2nd International Conference on Modeling, Big Data Analytics and Simulation (MBDAS 2021)

Journal of Physics: Conference Series Volume 2179

Shenzhen, China
14 – 15 November 2021

ISBN: 978-1-7138-5148-6
ISSN: 1742-6588

Printed from e-media with permission by:

Curran Associates, Inc.
57 Morehouse Lane
Red Hook, NY 12571

Some format issues inherent in the e-media version may also appear in this print version.

Institute of Physics
Dirac House, Temple Back
Bristol BS1 6BE UK

Phone: 44 1 17 929 7481
Fax: 44 1 17 920 0979

techtracking@iop.org

Additional copies of this publication are available from:

Curran Associates, Inc.
57 Morehouse Lane
Red Hook, NY 12571 USA
Phone: 845-758-0400
Fax: 845-758-2633
Email: curran@proceedings.com
Web: www.proceedings.com

TABLE OF CONTENTS

Journal of Physics: Conference Series **2179** (2022) 011001 doi:10.1088/1742-6596/2179/1/011001

Preface

Great honor to have your supports to 2021 2nd International Conference on Modeling, Big Data Analytics and Simulation (MBDAS2021) and participation in this academic discussion mainly direct at modelling and simulation, simulation optimization, mathematical modeling, time series analysis, complex system modeling and simulation, big data analytics and metrics, big data quality and provenance control, intelligent computing, quantum computing and related areas.

Due to the severe impact of Covid-19 and the travel restrictions, most of the authors can't attend the conference physically. The organizing committee of MBDAS2021 actively response to the call for epidemic prevention and control and adopt virtual form to conduct this academic conference. MBDAS2021 is rescheduled to be held via Tencent Meeting, November 15, 2021. The conference staff send a formal email notification to all the authors and participants before the meeting, try our best to co-ordinate the meeting and possibly make sure that all the details are in place. We set test room for our experts and authors to have a test on their video, audio and screen sharing, etc. in order to avoid the network meeting uncertainty. This virtual mainly conduct in the forms of Keynote Speech, Oral presentations, Posters, Free discussion.

It is great honor to invite five senior experts at here to share their keynote speeches to us. The first expert is Prof. Hui-Ming Wee from College of Electrical Engineering & Computer Science, Chung Yuan Christian University, Taiwan. The second expert is Prof. Christophe Claramunt from Naval Academy Research Institute & Arts & Métiers Institute of Technology, France. The third expert is Prof. Cheng Siong Chin from Newcastle University in Singapore, Singapore. The fourth expert is Prof. Stefano Zedda from Department of Business and Economics, University of Cagliari, Italy. The fifth expert is Prof. T. Velmurugan from Dwaraka Doss Goverdhan Doss Vaishnav College, India. They each make keynote speech for 50 minutes. And then the authors make 10mins oral presentations separately. All of them has contribute to a huge success of MBDAS2021 online conference.

Here are 115 papers widely cover Modeling, Big Data Analytics and Simulation and related topics in MBDAS2021 Calls for Papers. All the submissions in MBDAS2021 have gone through a double-blind peer reviewing, the editors and reviewers were responsible to make sure the scientific equality, verification and high standard. Each paper was assigned to 2-3 reviewer with similar research interests for reviewing. After reviewing and manuscript similarity detecting via Ithenticate, here are 48 papers were selected and accepted in MBDAS2021 Proceedings. Based on the main topics and suggestions of publication chairs, MBDAS2021 Conference Proceedings include 2 parts as below:

1. Modeling and Simulation Technologies and Applications
2. Data Analysis, Computational Methods and Algorithms Optimization

In the end, thanks for all the excellent reviewer members in the MBDAS2021 Technical Program Committee with professional reviewing and prompt response, Enthusiastic authors and interested attendees. We truly hope you enjoy yourselves and learn more from this academic discussion. Looking forward for the future collaboration with you.

MBDAS2021 Organization Committees

General Chairs
Prof. Dharmendra Prasad Mahato, National Institute of Technology Hamirpur, India
Prof. Isidoros Perikos, University of Patras, Greece
Prof. Arnulfo Luévanos Rojas, Autonomous University of Coahuila, México
Prof. Cheng Siong Chin, Newcastle University in Singapore, Singapore

Prof. Yilun Shang, Northumbia University, UK

Publication Chairs
Prof. Isidoros Perikos, University of Patras, Greece
Prof. Arnulfo Luévanos Rojas, Autonomous University of Coahuila, México

Publication Co-Chairs
Prof. Yilun Shang, Northumbia University, UK
Prof. Chandrani Singh, MCA Sinhgad Institute of Management, India

Keynote Speakers
Prof. Hui-Ming Wee, College of Electrical Engineering & Computer Science, Chung Yuan Christian University, Taiwan
Prof. Christophe Claramunt, Naval Academy Research Institute & Arts & Métiers Institute of Technology, France
Prof. Cheng Siong Chin, Newcastle University in Singapore, Singapore
Prof. Stefano Zedda, Department of Business and Economics, University of Cagliari, Italy
Prof. T. Velmurugan, Dwaraka Doss Goverdhan Doss Vaishnav College, India

Technical Program Committees
Prof. Ashok Panchapakesan, Prince Shri Venkateshwara Padmavathy Engineering College, India
Prof. Ateeq ur Rehman, University of Agriculture, Pakistan
Dr. Chong Leong Gan, Micron Technology Inc., Taiwan
Prof. Daniela Litan, Hyperion University, Romania
Prof. Grigoras Gheorghe, "Gheorghe Asachi" Technical University of Iasi, Romania
Prof. Zhiyu Jiang, Northwestern Polytechnical University, China
Dr. Ciortea Elisabeta Mihaela, "1 December 1918" University of Alba Iulia, Romania
Prof. Ioannis Filippopoulos, Hellenic American College, Athens, Greece
Prof. Prem Kumar Singh, Gandhi Institute of Technology and Management-Visakhapatnam, India
Dr. M. Madhiarasan, Indian Institute of Technology, India
Prof. Zaychenko Yuriy, Institute for Applied System Analisys, Ukraine
Prof. Gianpaolo Di Bona, University of Cassino and Southern Lazio, Italy
Dr. Mihaela Andrei, "Dunareaea de Jos" University of Gal, Romania
Prof. Volodymyr Khilenko, National University of Life and Environmental Sciences of Ukraine, Ukraine
Prof. Shadi I. Abudalfa, University College of Applied Sciences, Palestine
Dr. Hedayat Omidvar, National Iranian Gas Company, Iran
Prof. Nidà MEDDOURI, Virtual University of Tunis & IBM, Tunisia
Prof. Waleed Mahmoud Ead, Beni-Suef University, Egypt
Dr. Marius Iulian MIHAILESCU, The University of Bucharest, Romania
Prof. Shangce GAO, University of Toyama, Japan
Prof. M. Vijaya Laxmi, Jawaharlal Nehru Technological University, India
Prof. Kalidasan Rathinam, Lovely Professional University, India
Prof. Pavlo Maruschak, Ternopil Ivan Pul'uj National Technical University, Ukraine
Prof. Ming An Chung, National Taipei University of Technology, Taiwan
Prof. Elżbieta Macioszek, Silesian University of Technology, Poland
Dr. Catherine Nayer Tadros, Alexandria University, Egypt
Dr. Shahram Babaie, Islamic Azad University, Iran
Dr. Kingsley Okoye, Tecnologico de Monterrey, Mexico
Prof. Kartlos Kachiashvili, Faculty of Informatics and Control Systems of Georgian Technical University, Georgia
Prof. Mohamed Abdelsabour Fahmy, Suez canal University, Egypt
Dr. Hoang Nguyen, Hanoi University of Science and Technology, Vietnam
Dr. Jamal Talbi, Faculty of sciences et techniques of Settat, Morocco
Prof. Farhad Qaderi, University of technology, Iran
Dr. Abdollah Esmaeili, National Iranian South Oil Company (NISOC), Iran

Peer review declaration

All papers published in this volume of Journal of Physics: Conference Series have been peer reviewed through processes administered by the Editors. Reviews were conducted by expert referees to the professional and scientific standards expected of a proceedings journal published by IOP Publishing.

- **Type of peer review: Double-blind**
- **Conference submission management system: All the papers were submitted papers via the conference email: mbdas2021@sina.com**
- **Number of submissions received: 115**
- **Number of submissions sent for review: 101**
- **Number of submissions accepted: 48**
- **Acceptance Rate (Number of Submissions Accepted / Number of Submissions Received X 100):41.7%**
- **Average number of reviews per paper: 2**
- **Total number of reviewers involved: 88**
- **Any additional info on review process: MBDAS2021 uses a software to detect instances of overlapping and similar text in submitted manuscripts.**
- **Contact person for queries:**
  Ms. Yolanda Xu, Advanced Science and Industry Research Center, Hong Kong, mbdas2021@sina.com

Dynamic Simulation and Analysis of Engagement Process for Logarithmic Spiral Bevel Gear Based on Accurate 3D Model

Yang Xiao[1,*], Shengyu Hou[1], Shuling Dai[2], Xinfeng Xu[1] and Jie Zhang[1]

[1]School of Mechanical Engineering, Southwest Petroleum University, Chengdu, China
[2]School of Automation Science and Electrical Engineering, Beihang University, Beijing, China

*Corresponding author email: xiaoyang@swpu.edu.cn

Abstract. Based on the accurate 3D model of logarithmic spiral bevel gear, the operation of spiral cone gear engagement is analyzed and simulated by ANSYS. Through the nonlinear static contact analysis of different positions, the stress change rule and the maximum contact stress is obtained. Compared with the theoretical contact stress strength rule, this paper also analyzes the dynamic operation situation of the gear pairs, obtains the dynamic contact situation rule at different engagement times and different rotating speeds, the influence of the rotating speed on the gear contact characteristics is compared too. The static, dynamic performance and change rules of the gear engagement is obtained, which lay a solid foundation for the further research, design and manufacture of spiral bevel gears.
Keywords: Dynamic simulation; Logarithmic spiral cone gear; Static contact analysis; Dynamic contact analysis; Gear engagement process.

1. Introduction

Gear drives, being currently the most common type of transmission device, mainly determine the accuracy characteristics of mechanisms. Therefore, when designing new equipment, in addition to the requirements for ensuring the reliability of the drive based on gears, there is a very important issue of ensuring its accuracy characteristics. Spiral cone gear is a basic component in the machine to achieve intersection axis transmission movement and power, with many applications. Their quality and precision will have a great impact on the quality and precision of the final machine work. Parametric vibrations of gears make a significant contribution to the stationary dynamic-kinematic transmission error, which can lead to shock loads and damage to bearings[1]. Due to the geometric characteristics of the teeth, the overlapping coefficient increases during the gear engagement[2]. Compared with the straight bevel gear, the large cone gear have high transmission ratio and the better gear engagement, the higher bearing capacity. The operation is more stable, and the gear life is also 4~10 times that of the general cone gear, and the vibration and noise during high speed operation are significantly reduced to a certain extent[3]. Currently this type of gear is widely used in aerospace, aviation, automobile and other precision, high-speed, heavy load transmission fields. The working situation analysis, manufacturing accuracy and quality of the screw cone gear directly affect the quality, efficiency, noise, movement accuracy and life of these equipment's life. Therefore, the motion and operation of the spiral cone gear mesh based on the accurate model, the contact spots, transmission error and the bending stress on the tooth surface during load have always attracted the research and attention of scholars from all countries, and has become a very important technical problem in the gear design and production[4][5]. The analysis and solution of these problems can greatly improve the design and manufacturing level of spiral cone gear and meet the demand of high-quality machine needs.

The wheel teeth of the logarithmic helical cone gear are generated in this way: the wheel teeth are surrounded by two left and right tooth profile surfaces and the cone top and the bottom of the cone with the front cone and the back cone surface. The left and right two tooth profile surfaces are formed with the spatial spherical involution line as the tooth profile curve, moving along the cone logarithmic helical line for the occurrence line, and the surface can have an accurate mathematical description. The 3D entity of a wheel tooth is established for the full gear. However, in the previous analysis of spiral cone gear, the 3D model of cone gear established or by point cloud reverse approximation, or based on a simplified approximate finite element model, the shape and geometry of the analysis gear model is biased, so the analysis results are inevitably inaccurate. Contrary to the discussion above, in this paper, the simulation analysis is based on accurate 3D model of exact logarithmic helical gears[6] . Static and dynamic contact analysis were conducted by ANSYS to obtain more accurate contact work rules[7][8] . Figure 1 shows an exact 3D model of logarithmic spiral cone gears[9].

Figure 1. Accurate 3D logarithmic spiral cone gears model.

2. Static Contact Analysis of the Spiral Cone Gear

During the operation of the gear, only very few pairs of wheel teeth are engaged, and there have no contact stress for the other teeth. Therefore, in the static contact analysis, the influence of the nonmesh wheel teeth of the big gear on the integrated stiffness matrix is ignored. The mesh model is reduced to six teeth for mesh for analysis. The static contact analysis shown in Figure 1 and the gear parameters are shown in Table 1.

Table 1. Parameters of spiral cone gear.

Name	Active wheel	Slave wheel
Large-end modulus m (mm)	6	6
Number of teeth Z	15	46
Width of teeth b (mm)	44	44
Normal pressure angle α_n	20°	20°
Midpoint helical angle β	35°	35°
Caper angle δ	18.06°	71.94°
Material	20CrMnTi	20CrMnTi
Tooth surface hardness	58~63HRC	58~63HRC

We analyzed four different engagement positions of 0 °, 0.72 °, 1.44 °, and 2.16 ° relative to the initial position.

2.1. Establishment of the Finite Element Model

According to Table 1, the material of the two engagement gears is 20CrMnTi, density=7.85×10^3kg/m^3 and the elastic modulus E=2.0675×10^5MPa, Poisson ratio μ=0.3.

ANSYS Mesh has several different meseshing methods, and the density of the grid directly affects the accuracy of the solution results, but the overdense mesh increases the computational cost. In order to obtain more accurate calculation results, while considering the complex structural characteristics of the spiral bevel gear and its boundary conditions and load, in this case, a mixed tetrahedral and hexahedral mesh division is adopted, and the gear contact tooth surface engaged in the engagement is locally refined for better results. Also in order to reduce the computational amount, only the six-tooth model of the driven gear is solved to ensure the calculation accuracy. The mesh model for the gear side contact is shown in Figure 2.

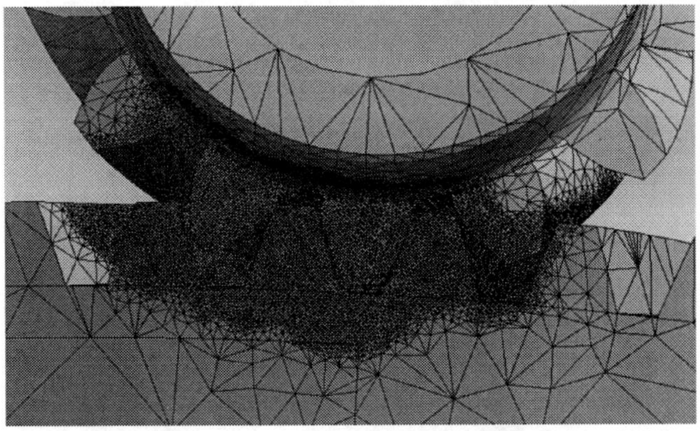

Figure 2. Contact area grid refinement.

For a contact analysis of the engagement gear using ANSYS, the contact area of the gear shall be defined. Due to the complexity of the spiral cone gear mesh surface, this paper adopts the automatic contact method in the face- surface contact to define the contact pair, and to judge and solve the contact surface with a defined contact relationship. In this example, the active wheel is defined as the contact surface, the slave wheel is defined as the target surface, and the defined contact pair is shown in the blue and red display areas in Figure 3. For nonlinear calculations of this partial contact, a penalty function is used in the Workbench platform to calculate the nonlinear contacts.

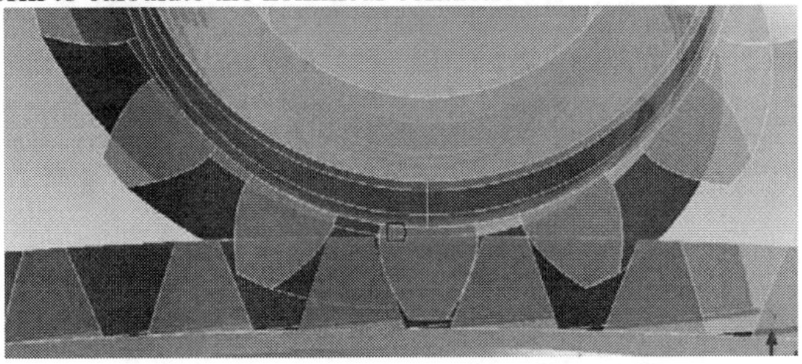

Figure 3. Definition of the contact pair.

All contact problems must define contact stiffness, including normal stiffness and tangential stiffness, and normal stiffness K_N is a key factor affecting the maximum penetration tolerance and pathological conditions. Considering computational convergence, the normal stiffness factor FKN=1.

FTOLN is the maximum penetration tolerance, and penetration exceeding this value makes new iterative calculations. The penetration tolerance value depends on the contact stiffness, which is too small to make it difficult to calculate convergence. Therefore, the maximum penetration tolerance selected in this article is FTOLN=0.1.

Considering the influence of friction factors, the friction coefficient between 20CrMnTi carburized steel gear contacts is 0.06[10].

2.2. Boundary Conditions and Load

Full constraints are placed on all the nodes of the inner ring surface of the big spiral cone gear, while the radial and axial degrees of freedom of the control nodes of the small gear inner ring are fixed, making the circumferential degrees of freedom rotating around the axis free. The gear is driven by a motor with a power of 500kW, for the high speed drive gear in this example the pinion spindle speed is 10000r/min and the calculated drive torque is T=300N·m. The large deformation switch is set to open to prevent the difficulty of calculation convergence.

2.3. Solution and Post-processing

In order to obtain the contact stress distribution of the spiral gear, the mesh model at different angles of the gear is analysed. During the spiral bevel gear engagement, the rotation angle ranges from 0° to 2.16 ° for each 0.72 °.By analysis, the contact stress at different locations of the spiral cone gear engagement model is shown in Figure 4, (a), (b), (c), (d). Through the results, the contact stress, strain distribution and strain value of the nodes in different directions are obtained.

The above method of calculating finite element contact analysis at several positions reflects the following characteristics:

- The analysis model of each engagement position can be achieved by the node rotation constraint of the active wheel, so this method enables the contact analysis at any instantaneous position in the gear engagement cycle.
- It's unnecessary to preset the contact load distribution, because the ANSYS can automatically capture and calculate the contact area and the contact load distribution.

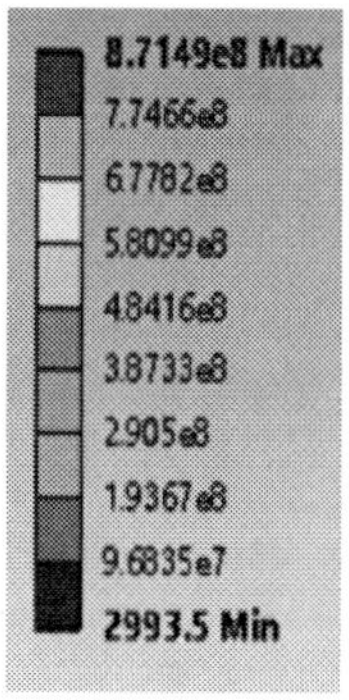

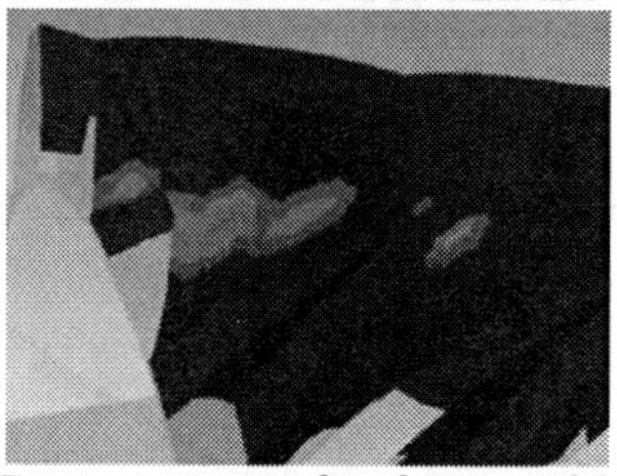

(a) Contact stress cloud map with 0 °

(b) Contact stress cloud map with 0.72 °

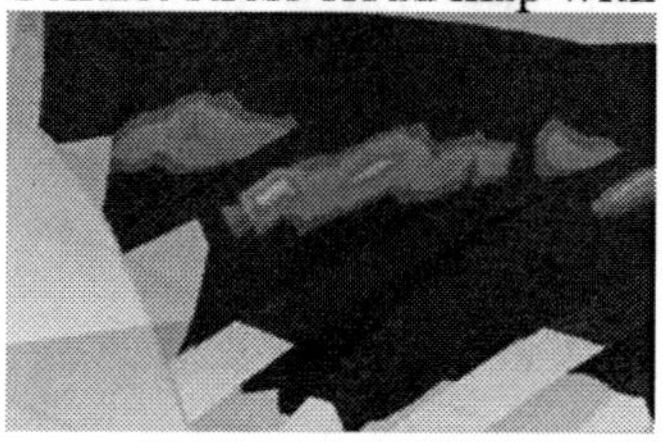

(c) Contact stress cloud map with 1.44 °

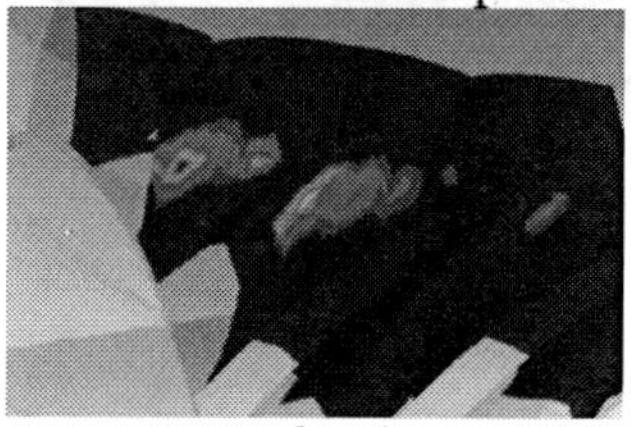

(d) Contact stress cloud map with 2.16 °

Figure 4. Cloud map of tooth surface contact stress during the engagement cycle.

As can be seen from Figure 4, during the engagement process, the gear gradually enters the engagement, the contact area gradually increases, and the contact stress on the tooth surface is also gradually increased. The wheel teeth have been in the engagement state, and the contact stress area is gradually increased, and the contact stress value is gradually increased too. Observation of the contact spot area can find that each contact spot area is linear in the distribution. In all stages of gear engagement, edge contact occurs near the tooth top, which adversely affects the arc tooth cone gear drive under high-speed heavy load conditions.

The results of the static contact analysis during the gear engagement drive cycle show that the trajectory of the helical cone gear is roughly linear in the direction in the contact spot area. During the gear engagement, the arc cone gear engagement is extended from the point contact to the surface area and distributed from its small end to the large end and tilted to the top.

During the engagement process, almost always three pairs of screw gear are in contact, so the screw gear has high bearing capacity, stable transmission and high transmission efficiency. During the engagement, the contact spot area gradually increases, and the maximum contact stress also increases. The teeth surface of the spiral cone gear will have edge contact, which affects the stability of the gear drive. During the gear engagement, the engagement contact spot area gradually decreases, and the maximum contact stress gradually decreases. In general, the contact stress of the tooth gear surface is increased and reduced during an engagement cycle.

When the processor solves the stress strain, displacement and contact stress of the assembly, Figure 5 shows the effect force cloud map of the gear contact at the initial position, and it can be seen that the maximum equivalent contact stress value is 871.1Mpa.

Figure 5. Gear initial contact position effect force cloud map.

Figs. 6 and 7 show the location of the maximum equivalent force of the active wheel. The maximum equivalent force of the slave and active wheels is 274.7Mpa and 366.3Mpa respectively, the maximum equivalent force is at the edge of the contact line and near the root, and pay attention to the stress concentration at the edge of the wheel teeth. The isoacting force of the active wheel engaged drive is greater relative to the driven wheel. This is due to the static contact analysis, where the slave wheel is a full constraint, where the moment is loaded on the active wheel, at the contact initial state, the active wheel contact impact is greater than the slave wheel.

Figure 6. Cloud map of effect force of slave wheel.

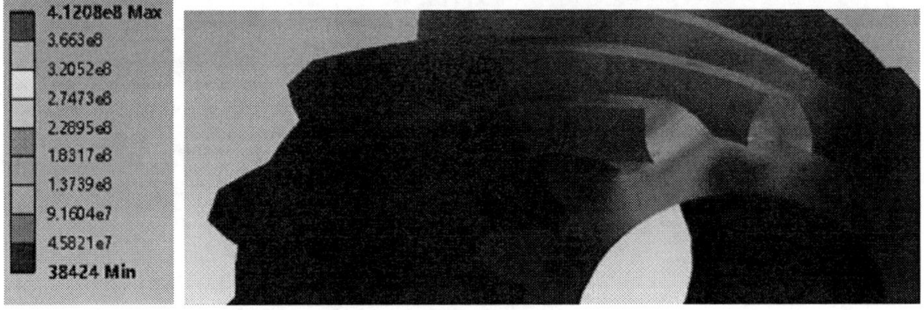

Figure 7. Cloud map of active wheel.

3. Dynamic Contact Analysis of Spiral Cone gear

In the process of gear transmission, the load is transmitted from gear by tooth, and the single and double teeth are always alternately engaged, resulting in vibration and impact of the gear mesh force and fluctuations of load and rotational speed, leading to periodic changes in the engagement stiffness. At the same time, there are nonlinear factors such as friction between the mesh surface, but only calculate the contact stress after multiple static contact at different engagement positions, but only consider the transient of the engagement process, and cannot accurately and comprehensively handle the mechanical characteristics of the reaction gear in their transmission process. Since the working environment of spiral cone gear is often in high-speed heavy load situations, in order to simulate the real gear continuous transmission process, it is necessary to conduct dynamic contact analysis of arc tooth cone gear[11]. Considering the influence of rotational speed, the finite element model of spiral cone gear is established in ANSYS Workbench, with transient dynamics analysis to obtain the dynamic contact stress of transmission process and the rule of dynamic engagement contact at different speeds of high, medium and low to provide reference for gear design[12].

3.1. Procedures and Processes of Dynamic Contact Analysis

During gear dynamic engagement, the motor applies an angular speed around the axis rotation to the active wheel through the coupling and shaft, while the driven wheel is driven around the central axis through the contact of the main and driven wheels. The driven wheel reaches the force balance under the action of the input torque and the load resistance torque. The angular velocity of the active wheel applied around the central axis is 1000rad/s , the resistance torque of the driven wheel is 120N. m[13]. In the transient analysis setting, the large deformation switch is opened, the load step end time is set to 0.0013s, the automatic time step switch is closed, the load step is 30, and the sparse matrix solver is used to solve the nonlinear contact transient. In order to analyse the contact situation at different times of the engagement process, the gear assembly contact state map at 6 different moments at high speed is selected, and the equivalent force cloud map is shown in Figure 8.

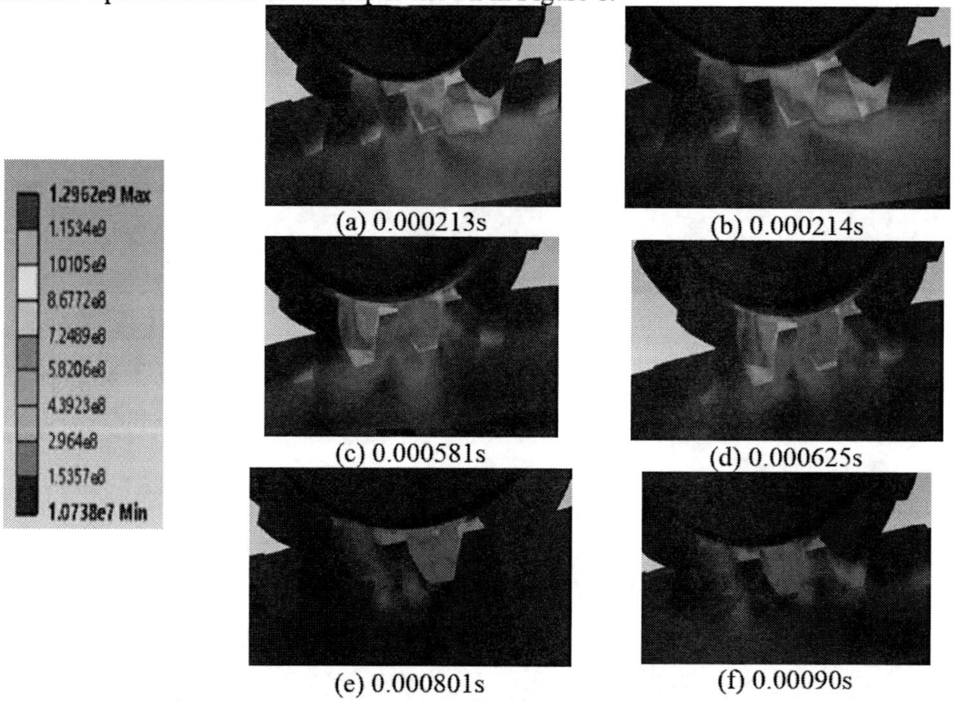

Figure 8. Cloud map of equal effect force at different time.

Select the maximum contact stress at different positions at different times in Figure 8 to obtain the stress results as shown in Table 2.

Table 2. Maximum contact stress at the different engagement positions of the gears.

Time	a	b	c	d	e	f
Engagement position	Single-tooth	Roout	Single-tooth	Roin	Double-teeth	Double-teeth
Maximum contact stress (Mpa)	1045	1093	1132	1240	859	642

Journal of Physics: Conference Series **2179** (2022) 012001 doi:10.1088/1742-6596/2179/1/012001

According to Figure 9 and Table 2, when the gear is in the high-speed rotation, the maximum contact stress of meshing in and out is greater than that of single tooth meshing. This is because the meshing in and out points deviate from the theoretical meshing in and out points, resulting in meshing in and out impact; The maximum contact stress of double tooth meshing is much less than that of single tooth meshing, which is due to the fact that the load is shared by multiple teeth, the area of contact area is increased, the comprehensive meshing stiffness is increased, the meshing contact deformation is smaller, and the solid contact stress is reduced.

3.2. Impact of Rotational Speed on Dynamic Contact

In order to analyse the influence of different speeds on the gear dynamic contact stress under different working conditions, we select three representative speeds of high, medium and low, and performs transient dynamic analysis of the engaging gear in ANSYS Workbench. The load boundary conditions are shown in Table 3. Working condition 1 is high speed, solved in front.

Table 3. Load working conditions.

Working condition No.	Angle speed (rad/s)	Load torque (N.M)
1	1000	120
2	100	120
3	10	120

In working condition 2, according to the engagement cycle corresponding to the speed, the automatic time step end time is set at 0.01 s, the load step is 40, extracting the equivalent force cloud map of the gear at different times at medium speed as shown in Figure 9.

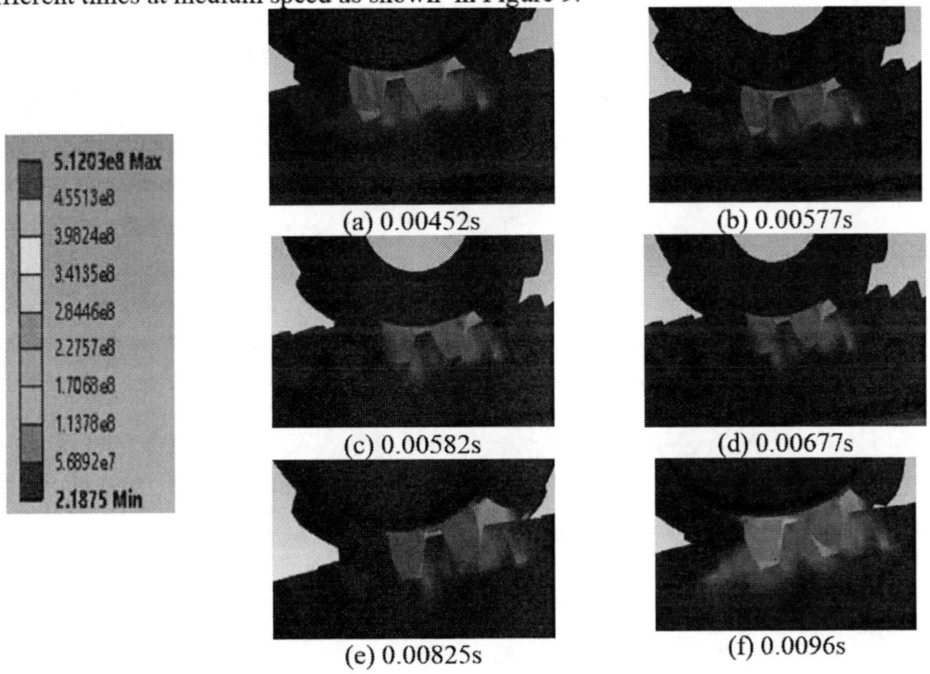

Figure 9. Cloud map of equivalent effect force at different times in working condition 2.

The maximum contact stress at different times in Figure 10 is selected to obtain the stress results as shown in Table 4.

Table 4. Working Conditions 2 Maximum contact stress in different engagement positions.

Time	a	b	c	d	e	f
Engagement position	Single-tooth	Roin	Double teeth	Double teeth	Roout	Single-tooth
Maximum contact stress (Mpa)	481	426	319	332	411	452

According to Figure 9 and Table 4, the maximum contact stress of the gear mesh in and out is slightly greater than that of the single tooth mesh, and also close to the single tooth, indicating that with the decreasing speed, the gear slowly enters the transmission stationary period, the load deformation is not changed, and the influence of the engagement impact of the gear is reduced.
Similarly, in working condition 3, the end time of the set load step is 0.08s, and the automatic time step switch is turned off, and the load step is set as 40, and the equivalent force cloud map of the gear engaged position at different times at low speed is shown in Figure 10.

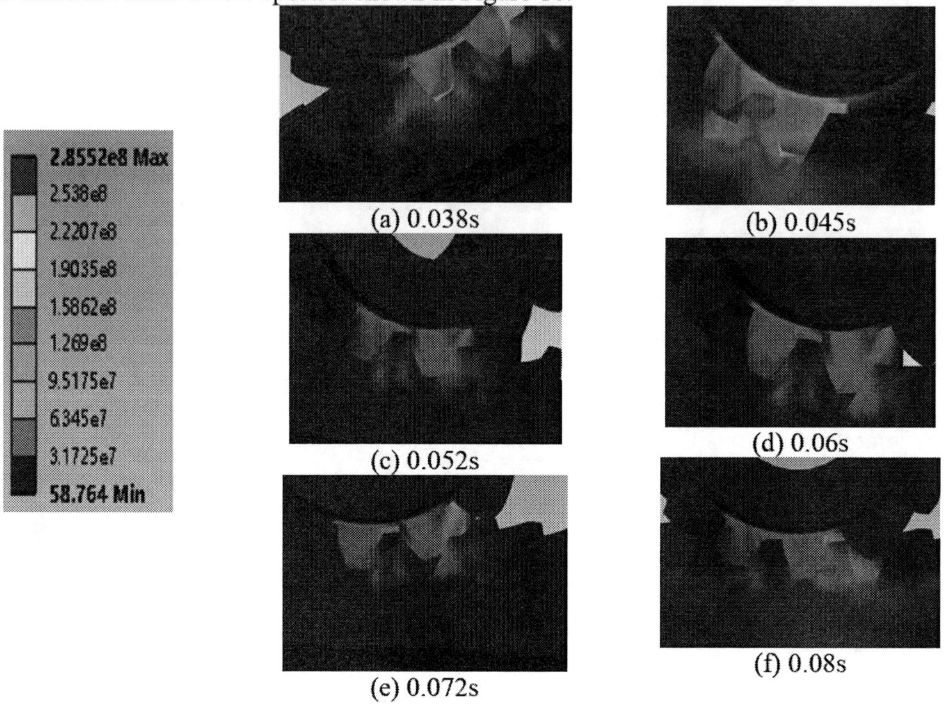

Figure 10. Cloud map of equivalent effect force at different times in working condition 3.

According to the analysis in Figure 10 and Table 5, at low speed, the maximum contact stress of the gear mesh in and out is equivalent to the contact stress of the two teeth, and the difference with the single teeth is getting smaller, which shows that the gear mesh moves smoothly at low speed and has good transmission performance.
According to the state at the low speed, the gear load deformation and impact are small. With the increase of the rotation speed, the load deformation and impact increase, and under the certain torque, the rotation speed is one of the leading factors in the amplitude of the contact force.

Table 5. Working Conditions 3 Maximum contact stress in different engagement positions.

Time	a	b	c	d	e	f
Engagement position	Single-tooth	Roin	Double teeth	Double teeth	Roout	Single-tooth
Maximum contact stress (Mpa)	265	173	206	213	168	254

4. Conclusion

With the same load torque of static and dynamic contact, from the equivalent cloud map, the dynamic simulation and analysis show static contact lines are straight lines along the small end to the large end and tilt towards the top of the tooth, the maximum equivalent force is at the edge of the contact line and near the root of the wheel. In the same load situation, the dynamic contact stress is larger than the static case, which is due to the load deformation of the gear during the mesh process and the large engagement impact under the internal excitation, the engagement point deviates from the theoretical position point, so that the mesh gear produces the line engagement, thus causing periodic vibration and impact. Compared with static contact analysis, dynamic analysis can easily solve the contact stress change rule at all moments of the whole engagement process, and be closer to the actual working condition[14].

Based on the exact 3D model of the spiral gear using ANSYS, through the nonlinear static contact analysis at different locations, the stress change rule is obtained and the maximum contact stress is solved in the ANSYS. Compared with the theoretical value, the results verify the accuracy of the contact problem by using the FEM method. Meanwhile, the dynamic contact situation of the gear at different engagement times is analysed, and the influence rule of the rotating speed on the contact characteristics is compared[15]. The simulation and analysis obtains the static and dynamic performance of gear work, which lays a solid foundation for further gear research, design and manufacturing.

Acknowledgments
The project is funded by the National Key Laboratory of Virtual Reality Technology and System's Open Research Project VRLAB 2020C07

References

[1] Sotnykov O., Rodionov M., Maruschak P., Brezinová J., Guzanová A., Apostol Y. Failure analysis of the hinge-lever mold oscillator bearings of the continuous casting machine. [J]. Strength, Fracture and Complexity 2014; 8 (3): 135-143.
[2] Fu Zheshao. Differential geometry and gear meshing principle [M]. Beijing: Petroleum University Press, 1999
[3] Litvin. Gear geometry and Application Theory [M]. Shanghai: Shanghai Science and Technology Publishing House, 2008.
[4] Guo Shuyi. 3D rapid modeling and Simulation for gears [M]. Beijing: Science Press, 2010
[5] Zeng Tao. Design and processing of spiral cone gear [M]. Harbin: Harbin University of Technology Press, 1989.
[6] Zhang Tingting. 3D Parametric Modeling and Application of Logical Bevel gears [D]. Southwest Petroleum University, 2019
[7] Wang Zhonghou. Dynamic performance simulation of thickness quasi-hyperbolic gear [M]. Beijing: Science Press, 2013.
[8] Xiao Yang et al. Accurate 3D parametric modeling of cylindrical helical gears based on Protoolkit [J] Journal of Graphics, 2017 (2): 180-184.
[9] Zhen Tao. Design and machining of spiral bevel gear [M]. Harbin: Harbin Institute of Technology Press, 1989.
[10] Su Zhijian et al. Solid styling method for Glisason arc bevel gear pair [J]. Computer Simulation, 2003 (8): 97-99.
[11] Zhu Xuewen, Zhu Guowen, He Xiaoxin. Parametric Model and Dynamic Analysis of Bevel gears [J]. Mechanical Transmission, 2006, 30 (2): 32-35.
[12] Love. Wilder Harber (translated by Zhang Zhixi). Principle of bevel gear and quasi-hyperbolic gear drive engagement [M]. Beijing: Machinery Industry Publishing House, 1958
[13] Wang Zhonghou, Wang Jie, Li Gang. Study on spiral cone gear drive error based on FEM [J]. Vibration and Impact, 2014 (14)
[14] Zhang Jinliang, Fang Zongde. Analysis of tooth surface contact stress of arc bevel gear [J]. Mechanical Science and Technology, 2007 (10)
[15] Wang Zhi, Fan Zhitao. Analysis of the research status of logarithmic spiral bevel gear [J]. Internal combustion engine and accessories, 2018 (23):76-77

Simulation Research on Automatic Compensation Technology of ADN Based on Neutral Point Active Compensation

Qihao Chen and Li'an Chen*

School of Electrical Engineering and Automation, Xiamen University of Technology, Xiamen, China

*Corresponding author email: chenla@xmut.edu.cn

Abstract. When the single-phase grounding fault occurs in the distribution network, the neutral point via the arc suppression coil grounding mode can effectively reduce the fault current and realize the self-extinguishing arc. However, as a large number of distributed generator and power electronic devices are connected to the distribution network, the structure of the network becomes more and more complex. In order to effectively reduce the fault current, the 10 kV active distribution network (ADN) was taken as the research object for simulation in this paper, and the support vector machines (SVM) algorithm were used for training by extracting the fault components, the neutral compensated current components and so on as eigenvalues. Used the controller to judge the system compensation state and complete the automatic compensation. The results show that under ideal conditions, the judgment time of the algorithm is 0.32 ms. And the algorithm can quickly respond to the fault component, improve the response speed and reduce the damage to electrical equipments.
Keywords: SVM; Simulation; Active compensation; ADN; Flexible grounding mode.

1. Introduction

Medium voltage distribution network is currently the most widely application scope, distribution network in power system in the operation of the indispensable to occupy the important position, with the level of urbanization gradually increase and the development of urban distribution network, power cable, transformation and upgrade, at the same time, the access of distributed generator (DG) and the utilization of electronic electronics, makes the grid structure complicated, The reactive power component, active power component and harmonic content in the grounding fault current are getting higher and higher, accounting for more than 10% of the total current[1]. As a kind of distributed flexible AC transmission system (DFACTS) distribution equipment, the active grounding mode of neutral point can effectively improve the power quality. The realization of arc suppression by active devices mainly has the following different sorts: Patent [2] invented a single-phase grounding fault current compensation arc suppression coil of active component device and method, by three-phase rectifier bridge according to controller commands into adjustable single phase alternating current, the transformer to power transformer neutral point injection system. Reference [1] proposed voltage arc suppression method based on flexible ground control technology , based on the fault phase voltage as control target, adopts the active inverter to inject grounding fault current zero sequence current compensation system, phase can be realized by changing the value of the injection current and zero sequence voltage to the system control, prompting the fault phase voltage zero. Reference [3-4] proposed a kind of magnetic-controlled reactor to replace the turning-regulated arc suppression coil as the master arc suppression coil in parallel with the active inverter, which is a master-slave arc suppression structure, to replace the full compensation of the ground fault current, and to solve the drawback that the turning-regulated arc suppression coil reactive current cannot be continuously adjusted. Reference [5] proposes a pure power electronic master-slave active arc suppression structure without arc suppression coil, and uses single-phase voltage source inverter to replace the arc suppression structure of the main arc suppression coil. However, due to the constant changes in the system state, the above methods require additional manual tuning or additional equipments in advance, resulting in increased costs. In this paper, 10 kV of ADN including photovoltaic is taken as the research object. The PV panel is connected to the main line, and power transmission is carried out through three branch lines. The load of branch lines is reflected in the form of total load. The single-phase grounding fault happened after the capacitive current and some components monitoring, fault component by comprehensive training SVM algorithm were monitoring the fault components characteristic value, at the same time in power transformer neutral point connected to the active arc suppression device through active control algorithm for adjusting arc suppression device, the fault compensation can be completed in time to improve the reliability and security of power supply in distribution network.

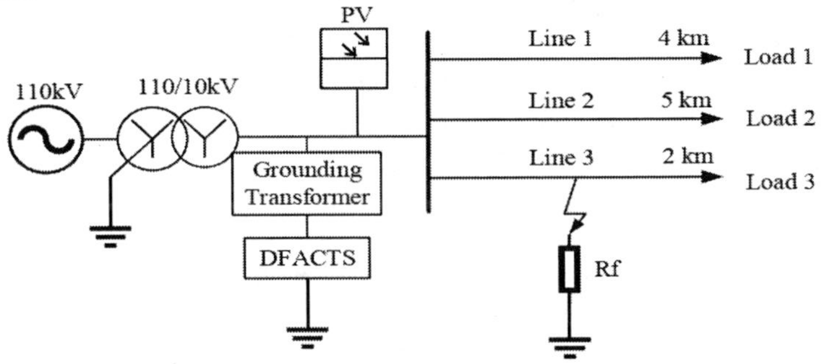

Figure 1. Structure diagram of simulation model.

Table 1. Wire parameters.

	R (Ω/km)	L (mH/km)	C (F/km)
Positive sequence parameters	0.092	0.277	0.261e-6
Zero sequence parameters	0.143	5.475	0.2e-6

2. Modeling and Simulation

The 10 kV ADN model is from the grid of Jiaxing, Zhejiang Province, whose neutral point grounding via the arc suppression coil was taken as an example in this paper, the network capacity current is 30 A, the capacity of arc suppression coil is 500 kVA, and the model is XHDCZ-500/10.5. Line structure is shown in figure 1. Single-phase grounding fault occurs in Line 3, and the wire parameters are shown in table 1, in which the insulation resistance to ground is set as 10 MΩ/km.

2.1. Distributed Generator

The photovoltaic generator is used as the DG to connected the ADN. The single PV cell can be equivalent to a PN junction in the mathematical model analysis. Figure 2 shows the equivalent circuit diagram.

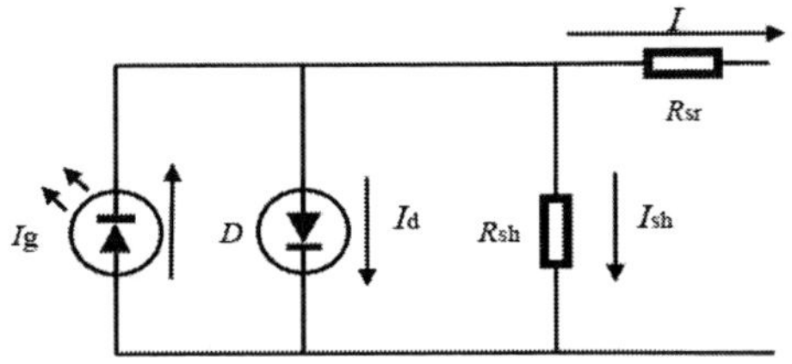

Figure 2. Equivalent circuit diagram of photovoltaic.

The equations are calculated as follows[6]:

$$I_g = I_{SCR} \frac{G}{G_R}[1 + \alpha_T (T_C - T_{CR})] \tag{1}$$

$$I_{sh} = (V + IR_{sr}) / R_{sh} \tag{2}$$

$$I_d = I_o[\exp(\frac{V + IR_{sr}}{nkT_C / q}) - 1] \tag{3}$$

$$I_o = I_{oR}(\frac{T_C^{\ 3}}{T_{CR}^{\ 3}})\exp[(\frac{1}{T_{CR}} - \frac{1}{T_C})\frac{qe_g}{nk}] \tag{4}$$

$$I = I_g - I_d - I_{sh} \tag{5}$$

Where, I_{SCR} - short-circuit current ; V_{OC} - open circuit voltage ; V_m - maximum working voltage ; I_m - maximum working current; T,G - temperature and light intensity; R_{sr},R_{sh} - series and parallel resistance of cell; I_o,I_d - saturation current and diode current; q,e_g,n,k - charge constant, band gap energy of materials, emission coefficient and boltzmann constant; α_T - temperature coefficient.

According to equations (1)-(5), an approximate simplified circuit model can be obtained, and the accuracy error can be controlled within 6 % in reference [6].

The amplitude, frequency and phase angle of DG need to be controlled synchronously in grid connection. The grid connection strategy of this model adopts the double loop decoupling control method of current inner loop and voltage outer loop based on dq coordinate transformation method[7-9]. Its principle is as follows: through the dq transformation method, the static coordinate system is transformed into the rotating coordinate system, and the PI controller is carried out in the dq axis of this coordinate system, and then the reference value of PWM carrier modulation signal is output through inverse change, so as to control the inverter. The harmonics of output are suppressed by LCL filter, which makes the system more stable. At the same time, the voltage outer loop monitors the phase angle frequency ω in the grid for phase locking and feedback. The control block diagram and parameters are identified in figure 3.

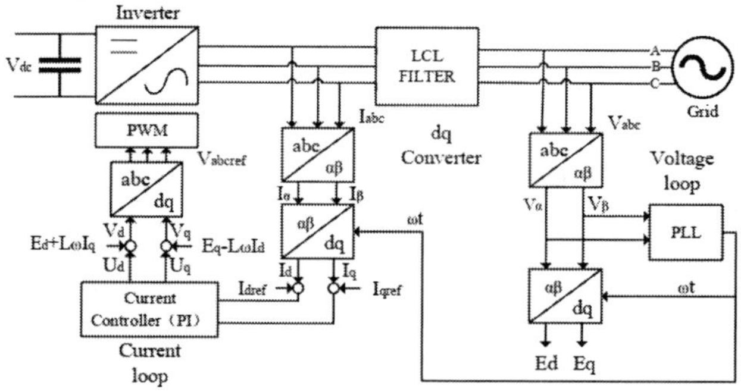

Figure 3. Block diagram of grid connected control strategy.

2.2. Active Arc Compensation Device

In this paper, a master-slave active arc suppression structure is adopted. The master part is the automatic turn adjustment type arc suppression coil, and the slave part is the converter[10-11]. The structure diagram of the device is shown in figure 4, and its operation principle is shown in figure 5.

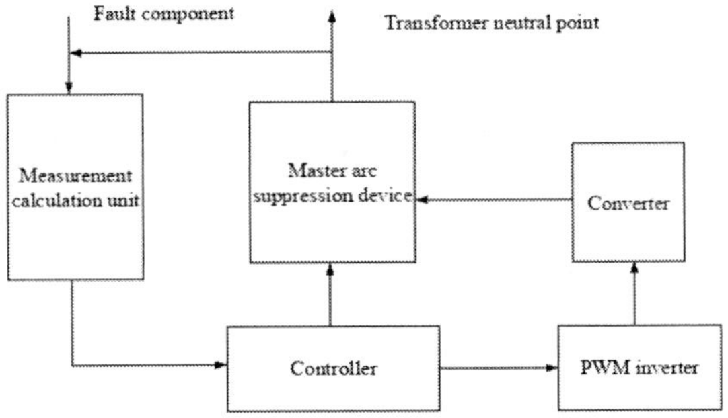

Figure 4. Structure diagram of active arc suppression device.

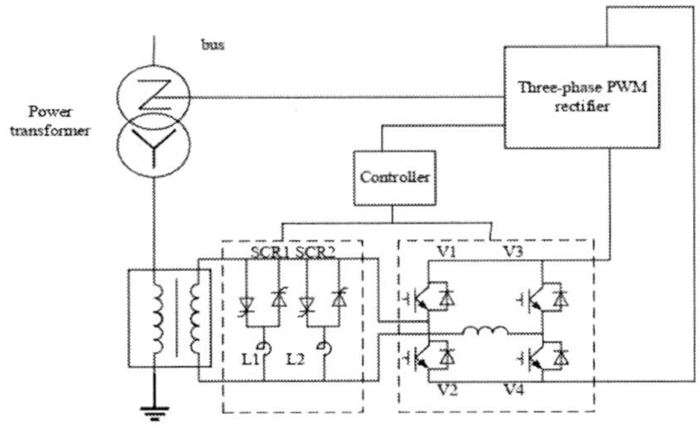

Figure 5. Operation schematic diagram of active arc suppression device.

When the single-phase grounding fault occurs, the device collects and transmits fault components through detection device of line-to-earth capacitive reactance and neutral current. Then the measuring unit calculate capacitive reactance of capacitive current, and the continuous signal is decomposed into discrete signal. By complete training SVM algorithm controller in advance, the controller can determine system operation state rapidly. The operation of the master-slave arc suppression device is adjusted to complete the full compensation operation dynamically in time. During the operation of the device, the master arc suppression device and the slave arc suppression device cooperate with each other to play a role. The master arc suppression coil adopts a new automatic turn adjustment type arc suppression coil to adjust the gear, which is used to compensate most of the grounding fault current. At the same time, the capacity of the slave arc suppression coil can be reduced, which is conducive to the rapid response of the converter compensation. The slave arc extinguisher part adopts an active device, which is used to accurately compensate the residual grounding current, mainly active power and harmonic components. The inverter uses a single-phase voltage source inverter, because the inverter only needs to compensate the active power and harmonic components, the required capacity configuration can be further appropriately reduced, so that the compensation scheme is more economical.

3. Data Extraction and Analysis

Support vector machine is a kind of binary classification model, which can be used for pattern classification and nonlinear regression like multilayer perceptron network and radial basis function network. The basic model definition of SVM is to establish the classification hyperplane as the decision surface, so that the maximum-margin hyperplane between the positive example and the negative example is maximized, and it is a linear classifier with the largest interval in the feature space, which is capable of binary classification of data. Figure 6 is the separating hyperplane diagram of SVM.

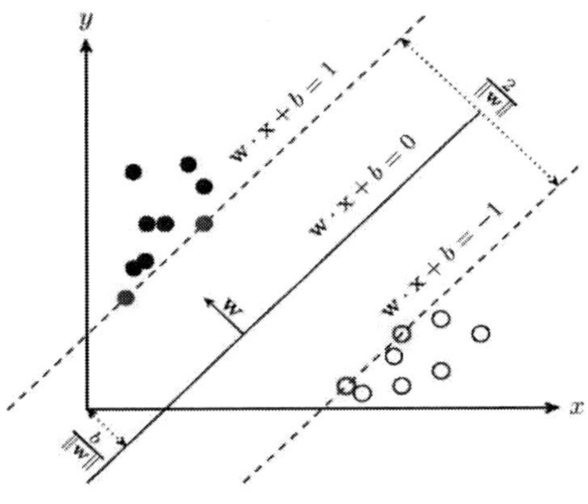

Figure 6. SVM separating hyperplane.

The inner product kernel between the "support vector" $X(i)$ and the vector of the input space X is the key to construct the learning algorithm of SVM, which is composed of the small subset extracted from the training data by the algorithm. Its architecture is shown in figure 7. Where K is the kernel function, and its types including linear kernel function, polynomial kernel function, radial basis kernel function, and two-layer perceptron kernel function.

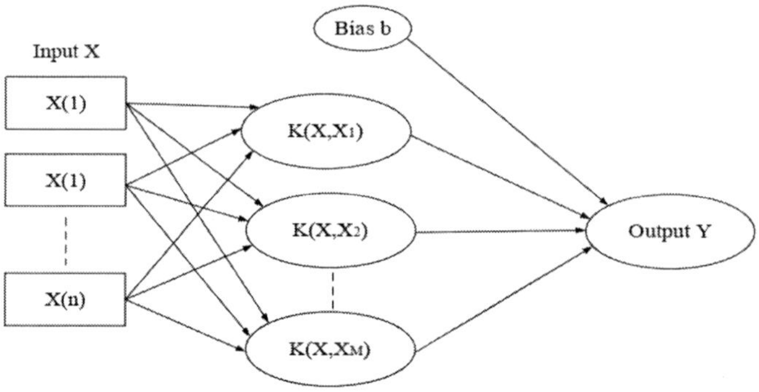

Figure 7. The architecture of SVM.

In order to ensure the correctness of the controller during device switching and reduce the classification errors caused by abnormal data interference, the SVM algorithm adopts one-to-one method to design the classifier. Figure 8 shows the step flow of classifier design.

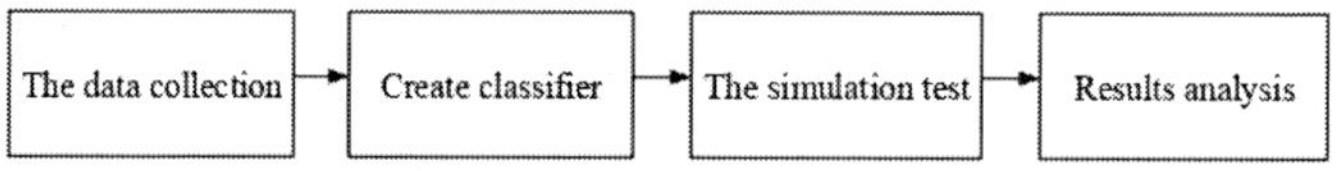

Figure 8. Step flow of classifier design.

3.1. The Training Results

During the creation and training of the classifier, in order to make the classification has higher reliability and classification effect, the data set collected in this paper should have a large information entropy content. The collected data are the three-phase voltage value V_{ABC} and current value I_{ABC} in feeder 3 of the fault line, as well as the earth short-circuit current value I_C in the case of grounding fault. There are 9 attributes including the voltage current component and neutral point current I_L. The system compensation status of the arc suppression coil is classified as 0 and 1. Where 0 represents the overcompensation state, while 1 represents the undercompensation and fully compensated state. The failure time t=0.2-0.5 s. There are samples of 14836 undercompensated and fully compensated states and 25164 overcompensated states. The decision plane of sample states is shown in figure 9.

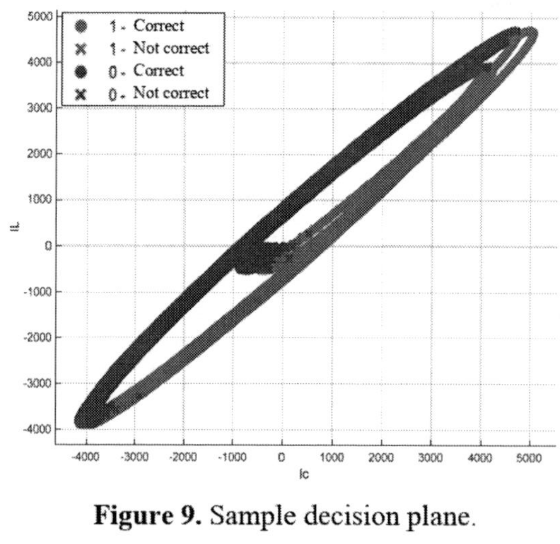

Figure 9. Sample decision plane.

The SVM adopts one-to-one classification, the kernel function is quadratic Gaussian type, the optimizer adopts Bayesian optimization, the iteration number is set as 30, and linear SVM, Gaussian SVM and optimizable SVM classifier are used. The classification effects are shown in figure 10-13. Receiver operating characteristic (ROC) curve means that smaller the distance from (1,1), the better the effect of the classifier. Area under ROC curve (AUC) is the area of the curve. The three types of classifier results are shown in table 2. It shows the result that the higher the number of iterations, the smaller the classification error.

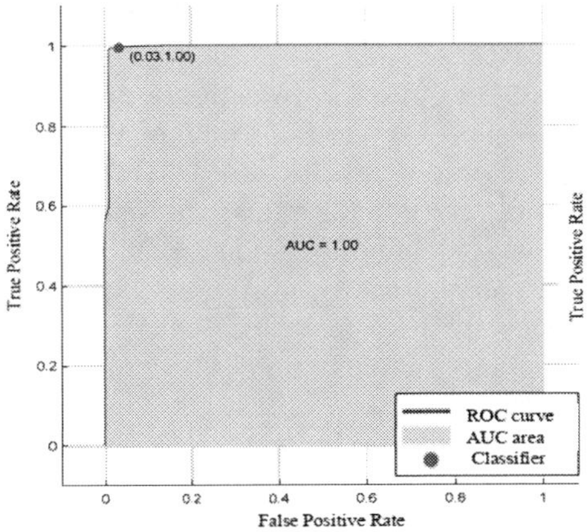

Figure 10. ROC curve of linear SVM.

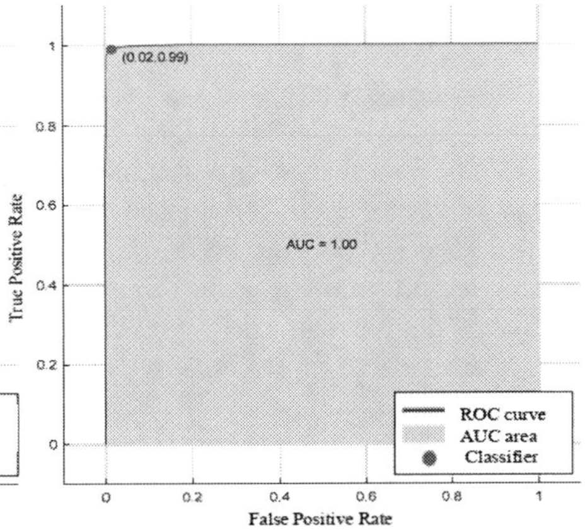

Figure 11. ROC curve of Gaussian SVM.

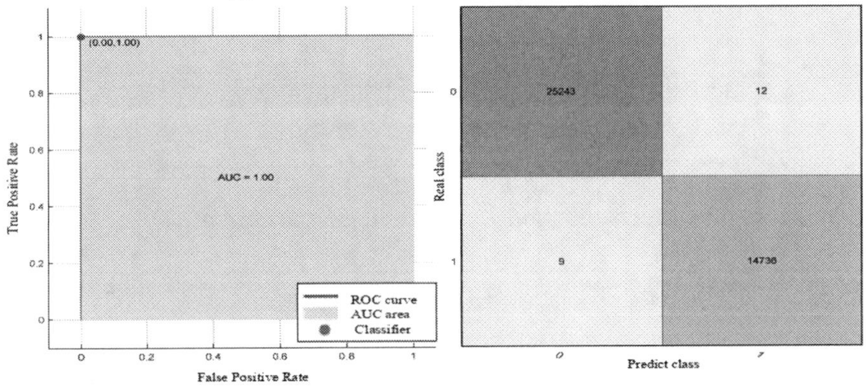

Figure 12. ROC curve and confusion matrix diagram of optimizable SVM.

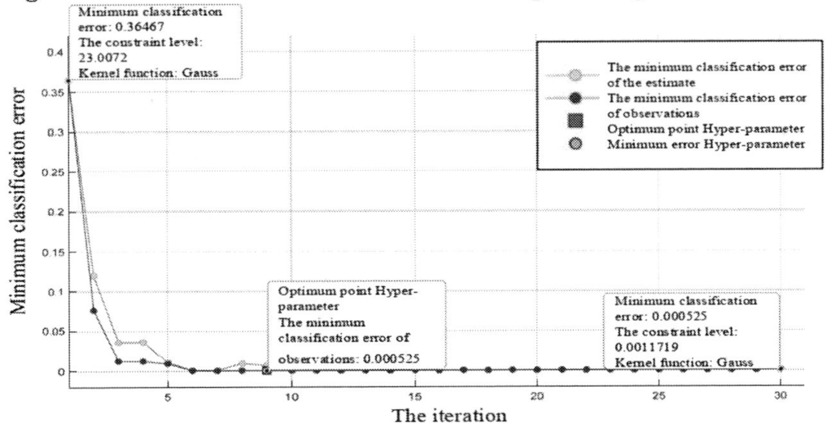

Figure 13. Minimization classification error of optimizable SVM.

Table 2. The judgment accuracy of SVM classifier on compensation state.

Algorithm type	Accuracy/%
Linear SVM	*98.5*
Gaussian SVM	*98.8*
Optimizable SVM	99.4

3.2. The Compensation Results

When t=0.2 s, it is assumed that A-phase grounding fault occurs. The waveforms of voltage and current in the grid are shown in figure 14. The voltage of A-phase decreases, while the voltage of non-fault B-phase and C-phase increases, and the line voltage remains symmetrical. When t=0.5 s, the fault ends, and the voltage and current in the distribution network return to a symmetrical state and operate normally. At the time t=0.3-0.5 s, the controller carries out switching operation on the active device, and the switching state and compensation effect of the device are shown in figure 15. It can be seen that the compensation device can track the current component of fault and realize the electrical component compensation effect of the system, and the calculate and judge time is 0.32 ms.

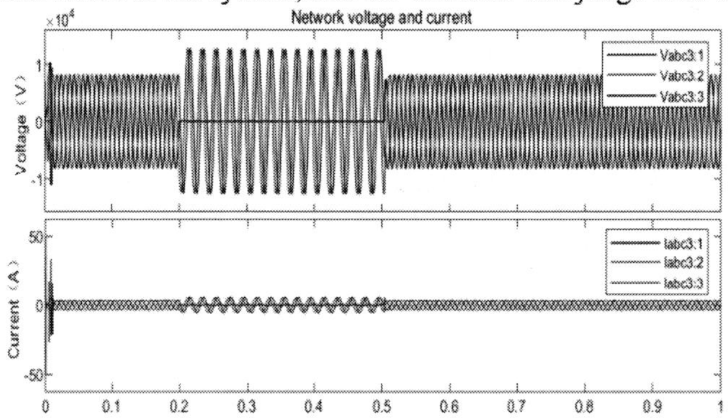

Figure 14. Distribution network voltage and current.

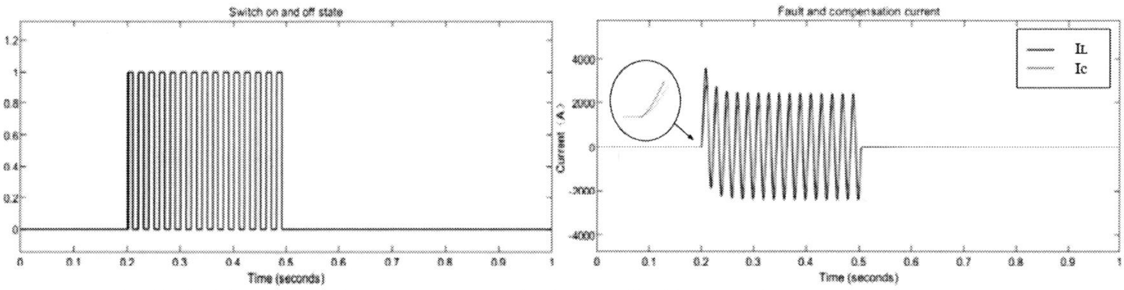

(a) Switching state of compensation device **(b)** Fault and compensation current

Figure 15. Compensation effect of active device.

4. Conclusion

In this paper, the model of 10 kV distribution network and the construction of active compensation device were built through the simulation software, and the compensation status of the system was determined by the control algorithm, and the switching and automatic compensation of the compensation device were carried out. The results show that by comparing three types of class SVM algorithm, the optimizable SVM algorithm has higher accuracy, and it can be used in the active compensation device, it can quickly recognize and compensate condition for compensation device through injection current of the neutral point to achieve automatic compensation effect. It also provides reference for the application of artificial intelligence algorithm in electrical equipments. In future studies, more diversified methods should be tried to compare, and the method with more economic and technical benefits should be selected.

References

[1] Zeng X, Wang Y, Li J, Xiong T. Novel Principle of Faults Arc Extinguishing & Feeder Protection Based on Flexible Grounding Control for Distribution Networks. Proceedings of the CSEE. 2012;32(16):137-43.
[2] Li X, Shi L, Wang C, Liu J, Dong X, Zhang D, et al., ^inventors; Arc suppression coil device and method for compensating active component of single-phase earth fault current. 2015-01-28.
[3] Chen B, Wang P, Shen W, Tian C, Wei L, Tu Z. Full compensation fault arc suppression principle and flexible control strategy of electromagnetic hybrid arc suppression coil. Transactions of China Electrotechnical Society. 2015;30(10):311-8.
[4] Engineering - Power Delivery; Study Data from Wuhan University Update Knowledge of Power Delivery (A Novel Neutral Electromagnetic Hybrid Flexible Grounding Method in Distribution Networks). Energy Weekly News. 2017.
[5] Chen Z, Zhang B. Non-inductive arc-suppression and active grounding compensation system based on master-slave inverters. Electric Power Automation Equipment. 2014;34(06):62-7.
[6] Su J, Yu S, Zhao W, Wu M, Shen Y. Investigation on engineering analytical model of silicon solar cells. Acta Energiae Solaris Sinica. 2001 2001-11-25(04):409-12.
[7] Guo L, Xue Y, Xu B, Li T, ^editors. Discussion about neutral grounding mode of Active Distribution Network. 2014 China International Conference on Electricity Distribution (CICED); 2014. Pub Place; Year Published.
[8] Che Y, Xu J, Yang Y, Zhou J, Zhao Y. Large Signal Modeling Method for AC/DC Independent Power System in dq-Coordinates. IEEE ACCESS. 2018:1.
[9] Shen X, Zheng J, Zhu S, Zhu L, Shi T, Qu L. A dq Axis Decoupling Parameter Identification Strategy for Grid-connected Inverter Controller of Photovoltaic Generation System. Automation of Electric Power Systems. 2014;38(04):38-43.
[10] Peng X, Jiang F, Tu C, Guo Q, Dai F, Fan R, et al. Dynamic Capacitive Compensator for Photovoltaic Connection and its Optimal Control. Power System Technology.2021:1-12.
[11] Li Q, Peng S, Ren C, Zhang Z, Zhao Q. Study on Full Compensation Arc Suppression of Active Current Injection for Single-phase Grounding Fault of Resonant Grounding System.Power Capacitor & Reactive Power Compensation. 2020;41(05):118-24.

Optimization Method of Overall Parameters of Dual Pulse Missile Based on Surrogate Model

Shenghui Cui*, Shifeng Zhang and Xibin Bai

College of Aerospace Science and Engineering, National University of Defense Technology, Changsha, Hunan, China

*Corresponding author email: 745032917@qq.com

Abstract. In this paper, the optimization method of missile overall parameters design based on surrogate model and the trajectory simulation system of three degree of freedom(3-DOF) dual pulse missile are established. An adaptive sampling point update method including three stages of potential feasible region location sampling, exploration sampling and potential optimal sampling is established. The adaptive sampling is realized through the constraint of distance. The convergence end criterion is established based on the accuracy of surrogate model, which can achieve efficient and reliable global optimization. By deriving the objective function of the optimization problem, the minimum take-off mass is determined, and reasonable constraints are set to ensure the speed and range requirements of the missile. The optimization algorithm based on surrogate model solves the problem of overall parameter design optimization of dual pulse missile and summarizes it.
Keywords: Surrogate model; Dual pulse missile; Overall parameter design optimization.

1. Introduction

The development of boost glide guidance weapons is of great significance to improve the strike ability. In order to improve the effective range, impact speed, maneuverability and strike accuracy of the aircraft, the overall parameters of the missile must be designed and optimized to meet the index constraints. The remarkable feature of hypersonic missile is its high flight speed and strong penetration attack ability. Therefore, for hypersonic missile, it is of great significance to ensure that it can achieve the whole hypersonic flight for its penetration ability.

Ego is one of the commonly used optimization algorithms based on surrogate model[1-2]. EGO (Efficient Global Optimization) is a global optimization method based on Kriging model. The convergence principle of ego algorithm is to make the objective function gradually converge to the global optimal solution[3] by constantly updating the sampling points and searching the design space. In order to solve the problem of high computational cost, bouhlel[4] and others constructed a new kernel function based on Kriging surrogate model and least square method, which has good computational efficiency for high-dimensional numerical examples. Li[5] et al. Proposed a bi-directional multi-surrogate global optimization method to solve the problems of too many samples and large amount of calculation in the optimization problem. Wang[6] et al. Proposed a multi-surrogate effective global optimization algorithm (MESGO) to improve the efficiency of the optimization algorithm.

Generally, there are many constraints in the practical engineering application of optimization algorithm, so the constraint optimization algorithm based on surrogate model is gradually developed. Basudhar[7] et al. Proposed a constraint effective global optimization method based on support vector machine to solve the problems caused by discontinuous or binary response, which greatly simplifies the constraint problem. Cho[8] and others proposed a unified sampling criterion to simplify the algorithm based on the sequential surrogate model optimization algorithm, so as to achieve the global optimal solution without any empirical constraints. Bagheri[9] et al. Proposed a new ego algorithm, which can well solve the constrained problems in the optimization problem and greatly reduce the calculation time cost. Antonio[10] et al. Proposed an optimization framework based on sequence model (SVM-CBO) to optimize a black box, multi extremum and complex objective function. The multidisciplinary optimization methods mentioned above are difficult to balance their exploration and development ability, which is difficult to ensure the efficiency and global convergence of the optimization method.

Therefore, in this paper, the overall parameter optimization method of dual pulse missile is established based on the surrogate model. The efficiency of the optimization method is greatly improved through the multi-stage adaptive sampling point update method. The missile simulation model is established to analyze the optimization results of dual pulse missile.

2. Mathematical Modeling of Overall Parameter Simulation

2.1. 3-DOF Trajectory Simulation Model

In the launch coordinate, the 3-DOF trajectory equation of the rocket[11] are as follows:

$$\begin{cases} m\dfrac{dV}{dt} = P\cos\alpha\cos\beta - X - mg\sin\theta\cos\sigma \\ mV\dfrac{d\theta}{dt} = P(\sin\alpha\cos\upsilon + \cos\alpha\sin\beta\sin\upsilon) + Y\cos\upsilon - Z\sin\upsilon - mg\cos\theta \\ -mV\cos\theta\dfrac{d\sigma}{dt} = P(\sin\alpha\sin\upsilon - \cos\alpha\sin\beta\cos\upsilon) + Y\sin\upsilon + Z\cos\upsilon \\ \dfrac{dx}{dt} = V\cos\theta\cos\sigma \\ \dfrac{dy}{dt} = V\sin\theta \\ \dfrac{dz}{dt} = -V\cos\theta\sin\sigma \end{cases} \tag{1}$$

Where, α,β are angle of attack and angle of sideslip; θ,σ,υ are flight path angle, heading angle and bank angle; p is motor thrust; X,Y,Z are drag, lift force and yawing force; x,y,z are the coordinate in the launching coordinate system.

2.2. Calculation Model of Dual Pulse Motor

The thrust model of dual pulse engine can be simplified to average thrust. Assuming that the i average thrust of the second stage pulse engine is P_i, the working time of the thrust is t_i and the specific impulse is defined as I_{sp}, it has the following relationship:

$$P_i t_i = I_{sp} g m_{pi} \tag{2}$$

For the single-stage dual pulse engine in this paper, the average thrust of the first pulse is P_1 and the working time is t_1; The average thrust of the second pulse is P_2 and the working time is t_2. When the charge quantity m_p is known, it has the following relationship:

$$P_1 t_1 + P_2 t_2 = I_{sp} g m_p \tag{3}$$

2.3. Aerodynamic Calculation Model

The main forces during flight are lift and drag, and the expression is as follows:

$$\begin{cases} D = \frac{1}{2} C_D \rho v^2 S_M \\ L = \frac{1}{2} C_L \rho v^2 S_M \end{cases} \tag{4}$$

In the above formula, C_D and C_L are drag and lift coefficients respectively; ρ is the atmospheric density; v is the speed; S_M is the aerodynamic reference area.

2.4. Flight Procedure Design

The flight sequence of hypersonic vehicle designed in this paper mainly includes take-off, negative angle of attack turn, turn end, engine first pulse flameout, maximum lift drag ratio glide, depression, engine second pulse ignition, terminal guidance and so on.

3. Mathematical Model of Overall Scheme Optimization

3.1. Design Variable

According to the above missile flight procedure, the optimized design variables are:

$$\begin{aligned} \mathbf{x} &= [x_1\ x_2\ x_3\ x_4\ x_5\ x_6\ x_7\ x_8\ x_9\ x_{10}\ x_{11}\ x_{12}]^{\mathrm{T}} \\ &= [m_p\ F_1\ \Delta t_1\ F_2\ \theta_0\ \Delta t_{g12}\ \Delta t_3\ \Delta t_4\ \alpha_m\ \Delta t_g\ \Delta t_p\ \alpha_p]^{\mathrm{T}} \end{aligned} \tag{5}$$

In the above formula, F_1 and F_2 are the average thrust of the first pulse and the average thrust of the second pulse of the engine respectively; Δt_1 represents the working time of the first pulse, and the working time of the second pulse can be obtained from equation(3). θ_0 is the initial launch velocity inclination angle; t_{g12} is the interval between the first and second pulse ignition of the engine; Δt_3 is the starting time of negative angle of attack turning; Δt_4 is the turning duration of negative angle of attack; α_m is the maximum angle of attack for negative angle of attack turning; Δt_g is the maximum lift drag ratio time between gliding and first pulse flameout. Δt_p , α_p are the interval time and angle of attack between negative angle of attack depression and maximum lift drag ratio gliding. The value range of optimized design variables in this paper is shown in the table below:

Table 1. Upper and lower bounds of missile trajectory optimization design variables.

Variables	Lower limit	Upper limit
m_p	3000 kg	5000 kg
F_1	150 kN	300 kN
Δt_1	30 s	45 s
F_2	30 kN	150 kN
θ_0	45 deg	85 deg
t_{g12}	350 s	600 s
Δt_3	3 s	10 s
Δt_4	5 s	20 s
α_m	-5 deg	-0.5 deg
Δt_g	20 s	60 s
Δt_p	-50 s	40 s
α_p	-6 deg	-1 deg

3.2. Objective Function

The objective of trajectory optimization by optimization algorithm is to find the minimum takeoff mass, that is, the minimum charge quantity m_p under the condition of meeting the constraints. Therefore, this paper selects the minimum takeoff mass as the optimization objective function[12]:

$$obj\_fun = \min m_p(x) = \min m_0(x) \tag{6}$$

3.3. Constraint Condition

According to the requirements of missile strike capability, it is necessary to meet the constraints such as range, maximum flight altitude, terminal strike speed, terminal speed inclination and second pulse working time. The specific expression is as follows:

$$\begin{cases} Range \geq 1000\ km \\ h_{max} \leq 80\ km \\ V_f \geq 1700\ m/s \\ \theta_T \geq -70 \text{ deg} \\ \Delta t_2 \geq 12\ s \end{cases} \tag{7}$$

Where, $Range$ is the longitudinal range; h_{max} is the maximum flight altitude; V_f is the terminal impact velocity and θ_T is the terminal velocity inclination.

4. EGO Optimization Algorithm

The basic framework of EGO optimization algorithm in this paper is as follows:

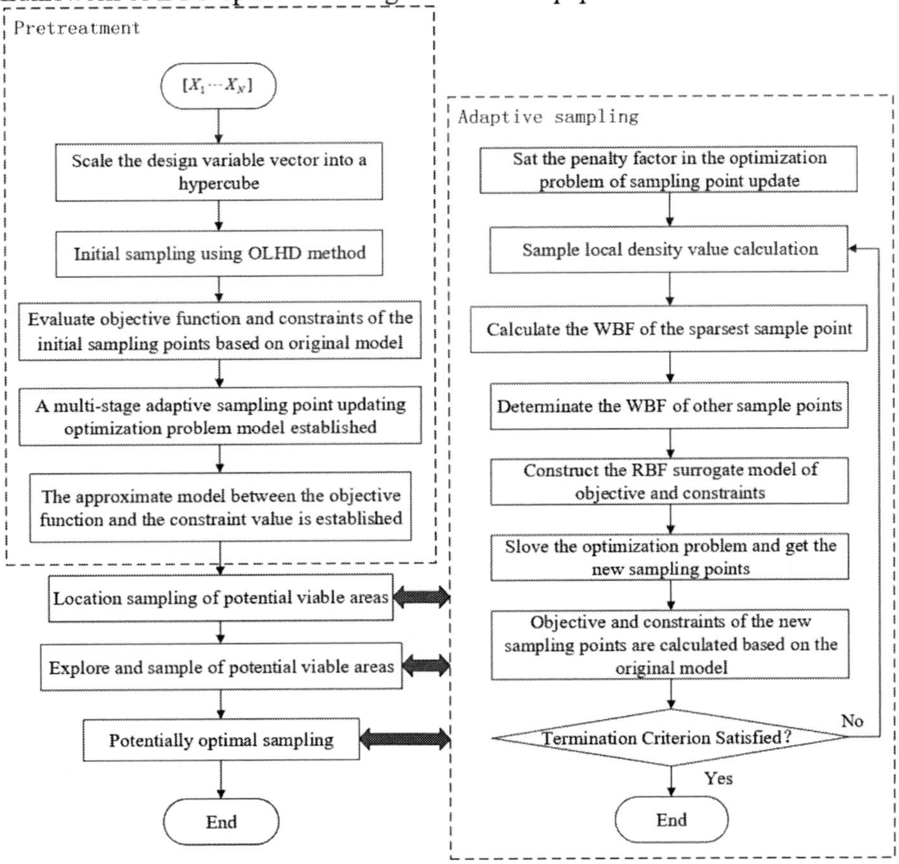

Figure 1. Flow chart of EGO optimization method.

(1) Initialization stage

The design space is mapped to the unit hypercube space, and the optimized Latin hypercube method is used to determine the initial samples. The number of samples is determined by the following formula:

$$N=\begin{cases}5\times dim, & dim\le 20\\ 100, & dim>20\end{cases} \tag{8}$$

Where, dim is the spatial dimension. The initial sample can be obtained by calculating the objective function and constraint value of the sample through the original model:

$$[x_i, y_i](i=1,2,\cdots,N) \tag{9}$$

Where, x_i represents the coordinates of sampling points, y_i represents the output response of the original model, which is generally a vector composed of objective function and all constraints.

(2) Approximation stage

This optimization method mainly uses the surrogate model established by the improved radial basis function to solve the global optimal solution of the approximate model.

(3) Convergence criterion

The convergence criterion is divided into three stages: location sampling, exploration sampling and potential optimal sampling.

(4) Sample point update stage

The adaptive addition strategy[12] proposed in this paper is as follows:

$$\begin{cases}\min\ \tilde{f}(x) & x_{\min}\le x\le x_{\max}\\ s.t.\quad \tilde{g}_i(x)\le 0 & i=0,1,2,\cdots,l\\ \qquad \tilde{h}_i(x)=0 & j=0,1,2,\cdots,m\\ \qquad P_d(x)\ge\delta & \end{cases} \tag{10}$$

Where, $f(x)$ is objective function, $g_i(x)$ and $h_j(x)$ are equality and inequality constraints. δ is the minimum distance constraint between sampling points. The constrained optimization problem is transformed into unconstrained optimization problem by penalty function method, as follows:

$$\min\quad f_p(x)=\tilde{f}(x)+\alpha\left\{\sum_i\varphi[\tilde{g}_i(x)]+\sum_j\psi\left[\tilde{h}_j(x)\right]\right\}+\beta\phi[p_d(x)] \tag{11}$$

Where, "~" refers to approximate values. α , β are penalty factors, φ , ψ , ϕ are continuous functions. In different sampling stages, the exploration and development ability is balanced by adjusting the value of penalty factors.

5. Optimization Results and an Analysis of Missile Overall

Taking the minimum charge quantity of dual pulse missile as the optimization objective, the overall parameters are optimized.

(I) The maximum flight altitude is less than or equal to 80km:

Table 2. Optimization value of missile design variables $h_{\max}\le$ 80 km.

Variables	F_1 /kN	Δt_1 /s	F_2 /kN	θ_0 /deg	Δt_{g12} /s	Δt_3 /s
Optimal value	218.38	34.38	30.00	62.45	429.07	9.56
Variables	Δt_4 /s	α_m /deg	Δt_g /s	Δt_p /s	α_p /deg	m_p /kg
Optimal value	9.78	-3.08	66.27	299.94	-3.97	3247.97

(II) The maximum flight altitude is less than or equal to 60km:

Table 3. Optimization value of missile design variables $h_{\max} \le 60\ km$.

Variables	F_1 /kN	Δt_1 /s	F_2 /kN	θ_0 /deg	Δt_{g12} /s	Δt_3 /s
Optimal value	233.89	39.24	30.46	60.51	393.53	11.41
Variables	Δt_4 /s	α_m /deg	Δt_g /s	Δt_p /s	α_p /deg	m_p /kg
Optimal value	9.99	-1.57	93.87	299.94	-4.90	3935.12

It can be seen from the above table that when the maximum flight altitude is reduced to 60km, the charge quantity of the missile increases by 705.15kg, and the thrust and working time of the first pulse of the corresponding engine increase. The average thrust of the second pulse of the engine remains basically unchanged, and its working time is obtained to be 15.19s, which is close to the value at the maximum flight altitude of 80km.

The objective function convergence results and trajectory optimization results of the missile are as follows:

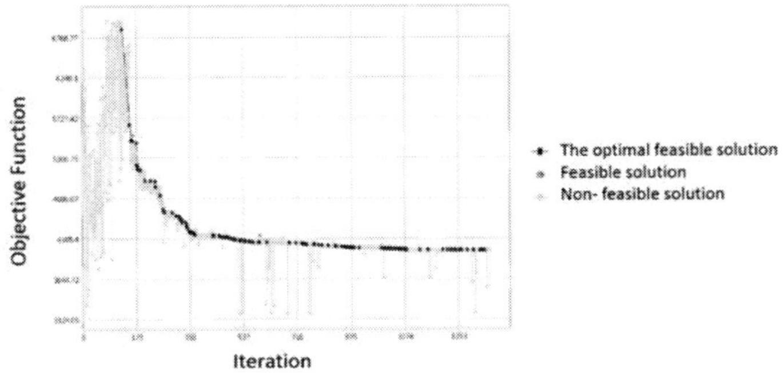

(a) 80km-Objective function convergence curve

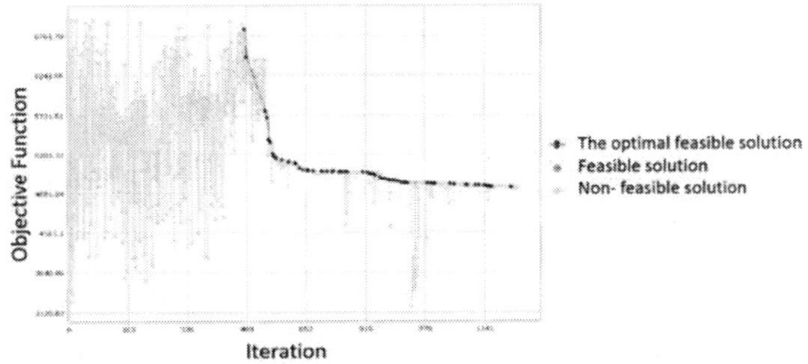

(b) 60km-Objective function convergence curve

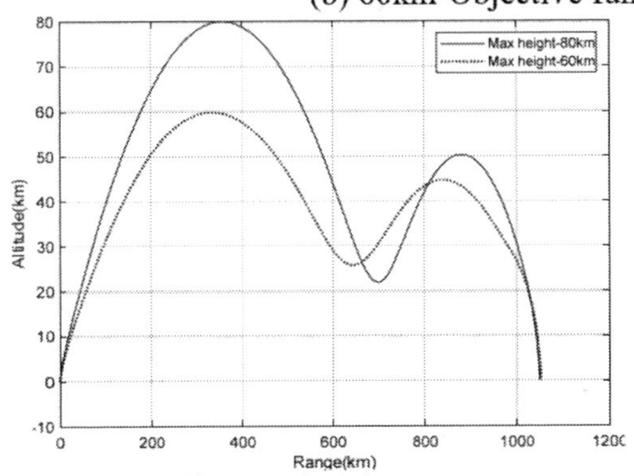

(c) Range-height curve

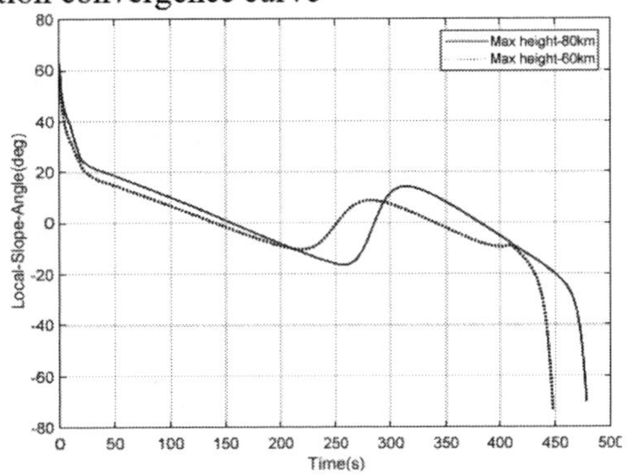

(d) Time-local flight path angle curve

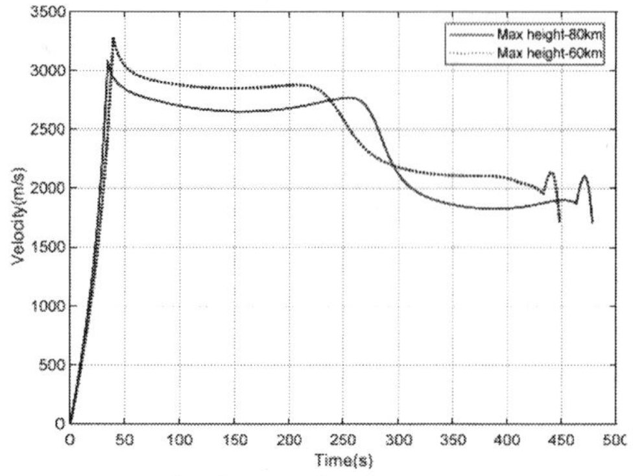

(e) Time-Velocity curve

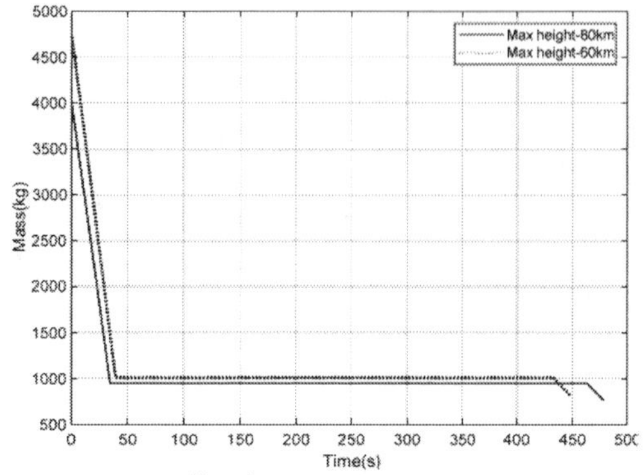

(f) Time-mass curve

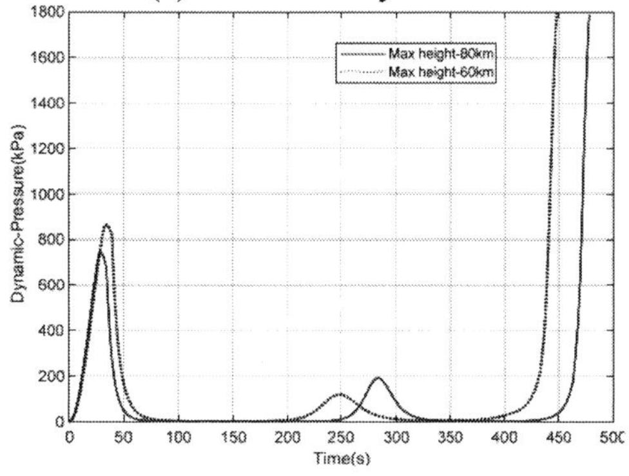

(i) Time-dynamic pressure curve

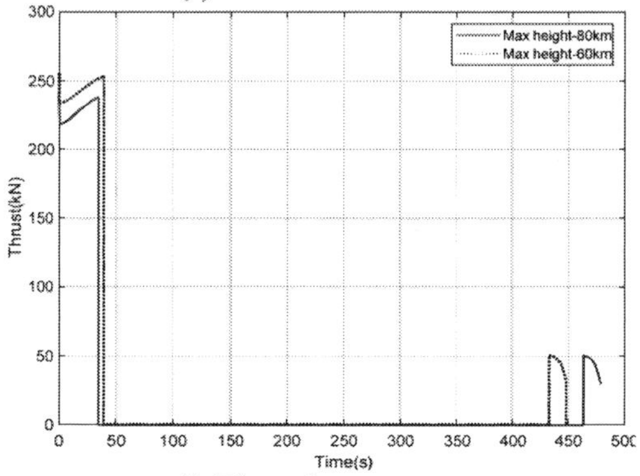

(j) Time-thrust curve

Figure 2. Optimization results.

From the optimal convergence curves at 80km and 60km, it can be seen that the missile takeoff mass curve converges when it is iterated to 1350 times and 1220 times, the optimal feasible solutions are 4042.17kg and 4754.92kg. The corresponding charges are 3247.97kg and 3935.12kg. The optimized longitudinal ranges are 1050.42km and 1056.09km, which meet the range requirements. It can be seen from the final impact speeds are 1707.47m/s and 1711.04m/s, meeting the terminal speed constraint. It can be seen from the local velocity inclination curve that the local velocity inclination is small and the projectile body is relatively horizontal during reentry, which is conducive to glide with the maximum lift drag ratio. The local velocity inclinations of the terminal are -70.32° and -73.38°, meeting the constraint of terminal inclination.

Through the optimization results, it can be seen that the optimization method converges after only calling the original model more than 1000 times. Compared with thousands of calls of traditional optimization methods, it has made great progress. The optimization method of the overall parameters of the dual pulse missile in this paper can provide reference for the development of hypersonic weapons and ensure its high-speed penetration ability.

6. Conclusion

The optimization algorithm based on surrogate model used in this paper has the characteristics of high efficiency and high precision of global optimization. It solves the main difficulties of multidisciplinary optimization algorithm, and the optimization of the overall parameters of dual pulse missile has a certain reference significance for its development.

Based on this method, the overall parameters of the dual pulse missile are designed and optimized. Taking the minimum takeoff mass (charge) as the objective function, the overall parameters of the missile are optimized through the value range of design variables and constraints, and the trajectory simulation of three degrees of freedom is carried out.

Acknowledgments
Thank Mr. Zhang for his guidance and the help of the senior brothers in the teaching and research office.

References
[1] Schonlau M. Computer experiments and global optimization: [Ph D Dissertation]. Waterloo, Ontario, Canada: University of Waterloo, 1997.
[2] Jones D R, Schonlau M, Welch W J. Efficient global optimization of expensive black box functions. *Journal of Global Optimization*, 1998, 13(4):455-492..
[3] Jones D R. A taxonomy of global optimization methods based on response surfaces. *Journal of Global Optimization*, 2001, 21(4):345-383.
[4] Mohamed A B, Nathalie B, Abdelkader O. Improving kriging surrogates of high-dimensional design models by Partial Least Squares dimension reduction. *Structural and Multidisciplinary Optimization,* 2016, Vol. 53(5): 935-952.
[5] Li, EY, Wang, H. Bi-direction multi-surrogate assisted global optimization. *Engineer Computers,* 2016, Vol. 33(3): 646-666..
[6] Wang, H, Ye, F, Li, EY, Li, GY. A comparative study of expected improvement-assisted global optimization with different surrogates. *Engineering Optimization,* 2016, Vol. 48(8): 1432-1458..
[7] Anirban B, Christoph D, Sylvain L. Constrained efficient global optimization with support vector machines. *Structural and Multidisciplinary Optimization,* 2012, Vol. 46(2): 201-221..
[8] Cho, S-G, Jang, J, Kim, J. Statistical surrogate model based sampling criterion for stochastic global optimization of problems with constraints(Article). *Journal of Mechanical Science and Technology,* 2015, Vol. 29(4): 1421-1427.
[9] Samineh B, Jurgen B, Wolfgang K. Constraint handling in efficient global optimization. *2017 Genetic and Evolutionary Computation Conference,* GECCO 2017.
[10] Antonio, C. Sequential model based optimization of partially defined functions under unknown constraints. *Journal of Global Optimization,* 2019, Vol. 79(2): 1-23.
[11] Kejun C, Luhua L, Yunhe M. Launch vehicle flight dynamics and guidance. Beijing: National De-fense Industry Press; 2014
[12] Masters DA, Taylor NJ, Rendall TCS, Allen CB, Poole DJ. Geometric Comparison of Aerofoil Shape Parameterization Methods. *AIAA Journal*. 2017;Vol.55(No.5):1575-89.

Research on Chain Fault Prevention Model Considering Voltage Stability

Huiqiong Deng and Xinyun Chen*

School of Electronic Electrical and Physics, Fujian University of Technology, FuZhou 350118, Fujian Province, China

*Corresponding author email: cxy_edu@126.com

Abstract. In this paper, we aim to prevent and control the chain tripping phenomenon in the pre-chain fault stage of the power grid, and use the L index and node injection power as the safety index to indicate the voltage stability of the power grid and the safety level of the current operation state of the power grid by adjusting the power output of the generating units. Based on this index and considering the various constraints of the grid, a preventive model is given to make the grid have the highest possible safety level against chain tripping and at the same time improve the voltage stability. In this paper, the prevention model is established based on the particle swarm algorithm, and the simulation results prove the feasibility and effectiveness of the proposed prevention model on the IEEE39 node system.
Keywords: Chain tripping; Prevention model; L index; Safety index.

1. Introduction

The power network is one of the most extensive and structurally complex man-made systems in the world today. As the backbone of the national economy, the safe operation of the power grid is closely related to the people's livelihood, and the frequent occurrence of large-scale power outages in recent years has fully exposed the vulnerability of large interconnected power grids [1]. The occurrence of large-scale power outages is often triggered by grid interlocking faults, and therefore, grid interlocking faults have long received the attention and attention of many researchers.

In recent years, research workers have proposed many new perspectives on the study of grid interlocking faults, providing many useful insights for the in-depth study of grid interlocking faults, for example, literature [2] focuses on wind power generation, a new energy source that is currently being promoted on a large scale in China, to study the impact of wind power access on grid interlocking faults.The literature [3] calculates the low voltage, frequency, and transmission line consequence impacts due to line faults by analyzing the evolution process of chain faults, calculates the risk value of each transmission line, conducts vulnerability assessment of each transmission line, and establishes a preventive control model. The literature [4] considers the interaction between system faults and changes in the operating state of components, and uses a Bayesian network approach to chain fault path search, which enables a more comprehensive analysis of the impact of each fault path on the system as a whole and provides a basis for making chain fault prevention decisions. Some researchers have also proposed to use the active power of the generator as the control means in the development of chain fault prevention, consider the time requirements of the preventive control means, and use the constraint degree and economic optimality of the grid tolerable as the optimization objectives [5-7]. In the literature [8], a nonlinear minimization model is used to establish the update fault

set based on the PCRA principle, so that the update voltage risk and countermeasures can be calculated until the risk is no longer reduced for the expected unexpected events.
This article first analyzes the power grid cascading tripping. Based on the current-type backup protection, combined with the various constraints of the power grid, a prevention model is established to calculate the stability index of the voltage of all nodes in the power network. The L-index analysis method is used to evaluate the voltage stability of all nodes. By sorting the voltage stability data of all nodes, find the nodes with the most vulnerable voltage stability, and at the same time inject power into the power network for the safety of node injecting power .The margin is constrained and optimized, and the shortest distance between the most vulnerable node and the current operating state and critical state of the grid is used as the objective function for tidal calculation, thereby reducing the risk of power network cascading failures and improving the operating stability of the power network.

2. The Establishment of the Model to Prevent Chain Tripping

2.1. Safety Margin to Prevent Chain Tripping
As a rule, if the grid is operated according to a pre-planned state, the grid is generally in a safe state at first, but after a period of operation, the grid may adjust its operating state due to load disturbances, and the grid may reach an unsafe state that can trigger a chain fault. To avoid this, preventive control should increase the distance between the grid and the unsafe state as much as possible.
When the first failure in the power network is removed, whether the remaining portion of the grid will have a chain trip in the grid due to a fault in any of the circuits can be measured by Eq. (1):

$$m(S) = \min \ (|\widetilde{S}| - |\widetilde{S}_{dist}|) \tag{1}$$

Eq. (1) $\widetilde{S}$ is the operating state of the power network at a certain moment, and $\widetilde{S}_{dist}$ is the security level boundary of the power network, and m(S) represents the measurement of the distance between them.
From the above analysis, it can be seen that in order to achieve the goal that the power grid is in operation and the power grid will not occur cascading tripping after the initial fault impact, the larger the value, the safer the power network. Therefore, in order to prevent the occurrence of cascading trips, it should be used as a safety constraint to determine whether the power grid is safe or not. Eq. (2) can be obtained:

$$f = \max(m) \tag{2}$$

2.2. Voltage Stability Index to Prevent Chain Tripping
This paper draws on the literature [9-10] and uses the L index as a safety index to evaluate the node voltage stability of the grid node. The L index was originally proposed by Kessel [12] and others, and was derived from a two-node system. When applied in a multi-node system (that is, a system with more than 2 nodes), the system needs to be divided into two groups: one group contains all PV nodes and is defined as N_P; the other group contains all PQ nodes and is defined as N_Q.
Literature [11] defines the local voltage stability index L_j of PQ node j as:

$$L_j = \frac{1}{V_j^2 Y_{jj}} \left\| \sum_{i \in T_Q} \frac{Z_{ji}^* S^+{}_i}{Z_{jj}^* \dot{V}_i} \right| \dot{V}_j \Big| = \frac{1}{V_j} \left| \sum_{i \in TQ} \frac{Z_{ji}^* S^+{}_i}{\dot{V}_i} \right| \tag{3}$$

In Eq. (3), $\dot{V}_i$ and $\dot{V}_j$ are the node voltage vectors of the PQ node respectively; Z_{ji}^* is the conjugate transimpedance matrix between PQ nodes j and i; Z_{jj}^* is the conjugate self-impedance matrix of PQ node j; $S^+{}_i$ is the equivalent load of the grid to node i; Y_{jj} is the amplitude of the Y_{QQ} self-admittance matrix.
For a PQ node j in a multi-node system, if the node voltage is stable, the value range of the L index of the node is (0,1); if the node voltage of the node is in a critical state, then the L index at this time

The value of is 1; if the node voltage of the node has been unstable, the value of the L index is greater than 1.
Therefore, the smaller the value of L index in Eq. (3), the more stable the power network , and the closer the value of L index to 1, the more unstable the power network . For the entire power grid, the L index values calculated by all PQ nodes are sorted from small to large, and the node with the largest L index value is the most vulnerable node in the power network. Because in the entire power grid, any voltage instability at any PQ node may cause the power grid to fail or even cascading trips. Therefore, in this paper, the most vulnerable node is used to represent the overall stability of the power network, that is, the weakest in the power network. The node voltage stability node determines the voltage stability of the overall grid, as shown in Eq. (4):

$$L = \max_{j \in T_Q} \left(L_j \right), j \in T_Q \tag{4}$$

In order to facilitate the optimization of the subsequent model, the value range of Eq.(4) should be preprocessed, and Eq.(4) should be transformed:

$$L' = 1 - L \tag{5}$$

The value range of the L index after processing is $(-\infty,1)$.When the node voltage stability of the power grid increases, the closer it is to 1, and when the node voltage stability decreases, the closer it is to negative infinity.

2.3. Optimization Model to Prevent Chain Tripping

In summary, it can be seen that the objective function obtained by combining the operating state safety distance of the power network and the L indicator reflecting voltage stability can be used as an indicator to measure the security of the power network. In the operation of the power network , if the processing of the generator set of the power network can be optimized and adjusted, under the premise of ensuring that the load demand is met, ensure that L' in Eq. (5) is close to 1 and f in Eq. (2) is greater than zero, and the value is greater than zero. Larger, the more stable the grid operation is guaranteed, the chain tripping of the grid will be effectively prevented. Therefore, Eq. (2) and Eq. (5) are integrated into the objective function Eq. (6):

$$F = \alpha_1 f + \alpha_2 L' \tag{6}$$

In Eq. (6), a_1, a_2 is the weight coefficient, $\alpha_1 = \alpha_2 = 0.5$, and the output power of the generator is to be optimized. From Eq. (2) and Eq.(5), it can be seen that both equations are to obtain their corresponding maximum values. Therefore, Eq. (6) should also be integrated as:

$$\mathrm{N} = \max (\mathrm{F}) \tag{7}$$

Adding the objective function of Eq. (7) to the various electrical constraints required by the power grid, a complete prevention model can be obtained.
Electrical constraints include two types of equations and inequalities.In these formulas, the equation constraints mainly include the power flow constraints of the power network before the initial failure and constraints after failure. Constraints can be abbreviated as Eq. (8) and Eq. (9):

$$\mathrm{H}^1(x) = 0 \tag{8}$$

$$\mathrm{H}^0(x) = 0 \tag{9}$$

In this prevention model, the inequality constraints that need to be considered are as follows:

$$\begin{cases} P_{Gt} \in \left[P_{Gt\cdot\min}, P_{Gt\cdot\max}\right], t=1,\cdots,N1 \\ Q_{Gt} \in \left[Q_{Gt\cdot\min}, Q_{Gt\cdot\max}\right], t=1,\cdots,N1 \\ U_k \in \left[U_{k\cdot\min}, U_{k\cdot\max}\right], k=1,2,\cdots,N2 \end{cases} \tag{10}$$

In Eq.(10), P_{Gt} is the active power output of the t-th generator in the power network; Q_{Gt} is the reactive power output of the t-th generator in the power network .Write Eq. (10) in abbreviated form can be expressed in the form of Eq. (11):

$$\mathrm{M}_0(x) \le 0 \tag{11}$$

Combining Eq.(2) and Eq. (7) to (11), the prevention model Eq. (12) that optimizes the power grid safety margin and voltage stability can be obtained by adjusting the active output of the generator:

$$\begin{cases} \mathrm{N} = \max(\mathrm{F}) \\ s.t.\ \ \mathrm{H}^1(x) = 0 \\ \quad\quad \mathrm{H}^0(x) = 0 \\ \quad\quad M^0(x) \le 0 \\ \quad\quad f \ge 0 \end{cases} \tag{12}$$

3. Prevention Model Calculation Process

This paper uses PSO to solve the prevention model. Compared with the conventional algorithm, the PSO has the advantages of good convergence, fast calculation speed, and not limited by the dimension of the problem [12], and the PSO also has the advantages of simple principle and easy to achieve optimization goals [13].

For the optimization problem shown in Eq. (12), in order to make the PSO serve the prevention model, this article should first simplify it appropriately:

(1) For each trend of the constraint in Eq.(12), this paper uses power flow calculation of power system for iterative calculation processing. If trend of the constraint is not satisfied, new particles will be generated in the particle swarm to replace it, so that the calculation conforms to the power flow constraint.

(2) In order to enable the prevention model in Eq. (12) to be better processed by particle swarm optimization, this paper transforms the objective function in Eq.(12) into the form of penalty function shown in Eq. (13):

$$\mathrm{NN} = \mathrm{N} + \sum_k \frac{1}{\beta_k}\left[\min\left(0, -g^0(x)\right)\right]^2 + \frac{1}{\gamma}(f)^2 \tag{13}$$

In Eq. (13), β_k and γ are penalty factors, and k is the inequality shown in Eq. (12).

4. Example Analysis

This article takes the IEEE-39 node system shown in Figure 2 as an example, and performs programming in the MATLAB environment. The component parameters in the program and the objective function Eq. (7) are all standard unit values. The reference capacity of the power network is 100MVA. Set the virtual current type backup protection to 5.77 per unit.

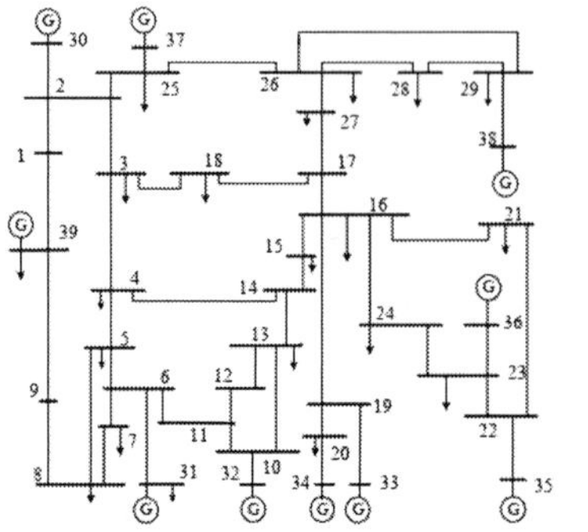

Figure 1. Diagram of the example system.

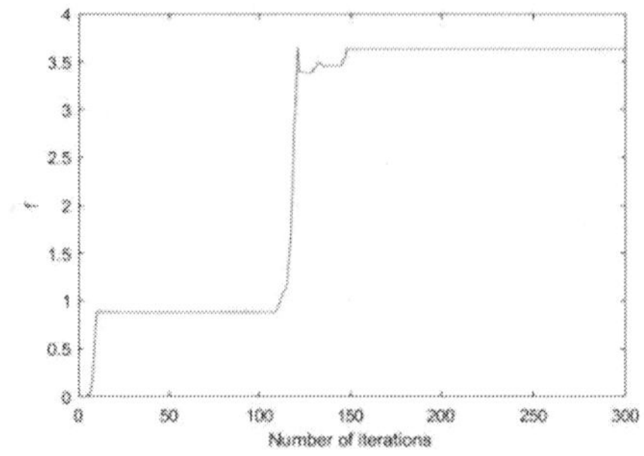

Figure 2. Grid operating state security constraint foptimization diagram.

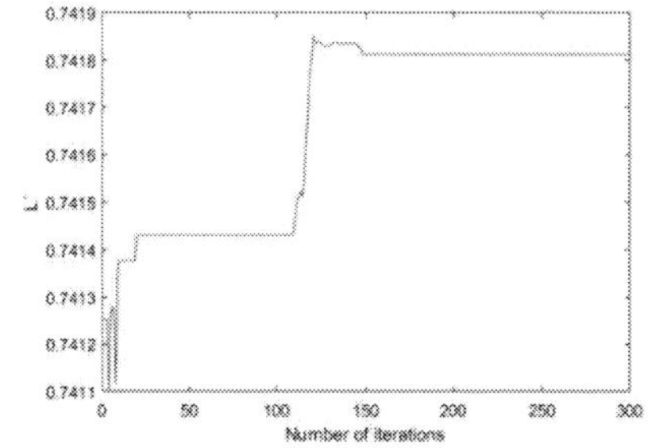

Figure 3. L index optimization diagram.

The y-axis values in Fig. 2 are the grid operation state safety distance values, and the x-axis is the number of iterations. As can be seen from Figure 2, the curve in the graph tends to rise during 300 optimization iterations of the particle swarm optimization algorithm, and when the particle swarm algorithm no longer rises after the 150th iteration, the final f value improves from 0.062 to 3.632.At this time, the distance between the current operating state of the grid and the critical state of the grid in which the chain tripping occurs is effectively increased, improving the safety level of the grid during operation. The y-axis in Figure 3 indicates the voltage stability index-- L', which improved from the initial 0.7412 to 0.7418 during 300 iterations, improving the voltage stability of the grid to some extent.

From the above results, it can be proved that the prevention model proposed in this paper (12) can effectively improve the safety level of the grid operation status and the overall voltage stability of the grid, and achieve the effect of preventing chain failures. Moreover, the prevention model proposed in this paper positively influences each other in terms of the safety level of grid operation and L index, and there is no conflict or mutual constraint.

5. Conclusion

This paper proposes an optimization model for the prevention of chain tripping in the power grid, and achieves the research results of preventing chain tripping in many aspects by improving the safety distance of the power grid operation state and enhancing the voltage stability of the power grid. The main conclusions of the article are as follows: The safety distance and L index of the grid operation state proposed in this paper quantifies the dominant factors affecting the level of grid safety, making the degree of influence have a clear physical meaning, and providing researchers with references when further studying and analyzing the complex operation state of the grid. The optimization model proposed in this paper can be used to test the grid and get the area with the shortest safety distance and fragile voltage stability in the operating state, which can be used to adjust the grid dispatching scheme, enhance the anti-disturbance capability of the grid and prevent the occurrence of chain tripping.

References

[1] Power, I. . (2013). 1 Causes of the 2003 Major Grid Blackouts in North America and Europe, and Recommended Means to Improve System Dynamic Performance Administrative Committee of the Power System Dynamic Performance Committee.
[2] Wang Weizhou, Liu Qian, Dan Yangqing, Wang Geng & Liu Wenying. (2015). Research on cascading failure prevention and control strategies for large-scale new energy integration into the power grid. New Technology of Electrical Engineering and Energy (03), 12-17.
[3] Zheng Qiao, Qiu Xiaoyan, Liu Nian, Sun Bin, Liu Ming & Zhu Lifang. (2014). Cascading failure prevention and control based on vulnerability assessment of transmission lines. Journal of Electric Power System and Automation (02), 32-36+76.
[4] Fu Rong, Jiang Guoping & Wang Baoyun. (2011). Grid preventive control considering the risk of system cascading failures. Power system protection and control (03), 12-17.
[5] Jiang Shengbo, Yang Jun, Wang Jianxiong, Yuan Wen, Xie Peiyuan, Luo Chao & Sun Yuanzhang. (2019). Large power grid cascading failure defense strategy based on prevention-emergency coordinated control. Electric Power Automation Equipment (12), 148-154.
[6] Yang Jun, Jiang Kai & Li Yang. (2017). Power system prevention-correction coordinated control strategy based on operational risk. Control and Decision (07),1301-1305.
[7] Li Yang. (2015). Research on the prevention-correction coordinated control based on operational risk (Master's thesis, Northeastern University).
[8] Tan, Y. , Li, Y. , Cao, Y. , Keune, B. , Jiang, L. ,& Li, S.(2016). Comprehensive decision-making method considering voltage risk for preventive and corrective control of power system. Iet Generation Transmission & Distribution.
[9] Jiang Tao, Chen Houhe & Li Guoqing. (2012). Voltage/reactive power zone regulation method based on local voltage stability index. Power System Technology (07), 207-213.
[10] Jia Hongjie, Sun Xiaoyan & Zhang Pei. (2006). Optimal power flow under voltage stability constraints based on L index. Journal of Electric Power System and Automation (01), 34-38+111.
[11] Kessel, P., Glavitsch, & H. (1986). Estimating the voltage stability of a power system. Power Delivery IEEE Transactions on.
[12] Wang Ansi. (2010). Power grid vulnerability assessment and stability control based on the accident chain (PhD dissertation, Huazhong University of Science and Technology).
[13] Wu Wenke, Wen Fushuan, Xue Yusheng, Zhao Xuanyu & Hu Rong. (2013). Power system cascading failure prediction based on Markov chain. Power System Automation (05), 29-37.

Effect of the Electromagnetic Susceptibility of Metal Interconnects: A Simulation Study

Jinfu Lin, Shulong Wang, Baowei Nie and Hongxia Liu*

Department Wide Bandgap Semiconductor Technology Disciplines State Key Laboratory, Xidian University, Xi'an, China

*Corresponding author email: hxliu@mail.xidian.edu.cn

Abstract. Large induced electromotive force will be generated on the metal interconnections under the action of strong electromagnetic pulse, thereby generating a large induced current. For copper metal interconnects commonly used in modern integrated circuits, large currents will cause electromigration, resulting in short or open in the circuit, causing damage to devices and circuits. In order to study the damage of metal interconnects under the action of electromagnetic pulse, in this paper, the induced electromotive force of metal interconnects under the action of HPM is derived through theoretical analysis, then analyzing the influence of different electromagnetic pulse parameters on the induced electromotive force of interconnects by the use of electromagnetic simulation software COMSOL. Using the established metal interconnect model, the electromagnetic sensitivity effect of metal interconnect is studied by analyzing the electric field distribution, current density and temperature distribution on the interconnect under electromagnetic pulse.
Keywords: Strong electromagnetic pulse; Metal interconnect; Induced electromotive force.

1. Introduction

Electromagnetic pulse (EMP) is a type of powerful energy field produced in an instant. This powerful energy field can be coupled into the electronic equipment through antennas, cable connections and metal pipes, resulting in a large induced voltage and current that causes the serious damage in the electronic equipment. The damage and interference of electromagnetic pulse to integrated circuits are mainly reflected in the damage of devices and metal interconnects [1-2]. For the MOS transistor circuit, the EMP damage is mainly reflected in the breakdown of the gate and the burning of the peripheral protection circuit [3], which leads to the latch-up effect of the digital CMOS circuit. For diodes, it is mainly the breakdown of the PN junction [4]. For metal interconnection lines, it is electromigration and fusing, which leads to short circuit and open circuit [5-7].

With the increase of interconnect density, the problem of electromagnetic reliability is becoming more and more serious. Electromagnetic pulse can transfer the coupling current or coupling voltage into the integrated circuit through the interconnect, resulting in a sharp rise in the temperature of the interconnect. Therefore, it is of great significance to study the response characteristics of interconnects and different EMP parameters. In recent years, much attention has been paid to the reliability of interconnection [8-12], especially the coupling between electromagnetic pulse and wire.

In this paper, the influence of electromagnetic pulse on metal interconnection is studied. The induced electromotive force of metal interconnection is solved by differential Maxwell equation and wave equation combined with boundary conditions. Then the influence of electromagnetic pulse period and pulse amplitude on induced electromotive force is analyzed by COMSOL simulation software version 5.2. In this paper, the copper interconnect model is established by COMSOL, and the electromagnetic excitation signal is applied to the interconnect port. By analyzing the current density and temperature field distribution of the metal interconnect under the strong electromagnetic pulse, the metal wire damage under the strong electromagnetic action is studied.

2. Model

In this paper, the hybrid physical field simulation software COMSOL is used to model and simulate the interconnect. Two physical fields, solid heat transfer and electromagnetic wave, are selected to analyze the electrothermal field distribution of the metal wire. Copper is selected as the metal interconnect material. In the model, the length is 20 μm, the width is 0.3 μm, the thickness is 0.18 μm, and the cross-sectional area is 0.054 μm^2. In order to simplify the problem, we assume that the electromagnetic pulse acts directly on the metal wire port through the air medium, and accumulates on the metal wire for a period of time to complete the heat transmission. The initial lattice temperature of the whole device is 300 K. The media and their corresponding parameters in the simulation are shown in the following table:

Table 1. The medium and its parameters.

Medium	Air	Copper(Cu)
Relative Permittivity (F/m)	1.0	1.0
Relative Permeability (H/m)	1.0	1.0
Electric Conductivity (S/m)	-	5.998×10^7
Density (Kg/m^3)	-	8960
Specific Heat Capacity(J/Kg·K)	1.4	385
Thermal Conductivity (W/m·K)	0.026	400
Melting Point (K)	-	1356

In order to facilitate the research, we use Gaussian pulse to describe high power microwave (HPM), and its electric field is:

$$E = E_0 \exp\left[-\frac{4\pi(t-t_0)^2}{\tau^2}\right] \tag{1}$$

Its magnetic field is:

$$H = E/\eta_0 , \eta_0 = 120\pi \tag{2}$$

The finite element method is used to solve the boundary value problem, which is a numerical method based on the differential equation of the electromagnetic field problem. Its principle is that many subdomains are used to replace the whole continuous region. In the subdomain, the unknown function is expressed by a simple interpolation function with unknown quotient, which transforms the original boundary value problem of infinite degrees of freedom into finite degrees of freedom. The finite element method is applicable to all kinds of physical fields described by any differential equation, and it is also suitable for solving time-varying fields, nonlinear fields and electromagnetic fields in complex media.

3. Results and Discussion

Electromagnetic pulse mainly enters the integrated circuit system through energy coupling, so the EMP effect is mainly electrical effect and thermal effect. Electric effect means that electromagnetic pulse forms induced voltage and induced current through coupling on metal connecting wire, and then acts on components. It will cause the performance of devices in the circuit to decline, the working state of devices to turn over, and the breakdown of semiconductor junctions. Thermal effect means that under the action of electromagnetic pulse, the temperature of the system will rise rapidly, and the heating rate is much higher than its heat dissipation rate, which is an approximate adiabatic process. Thermal effect can make the electronic components of an integrated circuit system or circuits that is very sensitive to temperature rise sharply, causing local thermal damage. If the temperature is too high in a short time, it will also cause local melting, resulting in permanent damage to the circuits.

3.1. Calculation and Simulation of Induced Electromotive Force

The second unit gives the electric and magnetic field equations of HPW to be studied. According to the Maxwell equation, the wave equation and the boundary condition, the formula for calculating the induced electromotive force on the interconnect can be obtained as follows:

$$\varepsilon = \frac{\partial \Phi}{\partial t} = \frac{\mu}{120\pi} E_0 \bullet \left(-\frac{4\pi}{\tau^2}\right) \bullet 2\left(t-t_0\right) \bullet \exp\left[-\frac{4\pi\left(t-t_0\right)^2}{\tau^2}\right] \bullet S \tag{3}$$

Among them, E_0 is the peak electric field intensity, t_0 is the delay time and τ is the time attenuation constant.

According to the formula for calculating the induced electromotive force on the interconnect obtained above, figure 1 shows the waveform of the induced electromotive force generated on the metal interconnect with different peak electric field intensity. The delay time is 500 ps and the time attenuation constant is 1 ns. It can be seen that when the electromagnetic pulse delay time and time decay constant are constant, the induced electromotive force on the metal interconnection line increases with the increase of the amplitude of the incident electric field intensity. When the peak value of incident field strength reaches 7.8×10^4 V/m, the maximum value of the induced electromotive force is about 4.5 V. The larger the incident field of the electromagnetic pulse is, the greater the induced electromotive force of the interconnect is.

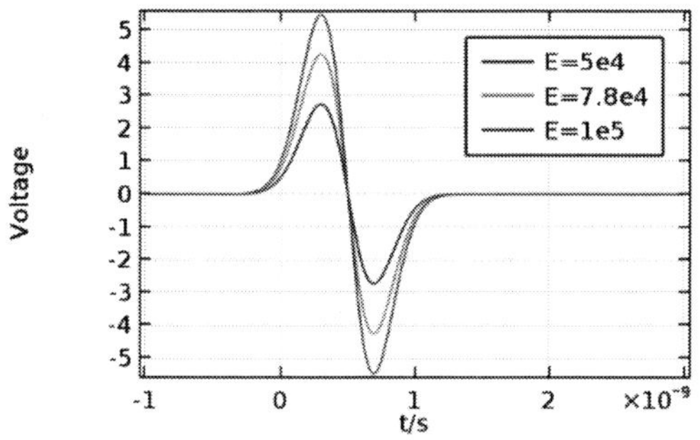

Figure 1. Effect of electric field intensity on induced electromotive force of interconnect.

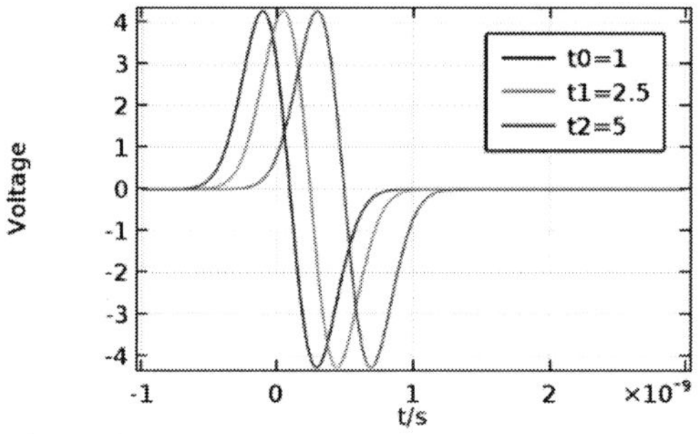

Figure 2. Effect of delay time on induced electromotive force of interconnect.

When the peak value of the incident field intensity is 7.8×10^4 V/m and the time attenuation constant is 1 ns, the waveform of the induced electromotive force on the wire under different delay time is shown in figure 2. It can be seen that the change of delay time has no effect on the peak value of induced electromotive force on metal interconnection lines, but has an effect on the time to reach the peak value. When the delay time changes from 100 ps to 500 ps, with the increase of the delay time, the time for the induced electromotive force on the metal interconnection line to reach the forward peak increases, and the time for reaching the reverse peak also increases accordingly.

When the peak value of the incident field strength is 7.8×10^4 V/m and the delay time is 500 ps, the waveform of induced electromotive force generated on the metal wire under different time attenuation constants is shown in Figure 3. It can be seen that the change of time attenuation constant will affect both the peak value and the period of induced electromotive force. With the increase of time attenuation constant, the period of induced electromotive force increases, and the time to reach the forward peak is the shortest and the time to reach the reverse peak is the longest. At the same time, the peak value decreases with the increase of time attenuation constant.

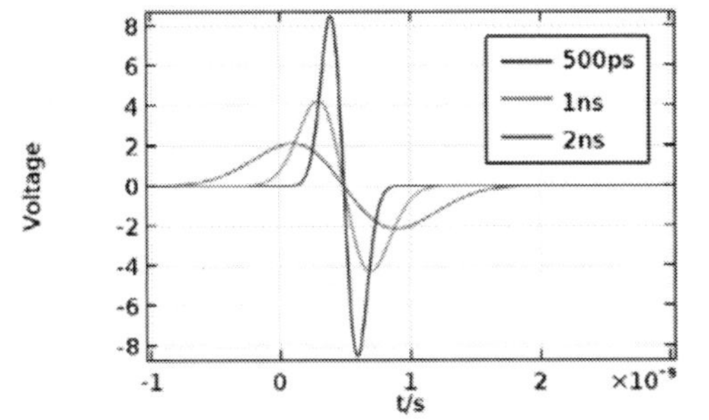

Figure 3. Effect of time attenuation constants on induced electromotive force of interconnect.

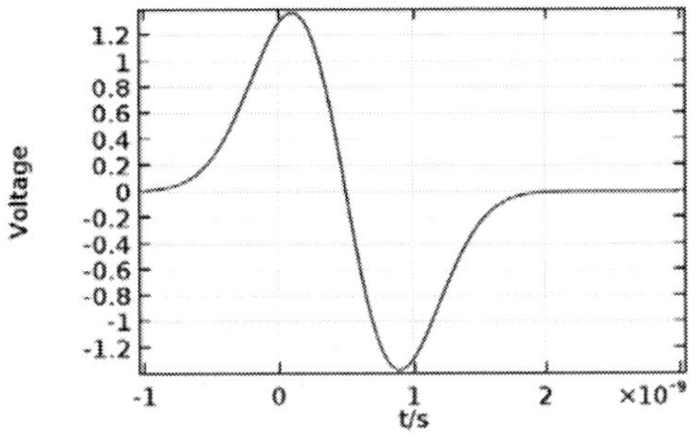

Figure 4. Inductive electromotive force corresponding to optimal EMP parameters.

From the above simulation results of the relationship between the parameters and the induced electromotive force of the interconnect, It can be seen that when the HPM is in the form of Gaussian pulse, the waveform of the induced electromotive force on the interconnect is very similar to that of the sine wave. The electric field strength of the HPW will affect the amplitude of the induced electromotive force. The larger the amplitude of the pulse, the higher the level of the pulse after attenuation, so that the induced electromotive force can be kept at a high value when coupled into the system. The delay time of HPM will affect the peak time of induced electromotive force. The larger the delay time is, the larger the peak time of electromotive force caused by interconnect coupling is. The HPM time attenuation constant will affect the peak value and the time of reaching the peak value of the induced electromotive force. The larger the time attenuation constant is, the smaller the peak value of the induced electromotive force is, the larger the period of the induced electromotive force is, and the shorter the time it takes to reach the positive peak value. Therefore, to minimize the peak induced electromotive force on the interconnect, the HPM electric field intensity is reduced and the time attenuation constant is increased. Figure 4 shows the calculated minimum induced electromotive force. In this case, the HPM electric field intensity is 5×10^4 V/m, the delay time is 500 ps, and the time attenuation constant is 2 ns. It can be seen that the peak value of the induced electromotive force on the interconnect is about 1.35 V.

3.2. Simulation of the Electrothermal Field Distribution

Then the electric field and thermal distribution of metal interconnects under the action of HPM is studied. At one end of the metal interconnect, a HPM wave with a peak electric field intensity of 7.8×10^4 V/m and a frequency of 1.5 GHz is applied. The induced current density and temperature distribution on the metal interconnect are shown in the following figure.

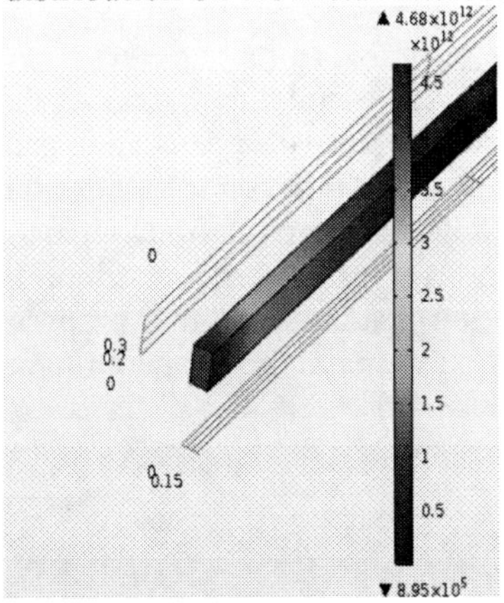

Figure 5. Induced current density distribution of metal interconnects under HPM.

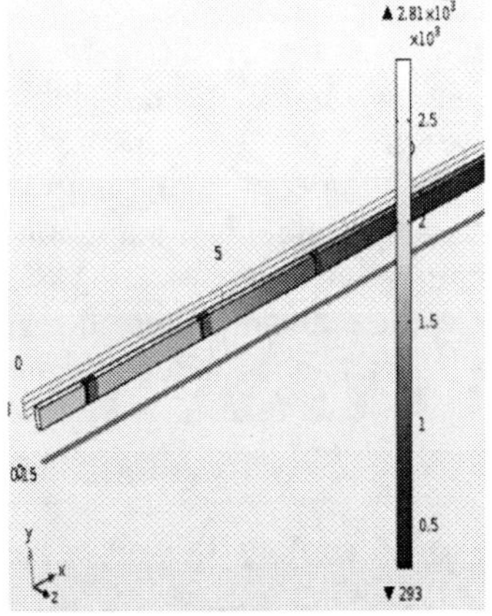

Figure 6. Temperature distribution of metal interconnect under HPM.

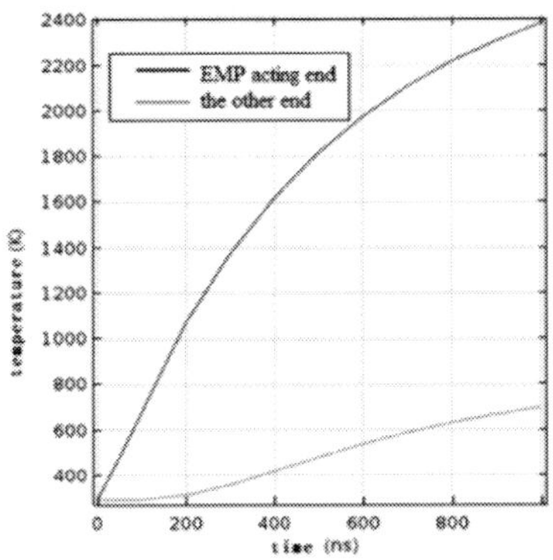

Figure 7. The temperature variation curve of different nodes.

In the very short time to the action of electromagnetic pulse, a great current will be produced on the wire. As can be seen from figure 5, the peak current density is 4.68×10^{12} A/m^2, and the peak induced current on the interconnect is 252.72 mA. Under the action of large current, the electromigration phenomenon of metal wire is very easy to occur. It can be seen from figure 6 that in a very short time under the action of electromagnetic pulse, the temperature on the wire reaches the melting point temperature of the copper material, causing the wire to fuse. From the figure 7 it can be clearly seen that the temperature rises faster as it gets closer to the end of the HPM action. This is due to the fact that under the action of HPM, the heat dissipation rate near the end face is much less than the heat generation rate, thus forming a large temperature gradient on the metal line. Due to the increase of temperature, the diffusion speed of atoms is accelerated, and the driving ability of electromigration is significantly improved. Finally, the electromigration phenomenon is formed rapidly.

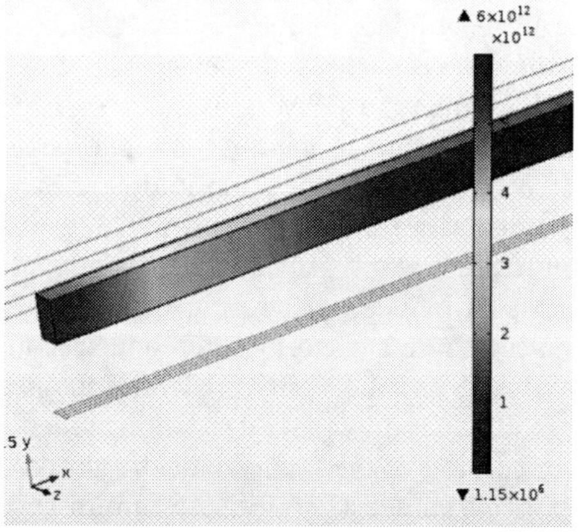

Figure 8. Induced current density distribution of metal interconnects under 1 GHz HPM.

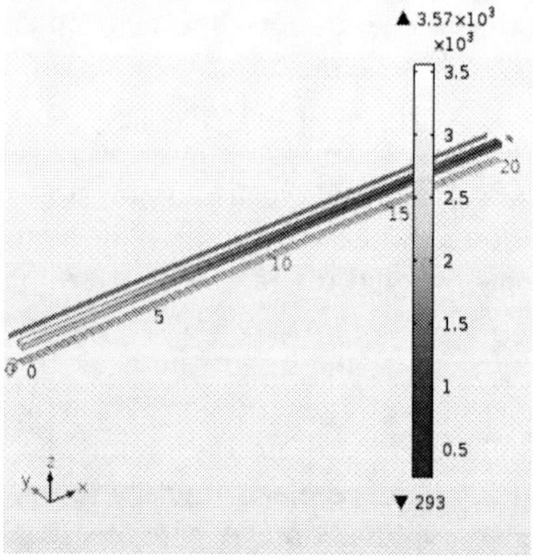

Figure 9. Temperature distribution of metal interconnects under 1 GHz HPM.

Keeping the electric field intensity constant, the pulse frequency of the electromagnetic pulse is reduced to 1 GHz, and the induced current density distribution and temperature distribution on the metal interconnect are shown in the figure above. It can be seen that with the increase of the pulse frequency, the value of the induced current on the wire decreases and the temperature on the wire decreases, which indicates that the higher the frequency, the more the number of pulses per unit time. However, because the power of the pulse source is certain, the amplitude of a single pulse decreases, which leads to the decrease of the induced current and the reduction of heat generated in the wire.

4. Conclusion

In this paper, we first derive the expression of the induced electromotive force on the metal interconnect under the action of a Gaussian pulse. Then, combined with the simulation software, the relationship between electric field intensity, pulse period and induced electromotive force is analyzed. In order to solve the current density and temperature distribution on the metal interconnect under the action of electromagnetic pulse, we establish the metal interconnect model based on the finite element method and the electromagnetic simulation software COMSOL, and get its specific distribution through three-dimensional simulation. At the same time, the simulation results show that properly increasing the electromagnetic pulse frequency can reduce the induced current and the temperature on the interconnect.

Acknowledgments

This paper is supported by Science and Technology Project of Headquarters of State Grid Corporation of China(5700-201941314A-0-0-00).

References

[1] Prasad D, Pan C, Naeemi A Impact of interconnect variability on circuit performance in advanced technology nodes. 17th ISQED, Mar 2016, Santa Clara, USA.

[2] Cheng Y, Chang Y, Len J, et al. Effect of layout on electromigration characteristics in copper dual damascene interconnects. Microelectronic Engineering, 2014, 128(5):19-23.
[3] Lin J, Liu H, Long S The MOSFET performance degrading induced by electric heating coupling under electromagnetic pulse environment. EDSSC, June 2018, Shenzhen, China.
[4] Dobykin VD Development of the theory for thermal damage of semiconductor structures by high-power electromagnetic radiation. J. Commun. Technol. Electron., 2008, 53(1):100-103.
[5] Wong T Time Dependent Dielectric Breakdown in Copper Low-k Interconnects: Mechanisms and Reliability Models. Materials, 2012, 5(9):1602-1625.
[6] Xu J Topological Structure and Analysis of Interconnection Networks. Springer Berlin, 2001, 7(3):969-970.
[7] Chang Y, Chen C, Chang T, et al. Fast phase transformation due to electromigration of 18 μm microbumps in three-dimensional integrated-circuit integration. Materials Letters, 2014, 137(15):136-138.
[8] Sandberg H and Murray R Model reduction of interconnected linear systems. Optimal Control Applications and Methods, 2009, 30(3):225-245.
[9] Fu C, Tan C, Wu S, et al. Width dependence of the effectiveness of reservoir length in improving electromigration for Cu/low-k interconnects. Microelectron Reliab, 2010, 50:1332-1335.
[10] Liang T and Xie Y Analytical Electromagnetic Field Coupling to Transmission Line Model Exploiting the Reciprocity Theorem. IEEE Transactions on Magnetics. 2021, 1-1. 10.1109/TMAG.2021.3079967
[11] Zhong X, Yu s, Chen L, et al. Test methods for electrochemical migration: a review. Journal of Materials Science: Materials in Electronics, 2017, 28(2):2279-2289.
[12] Olsen R and Tarditi A EMP Coupling to a Straight Conductor Above Ground: Transmission Line Formulation Based on Electromagnetic Reciprocity. IEEE Transactions on Electromagnetic Compatibility, 2019, 61(3):919-927.

Research on Modelling Method of Natural Illuminance Based on RBFNN

Yujie Zhang and Jing Guo*

School of Electrical and Control, Shaanxi University of Science and Technology, Xi'an, China

*Corresponding author email: 254874696@qq.com

Abstract. Aiming at the problem that the indoor natural illuminance is difficult to calculate accurately due to many complex factors in the environment. This paper proposes a method to establish the natural illuminance model by using radial basis function neural network (RBFNN). The RBFNN was trained by collecting indoor natural light data to obtain the benchmark model for calculating the natural illuminance distribution. The illuminance sensor arranged indoors was used to measure the real-time natural illuminance, and the output of the benchmark model was modified accordingly, so as to obtained a model that could calculate the real-time natural illuminance at any point in the room. The experimental results show that the maximum normalized error between the calculated and measured illuminance values of the model in this paper is no more than 15%, and the error of most test data is less than 8%, which solves the problem of accurate calculation of indoor natural illuminance, so as to provide support for indoor lighting control system.
Keywords: Intelligent lighting; Radial basis function neural network; Natural illuminance model.

1. Introduction

In the indoor intelligent lighting control system, reasonable use of natural light can greatly reduce the energy consumption of lighting [1]. Measuring or estimating the indoor natural illumination distribution0020is the premise of rational use of natural light. At present, photoelectric sensor is generally used for indoor measurement of illuminance, but it can only indirectly reflect the average illuminance in a large range of space. For the intelligent lighting control system that needs to obtain indoor illumination distribution, photoelectric sensor can't meet its accurate lighting control requirements.

To solve these problems, some researchers proposed a calculation model of indoor natural illuminance based on CIE sky type. The 15 types of sky published by CIE are used as the data source for calculating indoor natural illuminance. The distribution of sky brightness is obtained by taking the longitude and latitude of the building, the alternation of four seasons, the alternation of day and night, the window opening of the building, the orientation of the building, and the indoor reflectivity as variables, and the calculation model of indoor illuminance is established by using optical principles [2-6]. However, due to the use of expensive equipment to obtain the light climate parameters, this method brings a large cost to the data acquisition. The actual lighting environment is idealized in the modelling process, resulting in low accuracy of calculation results. In recent years, with the rise of artificial neural network, it is widely used in fault prediction and diagnosis, medical image classification, speech recognition and so on. Therefore, some scholars use BP neural network to estimate natural illumination [7-8], but the input of these neural networks is a dozen variables including time, outdoor temperature, clear sky index, solar radiation, building orientation, number of Windows, window orientation and so on, resulting in the time span of collecting training data for several months or even a whole year. It brings great cost to the establishment of the model.

In short, the existing methods consider too many input variables when modelling natural illuminance, resulting in high data acquisition cost of the model. In this paper, a natural illuminance modelling method based on radial basis function neural network (RBFNN) is proposed. It does not take the above complex and changeable environmental variables as parameters, but first uses radial basis function to establish a reference model with indoor position as input and natural illuminance as output. Then, a few sensors are arranged indoors to comprehensively reflect the impact of complex environment on illuminance distribution by using the changes of real-time measured natural illuminance, so as to correct the output of the benchmark model and form a complete natural illuminance model. Therefore, the method proposed in this paper avoids the high cost of data acquisition in natural light modelling, and brings great convenience to engineering implementation.

2. Radial Basis Function Neural Network

Radial basis function is a real valued function whose value only depends on the distance to a certain centre C, that is, $\phi(X)=\phi(\|X-C\|)$.

Radial basis function neural network (RBFNN) is a three-layer neural network composed of input layer, hidden layer and output layer [9-10]. The hidden layer is also called radial base layer. The activation function of each neuron is a radial basis function, and the structure is shown in Figure 1.

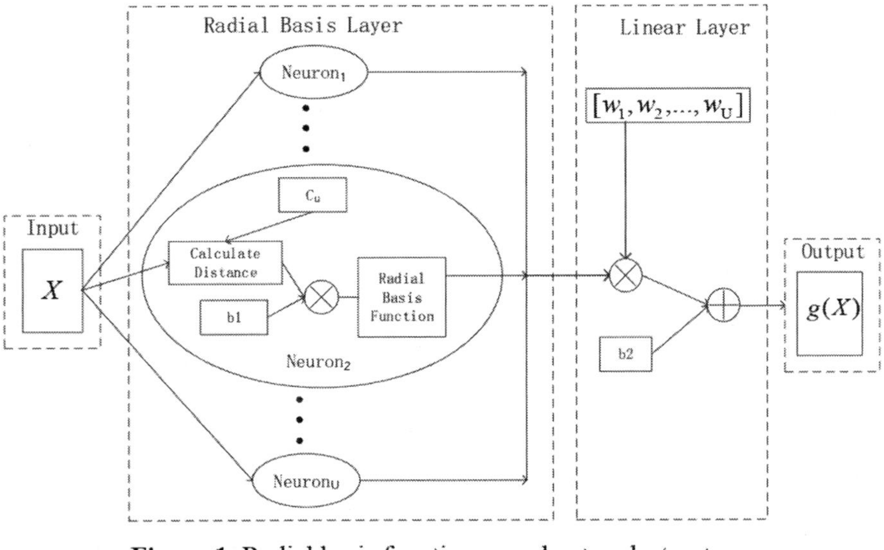

Figure 1. Radial basis function neural network structure.

RBFNN can generally be expressed as:

$$g(X)=\sum_{u=1}^{U}(w_u \times e^{-(b1\times\|X-C_u\|)^2})+b2 \tag{1}$$

Radial basis function neural network can mainly be used to solve regression problems and classification problems. Its approximation ability, classification ability and learning speed are better than BP (back propagation) neural network. It has simple structure, fast learning convergence speed, can approximate any nonlinear function and overcome the local minimum problem. Taking the position coordinates of the training data set as the input and the illuminance value as the output, this paper uses the basis function neural network to establish the indoor natural illuminance benchmark model.

3. Natural Illuminance Modelling

Indoor personnel generally pay attention to a certain horizontal working surface, so the natural illuminance model in this paper is to determine the relationship between each position on the indoor horizontal surface and the natural illuminance. The indoor natural light illumination modelling is divided into two steps: the first step is to select a certain period of the day (under the condition of natural light) with an illuminance meter to obtain the illuminance values at each measuring point on the indoor working face in this period. The RBF neural network takes the position information of the measuring point as the input and the illuminance value as the output to train the network to obtain the natural illuminance benchmark model based on this period; In the second step, on the basis of the benchmark model, the output of the benchmark model is corrected through the illuminance data of a few key points measured in real time, so as to achieve the purpose of calculating the real-time illuminance value at any position on the indoor working face.

3.1. Natural Illuminance Benchmark Model

The indoor horizontal plane is taken as the xoy plane of the three-dimensional coordinate system. The z-axis represents the natural illuminance on the horizontal plane. Then the illuminance distribution surface of natural light in the room is the result of the joint influence of many factors:

1) The internal factors affecting the indoor natural light intensity distribution include: the geographical location of the building, the orientation of the building, the location, size and number of windows, the reflective characteristics of indoor walls and object materials, the greater the depth from the window to the indoor, the weaker the natural light intensity, and so on [11].

2) The external factors affecting the distribution of indoor natural illuminance mainly include: with the rise and fall of the sun in the East and West, the radiation intensity of natural light changes from weak to strong and then weak, and the peak area of indoor illuminance also shifts from east to west; For the northern hemisphere, the light is strong in summer and weak in winter; On cloudy and rainy days, the direct sunlight is blocked by clouds, and the indoor illumination is insufficient, while on sunny days, the indoor illumination is sufficient [12-14].

As mentioned above, the indoor natural illumination distribution is affected by many factors, especially the randomness affected by weather changes. The current research is considering these factors comprehensively. First, calculate the sky brightness, and then take the sky as an area light source to establish the illuminance model. In this way, the modelling is difficult, and various professional meteorological equipment are needed to monitor the meteorological data, which is very inconvenient in the engineering implementation.

Therefore, this paper assumes that the internal factors affecting the indoor natural illuminance distribution determine the approximate shape of the indoor natural illuminance distribution surface. That is, for a given indoor environment, when there is a fixed outdoor light source, the value of the indoor illuminance distribution surface changes only with the change of the radiation intensity of the light source, while the shape of the illuminance distribution surface is basically unchanged. The weather conditions will not change suddenly in a short time most of the time. Therefore, the illuminance meter is used to collect the horizontal natural illuminance data in a short period of time, The illuminance data and the position data of the measuring points are made into a data set. The RBFNN trained by the training set can calculate the indoor natural illuminance distribution during the data acquisition period. The trained network is called the natural illuminance benchmark model.

3.2. Natural Illuminance Model

However, in order to obtain the real-time natural illuminance model, we must consider various external factors changing with time, so as to change the shape of the indoor natural illuminance distribution surface. Therefore, this paper proposes to reflect the dynamic change of illuminance distribution according to the monitored real-time illuminance data, and calculate the real-time natural illuminance by reasonably modifying the output of the natural illuminance benchmark model.

The illuminance value of natural illuminance distribution surface changes continuously with position. The smaller the distance between two points, the greater the correlation of illuminance value. Select a few key indoor locations to install illuminance sensors, and reflect the changes of natural illuminance in two periods at the same location through the ratio of the illuminance value measured in real time to the illuminance in the period of collecting training data. When calculating the real-time natural illuminance of the target point, according to the principle that the smaller the distance is, the greater the correlation between the illuminance values between the two points is, the natural illuminance changes of these few positions are applied with their respective influence correction weights according to the distance between them and the target point, and finally summed. Therefore, this method of correcting the illuminance of other points calculated by the benchmark model by using the natural illuminance changes of a few positions is actually to use the illuminance changes of a few positions to evaluate the dynamic changes of the whole indoor illuminance distribution according to the continuity of illuminance distribution, and then correct the model output to obtain the real-time indoor natural illuminance distribution.

Specifically, the correction formula is shown in formula (2):

$$\tilde{E} = E' \times \sum_{r=1}^{R} \frac{E_r w_r}{E_r' \sum_{r=1}^{R} w_r} \tag{2}$$

Where, w_r is the adjustment weight, which can be calculated as:

$$w_r = \frac{1}{\sqrt{(x - x_r)^2 + (y - y_r)^2}} \tag{3}$$

$[x, y]$ indicates the position of illuminance to be calculated (target position). $[x_r, y_r]$ defines the location of the sensor r. x_r and y_r share the same coordinate system with the working face. Equation (3) indicates that w_r is defined by the distance between the target position and the sensor r. The shorter the distance, the greater the influence of the real-time illumination change measured by the sensor on the illumination calculation of the target position. In equation (2), E' represents the output of the benchmark model at the target position. $\tilde{E}$ represents the real-time natural illuminance value obtained after correction. R represents that a total of R illuminance sensors are selected to measure the real-time illuminance value. $E_r (r = 1, 2, ..., R)$ represents the real-time natural illuminance value measured by sensor r. E_r' represents the illuminance value of the position during the training data acquisition period.

Therefore, the natural illuminance model is summarized as shown in Figure 2. The illuminance of the target position and the natural illuminance value of the sensor position in the training data acquisition period can be obtained through the RBFNN benchmark model. At the same time, the adjustment weight w_r is obtained from the coordinates of the target position and the sensor position. Finally, combined with the real-time natural illuminance E_r measured by the sensor, $\tilde{E}$ is obtained through the correction formula.

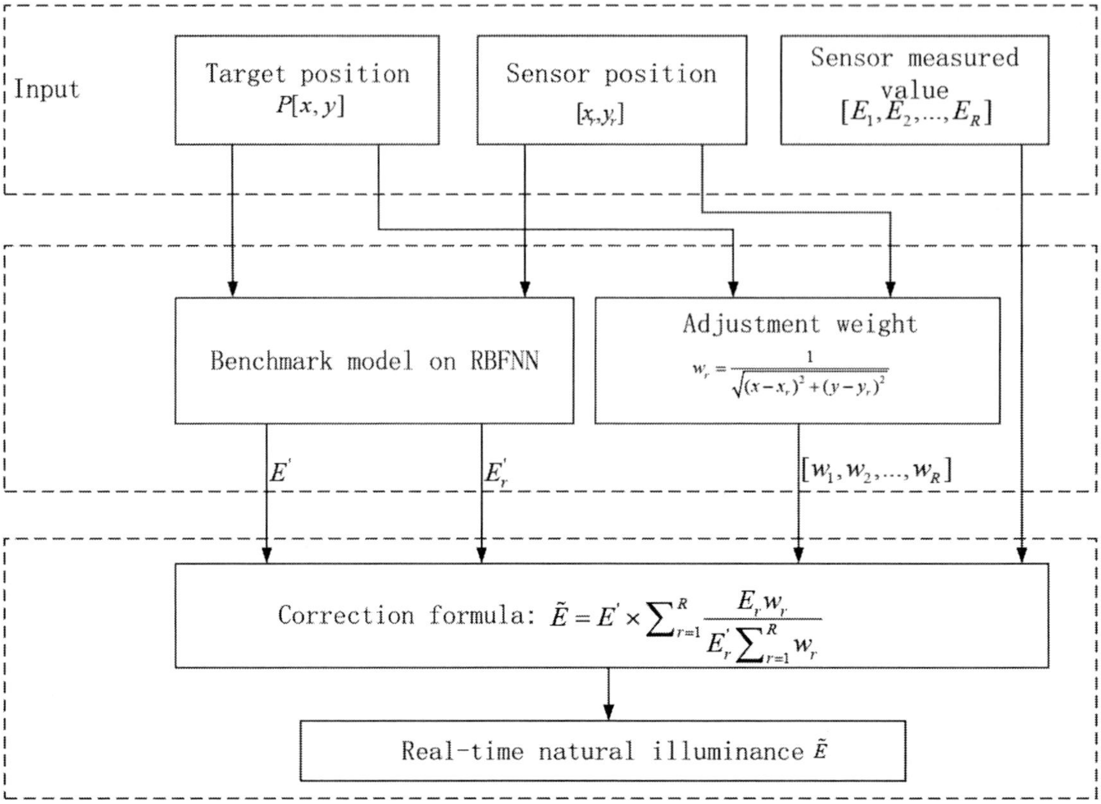

Figure 2. Frame diagram of natural illuminance model.

4. Experimental Verification

The experiment verifies the feasibility of the modelling method by comparing the illuminance calculated by the natural illuminance model with the verifying data. Since the measured value E of the key point sensor is an important part of the model correction formula (2), it is necessary to determine the position of the key point sensor first. Therefore, we propose three sensor layout schemes, compare their respective experimental results, and select the optimal scheme as the sensor layout scheme in this paper.

The experimental site is 2a505 laboratory in the experimental building of Shaanxi University of science and technology, and the building geographical coordinates are $108.92°$ east longitude and $34.35°$ north latitude. The laboratory is a $7.2m \times 7.8m \times 3.2m$ space, as shown in Figure 3. The indoor window faces south, the window height is $0.8m$, and the overall size of the window is $6m \times 1.35m$.

Figure 3. Schematic diagram of indoor three-dimensional space.

4.1. Data Acquisition and Network Training

On May 22, 2021, the weather was cloudy. The horizontal illuminance on the indoor desktop (height of 75cm) at multiple times of the day was measured with an illuminance meter. The horizontal coordinates of the measuring points and the measured illuminance value are used as the input and output of RBFNN to train the neural network. figure. 4 is an indoor top plan view. The dots in the figure represent the training data points measured from 9:00 to 9:05. A total of 24 groups of data are measured as the training data set of the benchmark model. The square represents three test points for verifying the illuminance value calculated by the model.

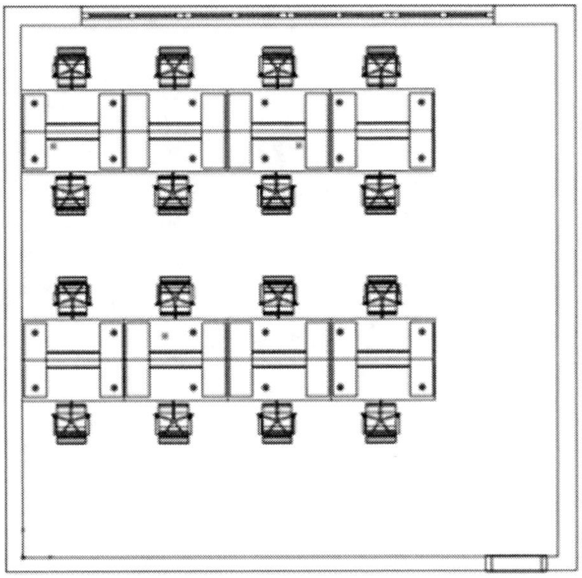

Figure 4. Indoor top plan.

Figure 5 (a), (b) and (c) are the layout schemes of three types of key point sensors. The triangle in the figure shows the location of the key point illumination sensor for obtaining real-time natural illumination.

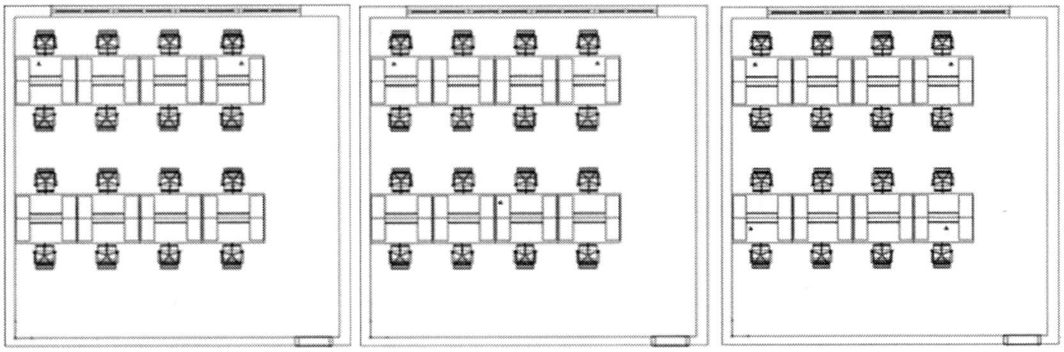

(a) (b) (c)

Figure 5. Schematic diagram of key point sensor position.

The training method of RBFNN in this experiment is "the centre is selected from the sample input": first initialize a radial basis function neural network without any hidden nodes, and then add hidden nodes to the network one by one. When the error meets the requirements, stop adding nodes. In order to minimize the number of hidden nodes, the node that can reduce the error the most will be selected as a new node every time.

The RBFNN trained on MATLAB is used as the natural illuminance benchmark model, and the illuminance distribution surface calculated by the benchmark model is shown in Figure 6.

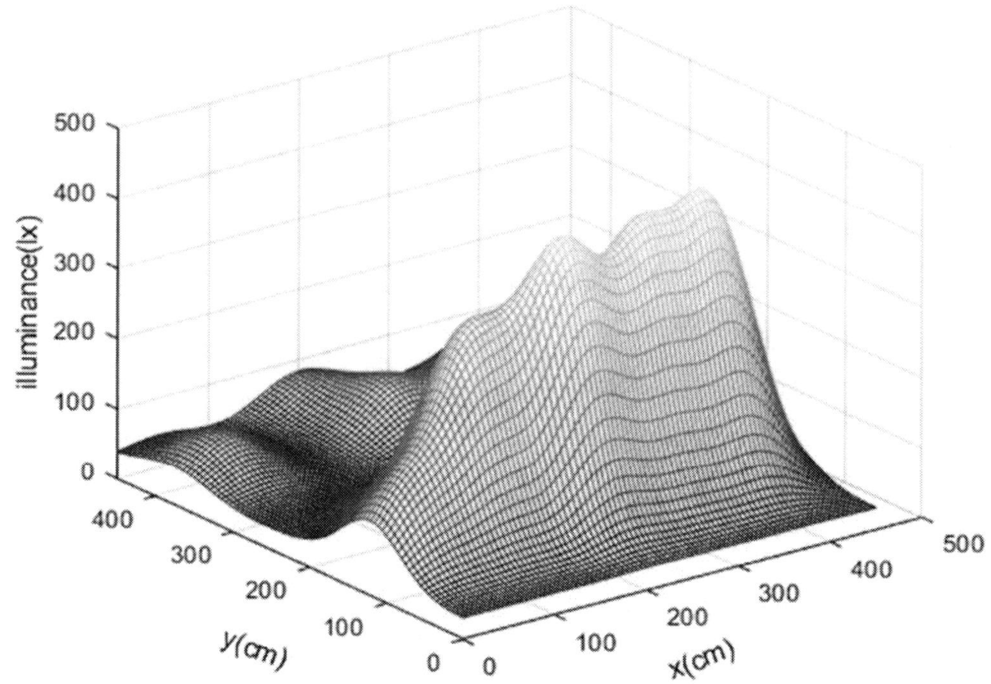

Figure 6. Illuminance distribution surface.

4.2. Analysis of Test Results

The experiment measures the illuminance of the test points shown in Figure 4 in each time period from 10:00 to 17:00 on the same day. At the same time, according to the real-time illuminance measured by the key point sensor, the illuminance calculation value is obtained by using the natural illuminance model described above. Compare the measured data with the illuminance value calculated by the model, and the statistical results are shown in the table below. Table 1 shows the experimental results of arranging two sensors shown in figure. 5 (a), table 2 shows the experimental results of arranging three sensors shown in figure. 5 (b), and table 3 shows the experimental results of arranging four sensors shown in figure. 5 (c).

Table 1. Calculation results of sensor layout scheme (a).

Time	Average error Δ / lx	Normalization error Δ'	standard deviation σ / lx
10:00~10:05	29.0	14.3%	28.6
11:00~11:05	15.3	9.9%	16.9
12:00~12:05	15.0	7.0%	13.2
13:00~13:05	18.1	23.3%	16.9
14:00~14:05	16.0	7.8%	17.2
15:00~15:05	9.2	7.9%	9.3
16:00~16:05	7.6	10.1%	7.5
17:00~17:05	5.6	33.0%	5.9

Table 2. Calculation results of sensor layout scheme (b).

Time	Average error Δ / lx	Normalization error Δ'	standard deviation σ / lx
10:00~10:05	16.0	7.9%	17.1
11:00~11:05	6.0	3.9%	6.1
12:00~12:05	12.0	5.6%	12.2
13:00~13:05	11.2	14.4%	10.1
14:00~14:05	10.0	4.9%	10.2
15:00~15:05	8.3	7.1%	8.4
16:00~16:05	6.0	8.0%	6.5
17:00~17:05	1.0	5.9%	1.3

Table 3. Calculation results of sensor layout scheme (c).

Time	Average error Δ / lx	Normalization error Δ'	standard deviation σ / lx
10:00~10:05	13.0	6.4%	14.1
11:00~11:05	4.0	2.6%	5.6
12:00~12:05	10.9	4.6%	11.3
13:00~13:05	12.2	15.7%	16.1
14:00~14:05	13.0	6.4%	13.4
15:00~15:05	4.5	3.9%	2.4
16:00~16:05	6.5	8.7%	4.9
17:00~17:05	0.8	4.7%	0.6

The average error Δ in the table is defined as the average of the absolute error between the illuminance $\tilde{E}$ calculated by the model and the verifying data E . The normalization error Δ' is defined as the normalization of the average error by the maximum value of verifying data. The standard deviation σ is defined as the standard deviation between the illuminance $\tilde{E}$ calculated by the model and the verifying data E .

Comparing the statistical results in the table, it can be seen that the errors of sensor schemes (b) and (c) are less than (a), (b) and (c) have little difference. Therefore, three sensor layout schemes (b) are selected as the optimal scheme in this paper. The maximum normalized error between the illuminance value calculated by the natural illuminance model and the actual measured natural illuminance value is no more than 15%, and most of the errors are within 8%, reaching the standard that personnel are not sensitive to environmental illuminance changes within 20%[15].

5. Conclusion

In this paper, a natural illuminance model based on RBFNN is proposed. Compared with the previous illuminance model, the model has the characteristics of less input parameters and simple structure. By comparing the experimental results of the three sensor layout schemes proposed in this paper, it is suggested that the location of the selected key point sensor can reflect the obvious change of natural illumination over time and can cover the working area in a wide range, such as the sensor layout scheme (b) shown in this paper. The illuminance model in this paper is only for natural light. The future research should comprehensively consider the indoor artificial lighting, and further establish the illuminance model of mixed light source.

References

[1] Pandharipande A, Caicedo D. Adaptive Illumination Rendering in LED Lighting Systems[J]. IEEE Transactions on Systems Man & Cybernetics Systems, 2013, 43(5):1052-1062.

[2] Hailong Chu, Yuning Zhang, Jian Wang. Indoor natural illuminance calculation model based on CIE sky model [J]. Optoelectronic technology, 2018,38 (03): 167-171 + 177.

[3] Younju Yoon,Jin Woo Moon,Sooyoung Kim. Development of annual daylight simulation algorithms for prediction of indoor daylight illuminance[J]. Energy & Buildings,2016,118.

[4] Hui Xiao, Jinguang Wang, Yi Zhang, et al. Study on actual sky model of general sky based on CIE standard [J]. Journal of lighting engineering, 2008 (01): 20-24.
[5] Yuxiang Wu, Chunyan Liu, Shangsheng Wen. Calculation of sky brightness distribution at any time based on CIE sky model [J]. Journal of optics, 2014,34 (11): 35-41.
[6] Hong Liu, Baizhan Li, Runming Yao. Dynamic calculation method of indoor illuminance of daylight lighting [J]. Journal of Chongqing University (NATURAL SCIENCE EDITION), 2007 (08): 85-88.
[7] Tuğçe Kazanasmaz,Murat Günaydin,Selcen Binol. Artificial neural networks to predict daylight illuminance in office buildings[J]. Building and Environment,2008,44(8).
[8] Varvara N. Katsanou,Minas C. Alexiadis,Dimitris P. Labridis. An ANN-based model for the prediction of internal lighting conditions and user actions in non-residential buildings[J]. Journal of Building Performance Simulation,2019,12(5).
[9] Shuhuai Guan, Yanxia Shen. Power load forecasting based on particle swarm optimization radial basis function neural network [J]. Sensors and Microsystems, 2021,40 (05): 128-131.
[10] Jingjing Li, Shaohua Xu. RBF process neural network based on combined kernel function and its application in indicator diagram diagnosis [J]. Computer engineering and science, 2021,43 (04): 746-752.
[11] Lei Pan. Research on indoor intelligent lighting control strategy combined with natural lighting [D]. Henan: Zhengzhou University, 2014.
[12] Ashish Pandharipande,David Caicedo. Daylight integrated illumination control of LED systems based on enhanced presence sensing[J]. Energy & Buildings,2010,43(4). Ashish Pandharipande,David Caicedo. Daylight integrated illumination control of LED systems based on enhanced presence sensing[J]. Energy & Buildings,2010,43(4).
[13] Vittorio Ferraro,Marilena Mele,Valerio Marinelli. Sky luminance measurements and comparisons with calculation models[J]. Journal of Atmospheric and Solar-Terrestrial Physics,2011,73(13).
[14] Yanna Ping. Research on indoor illumination control system based on natural daylighting [D]. Zhengzhou University, 2012.
[15] Yukio Akashi,Jason Neches. Detectability and Acceptability of Illuminance Reduction for Load Shedding[J]. Journal of the Illuminating Engineering Society,2004,33(1).

Modeling and Simulation Analysis of the Temperature Field of High Speed Missile Cabin

Long Yi*, Yuliang Li, Jindong Ouyang, Liyuan Zhu and Yaoqing Zhang

Jiangxi Hongdu Aviation Industry Group Co., Ltd., Nanchang, China

*Corresponding author email: 34016@nchu.edu.cn

Abstract. A temperature field analysis method is developed for the cabin of high-speed missile. The high fidelity three-dimensional model of the thermal protection system is build based on the finite element method, including structure and equipments inside the cabin. The model can express the heat conduction path, radiation relationship and the heat capacity of the equipment. The aero-heating and heat transfer are solved simultaneously. The convection coefficient relevant to experiment is used to simulate the heat transfer of the gas in the cabin. It can be used to large-scale engineering calculation for the thermal protection system of high-speed aircraft. The numerical example of a head cabin for high-speed missile shows that the obvious regional characteristics of the surface temperature outside the cabin is presented, which reflects the coupling effect of aero-heating and cabin structure. The temperature of the equipment in the cabin rises due to radiation and convection heat transfer. Heat capacity of the equipment and natural convection in the gap have obvious influence on the rate and amount of temperature rising. The developed APDL command stream can integrate commercial software of CAD and CAE. It is suitable for large and complex engineering structure because of the advantages such as high computational efficiency and strong portability.
Keywords: Temperature field; Thermal protection system; High-speed missile; APDL.

1. Introduction

When the aircraft is flying at high speed in the atmosphere, it will produce severe aerodynamic heating, which is also called "thermal barrier". The high temperature generated by aerodynamic heating will reduce the strength limit of the material and the load-bearing capacity of the structure, and cause thermal deformation of the structure, destroy the aerodynamic shape of the structure, and even directly cause the structure to burn. Therefore, a thermal protection structure must be used to dissipate, absorb or prevent heat transfer to the inside. On the other hand, the electronic equipment in the cabin will also generate a lot of heat when working in a confined space for a long time. Combination of the two will cause the temperature of the environment in the cabin to rise, which may affect the performance of the equipment[1-3]. Therefore, it is necessary to conduct a detailed analysis for the thermal protection of the aircraft[4].

Scholars have carried out a lot of theoretical research on the coupling problem between aerodynamic heat and the temperature of the structure[5-10]. In order to improve the calculation accuracy, CFD method is often used to calculate the external flow field, and finite element method is used to analyze the temperature field of the structure[11-17]. However, due to the huge cost of CFD method for solving aerodynamic heat, it is difficult to apply it to large-scale engineering structures. In order to improve the calculation efficiency, some researchers have completed the coupling analysis of aerodynamic heating / thermal response based on mature commercial software such as CFX, FLUENT and ANSYS[18]. Many researchers have also used engineering algorithm to calculate aerodynamic heating and commercial finite element software to calculate temperature field in the scheme design stage. However, the outer wall structure of the aircraft is mainly considered in the model, but the heating and heat storage capacity of the equipment in the cabin are less considered. According to the requirements of high-speed missile, such as reentry warhead, cruise missile and air-to-air missile, this paper develops a full model analysis method of the cabin temperature field. This method is based on the finite element method to perform high-fidelity three-dimensional modeling of the thermal protection system, cabin structure and cabin equipment. The model can express the heat conduction path, radiation relationship and the heat capacity of the equipment. The aero-heating and heat transfer are solved simultaneously. The convection coefficient relevant to experiment is used to simulate the heat transfer of the gas in the cabin. It can be used to large-scale engineering calculation for the thermal protection system of high-speed aircraft.

2. Aerodynamic Heat Calculation Method

First, the aerodynamic heat is calculated using the engineering formula. The heat flow formula for stagnation point developed by Fay-Riddell is suitable for high-speed missile warheads, which is:

$$q = 0.763\,\mathrm{Pr}^{-0.6}\left(\frac{\rho_w \mu_w}{\rho_s \mu_s}\right)^{0.1}\left(\rho_s \mu_s \left(\frac{du_e}{dx}\right)_s\right)^{0.5}\left[1+\left(Le^{0.52}-1\right)\frac{h_D}{h_s}\right]\left(h_s - h_w\right) \quad (1)$$

The reference enthalpy method developed by Echert is used to obtain the heat flux in the non-stagnation point region. For laminar flow:

$$q = 0.332\rho_e u_e \,\mathrm{Pr}^{-2/3}\,\mathrm{Re}_s^{-0.5}\left(\frac{\rho^* \mu^*}{\rho_e \mu_e}\right)^{0.5}\left(\frac{\rho_e \mu_e u_e r^{1.25} s}{\int_0^s \rho_e \mu_e u_e r^{1.25} ds}\right)^{0.5}\left(h_r - h_w\right) \quad (2)$$

For turbulent flow:

$$q = 0.0296\rho_e u_e \,\mathrm{Pr}^{-2/3}\,\mathrm{Re}_s^{-0.2}\left(\frac{\rho^*}{\rho_e}\right)^{0.8}\left(\frac{\mu^*}{\mu_e}\right)^{0.2}\left(\frac{\rho_e \mu_e u_e r^{1.25} s}{\int_0^s \rho_e \mu_e u_e r^{1.25} ds}\right)^{0.2}\left(h_r - h_w\right) \quad (3)$$

3. Temperature Field Calculation Method

Second, the temperature is calculated by finite element method. According to the law of conservation of energy and Fourier's law, the variables of the three-dimensional transient temperature field are functions of space coordinates and time:

$$\rho c\frac{\partial T}{\partial t} - \frac{\partial}{\partial x}\left(\mathrm{k}_x \frac{\partial T}{\partial x}\right) - \frac{\partial}{\partial y}\left(\mathrm{k}_y \frac{\partial T}{\partial y}\right) - \frac{\partial}{\partial z}\left(\mathrm{k}_z \frac{\partial T}{\partial z}\right) - \rho Q = 0 \quad (4)$$

Combined with the corresponding boundary conditions, the finite element method can be used to solve the above differential equations. In the design stage of the high-speed missile thermal protection scheme, the shape change of the missile body is not considered, and the adiabatic wall temperature boundary condition can be used in the analysis of the aero thermal environment. When calculating the temperature field, use the following formula to convert the heat flux and apply it to the outer wall of the cabin:

$$q_1 = \frac{q\left(h_r - h_{w1}\right)}{h_r - h_w} - \varepsilon\sigma T_{w1}^4 - \dot{m}\,h_{cs} \quad (5)$$

Where q , h_w and h_r are the heat flux, enthalpy and recovery enthalpy of the adiabatic wall, which are obtained from the analysis of aerodynamic thermal environment. q_1 , T_{w1} and h_{w1} represent the heat flux of loading, the temperature of outer wall and the corresponding enthalpy, which change with the calculation process in real time. ε is the external wall emissivity of the projectile, σ is the radiation coefficient of blackbody, both of which are constant. $\dot{m}$ and h_{cs} represent the mass loss rate and enthalpy of the material in the presence of ablation, respectively. The above formula has a non-uniform distribution on the surface of the projectile and needs to be calculated independently according to the space position.

In addition to the heat conduction of the installation joints, there is a gap between the cabin and the equipment, and radiation and convection heat exchange will occur. The heat accumulation effect cannot be ignored during long-term work. Due to the complexity of the space in the cabin and natural convection, it is difficult to simulate the fluid-solid coupling. This paper uses the Newton cooling formula to calculate:

$$q_2 = a(T_{w2} - T_f) \tag{6}$$

In the formula, q_2 represents the convective heat transfer heat flow, a represents the convective heat transfer coefficient, T_{w2} represents the wall temperature, and T_f represents the gas temperature. The above items all change in real time with the calculation process. Among them, the convective heat transfer coefficient a has a greater influence on the calculation results, and is related to the structure and temperature state, and can be interpolated according to the experimental data or the CFD calculation results.

When there is no shelter between the cabin and the equipment, there is radiant heat transfer. When the temperature of the inner wall of the cabin is higher, the influence becomes more significant. The radiant heat transfer follows the Stephen-Boltzmann law:

$$\phi = \varepsilon S \sigma T_{w2}^4 \tag{7}$$

In the formula, ϕ represents the radiant heat per unit time, and S represents the area of radiant surface. The higher the temperature, the more significant the radiation effect. In order to reduce the amount of calculation, indirect radiation or weak radiation can be ignored.

4. Finite Element Analysis Process

This article is based on APDL (ANSYS parametric design language) to establish the analysis process of a high-speed missile cabin temperature field:

1) A relatively simple structure can be established using APDL, but a large and complex structure can be established using CAD software such as CITIA, UG, SOLIDWORK, etc. The important parts like installation of joints and brackets must be handled carefully.

2) Input the material attributes such as thermal conductivity, specific heat capacity, density and other attributes. In order to load and post-process conveniently, the corresponding component need to be defined as required. Material attributes generally are different with temperature, which needs to be defined as temperature-dependent nonlinearity.

3) Import a mesh file or directly mesh the geometric model. For thin-walled structures such as skins, we can use the multi-layer shell element shell131, and solid70 for the solid structure. According to the requirements of accuracy, we can add the middle Node. Cover the surface effect element of surf152 at the boundary of the 3D model where convection and radiation need to be applied at the same time.

4) The thickness of thermal insulation layer can be defined parametrically, which is convenient for repeated evaluation and optimization design. At the same time, the temporary array can be defined parametrically to enhance the portability of command flow.

5) In order to reduce the times of reading the file, the load file is read and stored as a temporary array, which is applied to the outer surface after conversion. The relationship between convection and radiation heat transfer is defined for high temperature boundary and low temperature boundary, especially the loading surface and direction of shell element. The radiation direction of solid surface is

difficult to control, so the thin shell element can be covered on the surface to impose radiation boundary conditions. Volume heat generation rate must be defined for equipment heating.
6) Define the initial temperature field of the structure, and assign initial values to the corresponding parameters.
7) According to the time step, complete the loading and transient calculation. Using the temperature results of the previous time step to correct the heat flow load, and calculate the convective heat transfer coefficient and cabin air temperature by interpolation. The non-uniform net heat flow on the outer surface of the missile body is loaded in the form of concentrated load for every node. A smaller time step should be adopted to avoid divergence for load state with large gradient.
8) The temperature and other related quantities of the surface are extracted for the next step. The results of transient calculation and extraction need to be repeated until the end of calculation.

5. An Example of a High Speed Missile

The geometric structure of a head cabin for high-speed missile is shown in figure 1. The equipments in the cabin are installed with the cabin through the heat insulation gasket, and there is a gap between the equipments and the cabin. The research shows that the heat conduction of the equipment installation boss will cause the local temperature of the equipment shell to rise, and the insulation pad is generally used to block it. In this case, the convective and radiative heat transfer effects in the cabin are mainly considered, and the equipment installation boss is ignored. This example does not involve the ablation process. Shell131 shell element is used to mesh the cabin and insulation layer, and solid70 solid element is used to mesh the equipment, as shown in figure 2. Among them, the thickness of cabin and insulation layer and material performance are defined by layers, and equipment material is defined by equivalent property. Surf152 surface effect element is covered on the inner surface of the cabin and the outer surface of the equipment, which is convenient to apply convection and radiation heat transfer conditions at the same time. The degree of freedom of the surface element node on the inner surface of the cabin is defined as the inner wall temperature (TBOT).

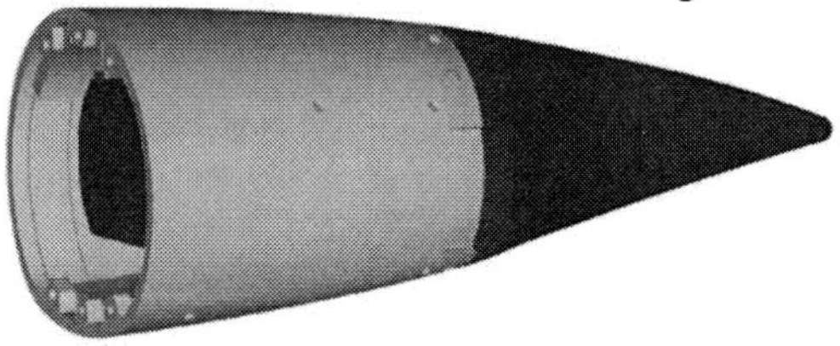

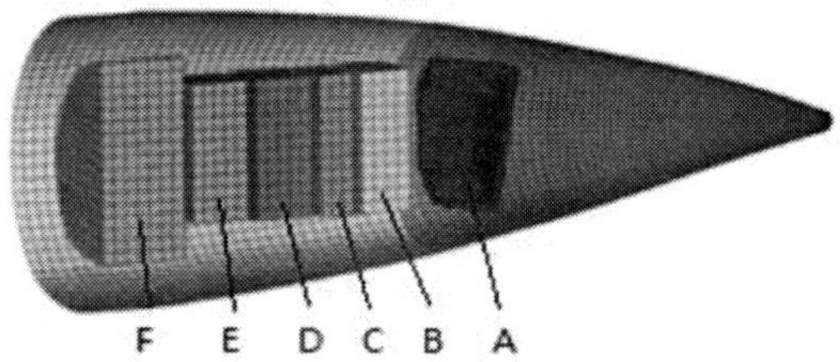

Figure 1. Geometry of a head cabin. **Figure 2.** Grid model of a head cabin.

Compile the APDL command stream, including the subroutine updating concentrated heat flow for independent nodes on the surface of the missile body. Calculate the transient temperature field according to the flight trajectory. The cloud diagram of temperature field for the structure at a typical time is shown in figure 3, and the history of surface temperature for the equipment in the cabin is shown in figure 4, in which A, B, C, D, E and F representing different equipment showing in figure 2. It can be seen that, affected by the distribution of aerodynamic heat, the material and thickness of the cabin structure, the outer surface temperature of the cabin has obvious regional characteristics. Radiation and convection heat transfer cause the temperature of the cabin equipment to rise slowly. Due to the difference in heat capacity and quality of the equipment, the amount of temperature rise is obvious different.

Figure 3. Schematic diagram of temperature field gradient.

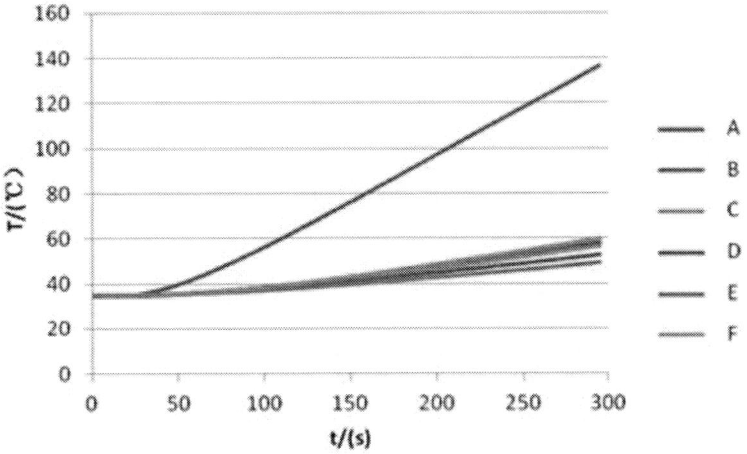

Figure 4. Device temperature time history.

6. Conclusion

In this paper, a high fidelity three-dimensional finite element model of the thermal protection system, cabin structure and cabin equipment is established. The heat conduction path, radiation relationship and heat sink effect of the equipment are reflected in detail. The process of aerodynamic heating and structural thermal protection is unified as the concentrated heat flow load related to the solution process. The heat transfer of the air in the cabin is simulated by the experimental correlation convection coefficient, which avoids the problem of fluid-heat-solid coupling. A numerical example of a high-speed missile shows that the surface temperature outside the cabin has obvious regional characteristics due to the influence of aerodynamic heat distribution, cabin structure material and thickness. Radiation and convection heat transfer cause the temperature of the equipment in the cabin to rise slowly. The developed APDL command stream can integrate mature software system of CAD and CAE. It has high computational efficiency and strong portability, and is suitable for large-scale engineering calculation.

References

[1] Rong K and Hong J2017*Mechanical environment test and evaluation of high performance system space vehicle*(*Strength and environment*vol 44) pp 1-7

[2] GUI Yewei and Liu Lei 2017*Research status and software development of fluid thermal solid coupling for hypersonic vehicles*(*ActaAeronauticaSinica*vol 8) pp 020844

[3] Chen Nashi 2018*Analysis of the integration of heat insulation and thermal control for the instrument cabin of high speed aircraft*(Harbin: Harbin Institute of technology) p 58

[4] Yang Yaqi and Wang Yaping 2020*Study on the thermal effect of the coupling between the interior and exterior of aircraft equipment*(*Journal of Southwest University of science and technology*vol35) pp 50-59

[5] WIETING A R and GUY R W 1976*Thermal-structural design/analysis of an airframe-integrated hydrogen-cooled scramjet*(*Journal of aircraft*vol 13) pp 192-197

[6] FALLEN D J and THORNTON E A 1983*Integrated thermal structural approach for shells of revolution*(*AIAA Journal*vol 21) pp 1475-1477
[7] Yang Zhibin and Qu Linfeng 2016*Temperature field coupling analysis of aircraft fuel tank under aerodynamic heating*(*Aeronautical computing technology*vol46) pp 31-34
[8] Li Zongyang and Li Yu 2020*Numerical analysis of thermal protection structure of aircraft hood considering multi field coupling effect*(*Aerospace defense*vol3) pp 8-15
[9] Huang Tang and Mao Guoliang 2000*Integrated numerical simulation of two-dimensional flow field, heat and structure*(*Acta aerodynamics Sinica*vol18) pp 115-119
[10] Xia Gang and Liu Xin Jian 2003*Numerical calculation of coupled hypersonic aerodynamic heating and structural heat transfer of bluff body*(*Journal of National University of Defense Science and technology*vol25) pp 35-39
[11] CULLER A J and MCNAMARA J J 2010*Studies on fluid-thermal-structural coupling for aerothermalelasticity in hypersonic flow* (*AIAA Journal*vol 48) pp 1721-1738
[12] Dong Weizhong and Gao Tiesuo 2015*Numerical study on surface temperature distribution and aerothermal coupling of hypersonic vehicle*(*ActaAeronauticaSinica*vol36) pp 311-324
[13] GUI yewei and Liu Lei 2015*Study on the calculation strategy and method of aerodynamic / thermal and structural multi field coupling*(*Journal of Engineering Thermophysics*vol36) pp 1047-1051
[14] Zhou Yinjia andMengSonghe 2016*Multi field coupled numerical study on thermal environment and structural heat transfer of hypersonic vehicle*(*ActaAeronauticaSinica*vol37) pp 2739-2748
[15] Yuan Kaihua and Cheng Meng 2017*hypersonic aerodynamics Heat Structural multi field coupling analysis*(*Tactical missile technology*vol5) pp 41-45
[16] Li Jiawei and Wang Jiangfeng 2019*Integrated calculation and analysis of fluid thermal solid coupling in "type IV" shock wave interference*(*ActaAeronauticaSinica*vol40) pp 123190
[17] Liu Jiawei 2016*Thermal structure fluid of hypersonic vehicle Structure Thermal multi field coupling simulation*(University of Defense Science and technology) p32
[18] Jiazhouxia and Wu Zhenqiang 2019 *Research on bi directional coupling of aerodynamic heat and structural heat transfer of aircraft*(*Strength and environment*vol46) pp 16-23

MD Simulation of the Laser Effects on Stopping for B^+ Ion in Plasmas

Pengfei Li, Guiqiu Wang*, Yi Duan, He Yi, Jian Wang and Shixuan Wang

College of Science, Dalian Maritime University, Dalian, China

*Corresponding author email: gqwang@dlmu.edu.cn

Abstract. In this work, the laser effects on the stopping for boron ion in plasmas are studied by molecular dynamics (MD) simulation. The expressions of the stopping power is derived based on the potential. The effects of different laser intensity, laser frequency, laser angle, plasma density and electron temperature on the stopping power of B^+ ion are discussed. The results tell that the stopping power of B^+ ion will decrease significantly with the increase of laser intensity, and the laser intensity has a strong modulation effect on the stopping power. On the other hand, when the included angle between the laser and the moving direction of boron ion is in the range of 0 to π / 2, the stopping power decreases gradually with the increase of the projectile angle. On the other hand, the influence of plasma parameters on the stopping power of boron ions is also very obvious. When other parameters are given, the increase of plasma temperature or the decrease of plasma density can decrease the value of stopping power.
Keywords: MD simulation; Stopping power; Laser.

1. Introduction

The stopping of ion beam in plasma under the radiation of laser is an important topic in the research of ion beam driven inertial confinement fusion (ICF) [1-3]. As per the literature it can be observed that laser-produced ion beam become most prominent tool for studying the high-energy-density physics [4-6]. Under the radiation of laser, the stopping of ion will be affected to some extent by the radiation of laser. In the existing reports, Arista [7] first adopted a general formula to describe the laser effects on the energy deposition of high-energy electrons propagating in degenerate electron gas. Nersisyan and Akopyan [8] found that the velocity of charged particles increased with the increase of laser field strength in solids. Roth et al [9] proposed that high current ion beam and high laser beam cooperate to drive ICF. The researchers reported the dynamic effect and energy deposition of incident ion beam in plasma under laser field [10]. Arista et al. [11] pointed out that the laser field can reduce the ion energy deposition for ions whose incident velocity was less than the electron thermal velocity in the plasma. Nersisyan et al. [12] reported that when ions pass through the plasma they obtained energy from the laser field and accelerate continuously. These researches are very helpful to understand the interaction mechanism of ion in plasma.

It is worthy to point out that the stopping power of ion in plasma is a very important quantity for ICF. In particular, the laser effect on ion stopping need to be well understood. Present work reports a comprehensive study on the influence of laser field on the stopping power of boron ions in plasma. This study will emphasis the laser effects and different plasma parameters such as laser intensity, laser frequency, laser angle, plasma density and electron temperature, etc.

2. Theoretical Model

Let us choose an ion with charge number Z, its velocity V, travelling in a plasma target with density n_0 and temperature T_e. Supposing the number of bound electrons of such an ion is n, and its charge density can be described as follows:

$$\rho_{ex}(r,t) = Ze\delta(r - vt) - \rho_e(r - vt) \quad (1)$$

The two terms to the right of the equal sign represent nuclear charge density and electron charge density respectively. We can only consider the contribution of electronic components in the plasma, because the thermal velocity of ions in the plasma is much smaller than that of ions. The laser field is $E_0(t)=E_0\sin\omega_0 t$, with the eqution of linearized Vlasov-Poisson, we can get the potenial $\Phi(r,t)$.

$$\left(\frac{\partial}{\partial t} + u\cdot\frac{\partial}{\partial r}\right)f'(r,u,t) = \frac{e}{m_e}\left(E_0\sin\omega_0 t - \frac{\partial\Phi}{\partial r}\right)\cdot\frac{\partial f_0(u)}{\partial u} \quad (2)$$

$$\nabla^2\Phi(r,t) = 4\pi\left[en'(r,t) - \rho_{ex}(r,t)\right] \quad (3)$$

where f_0 and f' are the undisturbed and disturbed electron distribution function respectively, and $n'(r,t) = \int du f'(r,u,t)$ is the disturbed density of electron .

$$\rho_e(R) = N\exp(-R/\Lambda)/\left(4\pi R\Lambda^2\right) \quad (4)$$

After the Fourier transform, we can get

$$\Phi(k,\omega) = \frac{4\pi\rho_{ex}(k,\omega)}{k^2\varepsilon(k,\omega)} \quad (5)$$

where the plasma dielectric function is

$$\varepsilon(k,\omega) = 1 + (k\lambda_D)^{-2}\left[X(\omega/kv_T) + iY(\omega/kv_T)\right] \quad (6)$$

where

$X(s) = 1 - 2se^{-s^2}\int_0^s e^{t^2}dt, Y(s) = \sqrt{\pi}se^{-s^2}$, $s = r - a_E\sin(\omega_0 t)$, and v_T is the electron thermal velocity $v_T = \sqrt{k_B T_e/m_e}$.

Using Fourier transform to the external charge density, we can get

$$\rho_{ex}(k,\omega) = Ze\int dt\exp\left[-ik\cdot a_E\sin(\omega_0 t) + i(\omega - k\cdot v)t\right] \quad (7)$$

$a_E = eE_0/(m_e\omega_0^2)$ is the transverse oscillation amplitude of the electron driven by the laser field. According to the Jacobi identity, $\exp[-ik\cdot a_E\sin(\omega_0 t)] = \sum J_n(k\cdot a_E)\exp(i\omega_0 t)$, we obtain

$$\rho_{ex}(k,\omega) = 2\pi\rho_\eta(k)\sum_{n=-\infty}^{\infty} Jn(\mathbf{k}\cdot\mathbf{a_E})\delta(\omega + n\omega_0 - k\cdot v) \quad (8)$$

$$\rho_\eta(k) = Ze\left(1 - \frac{\eta}{1 + (k\Lambda)^2}\right) \quad (9)$$

where $J_n(x)$ is the first kind of *n*th-order Bessel function. With Eqs.(5) and (6), we get the scalar potential as follows:

$$\Phi(r,t)=\frac{1}{2\pi^2}\sum_{n,m}\int\frac{d^3\vec{k}}{k^2}\rho_\eta(\vec{k})J_m(\vec{k}\cdot\vec{a}_E)\times J_n(\vec{k}\cdot\vec{a}_0)\frac{e^{i\vec{k}\cdot(\vec{r}-\vec{v}t)}e^{i(n-m)\omega_0 t}}{\varepsilon(k,\omega_n)} \quad (10)$$

where $\omega_n=kv-n\omega_0$ is frequency of the laser-harmonic Doppler shifted.
Over the laser period, let's take the time average to above expression. We can find that the quick oscillations with time *t* will be moved when *m=n*. Finally, the average interaction potential becomes time independent.
Assuming $\vec{E}_0$, $\vec{v}$ and $\vec{k}$ are all in the same plane, and α is the angle between $\vec{E}_0$ and $\vec{v}$. By introducing the variable $\omega=\vec{k}\cdot\vec{v}=kv\cos\theta$, we can define the quality

$$\chi(\vec{k},\omega)=\vec{k}\cdot\vec{a}_E=\frac{a_E\omega}{v}\cos\alpha-\kappa a_E\sin\alpha \quad (11)$$

The relationship of the three quantities can be found in Fig 1.

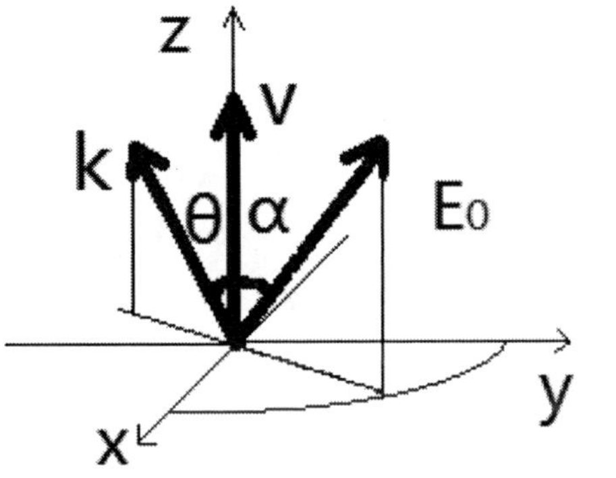

Figure 1. Relationship for three vectors $\vec{E}_0$, $\vec{v}$ and $\vec{k}$.
Using the potential, one can get the stopping power for ion, $S_Z(v)=-F_s(v)=\partial\Phi(r,t)/\partial r$.

3. Numerical Results Discussion

We choose B^+ (Z=5) as the study object. In the following sections, we discuss the influences of laser parameters and plasma parameters on the stopping power for B^+ in plasmas.
Figure 2 shows the stopping power *S* for Boron ion changing with the projectile velocity for different laser intensities. Other parameters are set as follows, the laser frequency $\omega_0=\omega_p$, the plasma density $n_0=10^{16}cm^{-3}$, the laser angle, i.e. the laser angle is π/3, and the electron temperature is T_e=3eV. The simulation results tell us that the stopping power shows a significant decline with the increase of laser intensity. Such a result indicates that the stopping power can be modulated by laser intensity.
Figure 3 shows the stopping power *S* for Boron ion changing with velocity for different laser frequencies. In the simulation, the laser intensity $a_E=3\lambda_D$, the plasma density $n_0=10^{16}cm^{-3}$, the laser angle $\alpha=\pi/3$, and the electron temperature is T_e=1eV. One can see that the stopping power changes obviously with laser frequency, especially for $\omega_0=\omega_p$, the stopping has more larger value than other frequencies. The changes are not very obvious between other higher laser frequencies.

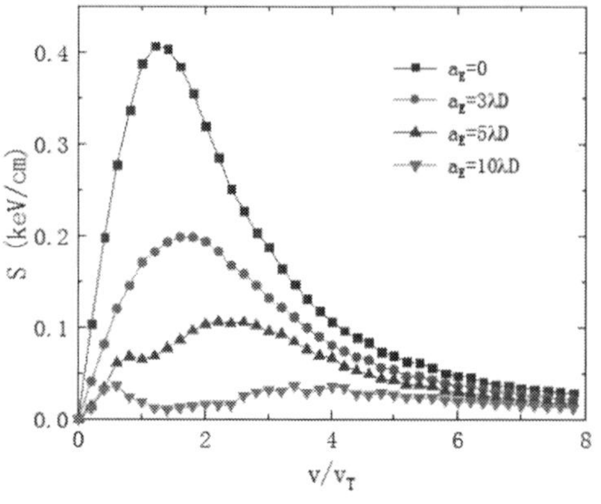

Figure 2. Influence of the laser intensity on the stopping power of Boron ion changing with the velocity ratio v/v$_T$.

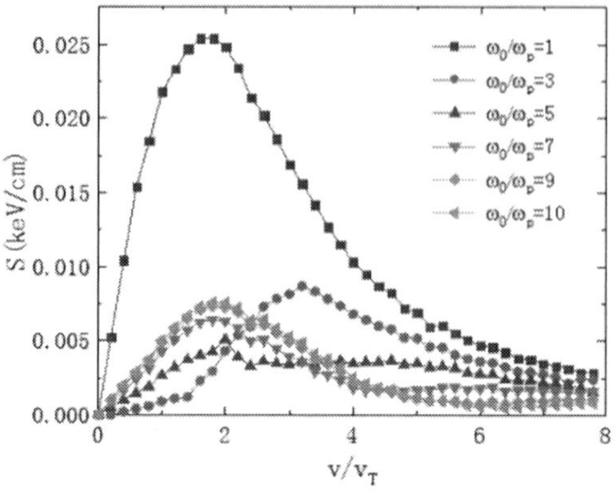

Figure 3. Influence of the laser frequency on the stopping power of Boron ion changing with the velocity ratio v/v$_T$.

Figure 4 shows the stopping power *S* for Boron ion changing with the projectile velocity for different laser angles. And the laser intensity $a_E = 3\lambda_D$, the plasma density $n_0 = 10^{16} cm^{-3}$, and the electron temperature is T$_e$=3eV. One can see that the magnitude of the stopping power decreases with laser angle increasing in the scale of 0 to $\pi/2$, the stopping power has maximum value at $v/v_T = 1.5$.

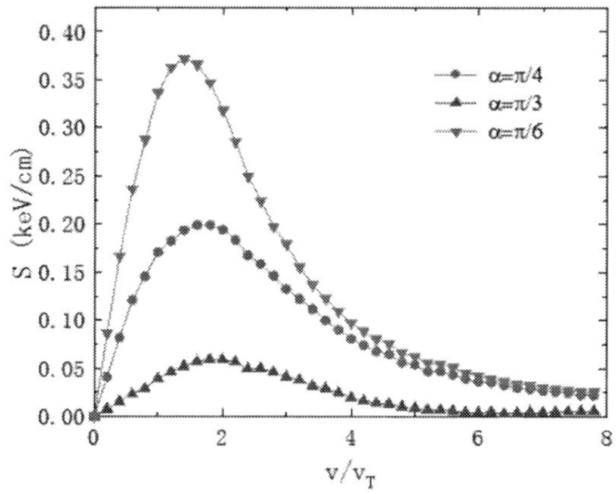

Figure 4. Influence of the laser angle on the stopping power of Boron ion changing with the velocity ratio v/v$_T$.

Figure 5 and Figure 6 show the stopping power *S* for Boron ion changing with the projectile velocity for different plasma densities and electron temperatures. In Fig.5, other parameters are set as follows, the laser intensity $a_E = 3\lambda_D$, the laser angle $\alpha = \pi/3$, the laser frequency $\omega_0 = \omega_p$, and the electron temperature is T$_e$=1eV. In Fig.6, the laser intensity, laser angle and laser frequency are set as same as in Fig. 5., the plasma density $n_0 = 10^{16} cm^{-3}$. One can see that the stopping power changes greatly with plasma density and electron temperature. Generally, the magnitude of stopping power has a larger value for higher plasma density and lower electron temperature.

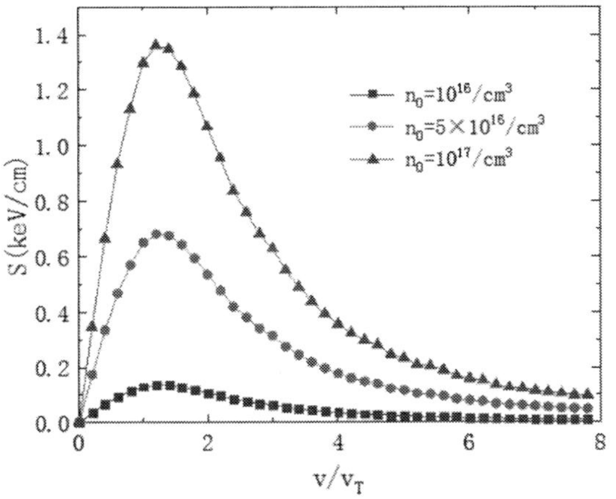

Figure 5. Influence of the plasma density on the stopping power of ion changing with the velocity ratio v/vT.

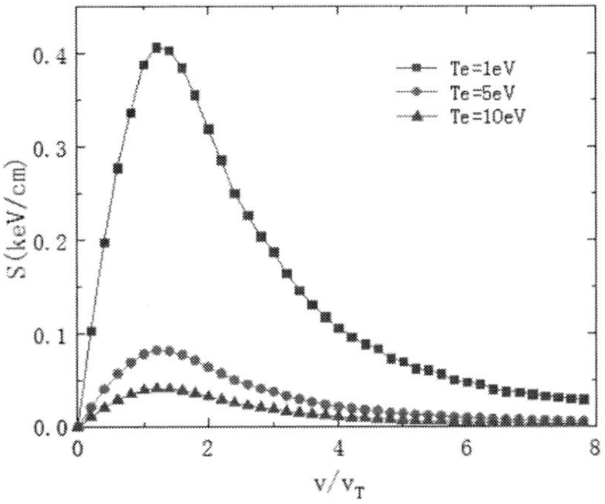

Figure 6. Influence of the electron temperature on the stopping Boron power of Boron ion changing with the velocity ratio v/vT.

4. Conclusion

Using the MD simulation method, the laser effects on the stopping for boron ion in plasmas have been studied. The focus is on the study of laser effects on the stopping power. By the theoretical deduction, we get the laser-assisted expressions for the potential and stopping power. Numerical results show that the laser-field intensity, frequency, and angle are important parameters affecting the interaction. The amplitude of stopping power decreases with increasing laser-field intensity, frequency, and laser angle. In addition, the plasma parameters are also important parameters affecting the stopping power, and the the amplitude of stopping power increase dramatically with the increasing plasma density and decreasing electron temperature.

Acknowledgments
This work is supported by the National Natural Science Foundation of China (11875096).

References

[1] Tahir N A, Lutz K J, Geb O, *et. al.* 1997 *Physics of Plasmas* 4 pp 796 -816
[2] Eliezer S, Murakami M, and Val J M M 2007 *Laser Particle Beams* 25 pp 585-592
[3] Barriga-Carrasco M D 2008 *Physics of Plasmas* 15 p 033103
[4] Kim J, Mcguffey C, Gautier D C, *et. al.* 2018 *Scientific reports* 8 p 17538
[5] Bang W, Albright B J, Bradley P A, *et. al.* 2015 *Scientific reports* 5 p 14318
[6] Ren J, Deng Z, Qi W, *et. al.* 2020 *Nature Communications* 20 p 18986
[7] Arista N R, Galvao R O M, Miranda L C M 1989 *Physics Review A* 40 pp 3808-3816
[8] Nersisyan H B, Akopyan E A 1999 *Physical Review A* 258 pp 323-328
[9] Roth M, Cowan T E, Key M H et al 2001 *Physical Review Letters* 86(30) pp 436-439
[10] Rrank A, Blazevic A, Grande P L et al 2010 *Physical Review E* 81 pp 230-239
[11] Arista N R, Galvao R, Miranda L C M 1990 *Journal of the Physical Society of Japan* 59(2) pp 544-552
[12] Nersisyan H B, Akopyan E A 1999 *Physical Review A* 258 pp 323-328

A Strabismus Surgery Parameter Design Model with WGAN-GP Data Enhancement Method

Renhao Tang[1], Wensi Wang[1,*], Qingyu Meng[2], Shuting Liang[2], Zequn Miao[2], Lili Guo[2] and Lejin Wang[2]

[1]School of Microelectronics, Faculty of Information Technology, Beijing University of Technology, Beijing, China

[2]Department of Ophthalmology, Peking University People's Hospital, Beijing, China

*Corresponding author email: wensi.wang@bjut.edu

Abstract. The purpose of this paper is a machine learning model that could predict the strabismus surgery parameter through the data of patients as accurately as possible. A strabismus surgery parameter design model's input is a Medical records and return is a surgical value. The Machine learning algorithms is difficult to get a desired result in this process because of the small amount and uneven distribution strabismus surgery data. This paper enhanced the data set through a WGAN-GP model to improve the performance of the LightGBM algorithm. The performance of model is increased from 69.32% to 84.52%.
Keywords: Strabismus surgery; Machine learning; Data enhancement.

1. Introduction

Strabismus is a congenital ophthalmological disease, which is mainly due to genetic inheritance or mutation. This disease causes abnormal and asymmetric developments of the eye muscles, which often lead to amblyopia and loss of stereoscopic vision. Early treatment is very important for strabismus, because the correction of amblyopia caused by strabismus generally needs to be conducted in childhood (under 12 years of age), and the formation of stereoscopic vision also only occurs in childhood. In China, the prevalence of strabismus is 1% to 3%, given the 1.4 billion population, the number of patients is approximately 30 million, and there are 400,000 new patients every year, which is one of the eye diseases with the highest incidence. Some patients will receive non-surgical treatment in childhood to improve their amblyopia, and undergo ocular muscle adjustment surgery in adulthood to improve the appearance of their eyes. Another part of the patients needs to undergo eye muscle adjustment surgery in childhood due to more severe strabismus. Among the more than 30 million strabismus patients, more and more patients have intention to undergo surgery to improve the symptoms of strabismus or to improve the appearance of the eyes. Strabismus surgery can be further divided into several different types, such as monocular strabismus, alternating strabismus, intermittent esotropia, and intermittent exotropia. It can also be divided into childhood strabismus surgery and adult strabismus surgery according to the age of surgery. The operation process requires sufficient experiences and skills to determine the Surgical Parameters, which is eye muscle adjustment amount and direction.

Due to the huge number of patients, many patients cannot choose to be treated in higher-level hospitals (Tier-III+) with better medical standards. Instead, they choose local general medical institutions or small ophthalmology clinics for surgery. In fact, more patients perform these surgeries in local hospitals than in the Tier-III+ hospitals in China. Due to technical and empirical issues in the

selection of surgical parameters, some patients undergoing strabismus surgery in local hospitals may result in insufficient strabismus correction or excessive strabismus correction. In comparison, the Tier III+ hospitals are less prone to such medical errors. In this paper, the data of more than one thousand successful strabismus operations in a Beijing Tier-III+ hospital was used to construct a deep learning model to help inexperienced doctors to train and assist them to give more accurate strabismus surgery parameters. In addition, the WGAN-GP [1] method is utilized to increase the number of simulated operations. By increasing the number of operations, the model can be better trained and improved for higher accuracy. The above situation has also been confirmed to exist in the experiments of other researchers. On Tan K C's UCI data set [2], the artificial intelligence algorithm has a significant effect on low-dimensional samples, and the accuracy can be maintained at about 99%. However, after Kamruzzaman SM increased the data dimension, the algorithm effect decreased significantly, down to 85% [3]; in the expert system of the Prasadl B team [4], the accuracy rate also fluctuated up and down 85%. At the same time, unbalanced data sets will also affect the accuracy of the medical diagnosis system. The team of Kamruzzaman SM processed the data sets by artificial means and found that the effect of artificial intelligence algorithms on unbalanced data sets is usually lower than In the original balanced data set, the neural network algorithm has the least impact, but in most cases the accuracy rate still drops significantly. It can be seen that the data dimension, data volume, and data balance of medical data sets all affect the performance of artificial intelligence algorithms [5].
This article aims to use GAN (Generative Adversarial Networks) technology to enhance medical data, increase the data volume of medical data, make the data volume compatible with the data dimension, improve the accuracy of artificial intelligence algorithm models, and complement the data of special cases. , Which makes the data distribution more uniform and improves the generalization ability of the algorithm. SSA B's team proposed the CEGAN[6] to resolve the data imbalance problem for the classification model. It was proved that the GAN is feasible for data enhancement. However, the original GAN network tends to converge slowly and the training process is unstable, and it needs to be adjusted based on the practical experience of engineering personnel. The introduction of Wasserstein distance [7] significantly improves the training convergence speed and training stability. Among them, WGAN-GP adds the Wasserstein distance and also adds gradient constraints to meet Lipschitz continuity. Therefore, this paper chose the WGAN-GP as the method of data enhancement because of the easier training process and simple structure
Using this model to train is more stable, speeds up the calculation process of the entire algorithm, and provides a reliable data enhancement method for the application of artificial intelligence technology in clinical medicine.

2. Materials and Methods

2.1. Dataset Description.

This article uses 1076 surgical cases of strabismus patients provided by Peking University People's Hospital. The data dimensions are 57 dimensions, including patient data such as the patient's age, dominant eye, near and distant exodeviation angle. It also contains the surgical value of the patient's strabismus correction surgery. Medical data is usually used for patient tracking and information management. The data in the data contains a variety of data formats and data types, which makes it impossible to directly input the data into the algorithm for training. Therefore, this paper filters out unqualified data through data cleaning and preprocessing, and unifies the data format into numerical data that can be processed by the computer.

2.2. Data Enhancement.

Medical data has uneven data distribution. Most patients have a high overall similarity, while the number of data on special cases is often very small. This makes the data of the 1,000 patients used in this article only have rare types of case data. A few, make the model under-fit the case data, which is not enough to make a correct diagnosis on these case data [8]. Therefore, through the WGAN-GP model, the data of rare cases is complemented, and the data distribution of the data set is improved, thereby improving the training effect of the final discriminant model.

In figure 1, the main three factors was chosen to show the distribution of the origin data. It shows that the distribution is uneven with the two concentrated areas. So it is not suitable for training a mathematical model to learn the feature of this data set. The uneven distribution could cause the training problem that is mentioned in the part of introduction.

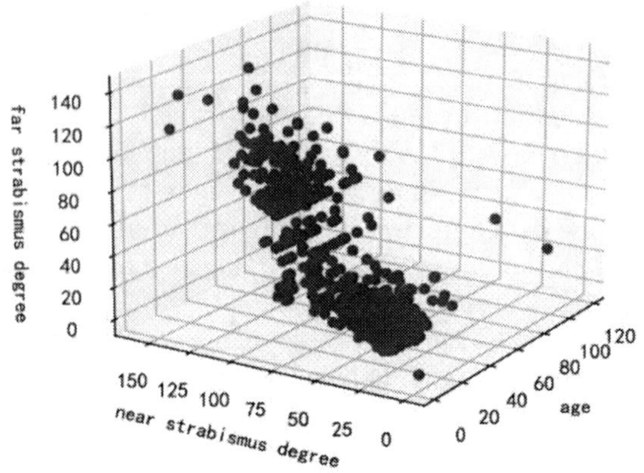

Figure 1. The distribution of the original data.

Figure 1 showed that the distribution of the origin data set was not even in the three primary factors(age, near strabismus degree and far strabismus degree).

Therefore, the data enhancement was into the consideration. The paper designed a workflow to generate the data via data enhancement method. It is shown by the flow figure:

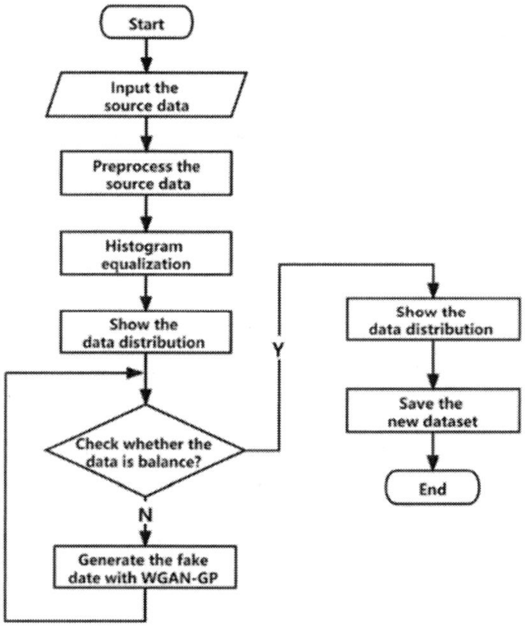

Figure 2. The workflow of the data enhancement.

The working focused on dividing the data into groups, training the generative model and testing the generated data set. The WGAN-GP model was trained to generate new data in the histogram equalization mapping space.

2.2.1. Histogram equalization. The principle of histogram equalization is to linearize the cumulative probability function of the function through a monotonic nonlinear mapping. The patient's case data can be mapped through the histogram equalization method. This classification method can more easily measure the difference between the cases and classify the cases.

Table 1. Data Mapping for the main factor.

Origin age	Mapping	Origin near strabismus	mapping	Origin far strabismus	mapping
4	10	22	10	16	10
5	20	27	20	19	20
6	30	30	30	20	30
8	40	35	40	24	40
9	50	37	50	28	50
10.5	60	43	60	34	60
14	70	72	70	55	70
21	80	85	80	75	80
29	90	94	90	82	90

The table 1 listed the value mapping which was calculated by histogram equalization algorithm for age, near strabismus degree and far strabismus degree. The origin data set could be mapping into a new space.

Through the histogram equalization algorithm and the distribution after data mapping, 10 is selected as the data grouping interval after mapping, and then the data group with 0 data amount is removed, leaving a total of 301 groups of data. Data mapping situation:

2.2.2. Train WGAN-GP model. GAN [9] is an unsupervised deep learning model. Common applications include image generative, image synthesis, and image generative from text. GAN training only needs a small amount of data to learn the distribution of data, so this article uses GAN to enhance the data of strabismus surgery. WGAN- GP is an improved model of GAN, and the speed of model training and convergence is faster. In this paper, WGAN-GP is improved, and DNN(a five-layer network, the generator and decision model are same) is used to replace WGAN-GP's CNN, thereby converting WGAN-GP from processing image data to enhanced numerical medical data. According to the grouping situation of the classified case data, training is carried out in a targeted manner. First, perform 3000 trainings on each set of data. After training, the cost is small and the Wasserstein distance is small. Finally, 301 WGAN-GP models are trained

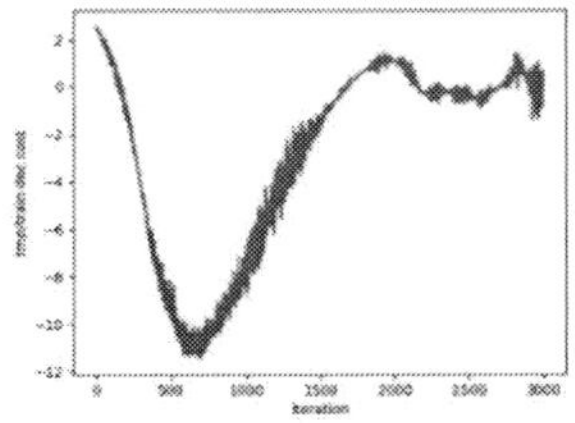

(a) training cost of group1 decision model

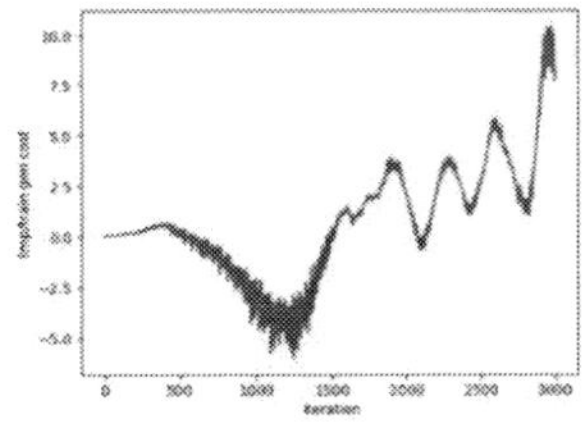

(b) training cost of group1 generative model

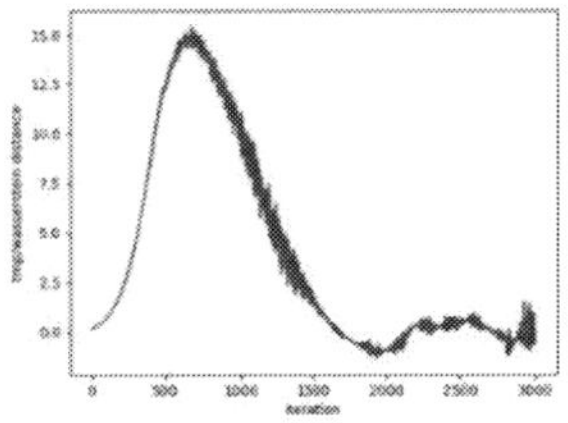

(c) wasserstein distance of group1

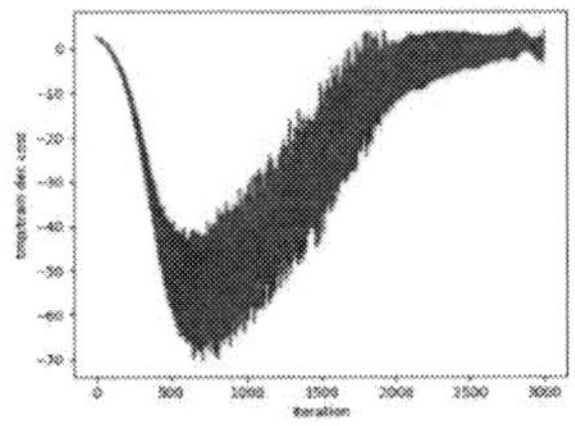

(d) training cost of group2 decision model

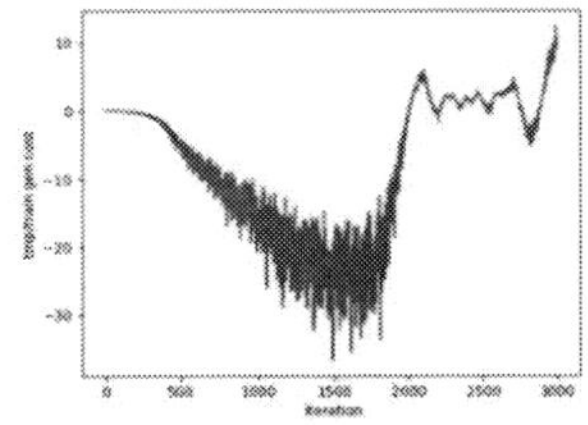

(e) training cost of group2 generative model

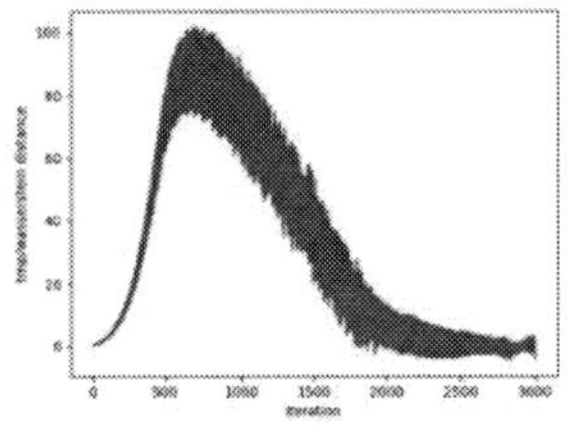

(f) wasserstein distance of group2

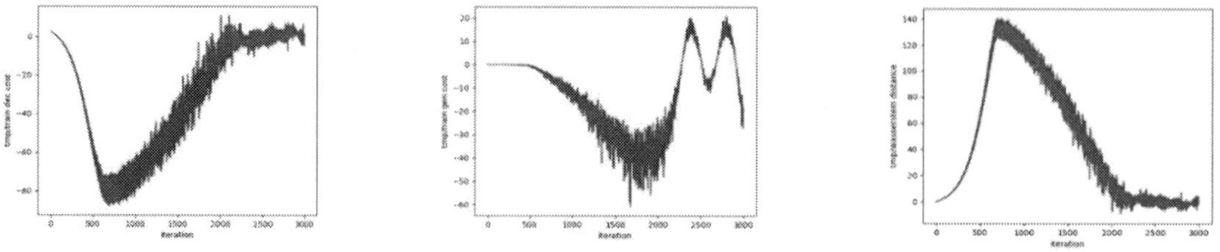

(g) training cost of group3 decision model (h) training cost of group3 generative model (i) wasserstein distance of group3

Figure 3. The result of training WGAN-GP model.

Figure 3 gave the examples for the training process from three groups.

This figure 3 listed the cost of decision model, the cost of generative model and wasserstein distance. The wasserstein distance is the standard to evaluate the balance between decision model and generative model. In these three sample group, the generative model was not in a stable position though the wasserstein distance is stable. Then the unstable group was adjusted in manual.

2.2.3. Test the performance of data set through LightGBM. LightGBM [10] is a decision tree model [11] based on ensemble learning. This article uses this model to test the performance of the data set. Since the LightGBM model has a good fitting effect on a small data set and a fast training speed, the LightGBM model is used to test the data set. The input of the model is the case data of the patient's examination, and the output of the model is the amount of operation corresponding to the patient's strabismus correction surgery. Test method: (1) In the control group, 80% of the data (800) are randomly selected from the original data as the training set, and then 10% of the data are used as the test set, and the remaining 10% are used as the validation set. Finally, use these data to compare LightGBM Model training; (2) Test the results from the enhanced data in the experimental group.

3. Result&Discussion

The Result was split to the two part, the distribution of generated data and the data performance in LightGBM.

3.1. Data Enhancement

Through the step of data enhancement, 0.3 Million medical records can be automatically generated. After comparison, it can be seen that the data of the insufficient data group has been supplemented, and the data distribution is more even.

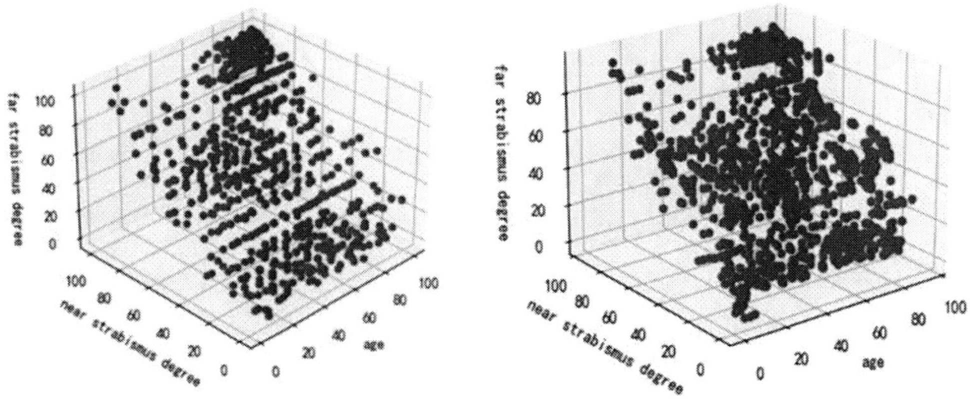

(a) Origin data distribution (b) Generated data distribution

Figure 4. The comparison between origin and generated data.

The figure 4 shows that the generated data set has more data point in the group and the distribution is more even than origin data set. The data in a group has passed the group filter which dropped the generated data outside the group. So it makes sure that the group data is generated by this group's WGAN-GP model. Although a few groups' generative model can not work obviously, the data enhancement has changed the distribution in the data set.

3.2. Testing in the LightGBM

Tested by LightGBM, the accuracy rate has been improved. After adding enhanced data, the over-fitting and under-fitting problems of the original model are improved. The model trained based on the Multiple classifier hyper-parameters uses the new data set to increase the accuracy of the data model from 70% to about 85%.

Table 2. AUC(area under the curve) of Testing Model.

	auc on training data set	*auc on testing data set*
Origin data	73.89%	69.32%
Generated data	84.91%	84.52%

The table 2 said the comparison of AUC between the model training by origin data set and the model training by generated data set.

4. Conclusion

In this paper, the case data of strabismus surgery is enhanced through the WGAN-GP network, so as to provide a better data set for improving subsequent modeling. The experimental results show that this method has a significant effect on improving the prediction accuracy of the prediction model with small amount of data and uneven distribution of samples. At the same time, the WGAN-GP model is still lacking in training stability. In the experiments of this article, the models of some data sets are not sufficiently trained under a fixed number of iterations. In the follow-up work, adjustments can be made to the insufficiently trained data sets. If this method is extended to other scenarios, it would be a better choice to make the GAN model adaptively train by dynamically adjusting the number of iterations. Through the histogram equalization method, the unevenly distributed data can be mapped to the new numerical space, which is convenient for visualization and grouping. However, in the data group with a small value range, the data generated by the WGAN-GP model is more difficult to fall back to the source data In the group, the data grouping method and the performance of the WGAN-GP model need to be further optimized.

References

[1] Ishaan Gulrajani, Faruk Ahmed, Martin Arjovsky, Vincent Dumoulin, and Aaron Courville. Improved training of wasserstein gans. arXiv preprint arXiv:1704.00028, 2017.
[2] Kay Chen Tan, Qiang Yu, CM Heng, and Tong Heng Lee. Evolutionary computing for knowledge discovery in medical diagnosis. Artificial Intelligence in Medicine, 27(2):129–154, 2003.
[3] SM Kamruzzaman, Ahmed Ryadh Hasan, Abu Bakar Siddiquee, Md Mazumder, and Ehsanul Hoque. Medical diagnosis using neural network. arXiv preprint arXiv:1009.4572, 2010.
[4] BDCN Prasadl, PESNK Prasad, and Y Sagar. An approach to develop expert systems in medical diagnosis using machine learning algorithms (asthma) and a performance study. International journal on soft computing (IJSC), 2(1):26–33, 2011.
[5] Luis Mena and Jesus A Gonzalez. Symbolic one-class learning from imbalanced datasets: application in medical diagnosis. International Journal on Artificial Intelligence Tools, 18(02):273–309, 2009.
[6] B Ssa, A Hl, C Plb, and Ola Yong. Cegan: Classification enhancement generative adversarial networks for unraveling data imbalance problems. Neural Networks, 133:69–86, 202
[7] Martin Arjovsky, Soumith Chintala, and L´eon Bottou. Wasserstein generative adversarial networks. In International conference on machine learning, pages 214–223. PMLR, 2017.

[8] William H Wolberg and Olvi L Mangasarian. Multisurface method of pattern separation for medical diagnosis applied to breast cytology. Proceedings of the national academy of sciences, 87(23):9193–9196, 1990.

[9] Ian Goodfellow, Jean Pouget-Abadie, Mehdi Mirza, Bing Xu, David Warde-Farley, Sherjil Ozair,Aaron Courville, and Yoshua Bengio. Generative adversarial nets. Advances in neural information processing systems, 27, 2014.

[10] Guolin Ke, Qi Meng, Thomas Finley, Taifeng Wang, Wei Chen, Weidong Ma, Qiwei Ye, and Tie-Yan Liu. Lightgbm: A highly efficient gradient boosting decision tree. Advances in neural information processing systems, 30:3146–3154, 2017

[11] Jyoti Soni, Ujma Ansari, Dipesh Sharma, and Sunita Soni. Predictive data mining for medical diagnosis: An overview of heart disease prediction. International Journal of Computer Applications, 17(8):43–48, 2011

Temperature Field Simulation and Structure Improvement of 12kV Switchgear

Yanjie Zhao, Li'an Chen*and Rui Qiao

School of Electrical Engineering and Automation, Xiamen University of Technology, Xiamen,China

*Corresponding author email: chenla@xmut.edu.cn

Abstract. Aiming at the heating problem of 12kV switchgear, the model of its main circuit module was established and simplified. Then, the temperature field and airflow field were simulated by finite element analysis software ANSYS. Besides, the distribution of temperature and airflow in the circuit breaker room was analyzed respectively and the heat dissipation structure of the switchgear was improved. Finally, two axial fans were installed at the top of the switchgear and the ventilation hole was designed at the bottom. The simulation results showed that this method can accelerate the air convection and reduce the temperature rise effectively. With the continuous increase of distribution network capacity, switchgear would have overheating problem during the long-term operation. If this problem cannot be solved, it would affect its insulation performance and even cause safety accidents. This study provides a reference for the design of switchgear and solves its overheating problem.
Keywords: Switchgear; Simulation; Finite element analysis; Temperature field; Airflow field; Structure improvement.

1. Introduction

In the power system, switchgear is a common terminal execution equipment and it is related to the safety of the power supply system. Since its compact structure and closed shell, heat cannot be dissipated effectively. Hence, when a large current flows through it, the temperature inside the switchgear will rise sharply, and cause the temperature rise to exceed the national standard[1-2]. With the development of switchgear in the direction of miniaturization and large capacity, its heating problem has become more and more serious. Therefore, it is necessary to conduct overheating analysis and improve its structure to reduce temperature rise.

Using Computational Fluid Dynamics (CFD) software to simulate the heating, temperature rise and heat dissipation process of the switchgear can provide guide for the design and optimization of the switchgear, thereby shortening the research and development cycle. In [3], COMSOL was used to solve and analyze the steady-state temperature field of the KYN28A-12 switchgear, which provides theoretical support for the high temperature warning of the switchgear. In [4], a three-dimensional model of the switchgear based on the finite element analysis method was established. Then, considering natural convection and forced air cooling, ANSYS was used to analyze and calculate the temperature field of the switchgear. It provided basis for temperature monitoring and control of the switchgear. In [5], CFD software Fluent and Maxwell were used to analyze the airflow field and temperature field distribution of the vacuum circuit breaker and current transformer for the long-term overheating problem of the switchgear. By changing the structure of the spring contact finger, the temperature rise was reduced by more than 4K.

This paper takes a certain type of 12kV switchgear as the research object, uses SolidWorks to establish a simplified model of the main circuit module of the switchgear, and then imports the ANSYS thermal simulation module Icepak to get the temperature field and airflow field distribution of the circuit breaker room under natural convection conditions. Subsequently, two axial fans are installed on the top of the switchgear and the ventilation hole at the bottom is improved. The simulation results show that the temperature field and airflow field distribution under forced air cooling are better than natural convection, which verifies the feasibility of the heat dissipation scheme.

2. Overheating Analysis

2.1. Basic Structure

The structure of a certain type of 12kV switchgear is shown in figure 1, it mainly includes 4 parts: busbar room, circuit breaker room, cable room and instrument room. It belongs to the indoor metal armored removable switchgear and its size is 800mm (width) × 1500mm (depth) × 2200mm (height).

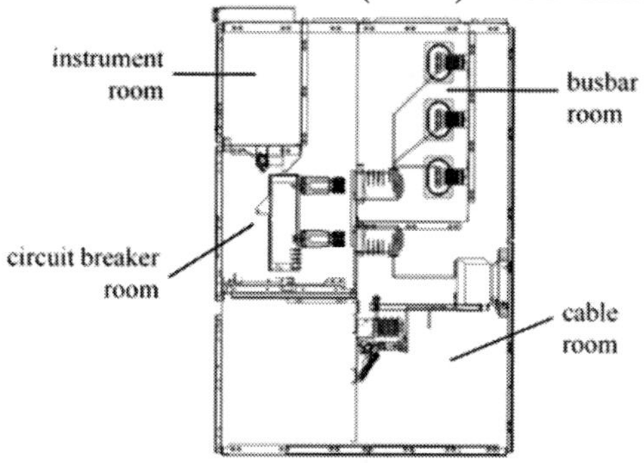

Figure 1. Basic structure of switchgear.

The switchgear adopts the *GB 3906-2006* standard, and the protection levels of the shell and compartment are IP4X and IP2X respectively. The main technical parameters are shown in table 1.

Table 1. Technical parameters of switchgear.

Parameters	Numerical value
Rated voltage(kV)	12
Rated current(A)	4000
Rated short-time withstand current(kA)	40/4s
Rated peak withstand current(kA)	100

2.2. Heat Source and Heat Transfer Analysis

The temperature rise inside the switchgear is mainly affected by two factors: heat source and heat transfer. The heat source of the switchgear mainly comes from the contact resistance and it is calculated by equation (1)[6]. It includes the Joule heat generated by the contacts, current-carrying conductors, busbars and their connections of the circuit breaker.

$$R_j = \frac{K_c}{(F_c/9.8)^m} \tag{1}$$

Here, R_j is the contact resistance; K_c is the coefficient related to the contact material; F_c is the pressure value of the contact position; m is the index related to the contact form.

In addition, the switchgear is a closed space with poor heat dissipation condition, so the heat transfer performance will also affect its temperature rise. The heat of the switchgear is mainly transferred through three methods: thermal conduction, thermal convection and thermal radiation[7]. Inside the switchgear, the busbar is the main heat source and it transfers energy to the surrounding solids in the form of thermal conduction. Thermal convection mainly occurs in the air inside the busbar room.

Journal of Physics: Conference Series **2179** (2022) 012010 doi:10.1088/1742-6596/2179/1/012010

Compared with the first two methods, the heat generated by thermal radiation is very small, so it can be ignored.

3. Simulation and Discussion

3.1. Geometry and Simplification
The actual model of the switchgear is very complicated. In order to facilitate the analysis, only the main circuit module is retained, as shown in figure 2(a). In the case of no influence on the path of the cooling circuit, the structure is simplified by using SolidWorks. During the process, the main heat sources of the switchgear, such as busbars, circuit breakers and current transformers are retained. For uneven surface parts, round holes and chamfers, model reconstruction is carried out. After simplification, the model is shown in figure 2(b).

(a) Before simplification (b) After simplification

Figure 2. Model of switchgear.

3.2. Meshing and Boundary Conditions
In order to facilitate calculation and analysis, the simplified model needs to be meshed. The quality of the meshing will affect the accuracy of the final simulation results[8]. The simplified model is imported into the ANSYS SpaceClaim module for preprocessing. Finally, the number of divided grids is 1321029, and the number of grid nodes is 1686712.

After meshing, boundary conditions are set by ANSYS thermal simulation module Icepak. The heat transfer method in the switchgear is mainly based on thermal conduction and thermal convection. The influence of thermal radiation is not considered, so natural convection is set as the main heat transfer method. Then, the convection heat transfer coefficient is set to $10\text{W}/(\text{m}^2{\cdot}\text{K})$ and the environment temperature is set to 40°C. The heat mainly comes from the busbar and its material is copper. Besides, the insulating material is epoxy resin and the propagation medium is air. The physical properties of each material are shown in table 2.

Table 2. Physical properties of each material.

Material	Thermal conductivity/ $[\text{W}{\cdot}(\text{m}{\cdot}\text{K})^{-1}]$	Density/ $(\text{g}{\cdot}\text{cm}^{-3})$	Specific heat capacity/ $[\text{J}{\cdot}(\text{g}{\cdot}\text{K})^{-1}]$	Resistivity/ $(\mu\Omega{\cdot}\text{m})$
Copper	392	8.9	0.39	0.018
Air	0.023	0.001	1.003	—
Epoxy resin	0.276	0.98	1.4	—

In addition, the skin effect of the conductor and the influence of Joule heating on the temperature rise are also considered. The skin effect is calculated by equation (2)[9]. Joule heating is reflected by setting the power of different conductive parts. The main circuit of the switchgear is connected with the current of 4000A, and the heating power of the A-phase, B-phase and C-phase busbars is all set to be 40W. The heating power of the circuit breaker and the current transformer are 180W and 30W respectively, and the total heating power is 330W.

$$K_{\mathrm{j}} = \frac{R_{\mathrm{f}}}{R_{0}} \tag{2}$$

Here: K_j is the skin effect coefficient of the conductor. R_f and R_0 are the resistance of AC circuit and DC circuit, respectively.

3.3. Simulation Results Analysis

The simulation is to study the operation of the switchgear under the extreme high temperature in a city in southeast China, and provide a reference for the design of the switchgear. Using the post-processing function of the CFD software Icepak, the simulation results of the switchgear can be obtained.

The temperature field distribution of the main circuit is shown in figure 3. As shown in the simulation results, the temperature at the junction of the main busbar and the branch busbars in the current-carrying circuit is highest, reaching 85.64°C. According to the provision of *GB/T11022-2011*, the temperature of bare copper during normal operation of the switchgear cannot exceed 75°C. The main reason for the excessive temperature rise is that the top ventilation holes are too small and the heat cannot be diffused in time. Therefore, the structure of the switchgear needs to be improved to meet the national standard.

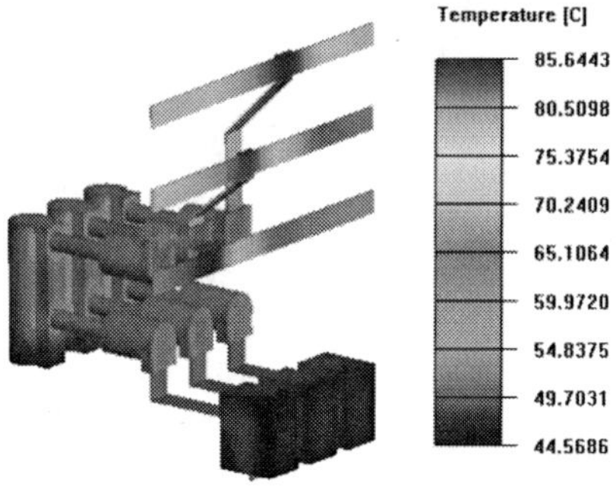

Figure 3. Temperature field distribution of the main circuit.

As can be seen from figure 4 and figure 5, the natural convection has little effect on the gas flow rate, which is no more than 0.44m/s. In addition, the circuit breaker room, as the main heat source, transfers heat to the air by means of thermal conduction and convection. The air expands by heat and transfers the heat to the outside of the switchgear along the vents. Eventually, this leads to a higher temperature in the upper part of the switchgear. The average flow rate of the gas in the cabinet is 0.16m/s. And in the circuit breaker room, since there is no ventilation hole, the gas flow rate is only 0.08m/s. Therefore, a good heat dissipation channel is not formed in the switchgear, which prevents the heat from being dissipated in time and accumulates the heat inside the switchgear. The temperature rise exceeds the standard, so the heat dissipation channel of the switchgear needs to be improved to reduce the temperature rise.

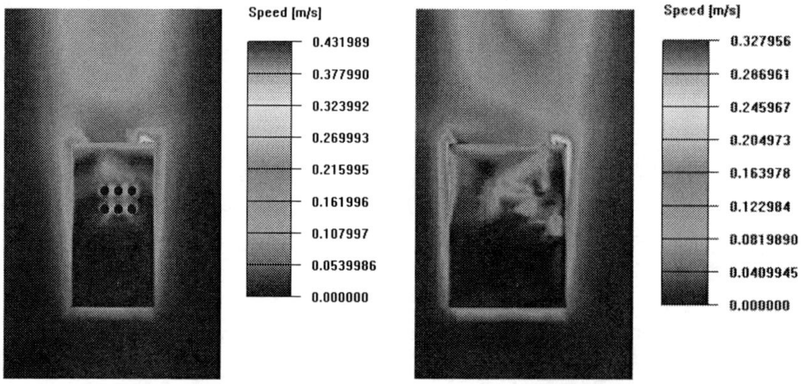

(a) Normal section (b) Side section

Figure 4. Air velocity distribution cross section of the switchgear.

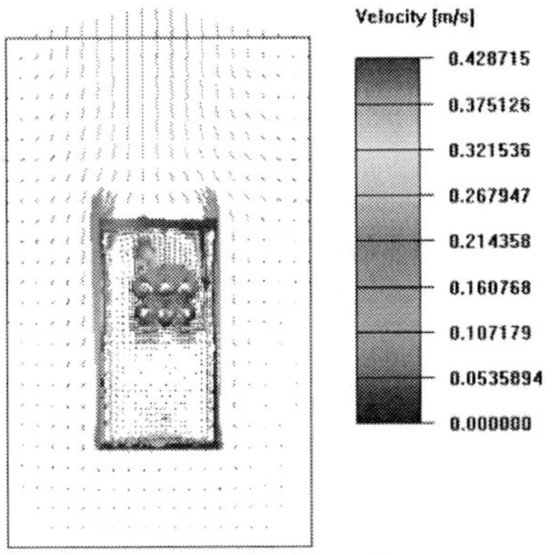

(a) Normal section

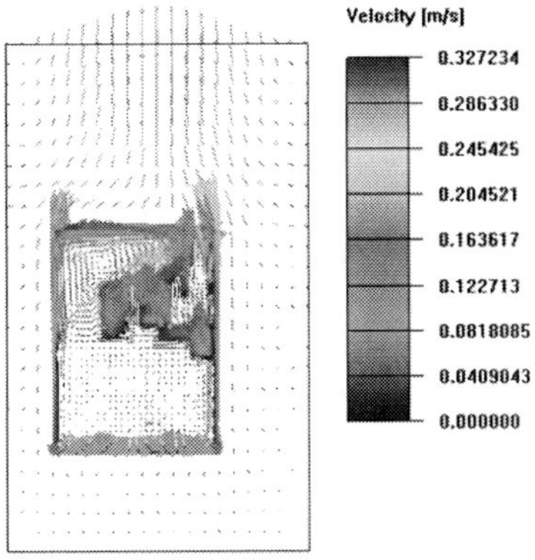

(b) Side section

Figure 5. Air velocity vector distribution of the switchgear.

4. Thermal Optimization

4.1. Structure Improvement

Aiming at the heat dissipation channel problem of the switchgear, its structure has been improved. At first, two axial fans with model AA1282HB-AW_A1_.1.2 are installed on the top of it for forced air cooling. Then, the number of ventilation holes at the bottom are added to increase the air intake, as shown in figure 6.

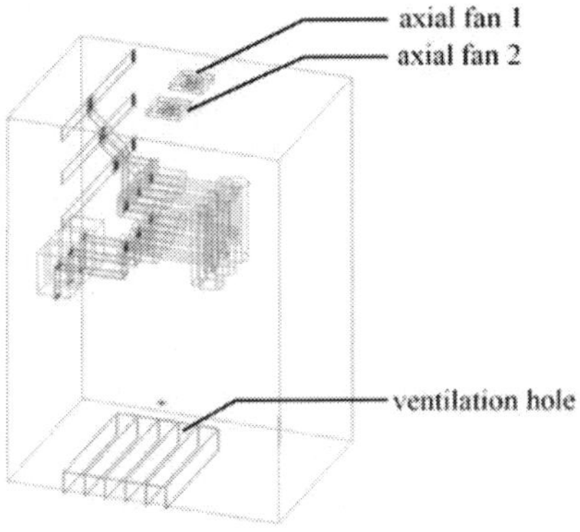

Figure 6. Switchgear after structure improvement.

4.2. Simulation

The improved switchgear model is re-imported into Icepak to divide the grid and set the boundary conditions. The physical properties of the material remain unchanged and the environment temperature is still set to 40℃. Then, the distribution of temperature field and airflow field of the switchgear are obtained.

The temperature field distribution of the improved switchgear is shown in figure 7. It can be seen that after the axial fans are added, the temperature rise of each part is significantly reduced. The connection between the main busbar and the branch busbar in the current-carrying circuit is still the highest temperature point, but compared with the value before the improvement, it has dropped to 74.77°C, a decrease of 12.69%. After improvement, temperature rise meets the requirements of *GB/T11022-2011*. It shows that the axial fan has a significant cooling effect.

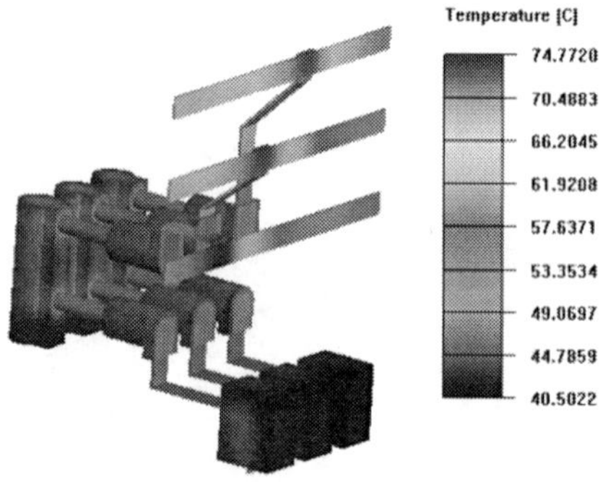

Figure 7. Temperature field distribution after improvement.

The simulation results of the airflow field distribution are shown in figure 8 and figure 9. It can be seen that, compared with natural convection, the two axial fans can expel the hot air in time, which significantly accelerates the gas flow in the switchgear. At the suction port of the axial fan, the gas flow rate reached 8.17m/s. In addition, due to the opening of ventilation holes at the bottom of the circuit breaker room, the average flow rate of the gas in the circuit breaker room is about 2.5m/s. In this way, an effective heat dissipation air duct is formed from top to bottom inside the switchgear. It can release the heat generated by the current-carrying circuit and then reduce the temperature rise of the switchgear.

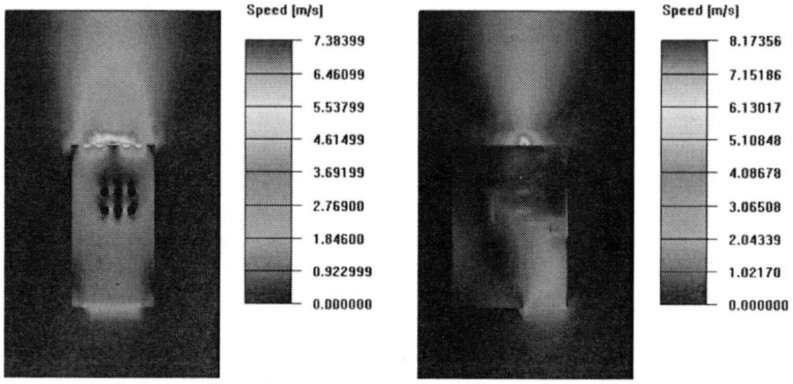

(a) Normal section (b) Side section

Figure 8. Air velocity distribution cross section after improvement.

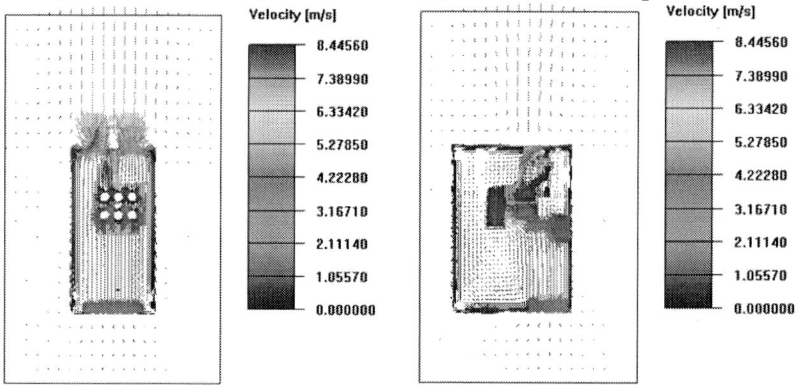

(a) Normal section (b) Side section

Figure 9. Air velocity vector distribution after improvement.

5. Conclusion

In this paper, two axial fans were arranged on the top of the 12kV switchgear for forced air cooling, and ventilation holes were designed at the bottom to improve the heat dissipation conditions of the switchgear. The simulation results showed that the improved temperature rise of the switchgear meets the national standard, and is reduced by 12.69% compared with the prior improvement. If this method is applied to switchgear products, it will bring certain economic benefits. Due to the limitation of conditions, this design was only carried out simulation analysis and temperature rise test verification was not carried out, and continuous improvement is needed in subsequent research.

References

[1] He R.-T. (2017) Thermal simulation and design improvements of the switchgear based on the natural convection. Electric Switchgear 55 (6):27-30
[2] Liu S.-J., Han W., Zhang H.-J. et al (2020) Research on temperature fluid field simulation and heat dissipation optimization of high voltage switchgear. High Voltage Apparatus 56 (10):63-69
[3] Deng C.-Z., Yang N., Deng H.-F. et al (2020) Simulation and analysis of temperature field of 10 kV high voltage switchgear based on COMSOL. Electrical Engineering Materials 47 (1):32-36
[4] Chen J.-Z., Du Z.-Y., Wang D. et al (2018) Numerical analysis of temperature field of high voltage switchgear. Advanced Technology of Electrical Engineering and Energy 37 (1):38-34
[5] Wang L.-J., Zheng W.-S., Wang L.-H. et al (2016) Electromagnetic-thermal-flow field coupling simulation of 12-kV medium-voltage switchgear. IEEE Transactions on Components, Packaging and Manufacturing Technology 6 (8):1208-1220
[6] Guo F.-Y., Chen Z.-H. (2008) Electric contact theory and its application technology. China Electric Power Press 37 (1):38-34
[7] Peng S.-Y., Ruan J.-J., Du Z.-Y. et al (2018) Influence of different heat dissipation conditions on temperature distribution in busbar room of switchgear. Advanced Technology of Electrical Engineering and Energy 37 (8):45-57
[8] Qu J.-Y., Chong F., Wu G. et al (2014) Application of computational fluid dynamics to predict the temperature-rise of low voltage switchgear compartment. IEEE 60th Holm Conference on Electrical Contacts (Holm) 2014 :1-5
[9] Dwight H B. (1922) Skin effect and proximity effect in tubular conductors. Journal of the American Institute of Electrical Engineers 41(3):203-209

Discrete-Time Modelling to Determine the Possible Operation Points of Incomplete-Resonance Wireless Power Transfer System

Daqing Luo[1], Jianfeng Hong[2] and Wenxiang Chen[1,*]

[1]Department of Instrumental and Electrical Engineering, Xiamen University, Xiamen, China
[2]School of Mechanicaland Automotive Engineering, Xiamen University of Technology, Xiamen, China

*Corresponding author email: wxchen@xmu.edu.cn

Abstract. The Independent energy injection and free resonance (IEIFR) system is a type of incomplete-resonance wireless power transfer system, which has a great tolerance with control strategy. This article uses a discrete-time modelling method to determine its all possible and rational operation points. This method samples the state variables with the switching cycle and once the system achieves steady-state, the mapping between those samples is a fixed point mapping. Combined with the known mode analysis, the rationality of those possible operation points can be verified. After the experiment verification, this method is helpful to analysis the incomplete-resonance wireless power transfer system.
Keywords: Wireless power transfer (WPT), Fixed point, Discrete-time modelling, Incomplete-resonance.

1. Introduction

Wireless power transfer (WPT) technology has wide application on electric vehicles, robots, portable devices and even underwater devices[1-5].

The fully-resonant type of WPT system is usually used with a variety of compensation networks, including series-series (SS), series-parallel (SP), parallel- series (PS) and parallel-parallel (PP)[6-10]. Their energy injection process and system resonance process are completed simultaneously. However, the switching frequency must be equal to the resonant frequency completely to meet the soft-switching condition. Hence there are few tolerances of the control strategy of the fully-resonant type. It is hard to complete the presented frequency tracking methods and impedance matching methods due to the high sampling frequency and short feedback time.

To promote the tolerance of the control strategy, an independent energy injection and free resonance (IEIFR) system has been proposed. Its power injection process and free resonance process are decoupling to be controlled independently. Due to its tolerance of the control strategy, it has a great advantage for feedback control with lower sampling frequency and longer feedback time. However, article[11] has not presented a rational modelling method to determine the operation point of IEIFR system. In article[11], the continuous-time domain modelling for one state variable is limited to calculate an approximate result about the studied state variable. The equations in article[11] is inadequate to determine the soft-switching operation point, hence the analysis is imperfect.

This paper uses discrete-time state space modelling to determine all possible soft-switching operation points. Concretely, the discrete-time state space modelling is a discrete-time mapping based on the interval of linear subsystem and the mapping of state variables in one switching cycle can be built via iteration. Then once the system achieves steady-state, the state variables are equal to its mapping in one switching cycle. Therefore, the mapping is a fixed point mapping and the possible operation points are determined with the proper soft-switching boundary condition.

2. Discrete-Time Modeling of IEIIFR System

The main circuit diagram of the IEIFR WPT system is shown in Figure 1. The diode D_0 is used to prevent current backflow. The switch S_1 is used to connect the DC voltage source u_{DC} to the circuit of the primary side. Besides, there is a switch S_2 control to access the resonant capacitor C_p on the primary side. The primary coil self-inductance L_p with internal resistance R_p and the secondary coil self-inductance L_s with internal resistance R_s constitute the mutual inductance coil, the main power transfer element. Additionally, between the primary coil and the secondary coil, M is the mutual inductance and k is the coupling coefficient. R_L is the system load resistance and C_s is the filter capacitor on the secondary side.
In IEIFR WPT system, according to the mode analysis in article[11], the processes of power injection and self-resonance are independent, which can be divided into two subsystems and whose equivalent circuits are shown in the Figure 2. During the processes of power injection.

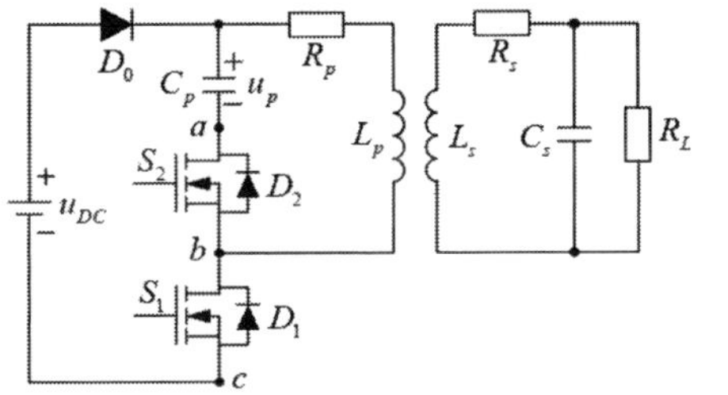

Figure 1. The IEIFR WPT system.

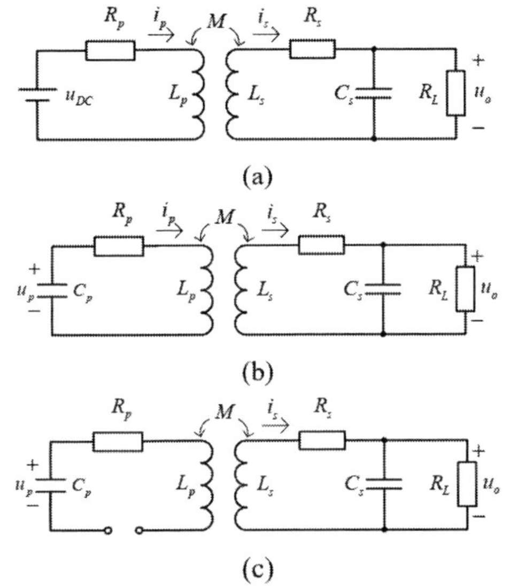

Figure 2. The equivalent circuits of the IEIFR WPT system (a) Power injection. (b) Self-resonance. (c) Discontinuous conduction mode.

Let $x=\begin{bmatrix} u_p & i_p & i_s & u_o \end{bmatrix}^T$ and $u=\begin{bmatrix} u_{DC} \end{bmatrix}$ be the state vector and the input vector of the system, respectively. In Figure 2, u_p, i_p, i_s and u_o are the voltage across capacitor C_p, the current through inductance L_p, the current through inductance L_s and the voltage across capacitor C_s, respectively
According to the Kirchhoff's voltage and current laws, the equivalent circuit shown in Figure 2(a) of the two subsystem can be described by the differential equations as follows:
Subsystem 1: Power injection

$$\dot{x} = A_1 x + B_1 u \tag{1}$$

Subsystem 2: Self-resonance

$$\dot{x} = A_2 x \tag{2}$$

Subsystem 3: Discontinuous conduction mode (DCM)

$$\dot{x} = A_3 x \tag{3}$$

where

$$A_1 = \begin{bmatrix} 0 & 0 & 0 & 0 \\ 0 & \frac{L_s R_p}{\Delta} & \frac{M R_s}{\Delta} & \frac{M}{\Delta} \\ 0 & \frac{M R_p}{\Delta} & \frac{L_p R_s}{\Delta} & \frac{L_p}{\Delta} \\ 0 & 0 & \frac{1}{C_s} & -\frac{1}{C_s R_L} \end{bmatrix}$$

$$B_1 = \begin{bmatrix} 0 & \frac{-L_s}{\Delta} & \frac{-M}{\Delta} & 0 \end{bmatrix}^T$$

$$A_2 = \begin{bmatrix} 0 & -\frac{1}{C_s} & 0 & 0 \\ \frac{-L_s}{\Delta} & \frac{L_s R_p}{\Delta} & \frac{M R_s}{\Delta} & \frac{M}{\Delta} \\ \frac{-M}{\Delta} & \frac{M R_p}{\Delta} & \frac{L_p R_s}{\Delta} & \frac{L_p}{\Delta} \\ 0 & 0 & \frac{1}{C_s} & -\frac{1}{C_s R_L} \end{bmatrix}$$

$$A_3 = \begin{bmatrix} 0 & 0 & 0 & 0 \\ 0 & 0 & 0 & 0 \\ 0 & 0 & -\frac{R_s}{L_s} & -\frac{1}{L_s} \\ 0 & 0 & \frac{1}{C_s} & -\frac{1}{C_s R_L} \end{bmatrix}$$

$$\Delta = M^2 - L_p L_s$$

$$M = k\sqrt{L_p L_s}$$

The analytical solution of each linearized mode can be presented as follows[12]:
Subsystem 1: Power injection

$$x(t) = f_1(t, x_0) = \Phi_1(t) x_0 + \int_0^t \Phi(t-\tau) B u_{DC} d\tau \tag{4}$$

Subsystem 2: Self-resonance

$$x(t) = f_2(t, x_0) = \Phi_2(t) x_0 \tag{5}$$

Subsystem 3: Discontinuous conduction state (DCM)

$$x(t) = f_3(t, x_0) = \Phi_3(t) x_0 \tag{6}$$

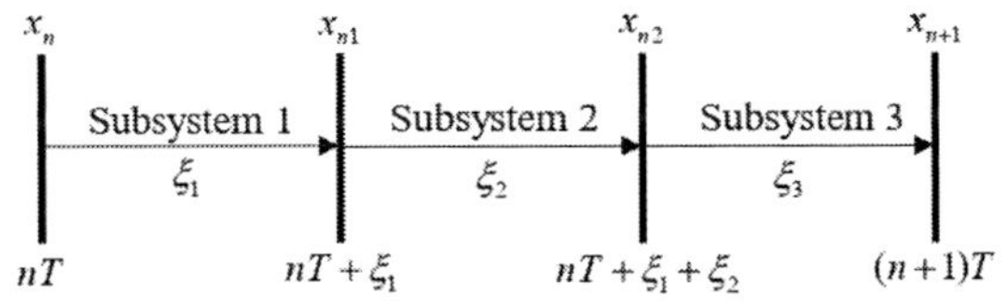

Figure 3. The mapping relationship in nth cycle.

Figure 3 shows the mapping relationship in the nth cycle. The intervals of those subsystem is ξ_1, ξ_2 and ξ_3 respectively and the corresponding mappings can be presented as follows:

$$x_{n1} = \Phi_1(\xi_1)x_n + \beta(\xi_1) \tag{7}$$

$$x_{n2} = \Phi_2(\xi_2)x_{n1} = \Phi_2(\xi_2)\Phi_1(\xi_1)x_n + \Phi_2(\xi_2)\beta(\xi_1) \tag{8}$$

$$x_{n+1} = \Phi_3(\xi_3)x_{n2} = \Phi_3(\xi_3)\Phi_2(\xi_2)\Phi_1(\xi_1)x_n + \Phi_3(\xi_3)\Phi_2(\xi_2)\beta(\xi_1) \tag{9}$$

where $\beta(\xi_1) = \int_0^{\xi_1}\Phi(\xi_1 - \tau)Bu_{DC}d\tau$.

Once the system achieves steady-state in the nth cycle, the initial state and the terminal state must be equal, namely, $x_{n+1} = x_n$. The mapping from x_n to x_{n+1} is a fixed point mapping and the fixed point is defined as $x^* = x_{n+1} = x_n$ which can be presented as follows:

$$x^* = [I - \Phi_3(\xi_3)\Phi_2(\xi_2)\Phi_1(\xi_1)]^{-1}\Phi_3(\xi_3)\Phi_2(\xi_2)\beta(\xi_1) \tag{10}$$

To meet the soft-switching conditions, the boundary of the subsystem must satisfied the zero-current switching (ZCS) condition. Hence the boundary condition can be presented as follows:

$$H_1 = C_1[I - \Phi_3(\xi_3)\Phi_2(\xi_2)\Phi_1(\xi_1)]^{-1}\Phi_3(\xi_3)\Phi_2(\xi_2)\beta(\xi_1) = 0 \tag{11}$$

Since ξ_1 and the ξ_3 are controllable but ξ_2 is uncontrollable, take a constant set of (ξ_1,ξ_3) as parameters of the fixed point equation (10) and the function with the respect of ξ_2 can be built to determine the value of ξ_2 as follows:

$$f_{\xi_1,\xi_3}(\xi_2) = C_1[I - \Phi_3(\xi_3)\Phi_2(\xi_2)\Phi_1(\xi_1)]^{-1}\Phi_3(\xi_3)\Phi_2(\xi_2)\beta(\xi_1) \tag{12}$$

The parameters of the IPIFR system are shown in Table 1 and the dc input voltage is 100V .The curve of $f_{\xi_1,\xi_3}(\xi_2)$ with the respect of ξ_2 is shown in Figure 4 and the intersection points of $f_{\xi_1,\xi_3}(\xi_2)$ with the horizontal line $f = 0$ are the possible values of ξ_2. In Figure 4, six intersection points 1)A, 2)B, 3)C, 4)D, 5)E and 6)F are 1) $78.4\mu s$, 2) $98.3\mu s$, 3) $163.1\mu s$, 4) $190.1\mu s$, 5) $249.1\mu s$ and 6) $280.7\mu s$, respectively. According to the shape of the curve shown in Figure 4, the six points can be divided into three groups based on the position of the three periodic peak. Then the curves of $i_p(t)$ and $u_p(t)$ are plotted in Figure 5 when ξ_2 takes $78.4\mu s$ and $98.3\mu s$ respectively. According to the performance of $i_p(t)$ and $u_p(t)$ shown in article[11], the fixed point of u_p must be less than zero, otherwise the soft-switching condition would not meet. The fixed point of u_p is 82.2V greater than zero when ξ_2 takes $78.4\mu s$ and other one is -81.0V when ξ_2 takes $98.3\mu s$. Therefore, point A is not a rational result of ξ_2 but point B is. In Table II, the fixed point of u_p are solved out with all the six points. The points A, C and E are not rational results of ξ_2 but the points B, D and F are.

Table 1. The parameters of IEIFR system.

u_{DC}(V)	L_p(μH)	C_p(μF)	R_p(Ω)	L_s(μH)	C_s(μF)	R_s(Ω)	k	R_L(Ω)
100.00	**660.0**	**0.4**	**0.2**	**585.0**	**0.4**	**0.2**	**0.5**	**10.0**

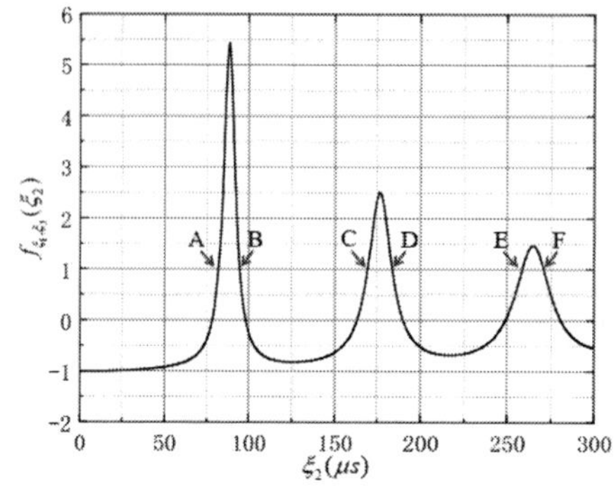

Figure 4. Curve of the fixed point function.

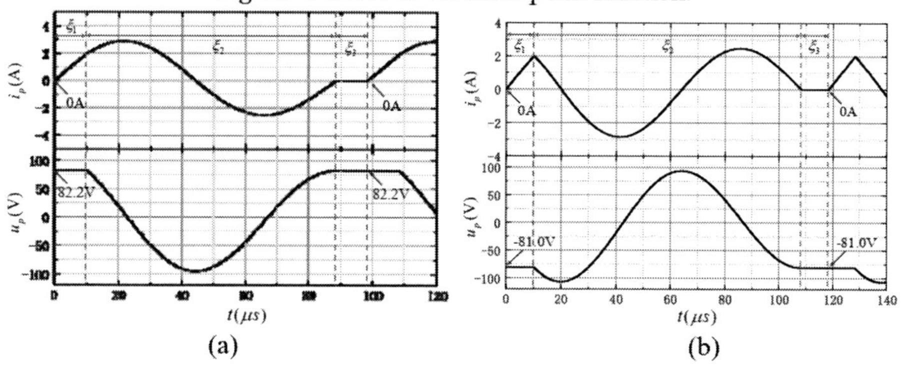

Figure 5. Curves of the simulation results under different operation point (a) Point A. (b) Point B.

Table 2. The results of operation points and the fixed point of u_p.

Point	$\xi_1(\mu s)$	$\xi_2(\mu s)$	$\xi_3(\mu s)$	u_p^*(V)	Is the point rational?
A	10	78.4	10	82.2	NO
B	10	98.3	10	-81.0	YES
C	10	163.1	10	50.2	NO
D	10	190.2	10	-47.9	YES
E	10	249.1	10	34.9	NO
F	10	280.7	10	-32.6	YES

3. Experiment Verification

The experimental waveforms of $i_p(t)$ and $u_p(t)$ under different operation points are shown in Figure 6 and the values of ξ_1, ξ_2 and ξ_3 in Figure 6 are listed in Table 3. The experimental results are very close to the theoretical results, therefore the modelling method is helpful to calculate the operation point and control the system.

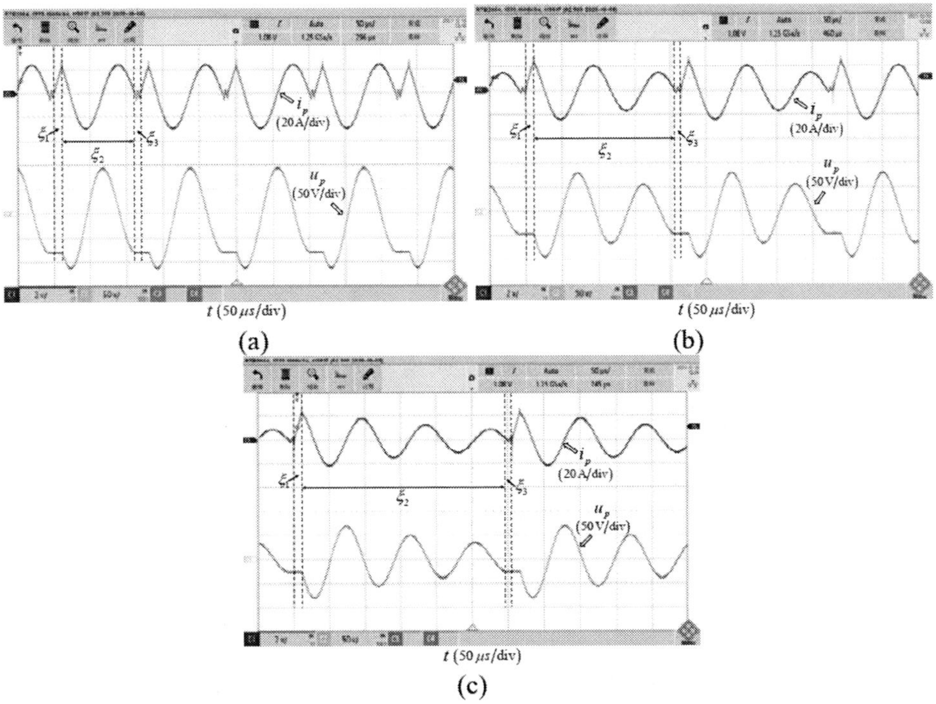

Figure 6. The waveforms of $i_p(t)$ and $u_p(t)$ under different operation points (a) Point B. (b) Point D. (c) Point F.

Table 3. Experimental results of operation points.

	$\xi_1(\mu s)$		$\xi_2(\mu s)$		$\xi_3(\mu s)$	
Point	Theoretical	Experimental	Theoretical	Experimental	Theoretical	Experimental
B	10	10.0	98.3	100.0	10	9.8
D	10	10.0	190.2	193.5	10	9.6
F	10	10.0	280.7	284.2	10	9.3

4. Conclusion

Since IEIFR system is different from the fully-resonance type of WPT system, the discrete-time modelling method is more suitable than the continuous-time domain modelling method or the ac impedance analysis modelling to analysis IEIFR system. The fixed point mapping in steady-state cycle has been built by iteration of analysis results of three linear subsystems. With the proper soft-switching boundary condition which is an analytical result of mode analyses, all possible operation points are determined precisely and the analytical results of all state variables can be calculated precisely either. It is well verified in experiment that the theoretical results is precise from the discrete-time model. Therefore, the discrete-time modelling method is helpful to analysis the incomplete-resonance system.

Acknowledgments

This work was supported in part by the Natural Science Foundation of China under Grant 51777177, and in part by the Science Foundation of the Fujian Province under Grant 2021J05262.

References

[1] Hunter H W, Aaron G, Kylee D S and Dan B 2012 *A High Efficiency 5 kW Inductive Charger for EVs Using Dual Side Control* (IEEE Transactions on Industrial Informatics) pp 585-595.

[2] Qingwei Z, Yun Z, Chenglin L, Yanjie G, Lifang W and Fang L 2017 *Experimental Study on Asymmetric Wireless Power Transfer System for Electric Vehicle Considering Ferrous Chassis*

(IEEE Transactions on Transportation Electrification) pp 427-433.
[3] Zheng Z and Bo Z 2020 *Omnidirectional and Efficient Wireless Power Transfer System for Logistic Robots* (IEEE Access) pp 13683-13693.
[4] Yan L, and Dongsheng B M 2016 *Wireless Power Transfer System Architectures for Portable or Implantable Applications* (Energies).
[5] Michael D F, Yates S and Paul R B 2001 *An interface system for autonomous undersea vehicles (IEEE Journal of Oceanic Engineering)* pp 522-525.
[6] Seonghwan K, Hyun-Ho P, Jonghoon K, Jingook K and Seungyoung A 2014 *Design and Analysis of a Resonant Reactive Shield for a Wireless Power Electric Vehicle* (IEEE Transactions on Microwave Theory and Techniques) pp 1057-1066.
[7] Yousu Y, Yijie W, Xiansheng L, Fanfan L and Dianguo X 2018 *A Novel Parameter Tuning Method for a Double-Sided LCL Compensated WPT System With Better Comprehensive Performance* (IEEE Transactions on Power Electronics) pp 8525-8536.
[8] Qingwei Z, Lifang W, Yanjie G, Chenglin L and Fang L 2016 *Applying LCC Compensation Network to Dynamic Wireless EV Charging System* (IEEE Transactions on Industrial Electronics) pp 6557-6567.
[9] Yeong H S, Bo H C, Eun S L, Gyu C L, Gyu-Hyeong C and Chun T R 2015 *General Unified Analyses of Two-Capacitor Inductive Power Transfer Systems: Equivalence of Current-Source SS and SP Compensations* (IEEE Transactions on Power Electronics) pp 6030-6045.
[10] Wenqi Z, Hao M, and Xiangning H 2009 *Investigation on Different Compensation Topologies in Inductively Coupled Power Transfer System* (Transactions of China Electrotechnical Society) pp 133-139.
[11] Lin C, Jianfeng H, Mingjie G, Zaifa L, Wenxiang C 2019 *A Converter Based on Independently Inductive Energy Injection and Free Resonance for Wireless Energy Transfer* (Energies).
[12] Chun S T, Yue S, Yu G S, Sing K N and Aiguo P H 2009 *Determining Multiple Steady-State ZCS Operating Points of a Switch-Mode Contactless Power Transfer System* (IEEE Transactions on Power Electronics) pp 416-425.

The CEEMD-LSTM-ARIMA Model and Its Application in Time Series Prediction

Chen Sun, Yingxiong Nong, Zhibin Chen, Dong Liang*, Ying Lu and Yishuang Qin
Information Center of China Tobacco Guangxi Industrial CO.,LTD., Nanning, Guangxi, China

*Corresponding author email: aloysiousliang@gmail.com

Abstract. Research shows that there are some data sets with strong timing in reality. After certain processing, they can predict the changes of transactions in the future to some extent and play a positive role in guiding production. However, they often have the characteristics of overall trend variability and seasonal fluctuation, and are also affected by many random factors. Therefore, the data show strong non stationarity. In this paper, a CEEMD-LSTM-ARIMA combined model is proposed. This method reconstructs variables by decomposing time series data, and adopts appropriate methods to deal with variables with different characteristics. In this paper, CEEMD-LSTM-ARIMA combined model is used to predict cigarette data. The experimental results show that the proposed combined model has faster convergence speed and higher accuracy than LSTM and ARIMA, and has more advantages in time series data prediction.
Keywords: Time series prediction; CEEMD; LSTM; ARIMA; Combined model.

1. Introduction

Time series is a series of data points obtained by a certain element according to the time sequence index, which reflects the change law of this element over time [1]. Time series analysis is to predict the development trend of future affairs or the changes in the future through the learning of these change laws. Therefore, it is a very effective data analysis method. Time series data exist in various fields such as meteorology, economy, medicine, electric power, science, transportation and so on. Through the study of time series data in various fields, we can mine a lot of information: the study of Precipitation Law is helpful to flood control and drought resistance; Forecast the economic trend and provide reference for formulating various policies; Mining the law of power consumption is helpful to power dispatching; Summarizing the historical traffic flow law is helpful to prevent congestion. With the continuous development of time series, attention to time series data in various fields is increasing.

In practical application, time series are often complex, nonlinear and non-stationary, and simple and stable time series are rare. Therefore, it is very important to find a method that can effectively deal with complex nonlinear and non-stationary time series and accurately predict them. The prediction methods of time series are roughly divided into simple linear time series prediction methods, nonlinear prediction methods based on machine learning and methods based on deep learning. Based on ARMA model, the processing method of geomagnetic declination missing number is studied in literature [2]. Literature [3] used BP neural network to predict the time series of water quality. Literature [4] studies how short-term wind power time series change based on CEEMD and random forest algorithm. Literature [5] has done research on prediction of aeroengine time series based on LSTM.

In order to explore more effective time series prediction methods, combined with complementary integrated empirical mode decomposition CEEMD [6], long-term and short-term memory network

LSTM [7] and differential integrated moving autoregressive average model ARIMA [8], this paper introduces a CEEMD-LSTM-ARIMA hybrid model, and applies the model to the prediction of cigarette time series data. By comparing the prediction results with LSTM method and ARIMA method, it is found that the combined model in this paper can divide and rule the characteristics of a single time series, so as to make full use of the explicit and invisible information of the data and effectively increase the accuracy of prediction.

2. The Basic Techniques of CEEMD-LSTM-ARIMA

2.1. CEEMD

The advantage of CEEMD is that the simultaneous addition of two Gaussian white noise with the same amplitude and the opposite phase solves the problem of CEEMD reconstruction error, while playing the inhibitory effect on the influence of mode mixing and residual white noise [9]. The specific decomposition steps of CEEMD are as follows:

(1) Add one set of opposite white noise sequences $n_1^+(t)$ and $n_1^-(t)$ to the original sequence $x(t)$to obtain sequences with positive noise $x_1^+(t)$ and sequences with negative noise $x_1^-(t)$.

$$\begin{cases} x_1^+(t) = x(t) + n_1^+(t) \\ x_1^-(t) = x(t) + n_1^-(t) \end{cases} \tag{1}$$

(2) The CEEMD algorithm breaks down the noise of the sequence to obtain the components I_{MF}, C_{1_j} and the trend surplus r_1.

(3) Similarly, using CEEMD decomposition to process the white noise sequence with opposite symbols in step (1) to obtain the components C_{-1_j} and r_{-1}.

(4) repeat the steps (1) ~ (3) n times to get n groups C_{1_j}, r_1, C_{-1_j}, r_{-1}.

(5) The mean of the components of 2 groups of residual positive and negative white noise obtained from multiple decomposition was selected as the final result, i. e.

$$\begin{cases} C_i(t) = \frac{1}{2n}\sum_{j=1}^{n}\left(C_{n_j} + C_{-n_j}\right) \\ r_n(t) = \frac{1}{2n}\sum_{j=1}^{n}(r_j + r_{-1}) \end{cases} \tag{2}$$

2.2. LSTM

LSTM is an RNN network for time series analysis, which introduces on the RNN the threshold structure (input gate, forgetting gate and output gate) to judge whether the information meets the requirements to control the cumulative speed of the information, so as to remember and update the new information with this structure and solve the problem of long-term dependence. The neural unit of each LSTM as shown in Figure 1 is composed of the cell state, namely the long-term state c_t, and the short-term state h_t, as well as the input gates i_t, the forgetting gates f_t, and the output gates o_t [10].

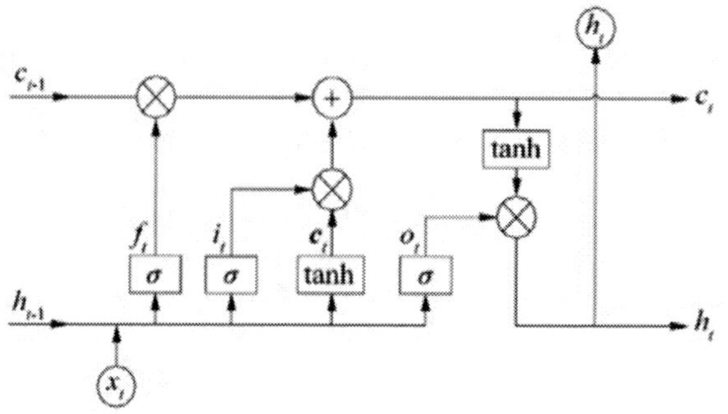

Figure 1. LSTM Structure Diagram.

The so-called cell state, which we can understand as a container for storing information, gradually changes and output the information in the container through the process control of the input, forgetting, and output doors. In each neural unit, the cell state experiences the forgetting gate, the input gate and the process of exporting information to the output gate. The input gate is copying the input information for processing the current neural unit. The entire input door contains 2 sections. The sigmoid activation function section determines what kind of input information will be updated, that is, a certain input information is ignored. The tanh section was used to construct a new candidate value vector, added to the current cell state. That is:

$$\begin{cases} i_t = \sigma(\boldsymbol{W}_i \cdot [x_t, h_{t-1}] + b_i) \\ c_t = \varphi(\boldsymbol{W}_c \cdot [x_t, h_{t-1}] + b_c) \end{cases} \tag{3}$$

Forgetting gate is used to determine what information the current state needs to be discarded, and LSTM learns to get the network to remember. That is [11]:

$$f_t = \sigma\left(\boldsymbol{W}_f \cdot [x_t, h_{t-1}] + b_f\right) \tag{4}$$

The output gate mainly controls the output information of the current hidden state. And there are:

$$\begin{cases} h_t = y_t = o_t \otimes \varphi(c_t) \\ c_t = i_t \otimes c_t + f_t \otimes c_{t-1} \\ o_t = \sigma(\boldsymbol{W}_o \cdot [x_t, h_{t-1}] + b_o) \end{cases} \tag{5}$$

In formula, x_t is the input and h_t is output of the network at time t [12]. c_t is the new candidate value vectors constructed for the tanh part at time t. f_t is Forget doors, i_t is input doors, ando_t is output gates. $\boldsymbol{W}_f, \boldsymbol{W}_i, \boldsymbol{W}_o, \boldsymbol{W}_c$ are the weight matrices for forgetting, input, output, and memory cells, respectively. b_f, b_i, b_o, b_c are the weight matrices for forgetting, input, output, and memory cells, respectively. σ is the sigmoid function, φ is for the Tanh function. Finally, the error and gradient calculation were performed by the time backpropagation (Bach Propagation Through Time, BPTT) algorithm to obtain the optimization parameters of the LSTM model [13].

2.3. ARIMA

ARIMA(p, d, q) is called the differential autoregressive moving mean model, which consists of a stochastic process consisting of autoregressive model AR(p) and moving mean model MA(q) [14].The specific forms are:

$$x_t = \varphi_1 x_{t-1} + \varphi_2 x_{t-2} + \cdots + \varphi_p x_{t-p} + \varepsilon - \theta_1 \varepsilon_{t-1} - \theta_2 \varepsilon_{t-2} \cdots - \theta_q \varepsilon_{t-q} \tag{6}$$

In formula: φ is autoregressive coefficient; p is autoregressive order; θ is moving average coefficient; q is moving average order; and $\{\varepsilon\}$ is white noise sequence.

3. Construction of CEEMD-LSTM-ARIMA Hybrid Model

This paper builds a prediction model based on the combination of complementary set empirical mode decomposition CEEMD, long-term and short-term memory network LSTM and differential integration moving autoregressive average model ARIMA. Because the historical data is extremely unstable and has great variability in the short term, firstly, the CEEMD modal decomposition method is used to decompose the historical data at different frequencies to enhance the timing characteristics reflected in the data and achieve the effect of noise reduction. And make full use of the advantages of LSTM and ARIMA in processing time series data to improve the prediction accuracy. The specific steps are as follows:

(1) Each time the opposite white noise signal is randomly added to the original sequence, the signal with the added white noise is decomposed into several components I_{MF} using CEEMD. After several repeats, the ones obtained after the decomposition were averaged separately, and the sequences were averaged as the final result. The j components I_{MF} and 1 residual component R can be decomposed.The X(t) is:

$$X(t) = \sum_{i=1}^{j} I_{\mathrm{MF}_i} + R \tag{7}$$

(2) According to the wave frequency and amplitude characteristics of the component image of I_{MF}, the decomposed components are classified as the high frequency components and the low frequency components.
(3) Data for predicting high frequency components using LSTM; low frequency components are predicted using ARIMA.
(4) The final prediction result is obtained by summing the predicted value of the high-frequency component and the predicted value of the low-frequency component in step (3).
To sum up, the model flow chart of CCEMD-LSTM-ARIMA is shown in Figure 2

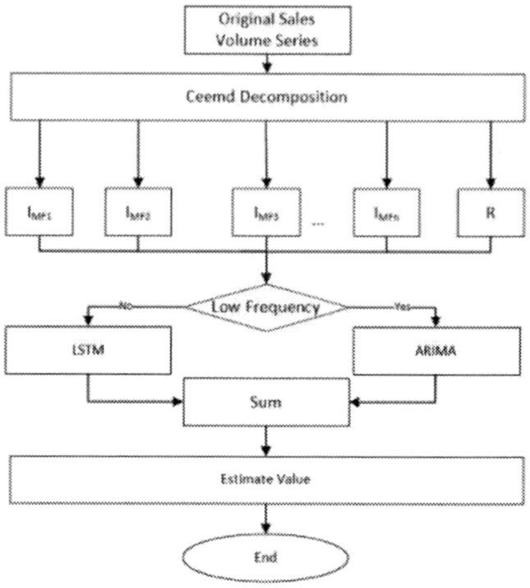

Figure 2. The CEEMD-LSTM-ARIMA prediction model flow.
To quantitatively evaluate the accuracy of the CEEMD-LSTM-ARIMA prediction model, two evaluation indexes were selected in this paper. Using the root-mean-square error (RMSE) and the average absolute percentage error (MAPE) to evaluate the final prediction performance of the model. The smaller value of the RMSE and the MAPE is, the higher the accuracy value is, which indicates that the prediction performance of the model is better.
(1) Root-Mean-Square Error

$$R_{\text{MSE}} = \sqrt{\frac{1}{n}\sum_{t=1}^{n}(X(t) - \hat{X}(t))^2} \tag{8}$$

(2) Average Absolute Percentage Error

$$M_{\text{APE}} = \frac{1}{n}\sum_{t=1}^{n}\frac{|X(t)-X(t)|}{|X(t)|} \tag{9}$$

In the formula, $X(t)$ is the actual data and $\hat{X}(t)$ is predicted data, respectively; n is the predicted number of samples.

4. Application of CEEMD-LSTM-ARIMA in Time Series Prediction

4.1. Time Series Data Set Division and Data Decomposition
This paper uses the monthly cigarette sales volume dataset of a city in China from 2009-2020 as the experimental number set, which divides them into training and testing. The data from 2009 to 2019 was used to perform training learning of predictive models. The data from January to December of 2020 was used as a test set to test the experimental effect of the prediction model.
The dataset are affected by a variety of factors, so the fluctuation trend presents irregular fluctuations with high amplitude. The original sequence was decomposed using the CEEMD model and 50 white noise sequences were added to the original time series with an amplitude set to 0.2. Six eigenmode functions I_{MF} and one residue term R are generated. The decomposition results are shown in Figure 3.

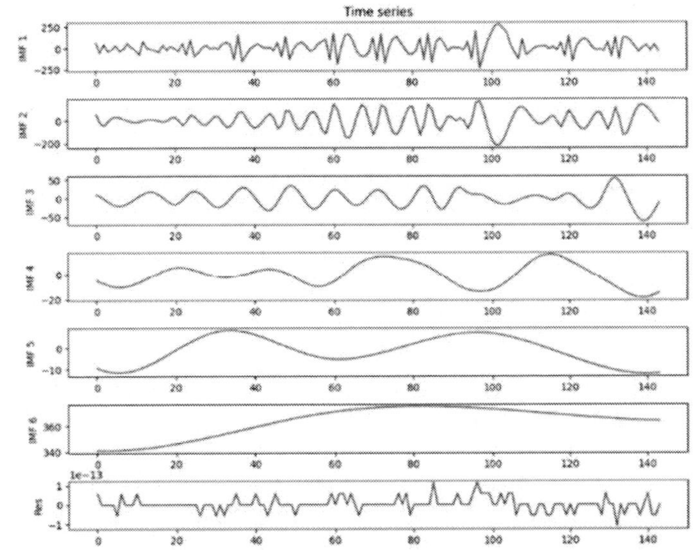

Figure 3. CEEMD exploded view.

The resulting six components are divided into two categories: high-frequency components and low-frequency components .The I_{MF1}~ I_{MF3} and Res components have short periods and large fluctuations, divided into high-frequency components and predicted using the LSTM algorithm. The I_{MF4}~ I_{MF6} fluctuations are small and gentle, with a cycle length over 2a, divided into low-frequency components and predicted using the ARIMA model. All component prediction values were summed for the results of prediction for the CEEMD-LSTM-ARIMA model.

4.2. Forecast Results and Comparative Analysis

To verify the effectiveness of the combined predictive models in this paper, the model is built using Python 3.6.6 programming. The results were also horizontally compared with the ARIMA model and the LSTM model, respectively [15].

The CEEMD-LSLM-ARIMA model predictions are shown in Figure 4. As can be seen from Figure 4, compared with the best results of other algorithms and the actual result, the overall fluctuation trend can be completely predicted. Comparisons of the error values for each model are shown in Table 1.

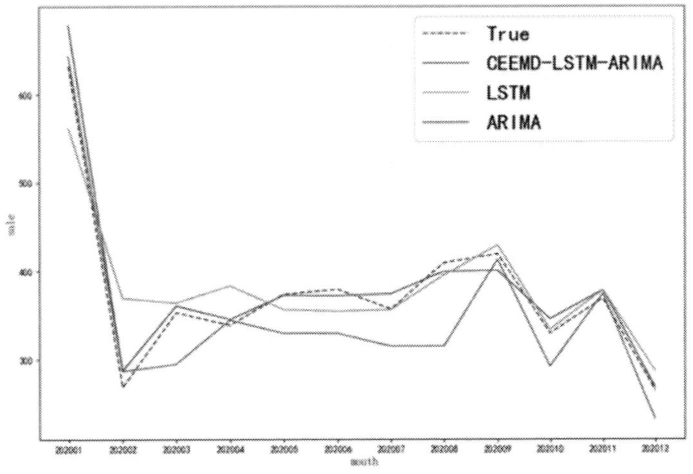

Figure 4. The comparison of prediction results.

Table 1. Error Analysis.

Model	MAPE	RMSE
CEEMD-LSTM-ARIMA	3.4366697993534587	17.5831155512592
LSTM	7.848120418623144	39.73578712316075
ARIMA	9.43125993005411	42.307496552960316

In this paper, combining the ARIMA and the LSTM model not only improves that the ARIMA model does not consider the influence of external factors, the number of samples should not be too much, and the weak prediction ability of nonlinear and unstable time series, but also combines the advantages of the LSTM model to fully excavate the data timing characteristics through its unique memory mode and forgetting mode. As shown in Table 1, both the CEEMD-LSTM-ARIMA model fits to the data and the data prediction error significantly outperformed the ARIMA and LSTM models.

5. Conclusion

In this paper, an intelligent model based on the combination of CEEMD, LSTM and ARIMA is proposed to explore a more effective means to deal with non-stationary data. Taking the cigarette data of a city in China as an example, the proposed method is verified by predicting the cigarette time series data. Firstly, CEEMD is used to decompose the cigarette time series data into several wave frequencies subsequences with inconsistent amplitudes. Then according to different wave frequencies, LSTM model and ARIMA model are used for prediction. Finally, the prediction results of the two methods are added to complete the final prediction of the whole combined model. The experimental results show that this method not only preserves the advantages of ARIMA model with high sensitivity to stable time series data and good at capturing linear relationships in data, but also combines the advantages of LSTM algorithm to fully mine data time series characteristics through its unique memory mode and forgetting mode. Compared with ARIMA and LSTM, CEEMD-LSTM-ARIMA in this paper has higher prediction accuracy, The method has good feasibility.

Acknowledgments

This research is funded by the Project of Guangxi China Tobacco Industry Co., Ltd. (No. 2021450000340063).

References

[1] Amal Saadallah, Matthias Jakobs, Katharina Morik. *Explainable Online Deep Neural Network Selection Using Adaptive Saliency Maps for Time Series Forecasting* [J]. Lecture Notes in Computer Science, 2021:404-420.

[2] Dong Baowei, Qian Qiuliang, Ren Yafei, et al. *Geomagnetic declination missing number processing method based on ARMA model* [J]. Geodesy and geodynamics, 2021,41 (11): 1152-1156.

[3] Zhang Mengdi, Xu Qing, et al. *Water quality time series prediction based on dynamic sliding window BP neural network* [J]. Journal of Environmental Engineering Technology, 2021: 1-11.

[4] LIU Qiang, HU Zhiqiang, ZHOU Yu, et al. *Short-term wind power forecasting based on CEEMD and random forest algorithm* [J]. Smart Power, 2019, 47 (6): 71-76.

[5] Lin Ruiguan, Wang Huawei, Che Changchang, et al. *Aeroengine predictive maintenance model based on LSTM classifier* [J]. Systems engineering and electronic technology, 2021: 1-10.

[6] Li Bingchen, Yu Huijun, Ding Huaxuan, Liu Jingyu. *Short term wind speed prediction based on CEEMD and LSTM-ARIMA* [J]. Energy Conversion and Management, 2021, 227: 1-16.

[7] Li Jie, Peng Qiyuan, Wen Chao. *Research on short-term passenger flow prediction of high-speed railway based on LSTM deep neural network* [J]. System engineering theory and practice, 2021,41 (10): 2669-2682.

[8] XU Xiaoyao, ZHANG Pengfei, JIANG Jian. *Dam deformation prediction based on EMD-GAELM-ARIMA algorithm* [J]. Computer and Modernization, 2020 (7):1-5.

[9] Bilin Shao, Maolin Li, Yu Zhao, Genqing Bian. *Nickel Price Forecast Based on the LSTM Neural Network Optimized by the Improved PSO Algorithm* [J]. Mathematical Problems in Engineering, 2019:1-15.

[10] Yu Qiong, Tian Xian. *Nonlinear time series prediction algorithm based on combined model* [J]. Computer engineering and science, 2021, 43 (10): 1817-1825.

[11] Chen Ding, Guizhi Wang, Xinyue Zhang, Qi Liu, Xiaodong Liu. *A hybrid CNN-LSTM model for predicting PM2.5 in Beijing based on spatiotemporal correlation* [J], Environmental and Ecological Statistics, 2021, 28(3):503-522.

[12] Zexian Sun, Shenshen Zhao, Jingxuan Zhang. *Short-Term Wind Power Forecasting on Multiple Scales Using VMD Decomposition, KMeans Clustering and LSTM Principal Computing* [J], IEEE Access, 2019: 166917-166929.

[13] Shang, Yilun. *A combinatorial necessary and sufficient condition for cluster consensus* [J]. Neurocomputing, 2016, 216: 611-616.

[14] Zhu Shaomin, Xia Hong, LV Xinzhi, Lu Chuan, Zhang Jiyu, Wang Zhichao, Yin Wenzhe. *Main pump state prediction of nuclear power plant based on ARIMA and LSTM combined model* [J]. Nuclear power engineering, 2021: 1-8.

[15] Longmin Shi, Yong Liu. *Soil phosphorus content prediction based on improved PSO-LSTM algorithm* [C], IEEE International Conference on Artificial Intelligence and Computer Applications (ICAICA), 2021:213-218.

Modeling and Simulation of Rock Bits Based on Mega Drilling Data

Zebing Wu, Mengjie Jiang*, Weixin Zheng, Yabing Gu, Xiping Zhai and Chenjuan Yang

School of Mechanical Engineering, Xi'an Shiyou University, Xi'an, Shaanxi, China

*Corresponding author email: 20211040455@stumail.xsyu.edu.cn

Abstract. Rock bits directly undertake the task of fracturing rock under earth during drilling operation. Accurate prediction of Rate of penetration (ROP) play an important role in reducing drilling costs and shortening the drilling cycle. This paper introduces ROP prediction methods including empirical mathematical modeling, computer simulation of bit-rock interaction and BP neural network model based on mega drilling data. The most significant factors affecting ROP are summarized as follows: weight on bit(WOB), rotation speed, rock formation properties, bit cutting structure, etc., which have a complex multi-parameter nonlinear relationship with ROP. The drilling data features complexity and big volume. Therefore machine learning is a best way to model the relationship between main factors and ROP. The advantages and disadvantages of the different ROP modeling methods are analyzed and compared. It shows that the BP neural network model based on mega drilling data demonstrates high accuracy, and is the future development direction in ROP prediction.
Keywords: Rock bits; ROP Modeling; Simulation; BP Neural Networks; Drilling data.

1. Introduction

Recently, more and more oil and gas resources are being consumed. In the future, oil resources will still be in an irreplaceable and important position. Inspired by the high demand for oil and gas resources and high oil prices, investment in oil development has been continuously increased worldwide, and the number of wells has been rising. However, as exploration and development continue to move toward deep, ultra-deep, and extended reach horizontal wells, the rate of mechanical drilling is low. Therefore, the long drilling and production cycle has become an important factor restricting the exploitation of deep-seated oil and gas resources In 2014, the oil industry was affected by the global economic recession. Oil prices fell to 35% before the recession. In 2020, oil prices fell to negative values. Due to extreme oil price fluctuations, the challenge of optimizing drilling performance is increasing. One of the main challenges is to improve rate of penetration (ROP). In 2021, benefiting from China's tremendous achievements in epidemic prevention and control, the oil and gas drilling industry is also vigorously developing and recovering.

There are many factors that control the performance of a drill bit, and the interaction of these factors is very complicated. One of the indicators of drill bit performance is ROP. For the past ten years, the ROP has been the focus of drilling optimization. The penetration rate of a well is a sign of drilling efficiency. High ROP means better drilling rig performance and production efficiency; ROP is a key parameter of the drilling optimization process, because it determines the drilling time and affects the drilling cost. In actual operation, predicting and optimizing ROP has always been a difficult problem. There are many factors that affect ROP, such as WOB, bit size, pore pressure gradient, ground machine pump equipment, rock formation properties, etc. These factors have an interactive effect on the ROP[1]. In order to improve the ROP, scholars worldwide have carried out a lot of research. Binghan 1964, Bourgoyne & Young 1974, Hareland and Rampersad 1994, Mohtari and Hareland 2010, etc[2-4] developed mathematical models successively under different factors affecting ROP. With the rapid development of computer simulation technology, Ma et al. conducted a simulation of the bit/rock interaction to predict the drilling speed of the drill bit[5]. The above methods have developed theoretical empirical formulas, numerical simulations, statistical data analysis, experimental research and other methods for ROP prediction technology, but these methods have certain defects and limitations. For example, with actual drilling operation, it is difficult to obtain real-time related parameters with the depth of the well. There is a certain difference between theoretical calculation and field measured data, which cannot fully meet the requirements of actual drilling operations.

In recent years, researchers in the field of drilling engineering have used artificial neural networks to predicting the influence of different factors on ROP through data analysis. Artificial neural networks have the ability to map and model complex nonlinear data and have a high degree of fit. R. Jahanbakhshi established a neural network ROP prediction model based on rock mechanical properties, bit structure, weight on bit, and bit speed. He conducted long-term investigations and data collection in an oil field in southern Iran, and the predicted ROP data output by the model and the data collected on site are compared. The comparison results show that the neural network model can efficiently forecast the ROP of this oilfield. Xia Hongquan et al. established a BP neural network ROP prediction model according to bit selection, and proved the accuracy of the model prediction based on actual drilling data[6]. Dong Qingqing et al. proposed a particle swarm optimization algorithm based on the conventional BP neural network rock drillability model, which further increased the precision of the neural network prediction model[7]. Wang Wen et al. established a ROP prediction neural network model for deep drilling in the Bohai Oilfield, and the output of the neural network is consistent with the actual ROP, and based on the prediction results of the model, a preferred plan for speed-up tools was proposed[8]. Based on actual drilling data, Du Yu et al. established three neural network models of BP, GA-BP and PSO-BP, and analysed the superiority of different models to the predictability of formation drillability[9].

This article first summarizes the main influencing factors that affect ROP, classifies many factors from controllability and uncontrollability, and summarizes the characteristics and degree of influence of different factors on ROP, and then based on the influence of different factors, the empirical mathematical model, the drill bit and rock interaction model and artificial neural network model analyse the performance of ROP prediction in detail, and finally put forward the development direction of ROP prediction application in actual drilling operations, such as BP neural network, deep learning, random forest decision tree and recurrent neural network.

2. Analysis of Factors Affecting ROP

The ROP is determined by controllable and uncontrollable factors. The former can be adjusted by equipment technology and technical means, such as bit structure parameters, rotary speed, hook load, drilling fluid properties, etc.; the latter is involved with factors that objectively exist and dynamically change with the drilling depth, such as rock properties, oil and gas reservoir depth, formation pressure[10], etc.

2.1. Controllable Factors

Drilling tools are the key factor affecting ROP. For different rock formations and drilling sections, the reasonable selection of drilling tools is the most direct factor to increase the ROP. Suitable drill bits, drill pipes and other drilling tools, as well as high-power drilling rigs can effectively save the time of drilling operations, running out and running in the pipe string and casing, thereby increasing the ROP[11]. High-efficiency coiled tubing drilling operations are also an important factor in reducing the trip time. The high-power and high-sensitivity drilling rig driving device and drawworks can effectively control the WOB, thereby improving the drilling efficiency. High-horsepower mud pumps can meet the requirements of different drilling sections for mud discharge rated pressure work. The actual working rate of this type of mud pump is low and the pump pressure frequently exceeds the working critical value, which seriously affects the ROP increase. Therefore, the development of high-horsepower and high-pressure pumping equipment is an urgent problem to be solved in our country[12]. High ROP requires drilling fluid tools and drilling fluids with high quality to carry cuttings, stop leaks, and ensure that the drilling fluid can continue to circulate to reduce the temperature of the bit, and continue to increase ROP while ensuring safe drilling[13]. In recent years, the selection of drilling speed-up tools and the development of new speed-up tools are the key research directions for improving ROP, such as composite percussion drilling tools[14], New-Drill speed-up tools[15], SL torsional impact tools[16], pulse jets-type hydraulic impact speed-up tool[17], whirling oscillating speed-up tool[18], whirling screw speed-up tool, etc.[19] Improving the rock breaking energy of the drill bit, the torque transmission efficiency of the drill string, and the conversion rate of mud hydraulic energy are considered as a starting point to improve ROP, these drilling speed-up tools have been successfully applied in major oil fields in China and have been continuously improved based on field feedback data. It should be noted that with the application of intelligent drilling rigs in the international oil and gas market, the concept of intelligent drilling and intelligent oilfields in the future will bring a new revolution to the oil and gas industry. Based on this premise, how to improve the automation level of the drill machine will be the key research topic for the future development.

2.2. Uncontrollable Factors

WOB is one of the most significant factors affecting ROP[20]. It determines the degree of interaction between the drill bit and the rock, the friction torque between the drill string and the borehole wall, and the rock-breaking volume. Normally, ROP increases with the WOB. It increases exponentially, and its index is generally in the range of 0-1. Different drill bit sizes have obvious differences in ROP. Large-sized drill bits have low rock breaking efficiency and slow ROP, but small-sized drill bits will result in too small boreholes, reduced casing layers, and poor adaptability to complex formations, Personalized design of drill bits for different oil and gas fields and specific drilling sections is an important research direction for future ROP speed increase. The nature of the drilling fluid has a very important impact on ROP and drilling safety. The density of the drilling fluid determines the pressure difference between the hydraulic pressure in the well and the pores of the formation. A suitable drilling fluid can balance the formation pressure, cool the drill bit, and carry cuttings. After the depth exceeds 6km, the bottom hole pressure difference can reach 60MPa and the temperature is higher than 200℃. As drilling fluid technology continues to evolve, the design concepts of high temperature resistant reagents such as SMC, SMP, and resin SPN have gradually improved[21]. Rock strength have a certain negative correlation with ROP[22]. The RPM and tooth wear directly affect the rock breaking efficiency of the drill bit. Therefore, it is necessary to reasonably control the drill speed and tooth profile and material selection.

3. Empirical ROP Models

To describe the relationship between each drilling factor and ROP more clearly, domestic and foreign scholars have proposed many empirical mathematical models, most of which are based on the formation properties combined with several controllable variables, and they have studied and analysed the mathematical relationship between ROP and various factors. Several classic ROP mathematical models recognized by experts and scholars are analysed as follows.

3.1. All Types of Empirical Models

Maurer, Bourgoyne derives a theoretical equation suitable for roller cone bits according to a large amount of experimental data obtained from single-tooth impact experiments, reflecting the relationship between ROP and WOB, rotation speed, bit diameter, and rock strength. The mathematical expression is:

$$ROP = \frac{KN(W - W_0)^2}{d_b^2 S^2} \tag{1}$$

The "Binghan Model" is modified on the basis of the "Maurer Model" and is suitable for situations where the WOB and the drilling rate are low. The model ignores the drilling depth, and its mathematical expression is:

$$ROP = K(\frac{W}{d_b})^{a_5} \tag{2}$$

In formulas (1) and (2), N is the rotation speed (r/min), S is the compressive strength of the rock (psi), K and a_5 are proportional constants, which are determined by experimental conditions, and W is the WOB (t), W_0 is the WOB threshold (t), and d_b is the drill diameter (mm). It can be seen from the formula that the predicted ROP is positively correlated with the WOB and rotation speed, and the rock strength is negatively correlated with the bit diameter.

"Bourgoyne and Young Model" introduces new parameters that affect the ROP, and determines the degree of influence of each parameter on the ROP, such as sand compaction, rock strength, pore pressure,WOB, rotary speed, hydraulic impact subjected by rock bit, tooth wear, etc., the mathematical expression of this model is:

$$ROP = f_1 + f_2 + f_3 + f_4 + f_5 + f_6 + f_7 + f_8 \tag{3}$$

$$f_1 = e^{2.303a_1} \tag{4}$$

$$f_2 = e^{2.303a_2(10000-D)} \tag{5}$$

$$f_3 = e^{2.303a_3D^{0.69}(g_p-9.0)} \tag{6}$$

$$f_4 = e^{2.303a_4D(g_p-\rho_c)} \tag{7}$$

$$f_5 = \left[\frac{(\frac{W}{d_b}) - (\frac{W}{d_b})_t}{4-(\frac{W}{d_b})_t}\right]^{a_5} \tag{8}$$

$$f_6 = \left(\frac{N}{60}\right)^{a_6} \tag{9}$$

$$f_7 = e^{-a_7h} \tag{10}$$

$$f_8 = \left(\frac{F_j}{1000}\right)^{a_8} \tag{11}$$

In the above formula, D is the vertical well depth (m), g_p is the pore pressure gradient (W/db), t is the minimum drill bit diameter (mm) that meets the drilling requirements, h is the tooth tip dullness, and a_1-a_8 are based on drilling conditions. The determined test constant, f_j is the hydraulic impact force (N) that the drill bit bears downhole.

"Warren Model" refers to "Maurer Model". In addition to factors such as rock strength, bit size, WOB, speed, etc., a large amount of experimental data has introduced new correction coefficients and determined new mathematical expressions:

$$ROP = \left(\frac{aS^2d_b^3}{N^bW^2} + \frac{c}{Nd_b}\right)^{-1} \tag{12}$$

In the above formula, S is the rock shear strength (psi), d_b is the drill bit diameter (mm), W is the bit weight (t), N is the drill speed (r/min), and a and c are dimensionless constants.
"Warren Model" is a more complete roller cone bit ROP mathematical model suitable for soft formations proposed by Warren by referring to the above empirical mathematical model. This model can predict the ROP under soft formations and low well pressure conditions. Considering the connection of drilling speed of WOB, drill pipe speed, rock strength and bit size, the response curve between ROP and each factor is analysed. At the same time, the rock cutting process from the beginning to the completion of rock cutting by the bit is also analysed. The influence of the amount of removal and the amount of bit wear on the ROP, the mathematical expression is:

$$ROP = \left(\frac{aS^2 d_b^3}{N^b W^2} + \frac{c}{N d_b} + \frac{c d_b^3 \gamma_f \mu}{F_{jm}} \right)^{-1} \tag{13}$$

In the above formula, F_{jm} is the impact force of the mud jet (t), γ_f is the specific gravity of the fluid, and μ is the plastic viscosity (mPa·s). The "Warren Model" can clearly describe the various factors in the continuous transition process of chip generation and chip stripping.
"Mohtari Model" puts forward a mathematical model of ROP based on WOB, tooth wear, bit size, and rock properties for PDC bits. Its expression is:

$$ROP = W_f G \frac{W^a N^b}{S D_b} \tag{14}$$

In the above formula, a, b, and G are the model constants, W_f is the bit wear coefficient (dimensionless), W is the WOB (t), D_b is the bit diameter (mm), and S is the compressive strength of the rock (psi), N is the drill speed (r/min).
"Hareland, Wu Model" focuses on the relationship between predicted ROP and rock compressive strength. This model considers the cutting structure of the bit in more detail, reflecting the effect of different cutting teeth and bit structure on ROP. Its expression is[13]:

$$ROP = K \frac{80 n_t m N^a}{D_b^2 \tan\varphi^2} \left[\frac{1}{c_2} \left(\frac{W}{100 n_t \sigma_p} - c_1 w \right) \right]^2 \left[1 - d \left(\frac{DG}{8} \right)^c \right] \tag{15}$$

In the above formula, σ_p is the strength of the rock under different pressure differences (psi), φ is the cutting angle between the tooth and the rock (°), W is the WOB (t), D_b is the diameter of the bit (mm), and m is one revolution of the bit. The number of effective cutting teeth, n_t is the number of teeth in contact with the rock at the bottom of the drill bit, w is the radial width of the inserted rock (mm), K, c, and d are all test coefficients, and N is the drill speed (r/min).

3.2. Summary to the Models

Summarizing the classic empirical mathematical model mentioned above, it can be seen that different ROP empirical mathematical models focus on different drilling factors. The parameters in the formula except for some model constants, such as a, b, G, etc. (usually obtained through test data), and the bit size D_b can be obtained before drilling operations; all other variables change with the change of drilling depth, such as WOB, rotation speed, rock strength, W_f bit wear coefficient, pore differential pressure, etc. These variables are difficult to obtain in the dynamic drilling process, so, the application of formulas in drilling operations has certain limitations, and the versatility is not strong. This paper takes the "Hareland, Wu Model" model as an example to compare the model with field drilling data. The ROP comparison curves for different well depths are shown in Figure 1.

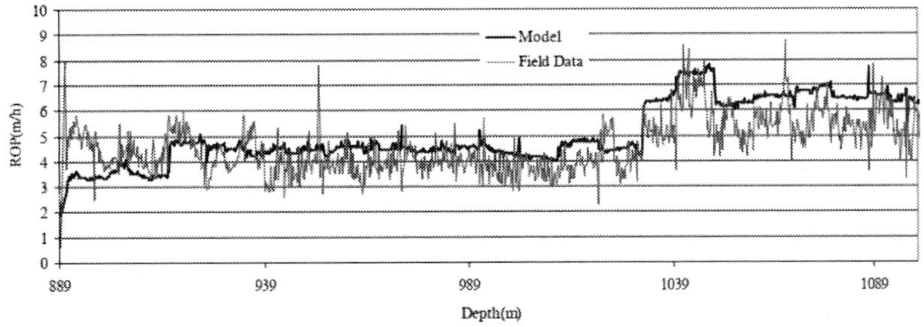

Figure 1. The comparison of ROP from the model and the field.

The comparison chart shows that the determined ROP mathematical model is in good agreement with the ROP of the actual drilling operation, and the overall trend is consistent. It should be noted that there are still large errors between the predicted ROP and the actual ROP at certain well depths, for example, the predicted ROP at 960m is about 4m/h, while the actual ROP is about 7.8m/h. This is due to the loose distribution of rock formations during actual drilling operations and the dynamic changes of working environment in the bottom hole.

4. Rock Bit/Rock Interaction Model

Low bit life and rock breaking efficiency are the main reasons that limit the increase of ROP. Therefore, in-depth study of the bit/rock interaction mechanism in drilling engineering is of great significance to improve the efficiency of oil and gas production[23]. The process of rock crushing by the drill bit is actually a comprehensive manifestation of the teeth on the bit interacted with the rock. Theoretical research can only solve some relatively simple, clear mechanism and strong regular problems, and it is difficult to comprehensively analyse the dynamic process of the bit teeth's breaking the rock; Experimental research can make people understand many phenomena with unclear mechanisms, but it sometimes takes a lot of the money and manpower, and the experiment cycle is long. Therefore, with the improvement of calculation and performance and finite element theory, combined with the in-depth study of the theory of the bit/rock interaction mechanism, computer simulation has become an important technical means for scholars to study the dynamic process of rock breaking by the bit teeth[24].

4.1. Bit/rock Interaction

Lian Zhanghua established a simulation model based on the parameters of the tooth profile and aimed at the size of the tooth crater at the bottom of the well[25]. Liu Qingyou et al. established a simulation model of the bit/rock interaction under the new tooth structure, and compared it with the rock breaking experiment of the single insert row, and obtained the simulation results with little error from the experimental data[26-27]. Yang Chunlei et al. used finite element software to establish a dynamic model of a single tooth row rock breaking system with compound motion, and a linear D-P model was used to simulate rock materials, which well reflected the impact indentation and rotating scraping in the process of rock fracture by the roller cone bit. This method has important reference significance for studying the interaction mechanism between different types of drill bits and rocks[28]. Kuang Yuchun et al. established a simulation model of the effect of bit structure, rock properties and weight-on-bit on ROP, and developed a digital simulation evaluation system for the interaction between the bit and rock[29-30]. Deng Minkai et al. established a simulation model of the PDC bit cutting teeth interacted with the rock, and studied the mechanical characteristics of the backrake angle of the cutting teeth on the bit/rock interaction and the influence on the efficiency of rock breaking[31]. Deng Yong et al. used the discrete element numerical simulation method to study the characteristics of crack formation and expansion during the dynamic fracture process of rock under impact load, as well as the relationship between the size of the crack formation and the impact velocity[32].

Heydarshahy Seyed Ali et al. designed 14 types of drill bits according to the International Association of Drillers (IADC) classification standards, established rock and drill bit models using the finite element method, and studied the rock breaking efficiency of drill bits of different drill bits. The results showed that: the height of the drill bit cone is a positive correlation between the area of the high-stress zone of the rock, and the rock cracks are mainly generated in the gauge area and the nose. The larger the cone height and gauge width, the longer the vertical cracks acting on the rock[33]. Zhu Xiaohua et al. established a three-dimensional finite element model of different cutting tooth cross-sections, based on different bottom hole confining pressures, studied the cutting tooth cross-sectional area, cutting tooth backrake angle and side rotation angle on the effect of rock breaking efficiency[34]. Yang Renqiang et al. proposed a multi-factor comprehensive optimization analysis method for drill bits. The specific idea is to apply the finite element method to comprehensively analyse the inherent characteristics, stress characteristics and temperature field characteristics of the PDC drill bit. Based on the comprehensive consideration of the three factors, the design and welding process are fully optimized, and the field test data is compared, and it is found that the life of the optimized bit is extended to 1.5 times that of the original bit[35]. In order to increase the market share of roller cone bits, this method has important reference significance for the tooth structure, nozzle design and internal bearing system optimization design of roller cone bits.

Wu Zebing et al. designed a cone-PDC hybrid drill bit. Since there are few researches on the temperature field and rock breaking characteristics of this kind of bit in the rock breaking process, a hybrid drill bit simulation model is established based on the finite element method and elastoplastic mechanics. The rock simulation model has studied the change of the temperature field of the bit and the rock crushing efficiency during the interaction process. The results show that: (1) the cone insert of the mixed bit crushes the rock first, and PDC cutter dominate the rock breaking when the crushing crater is formed; (2) the initial temperature of the mixed drill bit for breaking the rock rises sharply, and the temperature rises slowly after a period of time. The higher the WOB, the higher the temperature; (3)The hybrid bits have a large broken rock mass and a lower average temperature compared to conventional bits[36]. To shorten the drilling time of the hard rock formations, according to the characteristics of high brittleness and weak internal impact of hard rock, Peng Xu et al. used compound impact rock breaking technology to perform high-frequency axial impact and torsional impact on the rock with finite element method. A full-scale drill bit compound impact rock breaking model was established, and the effects of different impact frequencies, WOB and rotary speed on rock breaking efficiency were studied. The results showed that: (1) impact frequency and WOB have a greater effect on ROP; (2) compound impact can significantly increase ROP in hard rock formation[37]. These numerical simulation methods focus on the bit structure, rock properties and external loads, and study the mechanism of bit teeth penetration into the rock and the rock breaking characteristics. However, it can be noted that there have been many studies on PDC bit crushing rock by scholars in the past. There are few researches on the rock breaking mechanism of hybrid bits. Taking into account the different formation properties and ROP requirements of the actual drilling process, the simulation study of the drill bit interacted with the rock in the complex bottom hole environment still needs to be more detailed and in-depth with different kinds of drill bits.

4.2. Software Development for the Model

Take the roller cone bit as an example, given the structural parameters of the bit, the spatial position of any point on the roller cone bit can be determined in the compound cylindrical coordinate system. Combining the rotation and axial displacement movement of the bit, the bit space geometric equation is available where formula (16) represents[37], and regarding the spatial geometric equation of PDC bit, as early as 1993, Li derived the spatial geometric equation of PDC bit[38].

$$\begin{cases} \rho_p = \sqrt{\left(c_0 - h_p \cos\beta + r_p \sin\beta in\beta c_p\right)^2 + \left(r_p \sin\alpha_p - s\right)^2} \\ \theta_p = \theta_{0i} - \dfrac{2\pi\pi(-1)}{3} - \arctan\left(\dfrac{r_p \sin\alpha_p - s}{c_0 - h_p \cos\beta + r_p \sin\beta in\beta c_p}\right) \\ Z_p = z_0 - r_p \cos\beta os\beta c_p - h_p \cos\beta \end{cases} \tag{16}$$

The spatial geometric equations of roller cone bits and PDC bits can be combined with the bottom hole digital model, the well wall digital model, the rock breaking model, and the drill bit and rock interaction simulation mechanism for motion simulation and the drill bit breaking rock process simulation which this paper will not elaborate.

Based on the above equations and bottom hole model, a simulation software for the tri-cone bit the interacted with the rock was developed, and the model was calibrated through physical experiments. The ROP model based on the bit/rock interaction can predict the ROP under different lithology and different working conditions (WOB and RPM combination). Through simulation analysis, the ROP under different rotation speeds (low rotation speed and high rotation speed) and different rock formations (rock A and rock B) are studied. Figure 2, Figure 3, Figure 4, and Figure 5 are the comparison of field data and simulation results:

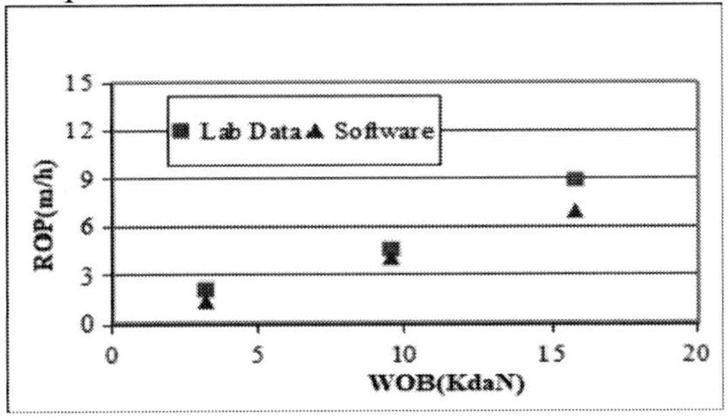

Figure 2. ROP comparison between software and lab data (IADC517, rock A, low RPM).

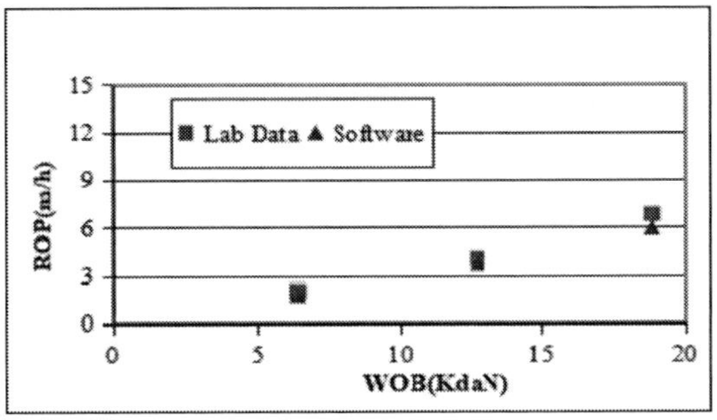

Figure 3. ROP comparison between software and lab data (IADC517, rock A, high RPM).

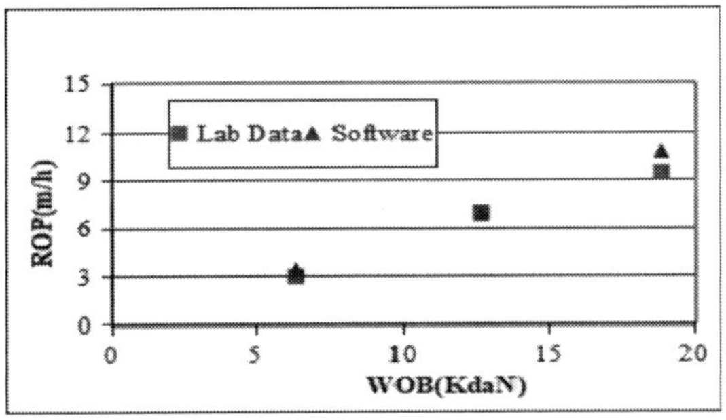

Figure 4. ROP comparison between software and field data (IADC517, rock B, low RPM).

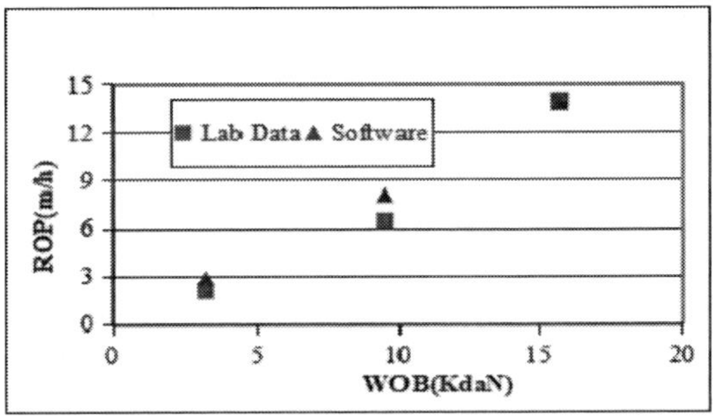

Figure 5. ROP comparison between software and lab data (IADC517, rock B, high RPM).

It can be seen from the comparison chart that the ROP simulation results of the tri-cone bit/rock interaction simulation software developed based on the bit space geometric equation and the bottom hole model are consistent with approximately 80% of the field data collected, and the overall trend is consistent. However, this should meet the premise that the drill bit central axis and the drilling axis coincide, and the rock texture is uniform. The simulation of the bit/rock interaction can not only simulate the load on the teeth, tooth row, and cone, but also predict the ROP of the bit under certain conditions including the cutting structure, rock type, weight-on-bit and speed.The main difference between the the ROP empirical mathematical model and the simulation model is that the latter has a very accurate description of the drill bit movement and cutting structure, so changing the tooth profile and the position of the tooth can reflect the change of ROP, so it can be used for subtle drill bit selection, but the former can't.

5. Machine Learning Model

In this paper, a ROP prediction model is established according to the BP neural network. BP neural network uses error back propagation algorithm, that is, through data feedback and error back propagation, the threshold between each neuron is continuously adjusted and corrected. The goal is to

minimize the variance of the output and obtain the weight coefficient. Judgment criteria are used to classify or predict unknown samples.
The selection of data samples is a very critical step in the neural network prediction process. There are many factors that affect ROP. Comprehensive consideration of the determined ROP mathematical model and the ROP model under rock-bit interaction, combined with the information available on site, the most effective factors affecting ROP are used as input layer neurons, including: Hole Depth, Hook Load, WOB, Standpipe Pressure, Surface Torque, Rotary RPM); the output is ROP. Therefore, the number of input points of the BP neural network is 6 and the output layer has 1 node. BP neural network predicts ROP model as shown in Figure 6.

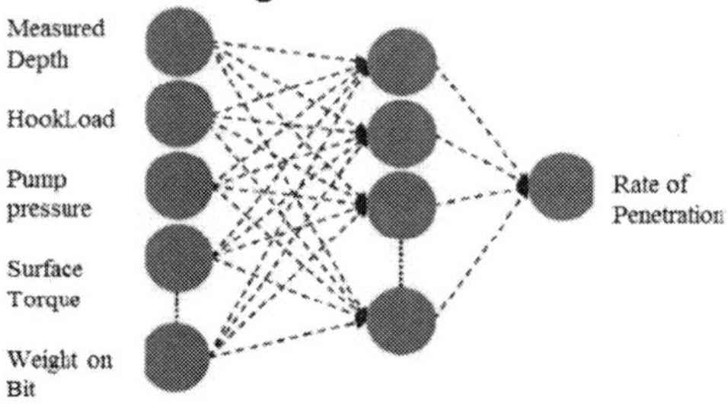

Figure 6. BP Neural Network ROP Model.

The quality of network training directly affects the accuracy of model prediction[17]. According to the field historical data collection system, the parameter data required by the model is derived, and invalid data is removed after preliminary screening. The established training database is shown in Table 1.

Table 1. Training Sample for the BP Neural Network.

MD/(m)	HL/(KDaN)	SPP/(KPa)	RPM/(r/min)	Tor/(NM)	WOB/(KDaN)	ROP(m/h)
2011.8	30	11330	26	5175	8.2	2.4
2012	30	11237	26	5241	8.3	2.3
2012.2	30	11366	26	5192	8.2	2.6
2012.4	30	11246	26	5126	8.3	2.4
2012.6	30	11326	26	5093	8.3	2.3
2012.8	30	11267	26	5159	8.3	2.4
2013	30	11354	26	5257	8.3	2.3
...	...	...		...	...	...

The data is normalized, and a three-layer BP network structure is selected. One hidden layer, and 5 neurons is present in the hidden layer. The BP neural network is used to train and predict ROP by using MATLAB software. In this experimental design, the entire training library has 290 sets of measured data, 264 sets of data are used to train the neural network model (Figure 8), and the remaining 16 sets of data are used to judge and predict the trained BP neural network model. The ROP is compared with the measured data, and the results are shown in Figures 7 and 9.

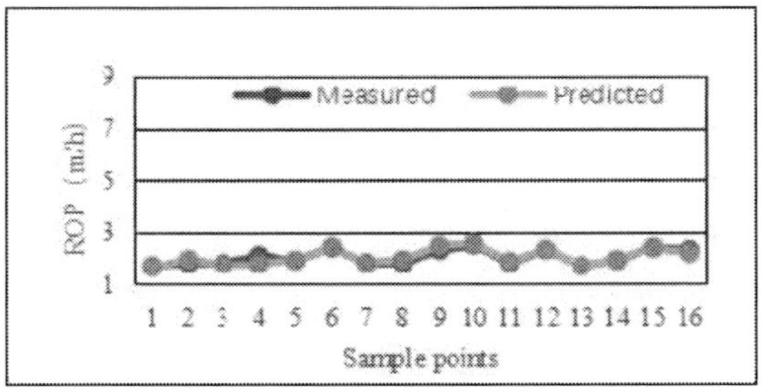

Figure 7. ROP Comparison between the measured and the predicted.

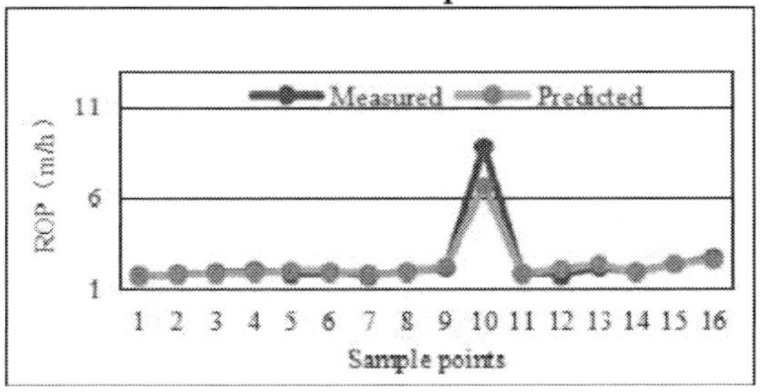

Figure 9. ROP Comparison between the measured and the predicted.

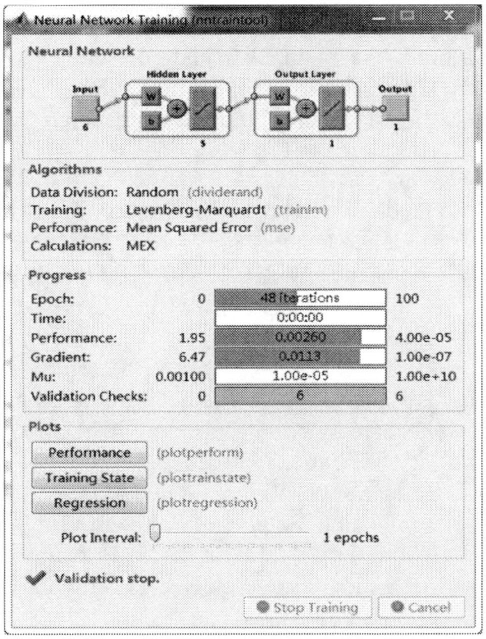

Figure 8. Neural network, algorithms and progress.

It can be seen that, except for the fourth sample point in Figure 7 and the tenth sample point in Figure 9, the predicted ROP is in very good agreement with the measured ROP, and the error fluctuation is small,the predicted ROP is consistent with approximately 94% of the measured ROP collected, that is, it meets the actual engineering requirements. It shows that the established ROP prediction model can predict the ROP of the corresponding formation relatively accurately.

6. Conclusions

(1) The deterministic ROP empirical mathematical model relies on specific environments and conditions, and only includes limited variables. If more uncontrollable variables are involved, the model will be very complicated, and some coefficients or constants in the model often require a lot of experiments and theoretical derivation. The process is very cumbersome; the advantage of the deterministic mathematical model is the fast calculation speed.

(2) In view of the shortcomings of the empirical mathematical model, the computer simulation study of bit/rock interaction has brought tremendous vitality. Through the establishment of accurate bit space geometric model and digital model of the bottom hole and borehole wall, combined with the rock breaking model and simulation mechanism, the drilling process of the bit can be simulated, and then the ROP of the bit can be predicted. The advantage is that the model can reflect the subtle changes of the bit cutting structure, so it can be well used in the selection of the bit; the disadvantage is that the model considers less factors affecting ROP, and the calculation speed is much slower than the empirical mathematical model.

(3) ROP while drilling is affected by many factors. The interaction of these factors is very complicated. Artificial neural network can reflect the complex relationship between multiple parameters. The prediction results of the present BP neural network model are in good agreement with the measured data. There are other machine learning models such as deep learning, recurrent neural networks, etc., which have good prediction results. Therefore, the use of neural network prediction technology for drill bit drilling performance optimization is the future development trend. Data-driven modelling and its applications to predict the bottom hole environment will prove to be the future of drilling operations.

Acknowledgments
We would like to thank Hundred-Talent Program Lab at Mechanical Engineering College, Xi'an Shiyou University for publication funding. Thanks are also given to the funding of the project (Z20257) from National Drilling Equipment Research Centre of Petro China.

References
[1] Hegde, C., Soares, C., & Gray, K. 2018. Rate of Penetration (ROP) Modeling Using Hybrid Models: Deterministic and Machine Learning. Unconventional Resources Technology Conference. doi:10.15530/ URTEC-2018.
[2] Bingham, M.G.1965. A New Approach to Interpreting Rock Drillability. Oil and Gas Journal, April 1965.
[3] Hareland, G., A. Wu, and B. Rashidi. 2010. A New Drilling Rate Model for Tricone Bits and Its Application to Predict Rock Compressive Strength. In Proceedings of the 44th US Rock Mechanics Symposium and 5th U.S.-Canada Rock Mechanics Symposium, held in Salt Lake City, UT June27–30, 2010.
[4] Hareland, G., Wu, A., & Rashidi, B. 2010. A Drilling Rate Model for Roller Cone Bits and Its Application. Society of Petroleum Engineers. doi:10.2118/129592-MS.
[5] Moran, D. P., Ibrahim, H. F., Purwanto, A., & Osmond, J. 2010. Sophisticated ROP Prediction Technology Based on Neural Network Delivers Accurate Results Sophisticated ROP Prediction Technology Based on Neural Network Delivers Accurate Results. Society of Petroleum Engineers. doi:10.2118/132010-MS.
[6] Cheng, Y.2020. Factors Affecting ROP and Analysis During Drilling Operation[J]. China Petroleum and Chemical Standard and Quality, 2020,40(03):18-19.
[7] Wang, D., Sun, J., Zhang, J., et al. 2021.Discussion on Technical Status and Development Direction of [J]. CHINA PETROLEUM MACHINERY, 2021, 49(01): 47-52.
[8] SU Z., TANG, Z., et al.2019. The Design and Field Testing of Two-Stage and Two-Speed Drilling Tools [J]. PETROLEUM DRILLING TECHNIQUES, 2019, 47(01):59-64.
[9] Sun, J., Zhao, Z., Bai, Y., et al. 2020.Progress in research of intelligent self-healing gel and its application prospects in the field of drilling fluid[J].ACTA PETROLET SINICA,2020,41(12): 1706-1718.
[10] ZHA, C., LIU, G., LI, J., et al. 2017.Development and Field Application of a Compound Percussive Jet[J]. PETROLEUM DRILLING TECHNIQUES, 2017, 45(01):57-61.
[11] QIAO, L.2016. Application of new accelerating tool NEW-DRILL in Yuanba Gas Field [J]. OIL DRILLING & PRODUCTION TECHNOLOGY, 2016, 38(01):14-17.
[12] Luo, X.2020. Application of SL Torsional Impact Tool in Halahatang Oilfield [J]. SINO-GLOBAL ENERGY, 2020, 25(S1): 144-150.
[13] Li, W., Li, S., Yan, L., et al.2018. Development and field application of a pulse–jet hydraulic impactor [J]. Natural Gas Industry, 2018, 38(05):87-93.
[14] CHEN, J., MOU, X., LI, H., et al.2020. Development and application of rotary-percussive and oscillatory drilling tool [J].FAULT-BLOCK OIL&GAS FIELD,2020,27(03):386-389.
[15] Tian, J., Yang, Y., Zhu, Z., et al.2019. Dwonhole dynamic characteristics of an actuator based on rotary screw[J]. ACTA PETROLET SINICA, 2019,40(02):224-231.
[16] WU, Z., GUO, L., PAN, Y., et al. 2018.Bottom hole WOB prediction in the process of horizontal well drilling and its application [J]. OIL DRILLING & PRODUCTION TECHNOLOGY, 2018, 40(01):9-13.
[17] Zhang, X. 2021.Brief analysis of high temperature and high pressure water-based drilling fluid technology Development [J]. West-China Exploration Engineering, 2021,33(05):27-28+32.
[18] CHEN, T., ZHANG, N., YANG, S., et al.2021. Research and application progress of low temperature drilling fluid [J/OL]. Applied Chemical Industry: 1-6[2021-05-13].
[19] Xue, Y., Li, Y. 2019.Influence Simulation of Parameters of Drill Bit on Rock Breaking Efficiency Based on Discrete Element Method [C].Proceedings of the 25th annual meeting of Society Beijing Mechanics, 2019:5.

[20] LIAN, Z., LIN, T., MENG, Y., et al. 2008.DYNAMIC SIMULATION OF ROCK-BREAKING PROCESS DURING GAS DRILLING [J]. Natural Gas Industry, 2008(06):88-89+152.
[21] Lian, Z.1994. Computer Simulation Model of Drill Bit's Breaking Rock[J].China Petroleum Machinery, 1994,22(8): 18-24.
[22] Liu, Q., Wu, Z., Ma, D.1998. Disc Bits and its Rock-breaking Mechanism[J]. Natural Gas Industry, 1998(01):67-69+10-11.
[23] Liu, Q., Wu, Z., Ma, D., et al. 1998.Simulation of rock breaking mechanism of disc bits[J]. China Petroleum Machinery, 1998(09):12-14+57.
[24] Yang, C., Li, B., Shi, G. 2007.Dynamic Characteristics of the Compound Rock Breaking System of Single Tooth Row[J]. Journal of Southwest Petroleum University, 2007(04): 138-140+199-200.
[25] Kuang, Y., Dong, Z., Wu, K., et al.2013. Development and Application of Numerical Simulation System for Three-cone Bit Drilling[J].Petroleum Drilling Techniques, 2013,41(01): 103-107.
[26] KUANG, Y., WU, K., YANG, Y. 2009.Simulation model of drilling process of three-cone bit [J]. Rock and Soil Mechanics, 2009, 30(S1):235-238.
[27] Deng, M., Wu, K., Hu, W., et al.2014. Rock-breaking Simulation and Experimental Analysis of PDC Bit Cutter [J]. China Petroleum Machinery, 2014,42(01):10-13.
[28] DENG, Y., CHEN, M., JIN, Y., et al.2016. Prediction Model and Numerical Simulation for Rock Fissure Length under Impact Load[J].Petroleum Drilling Techniques, 2016, 44(04):41-46.
[29] HEYDARSHAHY, S. A., KAREKAL, S. 2017.Influences of bit profiles on possible fracture modes [J]. PETROLEUM EXPLORATION AND DEVELOPMENT, 2017, 44(04): 630-637.
[30] ZHU, X., Yi, Q.2018. Influence of Cutting Shapes of PDC Bits on Rock Breaking Efficiency[J].China Mechanical Engineering, 2018, 29(07): 817-823.
[31] Yang, R., Xie, G., Wang, G.2019. Multi-factor Comprehensive Optimization Analysis of PDC Bit Based on Numerical Simulation [J]. Coal Mine Machinery, 2019, 40(07):71-75.
[32] WU, Z., LYU, L., WANG, Y., et al.2020. Rock-breaking characteristics and temperature field change of cone–PDC hybrid bits [J].Natural Gas Industry,2020,40(03):99-106.
[33] Wu, A., Hareland, G., & Lei, L. 2014. Computer Simulation and Analysis of Rock Breaking Process by Rock Bit Cutters. American Rock Mechanics Association.
[34] Li, S., D. Ma, and J. Hou. 1996. Study of PDC Bit Geometry, Journal of Southwest Petroleum University, 1996. Journal of Southwest Petroleum Institute. 18(2):82-86.
[35] Wang, K., Wei, F. 2003. Applications of Logging Information in Predicting Formation Anti-Drilling Parameters [J].Petroleum Drilling Techniques, 2003(05):61-62.
[36] DONG, Q., LIANG, X.2012. A Model for Predicting Formation Drillability Based on Optimized BP Neural Network [J].Exploration Engineering, 2012,39(11):26-28.
[37] WANG, W., LIU, X., DOU, P., et al.2018. A ROP prediction method based on neutral network for the deep layers [J]. OIL DRILLING & PRODUCTION TECHNOLOGY, 2018, 40(S1):121-124.
[38] Du, Y., Pan, Y.2020. Prediction Model of Rock Drillability Based on GA-BP Neural Network[J].Scientific Technology and Innovation,2020(25):57-59.

Phase-Variable Modeling and Comparative Study between a PMa-CPA and a CPA Alternator by Simulation Analysis

Ming-Fa Tsai*, Chung-Shi Tseng and Shu-Yui Bai

Department of Electrical Engineering, Minghsin University of Science and Technology, 1, Xinxing Rd., Xinfeng, Hsinchu, Taiwan

*Corresponding author email: mftsai@must.edu.tw

Abstract. This paper presents the phase-variable modeling of a permanent magnet-assisted claw-pole alternator in PSIM simulation software tool with the performance compared with a claw-pole alternator model, which is a special case of the permanent magnet-assisted claw-pole alternator model with the rotor permanent-magnet flux constant setting equal to zero. The constructed model can be simulated as a real alternator directly connected to a diode-bridge rectifier or switch-mode rectifier circuit for the integrated power application system simulation analysis. The proposed model contains two features, one of which is that the stator and rotor windings are circuit-based. The other one feature is that the mechanical torque input is based on a mathematical function. In this model, the rotor position-dependent variable inductor is replaced by a separate voltage source according to the voltage source absorption theorem. The simulation shows that the proposed model has faster torque step response and higher power efficiency than the claw-pole alternator model. The proposed model can be employed to the application of the control design and simulation study of voltage regulation in automobiles. The simulation results with 15-V command and the load sudden change verify the correction of the proposed model.

1. Introduction

The claw-pole alternator (CPA) is one type of wound-field synchronous machine with salient pole which can generate electricity in the stator coils with the rotor acting as an electromagnet. It has been widely used in automobiles for many years to provide the power through the diode-bridge rectifier, as shown in Figure 1, especially in the high speed engine operation, owing to the rugged structure in the rotor with a pair of claw poles secured around a cylindrical field winding [1-4]. However, because many mechanically driven loads, like ventilation fan, servo steering, breaks, and water pumps, in modern vehicles are gradually replaced by electrical drives, the power requirements in automobiles have been rising rapidly recently. The increasing electric load needs higher efficiency and generator power density. Despite their advantages, traditional CPAs suffer from some limitations such as the reliability, low power density, and low efficiency.

In order to meet the increasing power requirements, permanent magnet-assisted or hybrid-excited CPAs were proposed in the past [5-15]. Except the DC excited wound field, the permanent magnet-assisted CPA (PMa-CPA) has permanent magnet placing into the rotor, so that the torque density and power efficiency can be improved. In addition to connecting to the diode-bridge rectifier to supply power to vehicles, it has been shown that the switched-mode rectification with pulse-width modulation (PWM) coupled with some dedicated controls can provide great improvements in the transient response, efficiency, and power capability in previous works [16, 17]. A variety of integrated circuits (IC), such as Freescale TC80310, ONSEMI CS3361, ST L9409, and ST L9914, have been designed for the voltage regulation in various automobiles [18-21]. The simulation and test platform with real-time hardware-in-the-loop (HIL) has been also used to develop the control technology of the alternators, as shown in Figure 2, in which an induction motor under the CPA and battery rack is employed to substitute the automobile engine for driving the alternator and the speed reference and load change can be set by using the control and display panel together with an electrical load [22]. Prior to the HIL test, the modeling of the PMa-CPA for the system simulation analysis is essential in developing high-performance control schemes so as to optimize the efficiency.

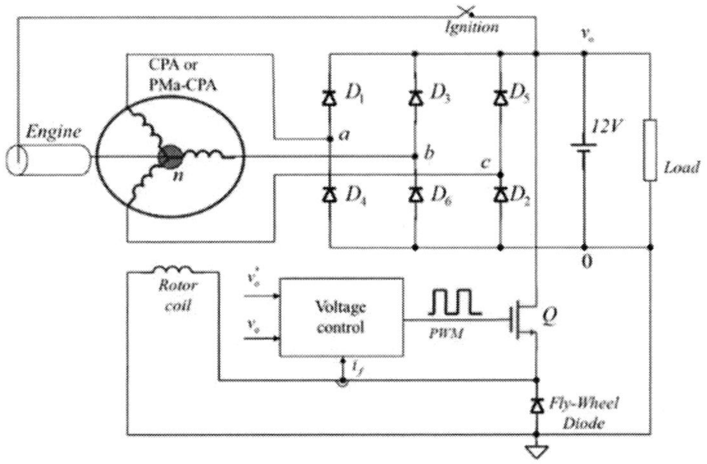

Figure 1. The structure for the voltage regulation in automobiles.

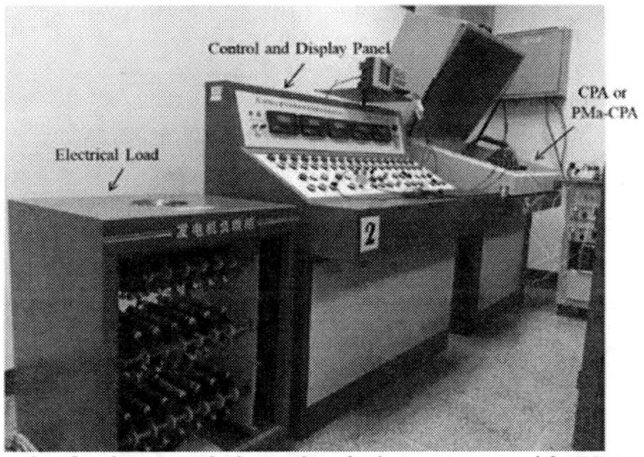

Figure 2. The hardware-in-the-loop real-time simulation system with CPA or PMa-CPA alternator.

For modeling three-phase AC machines including motors or generators, a variety of methods using software simulation tools, such as MATLAB/Simulink, PSIM, and PSPICE, have been studied and can be divided into two kinds of classification [23, 24]. One type is mathematic equation based model [25, 26] and the other one is classified as circuit component based model [27-29]. The problem of the equation-based model is that it cannot directly be coupled with external circuit such as the PWM inverter for integrated performance evaluation. The circuit component based model is then the candidate for the solution of the problem.

The modeling can be further classified as *d-q* modeling and the model in phase-variable form according to the different coordinate reference frames. The former is usually used for developing vector control algorithm in the synchronously rotating *d-q* reference frame [30, 31], and the latter which belongs to circuit component based model can be emulated as a real machine and connected directly to a power electronic converter for system integrated simulation analysis [32-34]. In the PSIM simulation tool, there has existed a synchronous machine element in the Motor Drive Module, which can be used as the CPA element for simulation analysis [35]. But, the element for the PMa-CPA model in PSIM is still unavailable.

The purpose of this paper is firstly to propose the PMa-CPA model as a sub-circuit block with phase-variable type in PSIM. One feature of the model is that the three-phase stator output windings and the rotor field winding are circuit-based. The equivalent circuit of each winding consists of a series connection of a resistor, an inductor, a rotor position-dependent variable inductor, and a dependent voltage source, respectively. For the construction of the model, the rotor position-dependent variable inductor is replaced by a separate voltage source according to the voltage source absorption theorem. The other feature of the model is that the mechanical torque input is based on a mathematical function. The proposed PMa-CPA model can be reduced to a CPA model by setting the rotor permanent-magnet flux constant to be zero. The performance between the PMa-CPA and CPA is then compared by the simulation analysis.

2. The Materials for the Proposed Model Construction

2.1. The Mathematical Model

The simplified structure of the PMa-CPA alternator with salient pole consists of three-phase stator windings, a rotor field winding, and permanent-magnet pairs mounted on the rotor. The rotor flux including the field-winding flux and the permanent magnet flux is aligned on the *d*-axis of the synchronously rotating *d-q* frame. Modeling of the PMa-CPA alternator can be divided into three parts which are the electromagnetic part, the electromechanical part, and the mechanical parts. Among them, the electromagnetic part is the most complicated, because it contains the magnetic coupling among the three-phase stator and the rotor flux linkages which are dependent on the rotor position angle described as follows.

Firstly, in the electromagnetic part, the stator and rotor voltage equations in the stationary *a–b–c* frame without considering the rotor damping cage can be expressed as follows:

$$\begin{bmatrix} v_a \\ v_b \\ v_c \\ v_f \end{bmatrix} = \frac{d}{dt}\begin{bmatrix} \psi_a \\ \psi_b \\ \psi_c \\ \psi_f \end{bmatrix} - \begin{bmatrix} R_s & 0 & 0 & 0 \\ 0 & R_s & 0 & 0 \\ 0 & 0 & R_s & 0 \\ 0 & 0 & 0 & R_f \end{bmatrix}\begin{bmatrix} i_a \\ i_b \\ i_c \\ -i_f \end{bmatrix} \tag{1}$$

where i_a, i_b, i_c are the stator currents flowing out of the generator; v_a, v_b, v_c are the three-phase stator voltages; i_f is the rotor exciting field current flowing into the rotor; ψ_a, ψ_b, ψ_c are the three-phase stator fluxes; ψ_f is the rotor field flux; R_s is the stator resistance, and R_f is the rotor field resistance. The stator fluxes ψ_a, ψ_b, ψ_c and rotor flux ψ_f, which are related to the stator and rotor field currents, the permanent magnet flux λ_{af}, and the rotor electrical angle θ_r, can be expressed as

$$\begin{bmatrix} \psi_a \\ \psi_b \\ \psi_c \\ \psi_f \end{bmatrix} = \begin{bmatrix} [L_s] & [L_{sr}] \\ [L_{sr}]^T & L_f \end{bmatrix}\begin{bmatrix} -i_a \\ -i_b \\ -i_c \\ i_f \end{bmatrix} + \lambda_{af}\begin{bmatrix} \cos\theta_r \\ \cos(\theta_r - \frac{2\pi}{3}) \\ \cos(\theta_r - \frac{2\pi}{3}) \\ 0 \end{bmatrix} \tag{2}$$

where

$$[L_s] = \begin{bmatrix} L_{aa} & L_{ab} & L_{ac} \\ L_{ba} & L_{bb} & L_{bc} \\ L_{ca} & L_{cb} & L_{cc} \end{bmatrix} \tag{3}$$

In (3), L_{aa}, L_{bb}, L_{cc} are the stator self-inductances in each phase and $L_{ab}, L_{ba}, L_{bc}, L_{cb}, L_{ca}, L_{ac}$ are the mutual inductances between different phase windings, which can be written as [36]

$$\begin{aligned}
L_{aa} &= L_{ls} + L_A - L_B \cos(2\theta_r) \\
L_{bb} &= L_{ls} + L_A - L_B \cos(2\theta_r + \frac{2\pi}{3}) \\
L_{aa} &= L_{ls} + L_A - L_B \cos(2\theta_r - \frac{2\pi}{3}) \\
L_{ab} &= L_{ba} = -\frac{L_A}{2} - L_B \cos(2\theta_r - \frac{2\pi}{3}) \\
L_{bc} &= L_{cb} = -\frac{L_A}{2} - L_B \cos(2\theta_r) \\
L_{ca} &= L_{ac} = -\frac{L_A}{2} - L_B \cos(2\theta_r + \frac{2\pi}{3})
\end{aligned} \tag{4}$$

where L_A is the stator mutual inductance, L_B is the mutual inductance between the stator and rotor windings due to rotor salient-pole effect, and L_{ls} is the stator leakage inductance. Furthermore, in (2),

$$[L_{sr}] = L_{sf} \begin{bmatrix} \cos\theta_r \\ \cos(\theta_r - \frac{2\pi}{3}) \\ \cos(\theta_r + \frac{2\pi}{3}) \end{bmatrix} \tag{5}$$

where L_{sf} is the magnetizing inductance between the stator and rotor-field windings. In (2) L_f is called the rotor self-inductance, which includes the rotor leakage inductance L_{lf} and the magnetizing inductance L_{sf}, and is written as

$$L_f = L_{lf} + L_{sf} \tag{6}$$

In addition, the rotor electrical speed and angle have the following relationship

$$\theta_r = \int \omega_r dt + \theta_{r0} \tag{7}$$

where θ_{r0} is the rotor initial position. It can be remarked that the model of the CPA is a special case of PMa-CPA. That is, the permanent-magnet flux λ_{af} in (2) is zero for the CPA.

Under the condition in which the three-phase currents are balanced, the stator and rotor voltage equations can be rewritten, respectively, by substituting (2) into (1) as follows.

$$\begin{bmatrix} v_a \\ v_b \\ v_c \end{bmatrix} = \begin{bmatrix} e_a \\ e_b \\ e_c \end{bmatrix} + \begin{bmatrix} e_{aL} \\ e_{bL} \\ e_{cL} \end{bmatrix} - R_s \begin{bmatrix} i_a \\ i_b \\ i_c \end{bmatrix} - L_s \begin{bmatrix} \frac{di_a}{dt} \\ \frac{di_b}{dt} \\ \frac{di_c}{dt} \end{bmatrix} \tag{8}$$

and

$$v_f = R_f i_f + L_f \frac{di_f}{dt} + e_{fL} + e_f \tag{9}$$

where

$$\begin{bmatrix} e_a \\ e_b \\ e_c \end{bmatrix} = -2\omega_r L_B \begin{bmatrix} \sin(2\theta_r) & \sin(2\theta_r - \frac{2\pi}{3}) & \sin(2\theta_r + \frac{2\pi}{3}) \\ \sin(2\theta_r - \frac{2\pi}{3}) & \sin(2\theta_r + \frac{2\pi}{3}) & \sin(2\theta_r) \\ \sin(2\theta_r + \frac{2\pi}{3}) & \sin(2\theta_r) & \sin(2\theta_r - \frac{2\pi}{3}) \end{bmatrix} \begin{bmatrix} i_a \\ i_b \\ i_c \end{bmatrix} - \omega_r(\lambda_{af} + L_{sf} i_f) \begin{bmatrix} \sin\theta_r \\ \sin(\theta_r - \frac{2\pi}{3}) \\ \sin(\theta_r + \frac{2\pi}{3}) \end{bmatrix} \tag{10}$$

$$\begin{bmatrix} e_{aL} \\ e_{bL} \\ e_{cL} \end{bmatrix} = L_B \begin{bmatrix} \cos(2\theta_r) & \cos(2\theta_r - \frac{2\pi}{3}) & \cos(2\theta_r + \frac{2\pi}{3}) \\ \cos(2\theta_r - \frac{2\pi}{3}) & \cos(2\theta_r + \frac{2\pi}{3}) & \cos(2\theta_r) \\ \cos(2\theta_r + \frac{2\pi}{3}) & \cos(2\theta_r) & \cos(2\theta_r - \frac{2\pi}{3}) \end{bmatrix} \begin{bmatrix} \frac{di_a}{dt} \\ \frac{di_b}{dt} \\ \frac{di_c}{dt} \end{bmatrix} + L_{sf} \begin{bmatrix} \cos\theta_r \\ \cos(\theta_r - \frac{2\pi}{3}) \\ \cos(\theta_r + \frac{2\pi}{3}) \end{bmatrix} \frac{di_f}{dt} \tag{11}$$

$$L_s = L_{ls} + \frac{3L_A}{2} \tag{12}$$

$$e_{fL} = -L_{sf} \left[\frac{di_a}{dt}\cos\theta_r + \frac{di_b}{dt}\cos(\theta_r - \frac{2\pi}{3}) + \frac{di_c}{dt}\cos(\theta_r + \frac{2\pi}{3}) \right] \tag{13}$$

and

$$e_f = \omega_r L_{sf} \left[i_a \sin\theta_r + i_b \sin(\theta_r - \frac{2\pi}{3}) + i_c \sin(\theta_r + \frac{2\pi}{3}) \right] \tag{14}$$

Secondly, the alternator electrical load torque in the electromechanical part can be obtained as follows.

$$T_e = \frac{dE_c}{d\theta_m} = \frac{P}{2}\frac{dE_c}{d\theta_r} \tag{15}$$

where P is the pole number, θ_m is the mechanical angle position of the rotor, and E_c is the coupling-field stored energy which can be expressed as [36]

$$E_c = \frac{1}{2}[i_a\ i_b\ i_c]([L_s] - L_{ls}I) \begin{bmatrix} i_a \\ i_b \\ i_c \end{bmatrix} + [i_a\ i_b\ i_c][L_{sr}] i_f + \frac{1}{2} L_{sf} i_f^2 + \lambda_{af} [i_a\ i_b\ i_c] \begin{bmatrix} \cos\theta_r \\ \cos(\theta_r - \frac{2\pi}{3}) \\ \cos(\theta_r + \frac{2\pi}{3}) \end{bmatrix} \tag{16}$$

with the identity matrix I. Since $[L_s]$ and $[L_{sr}]$ are variable due to θ_r, but L_{sf} is not function of θ_r, substituting (16) into (15) yields

$$T_e = \frac{-P}{2} L_B \begin{bmatrix} i_a & i_b & i_c \end{bmatrix} \begin{bmatrix} \sin(2\theta_r) & \sin(2\theta_r - \frac{2\pi}{3}) & \sin(2\theta_r + \frac{2\pi}{3}) \\ \sin(2\theta_r - \frac{2\pi}{3}) & \sin(2\theta_r + \frac{2\pi}{3}) & \sin(2\theta_r) \\ \sin(2\theta_r + \frac{2\pi}{3}) & \sin(2\theta_r) & \sin(2\theta_r - \frac{2\pi}{3}) \end{bmatrix} \begin{bmatrix} i_a \\ i_b \\ i_c \end{bmatrix} + \frac{2}{P}(\lambda_{af} + L_{sf} i_f) \left[i_a \sin\theta_r + i_b \sin(\theta_r - \frac{2\pi}{3}) + i_c \sin(\theta_r + \frac{2\pi}{3}) \right] \quad (17)$$

As can be seen, the alternator electrical load torque has two parts. The first term on the right-hand side of the equal sign in (17) is called the reluctance torque because of the salient pole effect. The second term is called the excitation torque which is owing to the rotor field-winding magnetizing flux and the permanent magnet flux.
Thirdly, the dynamics of the mechanical part can be written as follows

$$T_m - T_e = J\frac{d\omega_m}{dt} + B\omega_m \quad (18)$$

with the mechanical torque input T_m, the moment inertia J, the viscous coefficient B, and the mechanical rotation speed ω_m, which is related to the rotor electrical speed by

$$\omega_m = \frac{2}{P}\omega_r \quad (19)$$

Figure 3 illustrates the relationship between input torque and the alternator rotation speed with the transfer function block diagram as follows.

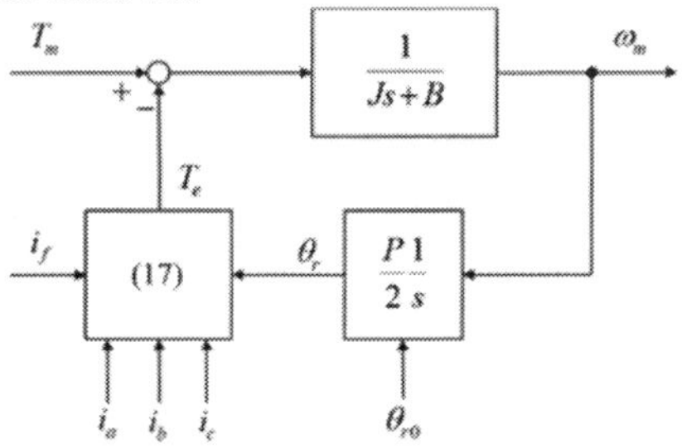

Figure 3. The transfer function block diagram from input torque to the alternator rotation speed.

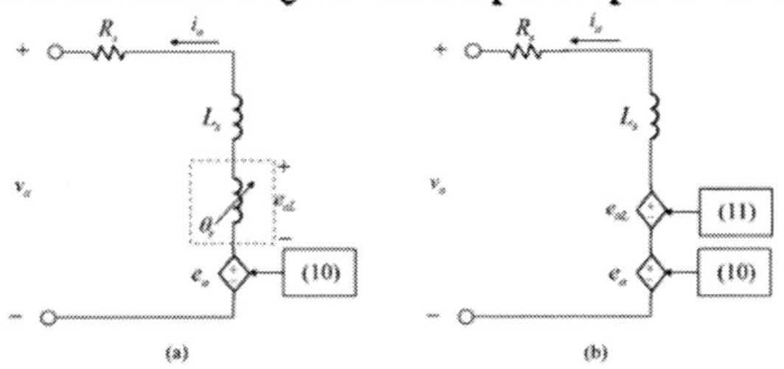

Figure 4. The equivalent circuit of the phase-a stator winding: **(a)** with variable inductor; **(b)** replaced by dependent voltage source.

2.2. Model Construction in PSIM

The e_a, e_b, e_c in (10) can be seen as the induced voltage sources in the three-phase stator windings. The e_{aL}, e_{bL}, e_{cL} in (11) can be seen as the voltage drops due to the inductance which is variable with the rotor rotation angle θ_r. The equivalent circuit of the phase-a stator with the variable inductance is shown in Figure 4(a). But in the PSIM simulation tool, no variable inductance components are available to construct the model. However, we can substitute the variable inductance with a dependent voltage source on the basis of (11) according to the voltage source absorption theorem [37]. The voltage drop by this dependent voltage source is equal to the voltage drop of the variable inductance we have replaced. Figure 4(b) illustrates the phase-a stator equivalent circuit with the proposed circuit element.

Similarly, the e_{fL} in (13) can be also seen as the voltage drops due to the inductance which is variable with the rotor rotation angle θ_r. The e_f in (14) can be seen as the back-emf in the rotor-field winding. The equivalent circuit of the rotor-field winding is shown in Figure 5.

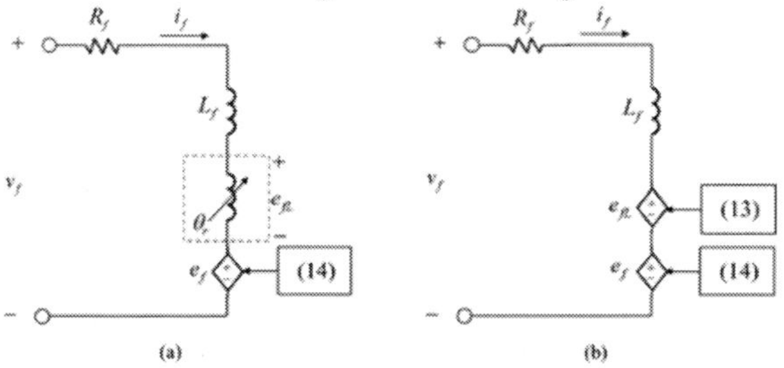

Figure 5. The equivalent circuit of the rotor field winding: **(a)** with variable inductor; **(b)** replaced by dependent voltage source.

The construction of the PMa-CPA alternator model in PSIM using the parameters listed in Table 1 is shown in Figure 6. The model is then reduced to be a CPA model as long as the rotor permanent-magnet flux constant λ_{af} in the blocks based on (10) and (17), respectively, is set to be zero.

As can be seen from Figure 6, the model block consists of four main parts based on the equations previously stated. These are an electrical part for the stator output currents based on (8) and rotor field input current based on (9), respectively, an electro-mechanical part for the alternator electrical load torque based on (17), a mechanic-to-electrical part, which produces the three-phase stator dependent voltage sources based on (10) and (11), the rotor field dependent voltage source based on (13) and back emf on the basis of (14), and a mechanical part according to Figure 3 for the alternator speed. The proposed model contains two features, one of which is that the stator and rotor field windings are circuit-based. The another feature is that the mechanical torque input is based on a mathematical function. This feature makes it easy to use.

The CPA model is firstly compared with the built-in synchronous machine (SM) element in the PSIM Motor Drive Module. However, some of the parameters used in the SM element, such as magnetizing inductance, must be set with the values in the *d*-*q* axes of the *d*-*q* reference frame. Thus, the *d*-axis magnetizing inductance L_{dm} and the *q*-axis magnetizing inductance L_{qm} should be obtained from the phase-variable value in Table 1.

Table 1. The PMa-CPA parameters without considering rotor damping cage.

Parameter	Value	Parameter	Value
R_s	0.1 Ω	L_{ls}	0.1 mH
L_A	2.6 mH	L_B	1.0 mH
L_s	4 mH	L_{sf}	2.4 mH
R_f	0.2 Ω	L_{lf}	5 mH
L_f	7.4 mH	λ_{af}	0.0579 Wb
J	$0.001\ \mathrm{kg \cdot m^2}$	B	0.0002 kg·m/rad/s
P	16		

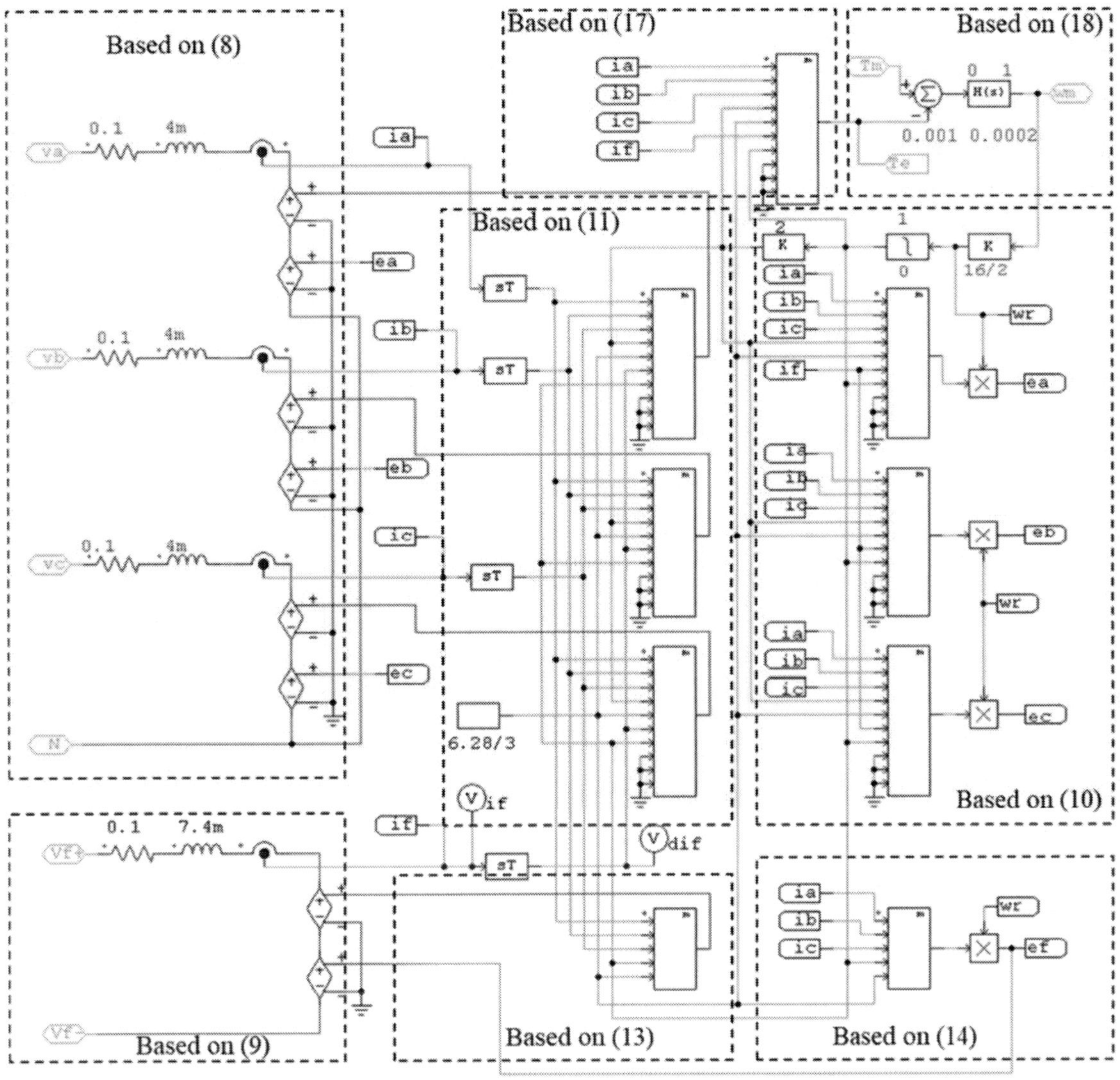

Figure 6. The PMa-CPA model constructed in PSIM.

Figure 7 (a) and (b) shows the stator and rotor magnetic fluxes in the coordinates between the a–b–c stationary reference frame and the stationary α–β reference frame and between the α–β reference frame and the d–q reference frame, respectively. By using the Clarke transformation in (20) and the Park transformation in (21) as follows, the stator d-axis magnetizing inductance L_{dm} and the q-axis magnetizing inductance L_{qm} can be obtained by setting $\theta_r = 0$ and $\theta_r = \pi/2$ in (4), respectively,

$$\begin{bmatrix} \psi_\alpha \\ \psi_\beta \end{bmatrix} = \begin{bmatrix} 1 & 0 & 0 \\ 0 & \frac{1}{\sqrt{3}} & \frac{-1}{\sqrt{3}} \end{bmatrix} \begin{bmatrix} \psi_a \\ \psi_b \\ \psi_c \end{bmatrix} \tag{20}$$

$$\begin{bmatrix} \psi_d \\ \psi_q \end{bmatrix} = \begin{bmatrix} \cos\theta_r & \sin\theta_r \\ -\sin\theta_r & \cos\theta_r \end{bmatrix} \begin{bmatrix} \psi_\alpha \\ \psi_\beta \end{bmatrix} \tag{21}$$

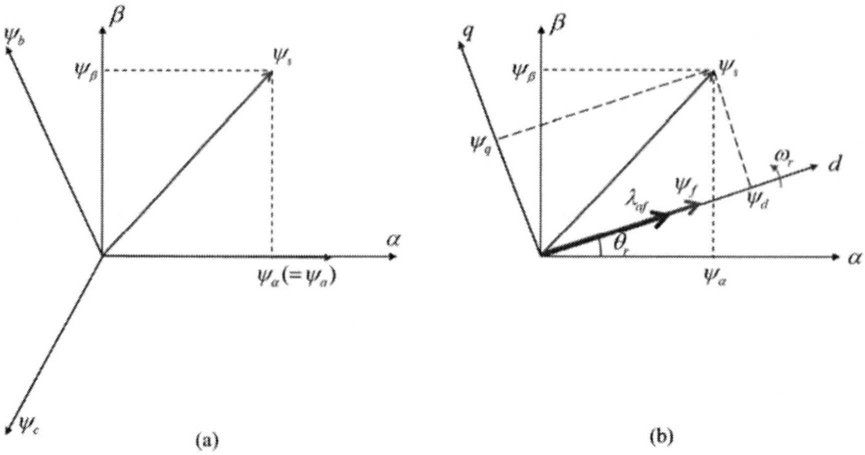

Figure 7. The magnetic fluxes: **(a)** in a–b–c and α–β frames; **(b)** in α–β and d–q frames.

where ψ_α and ψ_β are the stator flux components in the α–β frame, respectively, and ψ_d and ψ_q are the stator flux components in the d–q frame, respectively. By substituting (3) and (4) into (20) and then into (21), it follows that

$$\psi_d = (L_{ls} + \frac{3}{2}L_A - \frac{3}{2}L_B)i_d = (L_{ld} + L_{dm})i_d \tag{22}$$

and

$$\psi_q = (L_{ls} + \frac{3}{2}L_A - \frac{3}{2}L_B)i_q = (L_{lq} + L_{qm})i_q \tag{23}$$

where i_d and i_q are the d-axis and q-axis stator current components, respectively, and L_{ld} and L_{lq} are the d-axis and q-axis leakage inductance, respectively, written as

$$L_{ld} = L_{lq} = L_{ls} \tag{24}$$

and

$$L_{dm} = \frac{3}{2}L_A - \frac{3}{2}L_B \tag{25}$$

$$L_{qm} = \frac{3}{2}L_A + \frac{3}{2}L_B \tag{26}$$

In addition, because the rotor flux is aligned on the d-axis, the rotor field-winding magnetizing inductance L_{sf} in Table 1 is equal to the stator d-axis magnetizing inductance L_{dm}. From the parameter values of L_A and L_B in Table 1, one can obtain that $L_{dm} = 2.4\,\text{mH}$, $L_{qm} = 5.4\,\text{mH}$, and $L_{sf} = L_{dm} = 2.4\,\text{mH}$.

3. Results and Discussion

3.1. Comparison between the Constructed CPA and Built-in SM machine in PSIM

In the Motor Drive Module of PSIM, there has been a built-in synchronous machine (SM) model available. This built-in model can be acted as a CPA alternator by setting the appropriate parameters. For the verification, the constructed CPA model, which is obtained by setting the permanent magnet flux constant equal to zero in the PMa-CPA model, is used to compare with the PSIM built-in SM model. The comparison of the CPA model with the built-in SM model by using a 3-Nm torque step-command input, the same 10V voltage source for the same rotor exciting field current, and the same three-phase resistance load of $10\,\Omega$ in each phase is shown in Figure. 8(a). The SM machine parameters are shown in Table 2 by modifying the parameters of CPA model with $\lambda_{af}=0$ in Table 1. For without affecting the comparison between the two models, the rotor damping cage *d*-axis resistance R_{dr} and the rotor damping cage *q*-axis resistance R_{qr} are set to be a very large value of $100\,\text{M}\Omega$, respectively, and the rotor damping cage *d*-axis leakage inductance L_{drl} and the rotor damping cage *q*-axis leakage inductance L_{qrl} are set to be a very small value of $0.01\,\mu\text{H}$, respectively, in Table 2.

With the same 3-Nm torque step command, the alternator speed, stator current, and the rotor exciting field current responses between the two models are to be compared. Figure 8(b) illustrates the simulation waveforms. It can be seen that the speed, the stator current, and the rotor exciting field current responses of the CPA model are almost the same as those of the built-in SM model. This can verify that the constructed CPA model is correct. In addition, the 3-Nm torque step response of the CPA can be seen from the top of Figure 8(b), but is unavailable from the built-in SM model.

Table 2. The parameters of built-in SM.

Parameter	Value	Parameter	Value
R_s	$0.1\,\Omega$	L_{ls}	0.1 mH
L_{dm}	2.4 mH	L_{qm}	5.4 mH
R_f	$0.2\,\Omega$	L_{lf}	5 mH
R_{dr}	$100\,\text{M}\Omega$	R_{qr}	$100\,\text{M}\Omega$
L_{drl}	0.01 uH	L_{qrl}	0.01 uH
J	$0.001\,\text{kg}\cdot\text{m}^2$	B	$0.0002\,\text{kg}\cdot\text{m/rad/s}$
P	16	N_s/N_f	1

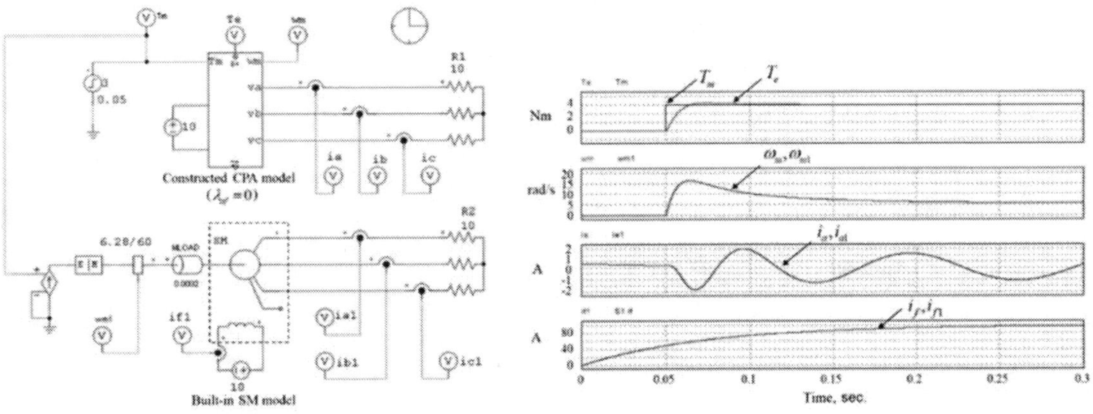

(a) (b)

Figure 8. Modeling verification of the CPA model: (a) Simulation model; **(b)** Simulation waveforms.

3.2. Comparison between the PMa-CPA and CPA

The comparison between the constructed PMa-CPA model with $\lambda_{af}=0.0579$ Wb and the CPA model with $\lambda_{af}=0$ by using the same 3-Nm torque step-command input, the same 10-V voltage supply for the same rotor exciting field current, and the same three-phase resistance load of $10\,\Omega$ in each phase is illustrated in Figure. 9(a). As can be seen, the PMa-CPA model has faster torque step response than the CPA model. However, the torque responses in the steady state are the same. In addition, the PMa-CPA model also has lower rotation speed response than the CPA model. This means that the PM-CPA model has less mechanical power than the CPA model to achieve the same steady-state torque response. This result indicates that the PMa-CPA has higher power efficiency than the CPA alternator.

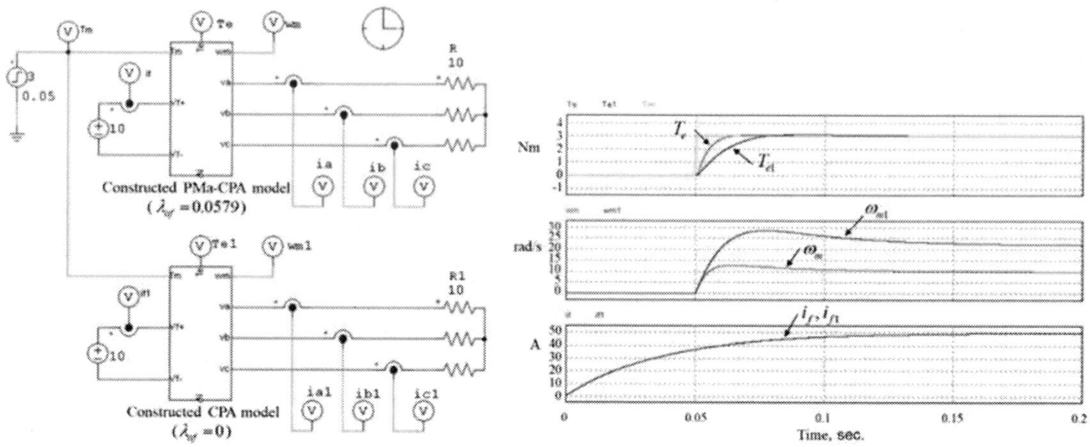

(a) (b)

Figure 9. Comparison between PMa-CPA and CPA: **(a)** The compared models **(b)** The compared waveforms.

3.3. Application to Voltage Regulation

The constructed PMa-CPA model can be applied to the controller design and simulation analysis for the voltage regulation in automobiles. The control structure of the voltage regulation system in PSIM with the primary resistance load of $1\,\Omega$ is shown in Figure 10(a), where the PMa-CPA model block is driven by a constructed induction motor (IM) model [32]. The IM motor model developed torque output is connected to the mechanical torque input of the PMa-CPA model block, in which the alternator toque output is connected back to the load torque input of the IM model. It has been obtained from Eq. (8), (10), and (11) stated in the previous Section that the phase voltage of each phase, which will be rectified to be a dc voltage through a diode-bridge rectifier, is not only related to rotor-field exciting current, but also has relationship with the stator current and the rotor rotation speed. The rectified output voltage can be then controlled in the closed-loop control system by regulating the rotor exciting field current, in which the load and speed variations can be considered as the disturbances to the system and compensated by a proportional-integral controller. The controller followed by a PWM signal generator is designed to control a transistor switch (Q1), through which the rotor exciting field current can flow from the battery to the rotor coil of the PMa-CPA model, while the switch is turned on. The exciting current continuously flow through the fry-wheel diode (D7) while the transistor switch is turned off.

By giving a 15-V step command at the time of 0.2 second, the simulation results are shown in Figure 10(b), which indicates that the output voltage is regulated at 15 V despite the sudden load change at the time 0.4 second and 0.6 second, respectively. As also can be seen, the load current achieves 15 A for the resistance load of $1\,\Omega$ before the load change of 0.4 second. After 0.4 second, the load resistance is changed to be about $1/3\,\Omega$ with the parallel connection of the $1\,\Omega$ resistor and the $0.5\,\Omega$ resistor, and hence the load is subsequently changed up to 45 A. After 0.6 second, the load resistance is changed back to $1\,\Omega$, and the load current is also subsequently changed down to 15 A. The alternator torque is gradually increasing from 10 Nm to about 35 Nm for the load current sudden change up to 45 A, and decreasing gradually down to 10 Nm for the load current sudden change down to 15 A, but the speed is kept near 40 rad/s for the load sudden changes.

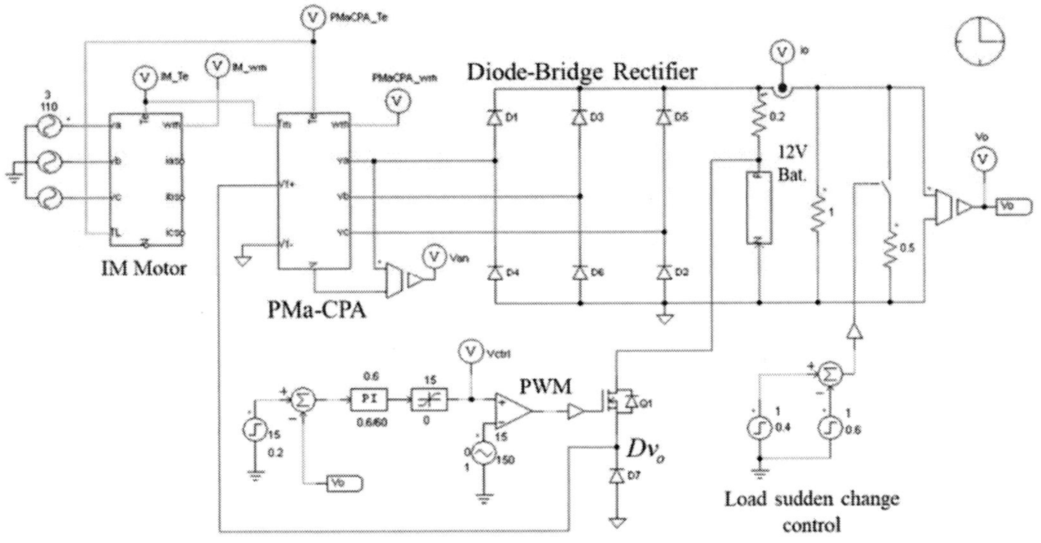

(a)

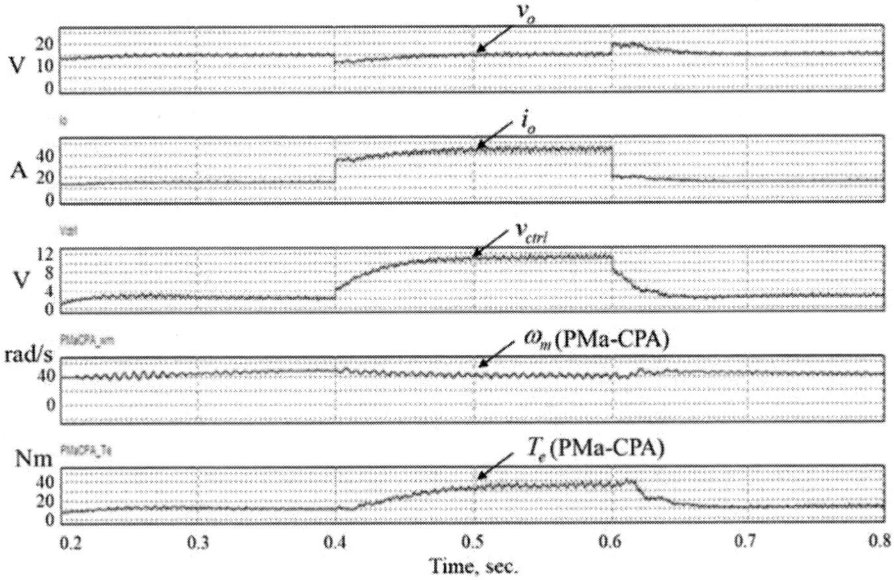

(b)

Figure 10. Voltage Regulation of PMa-CPA with load sudden change at 0.4 and 0.6 second: **(a)** Simulation model; **(b)** Simulation waveforms.

4. Conclusions

In this study, the modeling of a PMa-CPA alternator with phase-variable form in PSIM simulation software tool was proposed and the performance is compared with a CPA model, which is a special case of the PMa-CPA model with the rotor permanent-magnet flux constant setting to be zero. The constructed PMa-CPA model can be simulated as a real alternator directly connected to a diode-bridge rectifier or switch-mode rectifier circuit for the integrated power application system simulation analysis. The stator and rotor field windings are circuit-based. The equivalent circuit of each winding consists of a series connection of a resistor, an inductor, a rotor position-dependent variable inductor, and a dependent voltage source, respectively. For the construction of the model, the rotor position-dependent variable inductor is replaced by a dependent voltage source. In addition, the mechanical torque input is a mathematical function. The simulation shows that the PMa-CPA model has faster torque step response than the CPA model. The PMa-CPA model also has lower rotation speed than the CPA model. This means that the PM-CPA model has less mechanical power than the CPA model to achieve the same steady-state torque response. This result indicates that the PMa-CPA alternator has higher power efficiency than the CPA alternator.

The proposed PMa-CPA model can be applied to the controller design and simulation analysis voltage regulation in automobiles. The simulation results with 15-V command and the load sudden change also verify the correction of the proposed PMa-CPA model. Experiment verification for the voltage regulation will be the future work.

References

[1] Perreault, D. and Caliskan, V. Automotive power generation and control. *IEEE Trans. Power Electronics*, **2004**, 19, 618-630.

[2] Rees, S. and Ammann, U. A smart synchronous rectifier for 12 V automobile alternators. In Proceedings of IEEE 34th Annual Conference on Power Electronics Specialist (PESC'03), Acapulco, Mexico, 15-19 June **2003**, 1516-1521.

[3] Huang, L. Wang, J.; Fu, X. and Run, F. A soft start circuit for automobile voltage regulator. In Proceedings of 2011 International Conference on Electronics, Communications and Control (ICECC), Ningbo, China, 9-11 Sept. **2011**, 673-676.

[4] Unutulmaz, M.G. and Ergene, L.T. Excitation current control of a claw pole automobile alternator. In Proceedings of the 2013 3rd International Conference on Electric Power and Energy Conversion Systems, Istanbul, Turkey, 2-4 October 2013.

[5] Yang, C.; Lin, H. Guo, J. and Zhu, Z.Q. Design and analysis of a novel hybrid excitation synchronous machine with asymmetrically stagger permanent magnet. *IEEE Trans. on Magnetics*, **2008**, 44 , 4345-4356.

[6] Liu, X.; Lin, H. and Zhu, Z. Q. A novel dual-stator hybrid excited synchronous wind generator. *IEEE Trans. on Industry Applications*, **2009**, 45, 947-953.

[7] Kuroda, Y. and Morita, M. et al. Improvement of a claw pole motor using additional ferrite magnets for hybrid electric vehicles. In Proceedings of the International Conference on Electrical Machines 2010 (ICEM'2010), Rome, Italy 2-5 September 2010.

[8] Zhang, Q.; Huang, S. and Xie, G. Design and experimental verification of hybrid excitation machine with isolated magnetic paths. *IEEE Trans. on Energy Conversion,* **2010**, 25, 993-1000.

[9] Hagstedt, D. and Reinap, A. et al., Design and experimental evaluation of a compact hybrid excitation claw-pole rotor. In Proceedings of the 20th International Conference on Electrical Machines (ICEM'2012), Marseille, France, 2-5 September **2012**.

[10] Tutelea, L.; Ursu, D. and Boldea, I. IPM claw pole alternator system for more vehicle breaking energy recuperation. *Journal of Electrical Engineering*, **2012**, 12, 1-10.

[11] Rebhi, R.; Ibala, A. and Masmoudi, A. MEC-based sizing of a hybrid excited claw pole alternator, *IEEE Trans. on Industry Applications*, **2015**, 51, 211-223.

[12] Dajaku, G.; Lehner, B.; Dajaku, X.; Pretzer, A. and Gerling, D. Hybrid excited claw pole rotor for high power density automotive alternators. In Proceedings of the 2016 XXII International Conference on Electrical Machines (ICEM), Lausanne, Switzerland, 4-7 September **2016**, 2538-2545.

[13] Upadhayay, P.; Kedous-Leboucl, A.; Garbuiol, L.; Mipo, J.C. and Dubus, J. M. Design & comparison of a conventional and permanent magnet based claw-pole machine for automotive application. In Proceedings of the 15th International Conference on Electrical Machines, Drives and Power Systems (ELMA), Sofia, Bulgaria, June **2017**.

[14] Ye, C.; Liang, X.; Xiong, F.; Yang, J.; Xu, W. and Liu, Y. Design of an axial-flux PM-assisted claw-pole generator based on an equivalent magnetic circuit model. *IEEE Trans. Energy Conversion*, **2018**, 33, 2040-2049.

[15] Brahimi, N.; Tahi, S.;, Boudissa, E. and Bounekhla, M. Optimal rotor design of claw-pole alternator for performance improving. *Adv Syst Sci Appl*, **2019**, 02, 120-133.

[16] Rivas, J.M. and Keim, T.A. Performance improvement in alternators with switched-mode

rectifiers. *IEEE Trans. on Energy Conversation*, **2004**, 19, 561-568.
[17] Tang, S.Ch. and Otten, D.M. Design and evaluation of a 42-V automotive alternator with integrated switched-mode rectifier. *IEEE Trans. on Energy Conversation*, **2010**, 25, 983-993.
[18] TC80310 data sheet, Alternator regulator with LIN. rev. 5.0, Freescale Semiconductor, 2015.
[19] CS3361 data sheet, Alternator voltage regulator FET driver. rev. 11, ON Semiconductor, 2015.
[20] L9409 data sheet, Alternator voltage regulator with load response control. rev. 2, ST Semiconductor, 2013.
[21] L9914 data sheet, All silicon voltage regulator. rev. 5, ST Semiconductor, 2014.
[22] Tsai, M.-F. and Tseng, C.-S. An FPGA-based digital control IC for voltage regulation of automobile alternators. *Ta Yang Press*, L385, 2017.
[23] Zhao T. and Wang, Q. Application of MATLAB/SIMULINK and PSPICE simulation in teaching power electronics and electric drive system. In Proceedings of the IEEE International Conference of Electric Machines and Systems (ICEMS 2005), pp. 2037-2041, 2005.
[24] Tsai, M.-F.; Lee, T.-C.; Tseng, C.-S.; Chen, Y.-Y. and Lin, C.-H. Modeling and verification of an axial-flux permanent magnet motor with cogging torque ripple. In Proceedings of the 2013 IEEE TENCON (TENCON 2013), Xi'an, China, Oct. 22-25, 2013.
[25] Imayavaramban, M.; Chaithanya, A.V.K. and Fernandes, B.G. Analysis and mathematical modeling of matrix converter for adjustable speed ac drives. In Proceedings of the IEEE PSCE 2006 Conf., pp. 1113-1120, 2006.
[26] Apostoaia, C.M. AC machines and drives simulation platform. In Proceedings of the IEEE International Electric Machines & Drives Conference (IEMDC 2013), pp. 1295-1299, 2013.
[27] Mohammed, O.A.; Liu, S. and Liu, Z. A phase variable model of brushless dc motors based on finite element analysis and its coupling with external circuits. *IEEE Trans. Magnetics*, vol. 41, no. 5, pp. 1576-1579, May 2005.
[28] Mohammed, O.A.; Liu, S. and Liu, Z. Physical modeling of PM synchronous motors for integrated coupling with machine drives. *IEEE Trans. Magnetics*, 41, **2005**, pp. 1628-1631.
[29] Chen, J.; Guo, Y. and Zhu, J. An improved phase variable model based on electro-magnetic field coupled with its external circuits for performance evaluation of permanent magnet brushless dc motors. In Proceedings of the IEEE Conf. on Industrial Electronics and Applications, pp. 955-959, 2007.
[30] Rosa, F.C. and Bim, E.; A constrained non-linear model predictive controller for the rotor flux-oriented control of an induction motor. *Energies*, **2020**, 13, 3899.
[31] Tsai, M.-F.; Tseng, C.-S. and Lin, B.-Y. Phase voltage-oriented control of a PMSG wind generator for unity power factor correction. *Energies* 2020, 13, 5693.
[32] Tsai, M.-F.; Tseng, C.-S.; Chen, Y.-Y. and Peng, W.-Y. A new version of phase-variable modeling of an induction motor using PSIM. In Proceedings of the IEEE PEDS 2015, pp. 1047-1052, June 9~12, Sydney, Australia, 2015.
[33] Tsai, M.-F.; Tseng, C.-S.; Chen, C.-H.; Cheng, Y.-J. and Yang, C.-H. Phase-variable modeling of a synchronous reluctance motor using PSIM. In Proceedings of the IEEE SPEC 2016, 5-8 Dec. 2016, Auckland, New Zealand, 2016.
[34] Tsai, M.-F.; Tseng, C.-S. and Cheng, P.-J. Implementation of an FPGA-based current control and SVPWM ASIC with asymmetric five-segment switching scheme for AC motor drives. *Energies* **2021**, 14, 1462.
[35] Powersim, *PSIM User's Guide*. Version 9.0, Release 3. May 2010.
[36] Krause, P.C.; Wansynczuk, O. and Sudhoff, S.D. *Analysis of Electric Machine and Drive Systems*. 2nd Edition, Wiley Inter-Science, 2002.
[37] Sedra A.S. and Smith, K.C. *Microelectronic Circuits*. 5th edition, pp. E-3, Oxford, 2004.

Application of Linear Programming Modeling in Multiwave Missile Launch

Qitong Ou* and Chenhong Du

School of Applied Mathematics, Xiamen University of Technology, Xiamen, China

*Corresponding author email: ouqitong@xmut.edu.cn

Abstract. Linear programming modeling is mainly concerned with solving the maximum or minimum value of a linear function under a group of linear constraints. It has been widely used in many subjects and fields. As an application example of linear programming, this paper decomposed two-wave salvo into three steps, used Floyd algorithm and 0-1 programming to complete the task allocation model, and used Dijkstra algorithm to carry out path planning for the subsequent transfer of missile vehicles, and designed the transfer path with the highest survival rate.
Keywords: Mathematical modelling; Linear programming method; 0-1 programming; Floyd algorithm; Ant colony algorithm.

1. Introduction

Linear programming is mainly concerned with solving the maximum or minimum value of a linear function with a group of linear constraints. Linear programming is an early and widely used branch in operations research. The simplex method was proposed by G.B. Dentzing in 1947, and its successful implementation in computers made the problems solved by linear programming increase rapidly. It has been widely used in many subjects and fields such as national defense science and technology, economy, industry, agriculture, environmental protection, education and social science. This paper introduces the application of linear programming by taking a missile launch in national defense science and technology as an example.

In missile operations, the forms of attack and counterattack are obviously different under different military strength. If the military strength difference is large, then the defense strength and interception strength is relatively low, so the attacker only needs to use multiple waves of single missile launch to hit the enemy's core facilities or important road nodes. But the defense in the face of huge military strength, can only be small forces guerrilla and waiting for the opportunity to counterattack as the main means of response. However, if the difference in military strength is within the range of riposte, it indicates that the defense strength of the defending side is high, and the reduction of the forces of the attacking side to the first attack of the defending side is relatively limited, and the retention rate of the effective forces of the two sides will also have a significant impact on the trend of the subsequent battle. In 2018, Hao, Zhang and Sha [1] clearly pointed out that, considering the increasing importance of missile guidance system in the warfare, the reduction of the time for movement and the effective preservation of the combat force will be crucial for the army in the future. In 2018, Zhou and Qiu [2] designed a comprehensive algorithm based on cluster analysis and Floyd shortest path to transform multi-objective optimization problems into multiple single-objective optimization problems through linear weighting method, and verified the feasibility of this idea.

The core of the multi-wave missile launch planning study is to minimize the exposure time for vehicle launcher transfer. The minimum exposure time means a higher survival rate of the personnel, and at the same time, a higher survival rate of the launcher means a better preservation of the combat force. In 2014, Xie, Yang, Yin et al. [3] proposed the necessity of analyzing the factors influencing the single-shot killing probability of missile, and conducted a large number of simulation calculations. In 2013, Jin and Yu [4] put forward that penetration probability is the most important basic indicator that reflects the operational effectiveness of missile weapon system. After a series of studies, they reached the conclusion that in the battles between the military powers, considering the higher interception ratio, the single-wave time and multiple missile salvo fire is the best way to attack. For the defense side, how to defend the attack mode of single- wave multi- missile salvo will become the focus of research.

2. Problem Construction

2.1. Questions

Vehicle-mounted missile launchers usually lurk in the standby area. After receiving instructions from the headquarters, each vehicle-mounted missile launcher loads bombs from the standby area to the designated erectable area until all devices are in place and a salvo is fired. For a multi-wave salvo, vehicle-mounted missile launchers must travel to the logistics area to resupply the missiles between two rounds, or they cannot participate in the next salvo. Each settable area can only be used once between two successive launches.

Currently, one missile combat unit is engaged in combat operations. The unit has 24 vehicle-mounted missile launchers, which are divided into eagle, Leopard and Turtle types according to their functions. The corresponding numbers of the three vehicle-mounted missile launchers are 6, 6 and 12 respectively. The unit is deployed equally in two standby zones (D_1 , D_2) before engaging in combat operations. The operational area contains 6 logistics areas (Z01-Z06), and the number of erecting areas is 60 (F01-F60). Each erecting area can only permits at most one vehicle-mounted missile launcher at a time, and each logistics area can only permits at most two vehicle-mounted missile launchers at a time. Only one vehicle-mounted missile launcher can be reloaded for up to 10 minutes at a time. The amount of ammunition in the logistics area is absolutely sufficient and all three types of vehicle-mounted missile launchers can be replenished in any logistics area.

After receiving operational instructions, troops need to plan all routes for each vehicle-mounted missile launcher, including the allocation of erecting points, the route to erecting points, the choice of logistics area after salvo completion, the route to the designated logistics area, and the final requirement is the minimum overall exposure time.

The "exposure" is defined as: the vehicular missile launcher leaves the standby area, and the logistics time is not included (the logistics process is considered to be completely safe). The time of erection and removal of vehicle-mounted launchers located in an erecting area is excluded from consideration.

2.2. Map Construction

Suppose our map looks like Figure 1:

The area of operations in the image below is:

- 2 initial standby area for vehicle-mounted launchers: D_1 , D_2 ;
- 6 logistics areas for reload of vehicle-mounted launchers: Z01-Z06;
- 60 erecting operational areas for vehicle-mounted launchers: F01-F60;
- 62 road nodes: J01-J62.

(Roads marked with red lines are main roads, and other colored lines are ordinary roads) See appendix for detailed coordinates of each node and region.

2.3. Mode of Action of Vehicle-mounted Launcher

The vehicle-mounted launcher stands standby in the area D_1 or D_2 in situ when not on a mission. The standby area is shown in figure 2.

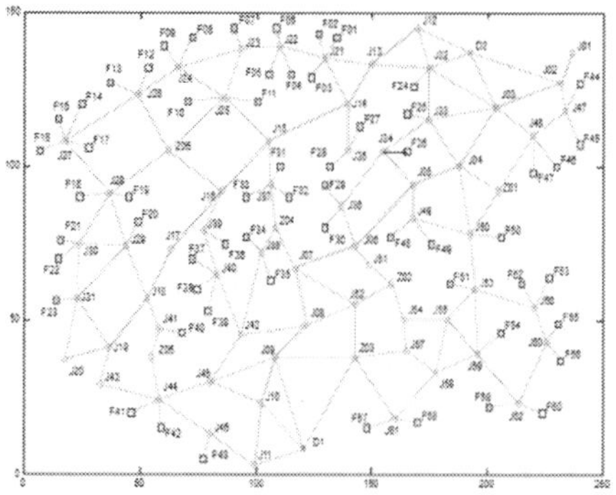

Figure 1. Route overview.

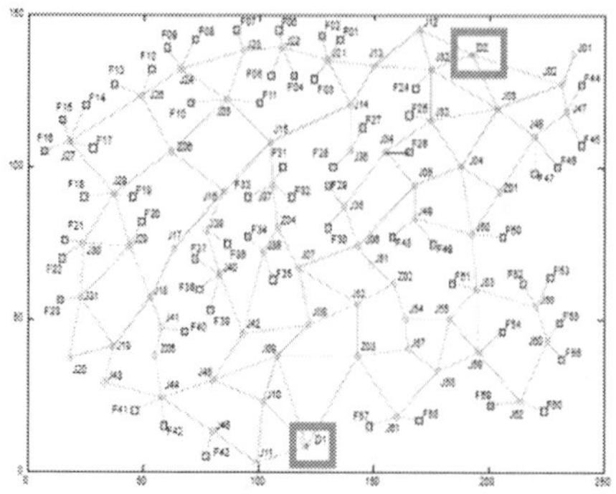

Figure 2. Schematic diagram of the standby area.

After receiving the combat instructions, each vehicle-mounted launcher loads bombs from the initial position D_i to the designated combat area F_j for erection along the designated road. When all vehicle-mounted launchers arrive at the designated position, a round of missile salvo starts. After the completion of a salvo, all vehicle-mounted launchers go to the designated logistics area Z_m for ammunition replacement, as shown in figure 3. After the transfer, they immediately go to the next designated location to wait for the next salvo.

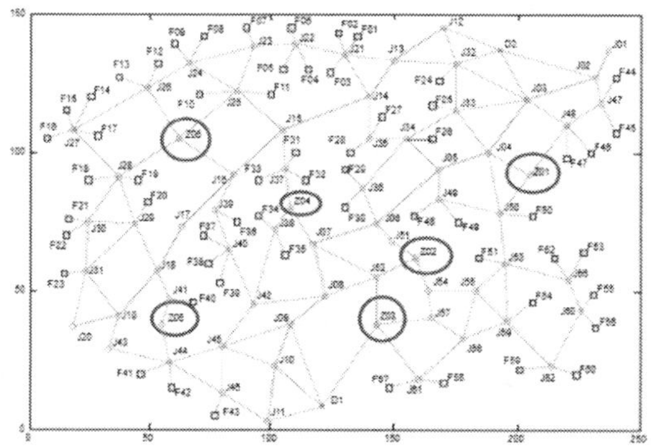

Figure 3. Logistics area.

2.4. Number and Type of Vehicle-mounted Launchers

There are 24 vehicle-mounted launchers, which can be divided into Eagle, Leopard and Turtle according to different types of bombs. The number of Eagle is 6, Leopard is 6 and Turtle is 12.

On main roads, the eagle vehicle-mounted launcher can travel at 70 km/h, the Leopard vehicle-mounted launcher at 60 km/h, and the Turtle vehicle-mounted launcher at 50 km/h. On ordinary roads, the Hawk can travel at 45 km/h, the Leopard at 35 km/h, and the Turtle at 30 km/h.

2.5. Conditional Assumptions

In order to ensure that the core of the study is the planning problem in multiwave missile launch, some conditions are assumed in this paper:

- Each operational unit may carry a maximum of one missile at a time;
- The same launch point is not used twice because the position will be exposed when the missile is launched;
- There may be only one device at the same time at each point where a vehicle-mounted launcher can be erected;
- The logistics area can accommodate a maximum of two launchers at the same time, and ammunition change cannot be performed at the same time. It takes 10 minutes for a single machine to change shells;

- Ammunition in the logistics area is sufficient, regardless of the situation that ammunition will be changed out, and the protective measures in the area are in place. The missile loading vehicle is considered to be absolutely safe in this area, so the exposure time is not included;
- The main road is relatively wide, allowing no more than two vehicle-mounted launchers to pass through at the same time, while the other common roads can only allow no more than one vehicle-mounted launcher to pass through;
- There is no capacity limit for road nodes;
- The technical time required for the erection of vehicle-mounted launchers is not taken into consideration in this paper;
- Weather conditions, road surface repairment and traffic are not taken into consideration.

3. Modeling and Solving

The exposure time should be as short as possible, so different requirements and restrictions will be added to better the study.

3.1. Model 1: Two Salutes

3.1.1. Conditional formulation. Assuming that the current operational requirements are received and two salvo missions are to be carried out, and all units must participate in each round of salvo, the overall minimum exposure time is calculated on this premise.

3.1.2. Problem analysis. The general process of the two-wave salvo mission can be summarized as follows: First, all vehicle-mounted missile launchers proceed from departure D_i to launch point F_j for the first salvo; After a salvo, the vehicle-mounted missile launchers proceed from the logistics area F_j to perform a reload operation Z_m, and from the logistics area maneuver Z_m to the next erection area F_j for a second salvo.

3.1.3. Model establishment. Since the real coordinates of each node and region are known, we can first construct a path matrix of 130*130. Since there are three types of vehicle-mounted missile launchers: Eagle, Leopard and Turtle, the speed of these three types of devices is different. Therefore, after substituting different speeds and using Floyd algorithm, we will get three 130*130 corresponding shortest path matrices T_Y, T_B, T_G, and respectively correspond to Eagle, Leopard and Turtle.
For the first question, we can first decompose it into three steps:

- Step 1: Vehicle missile launcher from departure D_i to erection area F_j.
- Step 2: After a salvo from the vehicle missile launcher, the vehicle moves from the erection area F_j to the logistics area Z_m.
- Step 3: After the vehicle missile launcher is reloaded, maneuver from the logistics area Z_m to the next erection area F_j.

After dividing the problem into three steps, 0-1 programming was used to constrain the problem. And in order to coordinate with the constraints of 0-1 planning for solving, the above shortest path matrix T_Y, T_B, T_G, and three new matrices are obtained after screening, and these three new matrices DF, FZ, ZF, correspond to the overall shortest exposure time of the three steps in sequence respectively.

3.1.4. Setting variables and symbols. In order to achieve the goal of shortest exposure time T, set the objective function:

$$\min T = X_u DF + X_o FZ + X_p ZF \tag{1}$$

0-1 variables are as follows:

(1) x_u indicates whether in the step 1, the vehicle-mounted missile launcher chooses to move from the initial area D_i to the launch site F_j,

$$x_u = \begin{cases} 0, \text{Do not travel from initial zone } D_i \text{ to erecting zone } F_j \\ 1, \text{Travel from initial zone } D_i \text{ to erecting zone } F_j \end{cases}, u = 1,2,\ldots,120.$$

The matrix of x_u is denoted as X_u

(2) x_o indicates whether in step 2 the vehicle missile launcher chooses to travel from the erection area F_j to the logistics area Z_m

$$x_o = \begin{cases} 0, \text{Do not travel from erection area } F_j \text{ to logistics area } Z_m, \\ 1, \text{Travel from erection area } F_j \text{ to logistics area } Z_m, \end{cases}, o = 1,2,\ldots,360.$$

The matrix of x_o is denoted as X_o

(3) x_p indicates whether the vehicle-mounted missile launcher chooses to travel from the logistics area Z_m to the launch site F_j in step 3

$$x_p = \begin{cases} 0, \text{Do not travel from logistics area } Z_m \text{ to erection area } F_j, \\ 1, \text{Travel from logistics area } Z_m \text{ to erection area } F_j, \end{cases}, p = 1,2,\ldots,360.$$

The matrix of x_p is denoted as X_p.

Define the following symbols:

- YD_i : number of Hawk vehicle-mounted launchers in the initial area D_i (i=1,2);
- BD_i : number of Leopard vehicle-mounted launchers in the initial area D_i (i=1,2);
- GD_i : number of Turtle vehicle-mounted launchers in the initial area D_i (i=1,2);
- $x_{D_iF_j}$: The number of vehicle-mounted launchers from the initial area D_i to the erection area F_j in step 1
- $x_{F_jZ_m}$: The number of vehicle-mounted launchers from the erection area F_j to the logistics area Z_m in step 2,
- $x_{Z_mF_j}$: the number of vehicle-mounted launchers from the logistics area Z_m to the erection area F_j in step 3.

3.1.5. Constraints. (1)Since only one vehicle-mounted launcher can be operated on ordinary roads at a time, vehicle-mounted launchers should initially be evenly distributed in two initial areas:

$$YD_1 = YD_2 = 3. \tag{2}$$

$$BD_1 = BD_2 = 3. \tag{3}$$

$$GD_1 = GD_2 = 6. \tag{4}$$

(2)In step 1, when the vehicle-mounted launchers are distributed to various emplacement points, each point can only have at most one device at the same time:

$$\sum x_{D_iF_j} \leq 1. \tag{5}$$

(3)In step 2, all vehicle-mounted launchers will have to travel to the logistics area to replenish the missiles:

$$\sum x_{F_j Z_m} = 24 . \tag{6}$$

(4)In step 2, when vehicle-mounted launchers go to the logistics area to replenish missiles, a maximum of two launchers may be present in each logistics area at the same time:

$$x_{F_j Z_m} \le 2 ,(\mathrm{m}=1,2,3,4,5,6). \tag{7}$$

(5)No more than one vehicle-mounted launcher may be allowed on ordinary roads at a time, and the number of vehicle-mounted launchers from each erecting area to the logistic area must be less than or equal to the number of vehicle-mounted launchers from each logistic area to each erecting area:

$$\sum_{z\in Z} x_{fz} \le \sum_{d\in D} x_{df}, f \in F . \tag{8}$$

(6)In step 3, The number of vehicle-mounted launchers of all types leaving the logistics area shall be equal to that entering in Step 2:

$$x_{F_j Z_m} = x_{Z_m F_j} ,(\mathrm{m}=1,2,3,4,5,6). \tag{9}$$

(7)In step 3, the vehicle-mounted launchers shall not repeat the previous ones when going to each erecting area, and the maximum number of launchers in each erecting area shall be 1:

$$\sum_{d\in D} x_{df} + \sum_{z\in Z} x_{zf} \le 1, f \in F . \tag{10}$$

3.1.6. Conclusions and optimization. After the program runs, considering the waiting time of vehicles on ordinary roads in the same direction, vehicles in reverse situations, vehicles in replenishment in logistics area, and taking into account the advantage of the main road for permitting the passing of two vehicle-mounted launchers at the same time, finally we get the distribution diagram of the following three stages. (Figure 4 - Figure 6)

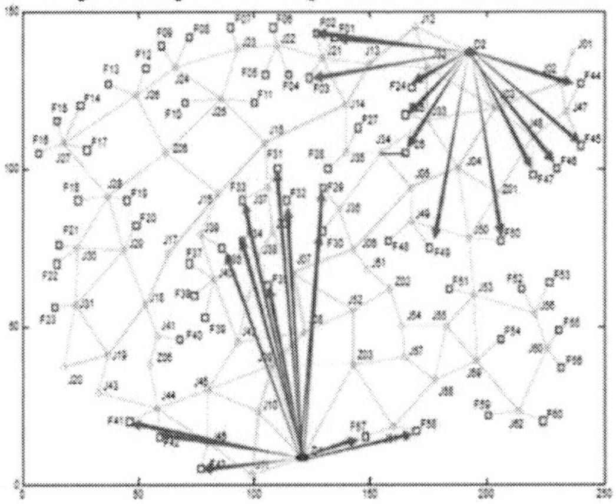

Figure 4. Schematic figure of step 1.

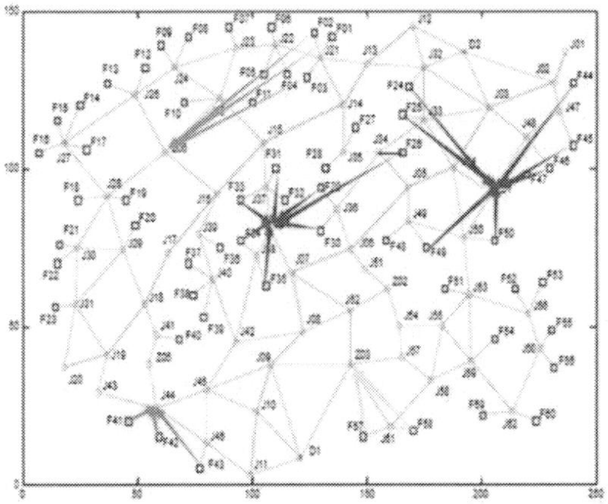

Figure 5. Schematic figure of step 2.

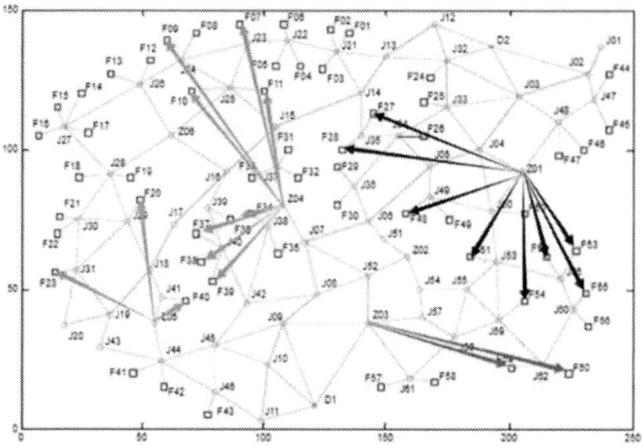

Figure 6. Logistics area.

Finally, the model solved the overall minimum exposure time of 24 vehicle-mounted launchers:

$$\min T = 8957.86\,\text{min}. \tag{11}$$

3.2. Model 2: Add Logistics Area

3.2.1. Conditional formulation. The proper layout of logistics areas will greatly enhance the survivability of combat units. Now two additional logistic areas can be added in road nodes J25, J34, J36, J42 and J49, in addition to the existing six logistics areas. The shortest exposure time that can be achieved after the addition of logistics area is studied.

3.2.2. Problem analysis. The number of additional logistics areas has no effect on step 1 of model 1, so the main research objects are the step 2 and step 3. Choosing 2 out of 5 to increase the logistics area results in 10 different situations compared to model 1. After comprehensive consideration, the ergodic sorting method can be used to minimize the total exposure time.

3.2.3. Model establishment. The two salutes are still divided into three steps, and T_i (i =1,2,3) is used to represent the total minimum exposure time of the three steps. Since the addition of logistics area will not affect the first step, the minimum time of the first step in model 1 can be used directly. The objective function of the other two steps is:

$$T_2 = \sum_{j=1}^{60}\sum_{m=1}^{8} t_{F_j Z_m} x_o \,. \tag{12}$$

$$T_3 = \sum_{m=1}^{8}\sum_{l=1}^{36} t_{Z_m F_l} x_p \,. \tag{13}$$

3.2.4. Setting variables and symbols. In order to facilitate more detailed constraints on each part, the following symbols are defined on the basis of Model 1:

- $t_{F_j Z_m}$: Minimum time for vehicle-mounted launchers to travel from the erection area F_j to the logistics area Z_m ;
- $t_{Z_m F_l}$: Minimum time for vehicle-mounted launchers to travel from the logistics area Z_m to the erection area F_l
- F_l : Set up area not used in the first round of salvo.

3.2.5. Constraints. (1) Erection points that were not accessed by vehicle-mounted launchers in the first step cannot appear out of nothing in the second step:

$$x_o \le x_u \,. \tag{14}$$

(2) Each vehicle-mounted launcher heads to the logistics area to reload after a salvo:

$$\sum_{j=1}^{60}\sum_{m=1}^{8} x_o = 1 \,. \tag{15}$$

(3) After loading, each vehicle-mounted device will proceed to the new erection area:

$$\sum_{m=1}^{8}\sum_{l=1}^{36} x_p = 24 \,. \tag{16}$$

(4) A maximum of 1 vehicle-mounted launcher per erection area at the same time:

$$\sum x_p \le 1 \,. \tag{17}$$

3.2.6. Conclusions and optimization. $\min T = \sum T_i$. In the solution process of Model 1, it can be obtained that T_1=2459.2min. $\text{Min}\, T_2 + T_3 = 5235.1\text{min}$ can be obtained after the solution is simulated by the program and the 10 optimal overall exposure time is sorted to the minimum. Therefore, after the addition of two new logistics areas, the shortest total exposure time is 7694.3min.

4. Conclusion

In this paper, after shielding some irrelevant conditions, based on some assumptions consistent with the real situation, with the help of 0-1 programming model, FLoyd algorithm, ant colony algorithm and other methods, a sufficient study of two-wave missile salvo is carried out. Via the study, multiple wave time programming problem is decomposed into several simple single objective 0-1 planning stage, and then by the method of linear weighted integration with multiple stages, the optimal solution can be found out. This will greatly reduce the amount of time to solve these problems, simplify the solving of multi-objective optimization problem.

Acknowledgments

The paper is supported by the major Teaching Reform Project of Fujian Province in 2021, Item no. FBJG20210093.

References

[1] Hao Liangliang, Zhang Jinsheng, Sha De. *Application of Rectangular Constraint Dijkstra Algorithm in Missile Launch Planning* [C]. National Security Geophysics Series no. 14 -- Resources, Environment and Geophysics, Special Data collection of Chinese Geophysical Society, 2018: 307-311.

[2] Zhou Shuang, Qiu Tao. *Planning problems in Multi-wave Missile Launch* [J]. Agricultural Equipment and Vehicle Engineering, 2018, 56(11): 52-60.

[3] Xie Bang-Rong, Yang Jian-ying, Yin Jian et al. *Analysis of Influencing Factors of Missile Single Kill Probability* [J]. Fire & Command Control, 2014,01(02): 60-64.

[4] Jin Hong, YU Yue. *Penetration Probability Calculation Model based on Multi-wave Salvo Interception* [C]. Proceedings of the 1st China Command and Control Conference, 2013: 816-821.

Appendix

Initial area coordinates

D1	121	8	**D2**	193	137

2. Logistics area coordinates

Z01	205	92	**Z04**	108	80
Z02	159	62	**Z05**	55	38
Z03	143	38	**Z06**	62	105

3. Point coordinates that can be set up

F01	135	142	**F31**	110	100
F02	127	143	**F32**	114	90
F03	124	129	**F33**	95	90
F04	115	130	**F34**	95	77
F05	105	130	**F35**	106	63
F06	108	145	**F36**	86	75
F07	90	145	**F37**	72	70
F08	72	142	**F38**	74	60
F09	60	139	**F39**	79	53
F10	70	121	**F40**	68	46
F11	100	121	**F41**	46	20
F12	53	132	**F42**	59	15
F13	37	127	**F43**	77	5
F14	25	120	**F44**	240	127
F15	15	115	**F45**	240	107
F16	7	105	**F46**	230	100
F17	28	106	**F47**	220	98
F18	24	90	**F48**	158	77
F19	45	90	**F49**	176	75
F20	49	82	**F50**	206	77
F21	16	76	**F51**	184	62
F22	15	70	**F52**	215	62
F23	14	56	**F53**	227	64
F24	168	126	**F54**	206	46
F25	165	117	**F55**	231	49
F26	165	105	**F56**	232	37
F27	145	113	**F57**	148	15
F28	132	100	**F58**	170	17
F29	130	94	**F59**	201	22
F30	130	80	**F60**	224	20

4. Coordinates of road nodes

J01	237	137	**J32**	175	132
J02	232	127	**J33**	175	115
J03	204	119	**J34**	155	105
J04	188	100	**J35**	140	105
J05	168	94	**J36**	137	87
J06	143	74	**J37**	106	94
J07	117	67	**J38**	102	72
J08	122	48	**J39**	77	79

J09	108	38	**J40**	83	65
J10	102	23	**J41**	58	47
J11	99	3	**J42**	93	45
J12	170	145	**J43**	33	29
J13	150	133	**J44**	58	24
J14	140	120	**J45**	80	30
J15	105	108	**J46**	80	13
J16	85	92	**J47**	234	118
J17	63	73	**J48**	220	110
J18	53	57	**J49**	168	83
J19	37	41	**J50**	193	78
J20	18	37	**J51**	149	68
J21	130	135	**J52**	143	55
J22	110	139	**J53**	195	60
J23	93	138	**J54**	164	50
J24	66	132	**J55**	183	50
J25	86	122	**J56**	221	54
J26	49	123	**J57**	165	40
J27	18	108	**J58**	178	33
J28	37	91	**J59**	196	39
J29	44	74	**J60**	226	43
J30	24	75	**J61**	160	18
J31	23	57	**J62**	214	23

The Optimization Simulation Analysis of Heavy NC Machine Tool Bed

Bing Luo[1,*] and Xin Zhao[2]

[1]School of Mechanical Engineering, Shenyang Ligong University, Shenyang 110159
[2]FuShun YongMao Construction Machinery CO.,LTD, China

*Corresponding author email: fzylingyf@163.com

Abstract. In order to reduce the high manufacturing cost of heavy numerical control (NC) machine tool bed, the paper first introduced the stiffness and strength calculation method of constant cross-section and constant span beam, and deduced the theoretical optimization model with the minimum cross-section area of the bed as the optimization objective. Using the interior point method completed the optimization analysis of the machine bed. Then, the ISIGHT software was used to realize the integration of design software and analysis software, and the optimization simulation analysis of the machine bed was completed under the same interface. Finally, the calculation accuracy of the theoretical model is verified by comparing the optimization results of the theoretical model and the simulation model. The weight can be reduced by more than 10% on the basis of ensuring the stiffness and strength of the machine bed. The method lays a theoretical foundation for the manufacture of the machine bed.
Keywords: Heavy NC machine tool; Bed; Optimization; ISIGHT; Simulation.

1. Introduction

Heavy NC machine tool bed is the main load-bearing component of the whole machine, and its quality is directly related to the machining accuracy and precision preserve of heavy NC machine tool. Therefore, scholars from all over the world pay great attention to the mechanical properties of heavy NC machine tool bed [1,2]. Shao Yinghe et al.[3] proposed an analysis method for the relationship between performance of the bed and whole machine. Liu Bing et al.[4] analyzed the static and dynamic characteristics of machine tool bed with APDL language provided by ANSYS. Liu Zhisong et al.[5] proposed a prediction algorithm of bed stress distribution based on distributed strain gauge, which is suitable for any distribution of measuring points. Bo Ruifeng et al.[6] did a dynamic analysis of deep hole machine bed based on multi frequency topology optimization. Cao Li et al.[7] analyzed the influence of bed material on dynamic characteristics of the machine tool based on ANSYS Workbench. Liu Kuo[8] analyzed the harmonic response of the machine tool bed. Kpczak N et al.[9] carried out the modal analysis method and experimental research of the machine tool bed. Ziyao W et al.[10] improved the structure of the machine bed based on the modal analysis results.

However, due to the huge quality of heavy machine tool bed, it brings great trouble in the process of assembly and debugging, and also causes the waste of resources. Therefore, it is necessary to carry out the optimization design research of heavy NC machine tool bed. In this paper, through the simplification of the machine bed, a theoretical method of optimization design is obtained which is easy to realize and operate, and the accuracy of the theoretical optimization method is verified by comparing with the simulation optimization method. The method lays a foundation for the optimization design of heavy NC machine tool bed.

2. The Optimization Model of the Machine Bed

2.1. Simplification of the Machine Bed Model

In this paper, the non-supporting parts of the model are removed, including the skirt which plays the role of guidance and technic, and the supporting part of the bed is completely retained. The structure of the model is regularized to facilitate the calculation of parameters such as the centroid and moment of inertia of the model. The specific simplified results are shown in Figure 1. Figure 1-(a) shows the actual structure, and figure 1-(b) shows the simplified model.

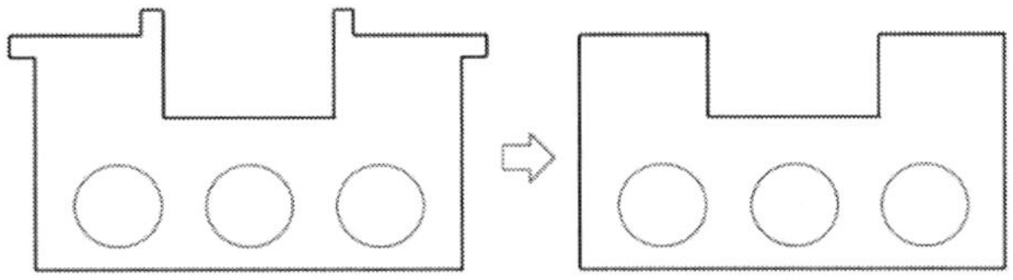

(a) (b)

Figure 1. Simplified process diagram of machine bed model.

2.2. Centroid Calculation of the Machine Bed Section

For heavy NC machine tool bed section, it can be regarded as a combination of multiple graphics. When calculating the section centroid, it can be combined and deduced. Because the algebraic sum of the static moment of each part to a certain axis is equal to the static moment of the whole to the same axis, there is a formula:

$$\begin{cases} S_z = \sum_{i=1}^{n} A_i \overline{y_i} \\ S_y = \sum_{i=1}^{n} A_i \overline{z_i} \end{cases} \tag{1}$$

where, $\overline{y_i}$, $\overline{z_i}$ —centroid coordinates of section for the i graph;

Ai —cross section area of the i graph;

S_y, S_z —static moment of the bed section respect to the coordinate axes.

2.3. Calculation of Inertia Moment of the Machine Bed

The cross section of the machine bed is reasonably partitioned to complete the calculation of the moment of inertia. The specific partition method is shown in Figure 2.

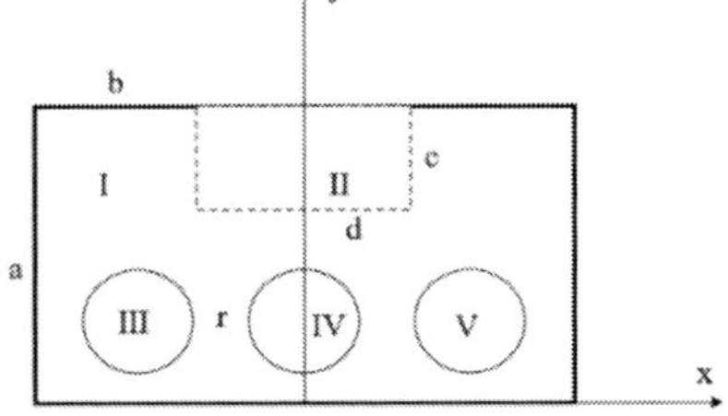

Figure 2. Partition method of machine bed.

According to the formula for translation of axis and moment of inertia, the moment of inertia of heavy machine tool bed can be obtained as follows:

$$I_z = I_z^I - I_z^{II} - I_z^{III} - I_z^{IV} - I_z^{V} \tag{2}$$

2.4. Mechanical Model of the Machine Bed

The machine bed is generally placed on the bed base, and its bottom is fixed by multiple bolts. Based on theoretical mechanics, the bed can be regarded as a continuous beam structure for analysis, and its force is shown in Figure 3.

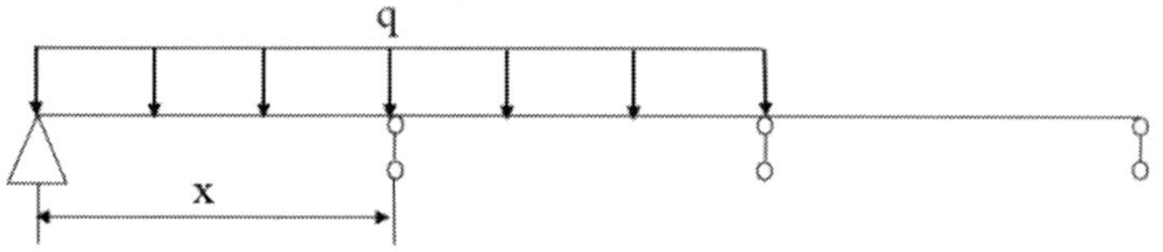

Figure 3. Force diagram of the machine bed.

In order to realize the optimal design of the machine bed, it is necessary to obtain the maximum load deformation and normal stress of the machine bed. When calculating the continuous beam with equal span and equal section, the following formula can be used [11]:

$$\begin{cases} \sigma_{max} = \dfrac{M_{max} y_{max}}{I_z} \\ f_{max} = \dfrac{\alpha q l^4}{100 E I_z} \end{cases} \tag{3}$$

where, σ_{max} —maximum stress;

M_{max} —maximum bending moment;

y_{max} —the distance from the furthest point of the cross section to the neutral axis;

f_{max} —maximum deformation;

q —uniform load;

l —length of continuous beam;

E —elastic modulus;

α —deflection coefficient;

I_z —moment of inertia

2.5. Establishing of Optimization Model

In this paper, the optimization model of heavy NC machine tool bed is established with the minimum section size as the optimization objective, and the section size, maximum load-bearing deformation and maximum stress as the constraint conditions. The optimization analysis of heavy NC machine tool bed is completed by using the interior point method.

The optimization model is as follows:

$$\min(f) = ab - cd - 4.712388r^2 \tag{4}$$

The constraint condition is as follows:

$$g_{1(x)} = 200000000 - \frac{87.75\times 10^3 \times \dfrac{ab\times 0.5a - cd\times(a-0.5c) - 0.8246679r^2}{ab - cd - 4.712388r^2}}{\dfrac{1}{3}ba^3 - (\dfrac{c^2}{3} + a^2 - ac)cd - 1.178097r^3 - 0.14431688r^2} \geq 0$$

$$g_{2(x)} = 0.0000025 - \frac{1.409016\times 10^{-7}}{\dfrac{1}{3}ba^3 - (\dfrac{c^2}{3} + a^2 - ac)cd - 1.178097r^3 - 0.14431688r^2} \geq 0$$

$$g_{3(x)} = a/0.6 - 1 \geq 0$$

$$g_{4(x)} = 1 - a \geq 0$$
$$g_{5(x)} = b / 1.2 - 1 \geq 0$$
$$g_{6(x)} = 1 - b / 2 \geq 0$$
$$g_{7(x)} = c / 0.2 - 1 \geq 0$$
$$g_{8(x)} = 1 - c / 0.4 \geq 0$$
$$g_{9(x)} = d / 0.6 - 1 \geq 0$$
$$g_{10(x)} = 1 - d \geq 0$$
$$g_{11(x)} = r / 0.09 - 1 \geq 0$$
$$g_{12(x)} = 1 - r / 0.15 \geq 0$$
$$g_{13(x)} = \frac{a-c}{2} - r \geq 0$$

2.6. Optimization Method Programming

Optimization Toolbox module in Matlab software is used to solve the problem in this paper. The specific parameter input is shown in the figure 4:

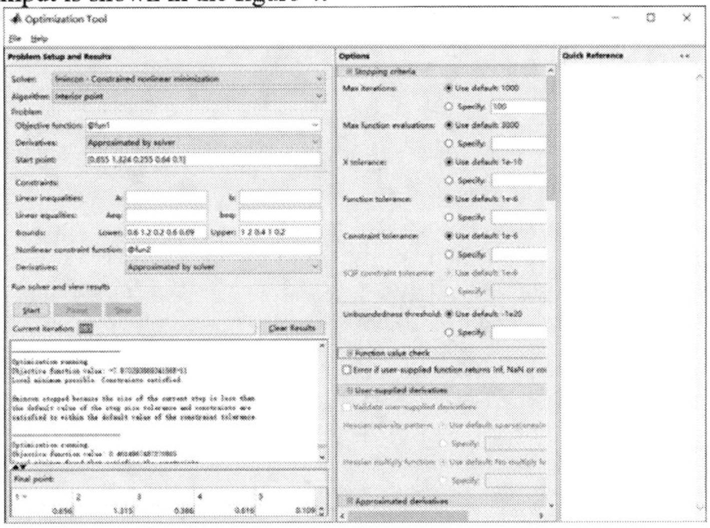

Figure 4. Optimization analysis process of the machine bed.

3. Finite Element Simulation Calculation of the Machine Bed

It is important to integrate the coupling structural parts, the rules of different disciplines and different CAE analysis software together for overall analysis and optimization, so that each optimization target can be fully analyzed and compared in the optimization process. In the process of completing the simulation calculation of heavy machine tool bed, the application of MATLAB (for optimization analysis), ANSYS (parametric modeling and mechanical simulation calculation of machine bed) and other software should be coordinated, so this paper realizes the optimization analysis of complex system based on ISIGHT software. Figure 5 shows the displacement and stress nephogram of heavy machine tool bed.

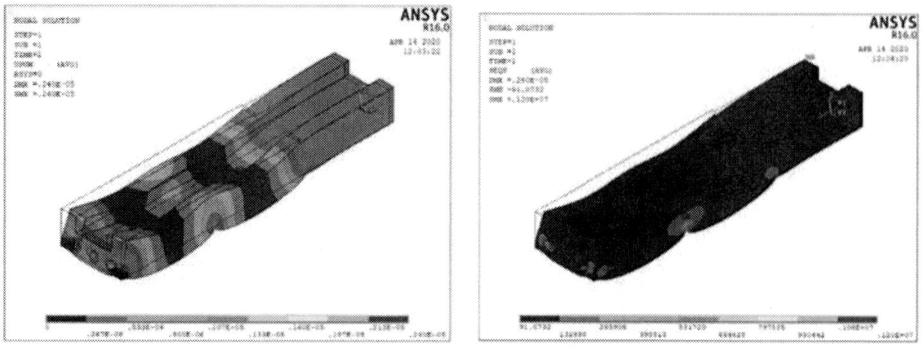

Figure 5. Analysis nephogram of the machine bed.

The paper completed parametric modeling based on ANSYS APDL, through which the strength and stiffness of the heavy NC machine tool bed can be checked. Using the seamless interface technology of ANSYS and ISIGHT, the paper integrated ANSYS under the ISIGHT interface to complete the optimization analysis of the bed. Figure 6 shows the optimization process of maximum deformation and stress.

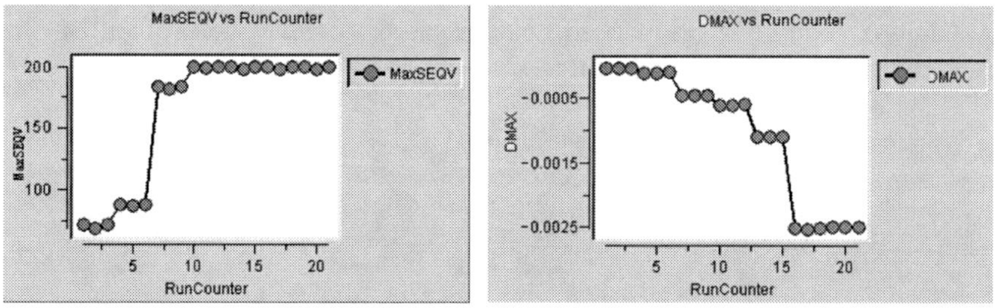

Figure 6. Optimization process of maximum deformation and stress.

The optimization results of the machine bed are shown in Figure 7. The cross-sectional dimensions of the bed are a = 0.662m, b = 1.233m, c = 0.364m, d = 0.652m, r = 0.112m. After optimization, the mass of machine tool is 2379429.15kg, which is reduced by 13.96% compared with the original mass of 2765615.5kg.

Parameter	Mode	Lower	Value	Upper	Objective	Target
*SET, a		0.6	0. 66245	1.0		
*SET, b		1.2	1. 2335	2.0		
*SET, c		0.2	0. 36466	0.4		
*SET, d		0.6	0. 65259	1.0		
*SET, r		0.09	0. 11235	0.15		
DMAX		-2.5E-6	-2. 4615E-			
MASS			2379400. 0		minimize	
SMAX			1. 9683E8	2.0E8		

Figure 7. Optimization results of the heavy machine bed.

4. Comparison between Theory and Simulation

In this paper, the optimization results of the theoretical model are compared with the simulation results, and the comparison results are shown in Table 1:

Table 1. Comparison of theoretical and simulation results.

Parameters	Optimization results of theoretical model	Optimization results of simulation
a	0.651	0.662
b	1.305	1.233
c	0.357	0.364
d	0.618	0.652
r	0.109	0.112

By comparing the data in Table 1, the error between the optimization results of the theoretical model and the simulation results of the machine bed is 9.2%, which proves the correctness of the theoretical model established in the paper. Because the theoretical model is simple in program and convenient in operation, it is easier to be applied in engineering, which provides a theoretical basis for guiding the design of the heavy NC machine tool bed.

5. Conclusion

In this paper, through the reasonable partition and parameter setting of the machine bed, the calculation of the centroid and moment of inertia of the machine bed is completed based on the theoretical mechanical knowledge. Based on this, the mechanical calculation model of equal span and equal section is deduced. Then, taking the minimum section size as the optimization objective, the optimization model of the machine bed is established, and the optimization analysis of the machine bed is realized by using the interior point method. The paper through ISIGHT software integrated with MATLAB and ANSYS software, sets optimization design platform of the heavy NC machine tool bed, and verifies the accuracy of the optimization model of the theory. The analysis shows that the optimized quality of the machine bed was reduced by 14%, and can greatly reduce the manufacturing cost of heavy NC machine tool bed. This method is easy to realize and master, and can be used as the main reference method for the structural design of the heavy NC machine tool.

References

[1] Tian Y Liu Z and Dong X 2018 Bearing deformation of heavy-duty machine tool-foundation systems and deformation detection methods *Archive Proceedings of the Institution of Mechanical Engineers Part C Journal of Mechanical Engineering Science* pp 3232-3245

[2] Tian Y Liu Z and Xu X et al 2019 Systematic review of research relating to heavy-duty machine tool foundation systems *Advances in Mechanical Engineering* pp 3232-3245

[3] SHAO Yinghe HONG Rongjing and YU Chunjian 2020 Static characteristics analysis of machine tool bed performance *Journal of Nanjing Tech University (Natural Science Edition)* pp 110-114

[4] Liu Bing Chen Cheng and Wang Yong et al 2019 Dynamic characteristic analysis and structure optimization of compound machine tool bed *Journal of Electronic Measurement and Instrumentation* pp 72-77

[5] LIU Zhisong WU Jiakun and KUANG Kang 2018 Research on stress distribution detection method of large CNC machine tool bed *Modular Machine Tool & Automatic Manufacturing Technique* pp 48-50

[6] Bo Ruifeng Yin Shuyu Yan Shuaiyin and Shen Xingquan 2015 Dynamic analysis of deep hole machine lathe bed based on the multiple frequencies topology optimization *Machine Design & Research* pp67-69+73

[7] CAO Li and ZHONG Jianlin 2015 Analysis of the effect of bed materials on machine tool's dynamic characteristics based on ansys workbench *Modular Machine Tool & Automatic Manufacturing Technique* pp 61-64

[8] Kuo L Chun-Shi L Jian-Feng L et al 2011 Harmonic response analysis for bed and column of VMC0540d machine tool *Machinery Design & Manufacture* pp 162-164

[9] Kpczak N and Pawowski W 2015 Theoretical and Experimental Modal Analysis of Machine Tool Bed *Mechanik* pp 713/194-713/198

[10] Ziyao W Xiangbo Z E Shoutang L V et al 2016 Modal Analysis and Structure Improvement of Machine Tool Bed *Modern Manufacturing Technology and Equipment*

[11] 1998 Manual for static calculation of building structures *China architecture & building press* pp 149-193

Numerical Simulation of Impact Resistance of Composite Honeycomb Sandwich Radome under High-Speed Impact

Yang Shen*

Nanjing Communications Construction & Investment Holding (Group) Co., Ltd., Nanjing, China

*Corresponding author email: 64600855@qq.com

Abstract. In order to study the impact resistance and dynamic response of the composite honeycomb sandwich radome under high-speed bird impact, a simulation model of the bird impacting composite honeycomb sandwich radome with glass fiber reinforced composite (GFRP) panel and meta-aromatic polyphthalamide honeycomb (NOMEX[R]) core is established through the finite element simulation software LS-DYNA. The effects of thickness distribution of the upper and lower panels of the radome, and the layering angle of the upper panel on the impact resistance of the radome are studied. The research results show that the increase in the thickness distribution of the upper panel is a positive gain for impact resistance of the radome. Inserting some other angled layers in the orthogonal layer can improve the impact resistance of the honeycomb sandwich structure.
Keywords: Radome; Honeycomb sandwich; Bird impact; Composite material; Finite element simulation.

1. Introduction

The civil aviation industry has developed rapidly, and aviation accidents have also attracted great public attention in recent years. As a threat to aircraft, bird impacts can cause serious damage to aircraft structures [1-2]. Due to its special geometric structure, the honeycomb sandwich structure material has excellent properties such as light weight, high strength, good heat insulation performance, and good shock absorption performance. In recent years, it has been widely used in automobiles, ships, high-speed rail, aerospace and other fields. Giglio [3] et al. used finite element simulation software to study the influence of the structural parameters of the Nomex honeycomb sandwich panel on the compression simulation on the basis of existing experiments. It was found that the more unit cells of the Nomex honeycomb core, the closer the compression simulation result is to the test result. Tolga [4] conducted an experimental study on the influence of the design parameters of the honeycomb sandwich composite material on the low-velocity impact dynamic response of the honeycomb sandwich composite material. The results show that the thickness of the honeycomb core has almost no effect on the strength of the composite material, and affects the impact performance of the sample. The panel thickness has the biggest effect. Akatay [5] used the drop weight test to repeatedly impact the honeycomb sandwich structure composed of unidirectional glass fiber reinforced epoxy composite panel and aluminum honeycomb core used inside the aircraft to determine the impact energy, maximum load and permanent deflection of the honeycomb sandwich structure, to evaluate its repeated impact resistance and damage resistance. Nishikawa [6] used numerical simulation methods to discuss the damage characteristics of the single-way composite plate, and carried out a simple modeling of the bird impacting the composite fan blades, and analyzed the bird impacting composite material with different speeds and different qualities of the bird body. The fiber failure process and matrix failure process of composite laminates are studied, and the change of contact force in the process of bird impacting the composite laminate is extracted.

Based on the above research background, aiming at the impact resistance of the aircraft radome with composite honeycomb sandwich structures, the LS-DYNA explicit dynamic analysis module is selected and the SPH method is used to study the dynamic response, energy absorption properties, and damage mechanism of the bird impacting the radome. The effects of different impact positions, impact angles, the thickness distribution of the upper and lower panels of the radome, and the layering angles of the upper panels of the radome on the dynamic response, damage and energy absorption of the radome are discussed. It has significant value for the structural design of aircraft radome.

2. Simulation Model

2.1. Geometry Structure

In this paper, the bird impacting radome is simulated in LS-DYNA based on the A320 aircraft radome. The actual sandwich structure is shown in Fig. 1. From the inside to the outside, there are 4 layers of 0.25mm thickness Glass fiber layup, honeycomb core, 5 layers of 0.13mm and 0.25mm mixed glass fiber layup.

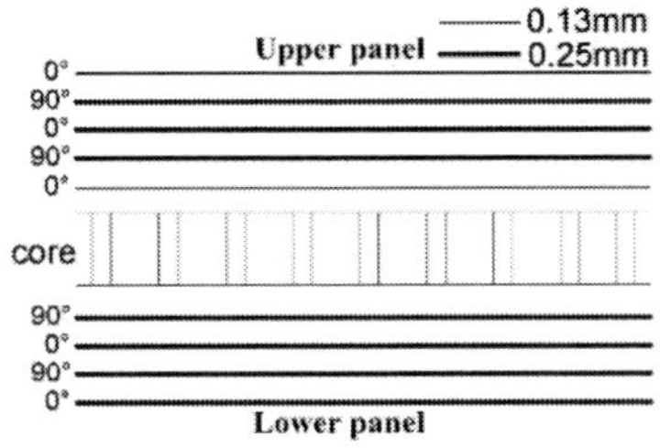

Figure 1. Composite honeycomb sandwich structure

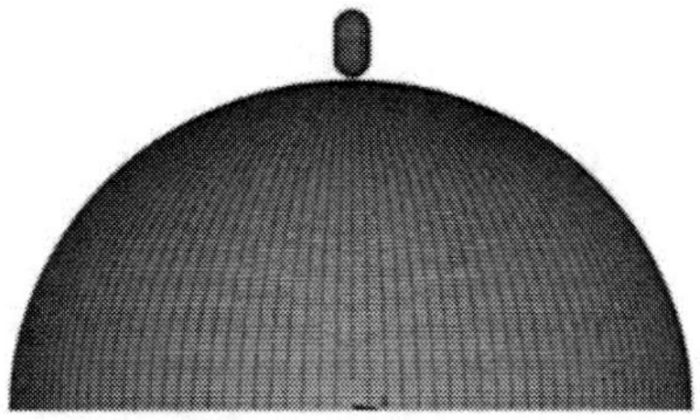

Figure 2. Finite element model.

The sandwich core of the radome is a regular hexagonal Nomex aramid paper honeycomb with a cell side length of 2.7mm, the honeycomb wall thickness t=0.27mm, the honeycomb height h=8mm, and the density ρ=48kg/m^3. The ply angle is designed to be [0°/90°/0°/90°/0°/core/90°/0°/90°/0°] from outside to inside. In the research, the radome model is simplified as a hemispherical shell structure with a certain thickness and a radius of 1100mm. The bird body is a hemispherical cylindrical structure with a diameter D=113mm and length L=226mm. The bird is simulated by SPH method, and the composite radome is simulated as solid elements [7]. The finite element model is shown in Fig. 2.

2.2. Material Properties

The constitutive model of the bird body material uses the keywords *MAT_NULL and *EOS_GRUNEISEN to provide the stress tensor for the material. The main parameters of the bird body material constitutive are shown in Table 1.

Table 1. The main parameters of bird body.

ρ/(kg·m^{-3})	C	S_1	S_2	γ_0
950	1.647	1.921	-0.096	0.35

Table 2. Material properties of GFRP laminate.

Parameter	value	parameter	value
ρ/(kg·m^{-3})	1570	v	0.34
E_{11}/Gpa	146	X_c/Gpa	1.28
E_{22}/Gpa	8.04	S_c/Mpa	220
E_{33}/Gpa	8.04	X_t/Gpa	1.59
G_{12}/Gpa	8.16	Y_t/Mpa	24
G_{23}=G_{13}/Gpa	2.34	Y_c/Mpa	120

The material model of the glass fiber composite panel is GF300/MTM28, and the keyword *MAT_ENHANCED_COMPOSITE_DAMAGE is used to define the material model. The damage judgment of this material model follows the Chang-Chang failure criterion. The material parameters are shown in Table 2.

The honeycomb core is simulated with the keyword *MAT_MODIFIED_HONEYCOMB. This material model can well simulate the anisotropic material of Nomex aramid paper honeycomb core. The equivalent material parameters are shown in Table 3.

Table 3. Equivalent material parameters of Nomex honeycomb.

Parameter	value	parameter	value
ρ/(kg·m^{-3})	48	E_3/Mpa	85
v	0.33	G_{12}/Mpa	42
E_1/Mpa	0.46	G_{23}/Mpa	26
E_2/Mpa	0.40	G_{13}/Mpa	0.31

2.3. Contact and Boundary Conditions

The SPH method is used to turn the bird body into a set of quality points. The contact between the radome and the bird is defined as *CONTACT_NODE_TO_SURFACE, and the *CONTACT_SINGLE_SURFACE contact is also defined to check whether penetration occurs. The movement and rotation of the radome is fixed in the x, y, and z directions.

3. Simulation Analysis of Bird Impacting Radome

After calculating the bird impact radome model, the stress distribution at some moments of the impact process is shown in Fig. 3.

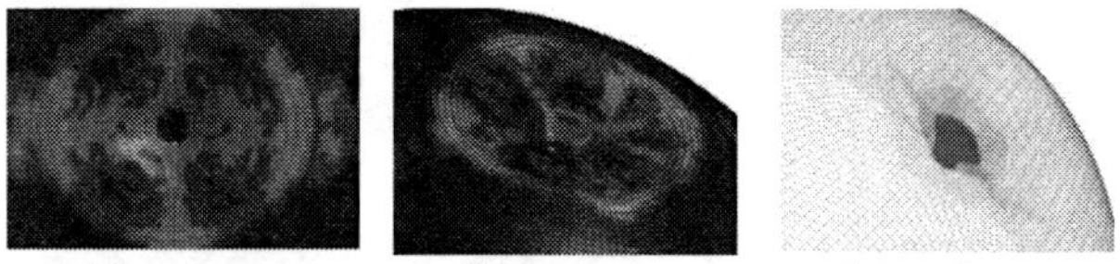

(a) t=1ms (b) t=2ms (c) t=3ms

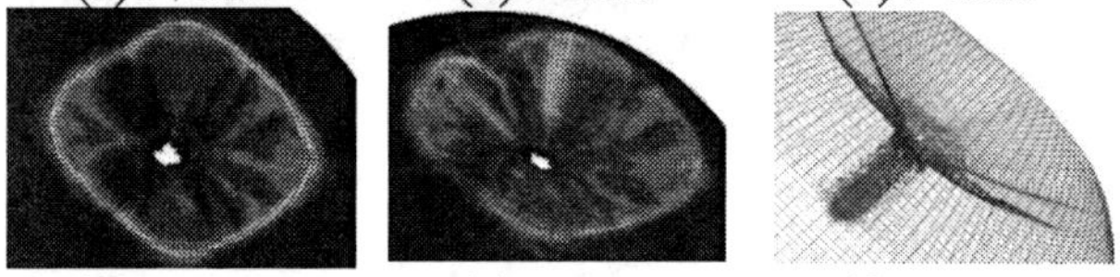

(d) t=4ms (e) t=6ms (f) t=8ms

Figure 3. Stress distribution of the radome.

The kinetic energy change history of the bird body is shown in Fig. 4, the time history of the contact force between the bird body and the honeycomb sandwich radome is shown in Fig. 5, and the time history of the displacement of radome center point is shown in Fig. 6.

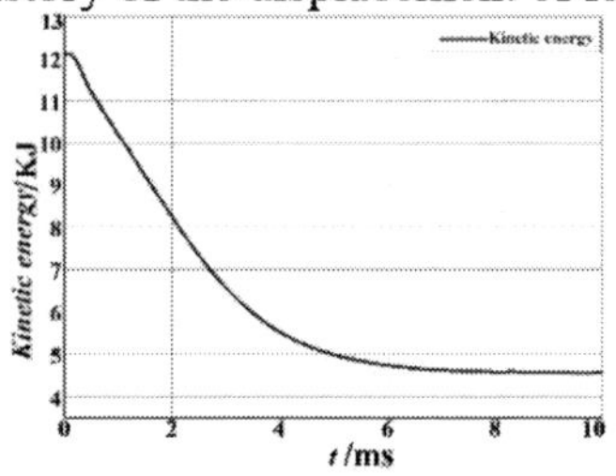

Figure 4. Kinetic energy of the bird.

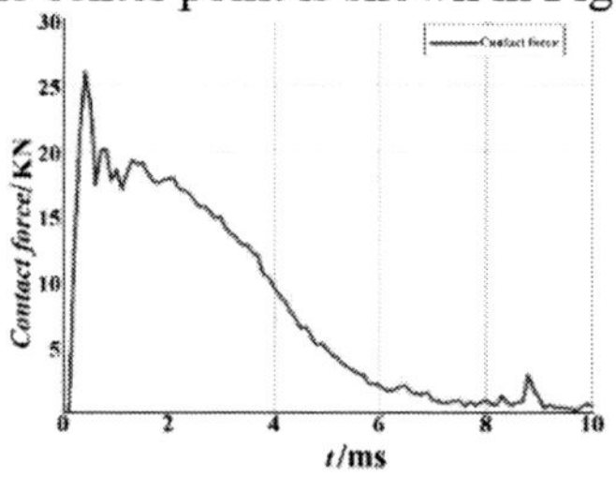

Figure 5. Contact force.

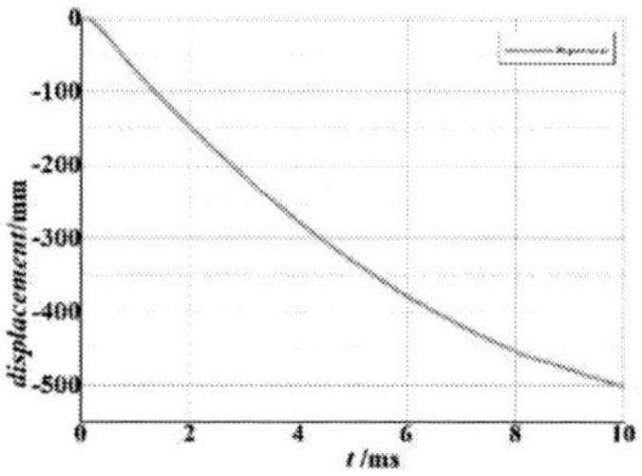

Figure 6. Displacement of radome center point.

From Figs. 3 to 6, it can be seen that the bird body produces a great contact force at the moment of contact with the honeycomb sandwich radome. The upper panel fails first, the contact force becomes smaller, and then the honeycomb core and lower panel failed. The contact force dropped further, and the bird body had penetrated the radome in 3ms. It can be seen from Fig. 4 that there is still contact force between the bird body and the radome, because the bird body deforms greatly during the impact. As shown in Fig. 3(c), the contact area with the radome increases. At 8ms, most of the bird body passed through the radome, and the downward trend of the bird's kinetic energy slowed down. In the process of the bird impacting the radome, the radome produced a huge deformation, and the center point displacement was close to 500mm at 10ms. The kinetic energy of the bird body was reduced from the initial kinetic energy of 12111J to 4559J, which was a decrease of 7552J.

4. Analysis of the Impact Resistance of the Radome

4.1. Thickness Distribution of Upper and Lower Panels

How to improve the impact resistance without increasing weight is an important content of bird impact research.

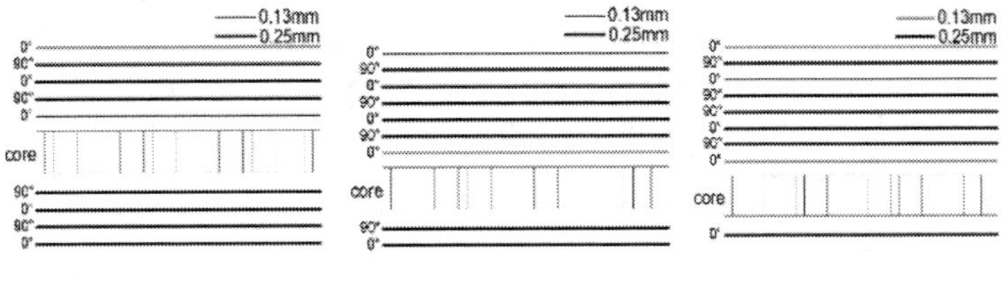

(a) K1 (b) K2 (c) K3

Figure 7. Schematic diagram of different thickness distribution.

The method of redistributing the thickness of the upper and lower panels of the radome is used to study the impact of its impact resistance performance without increasing the weight. Without increasing the weight or changing the thickness of the single-layer board, the sub-layers on the upper and lower panels are moved to change the thickness distribution. The three designs are K1: the thickness of the upper panel is 1.01mm, K2: the thickness of the upper panel is 1.51mm, and K3: the thickness of the upper panel is 1.76mm. In order to prevent the layup sequence from interfering with the simulation results, all orthogonal layups of 0° and 90° are adopted, as shown in Fig. 7.

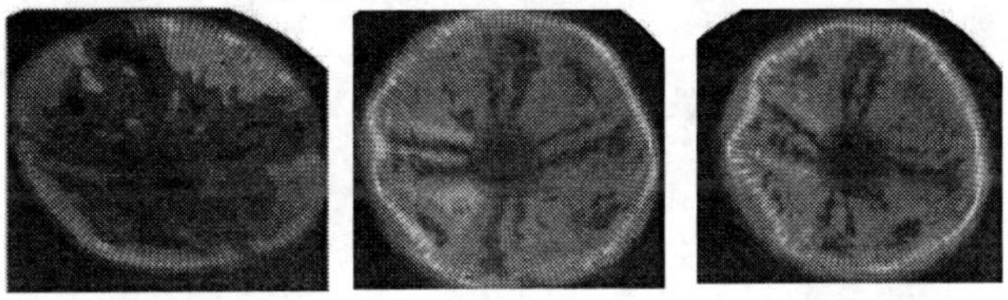

(a) 0° (b) 10° (c) 20°

Figure 8. Stress distribution of the radome.

The impact result is shown in Fig. 8. The time history of the kinetic energy of the bird is shown in Fig. 9. The maximum displacement of the impact point is shown in Fig. 10. The maximum deformation diameter of radome at 10ms is shown in Table 4.

Table 4. Maximum deformation diameter of the radome at 10ms.

Scheme	K1	K2	K3
Max diameter/mm	1242	1330	1274

From the impact results of the three schemes in Fig. 8, it can be seen that the damage degree of K1 is the largest, the upper panel of K2 scheme partially fails, the radome has not penetrated, and the bird body spreads around the surface of the radome. The degree of deformation of the radome of the K3 scheme is not as good as that of the K2. After the failure of some elements on the upper panel of the K3 scheme, the sandwich structure no longer fails, and the radome is not penetrated.

It can be seen from Fig. 9 that the energy absorption effects of the K2 and K3 schemes are relatively close, which are significantly better than the K1 scheme. It can be seen that the increase in the thickness distribution of the upper panel is a positive gain for the anti-bird impact performance of the composite honeycomb sandwich radome.

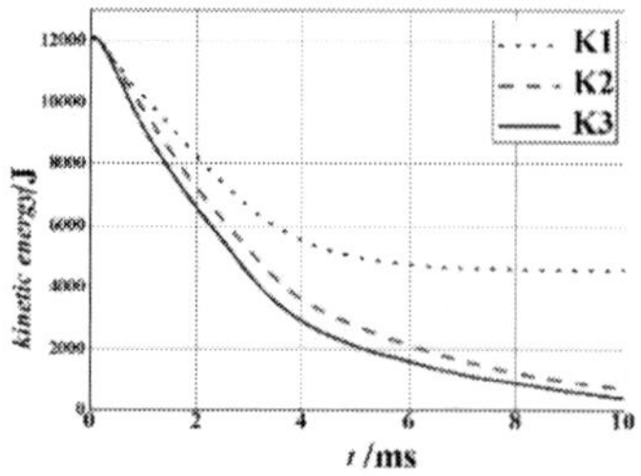

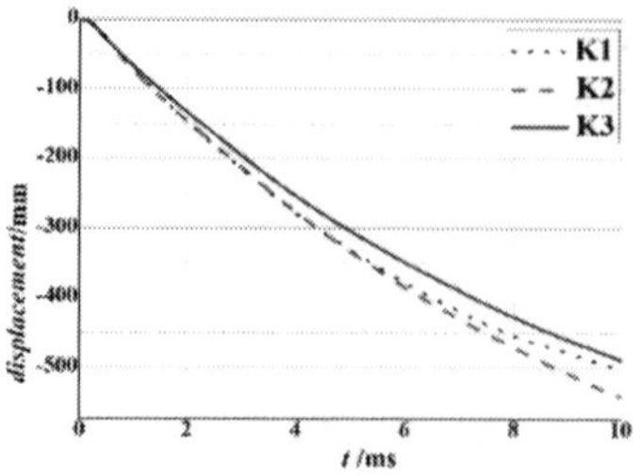

Figure 9. Kinetic energy of the bird. **Figure 10.** Displacement of radome center point.

From Figs. 9 to 10, and Table 4, the K2 scheme absorbs less kinetic energy than K3, but the impact point displacement in the K2 scheme is greater than K3, and the maximum diameter of the radome deformation is greater than K3. The solution K3 with the largest thickness of the upper panel has the best energy absorption effect, and has better impact resistance and resistance to deformation.

Therefore, more thickness can be allocated to the upper panel to improve its impact resistance and energy absorption effect. In the lightweight design, the overall thickness can be reduced, and the upper panel distribution thickness can be increased to meet the lightweight requirements.

4.2. Laying Angle of Upper Panel

It is also a feasible solution to improve the impact resistance of the composite sandwich structure without changing the weight, but changing the layup angle. The above panel is the research object, and the influence of the change of the layup angle of the upper panel on the dynamic response of the radome is analyzed. The order of the thickness of the sub-layer remains unchanged, and 3 types of layup schemes for the upper panel are designed: A1[0°/90°/0°/90°/0°], A2[0°/30°/90°/-30° /0°], A3[0°/45°/90°/-45°/0°].

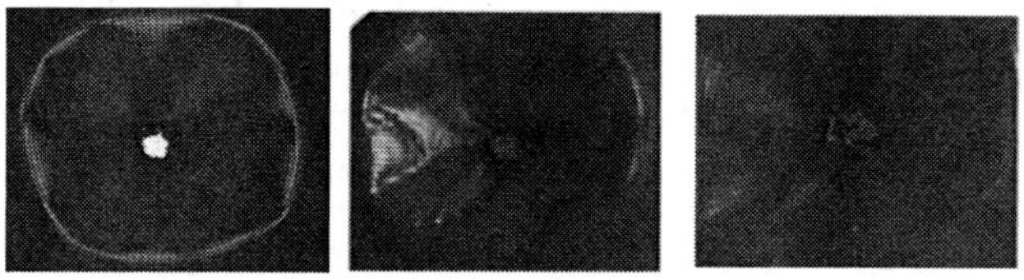

(a) A1 (b) A2 (c) A3

Figure 11. Deflection distribution of the radome.

The comparison of impact results is shown in Fig. 11. The maximum displacement of the impact point is shown in Fig. 12. The kinetic energy time history of the bird is shown in Fig. 13. The maximum deformation diameter of radome at 10ms is shown in Table 5.

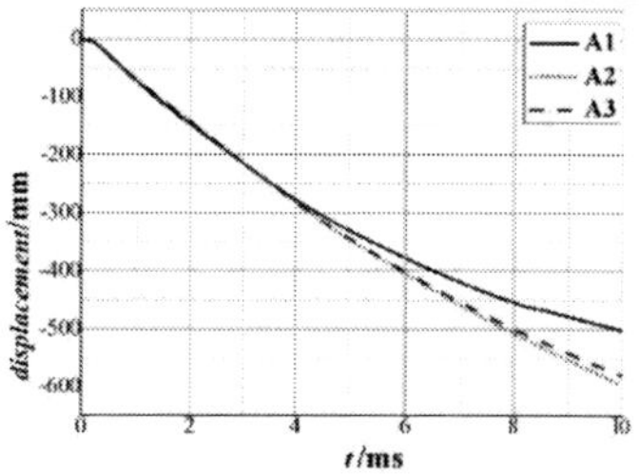

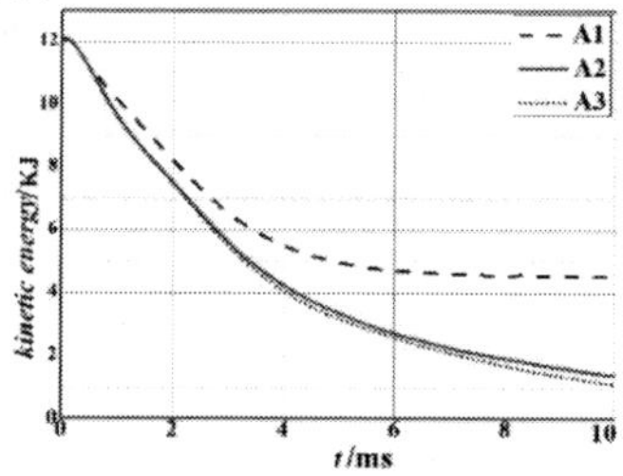

Figure 12. Displacement of radome center point. **Figure 13.** Kinetic energy of the bird.

Table 5. Maximum deformation diameter of the radome at 10ms.

Scheme	A1	A2	A3
Max diameter/mm	1242	1440	1362

From the impact results of the three layup schemes in Fig. 11, it can be seen that the radome of A1 scheme is penetrated, and the radome of A2 and A3 is not penetrated. The upper panel and honeycomb core of A2 failed, the lower panel did not fail, the upper panel of A3 failed, a small part of the honeycomb core unit failed, and the lower panel did not fail. The impact resistance of the layup scheme is A3>A2>A1.

Comparing A2 and A3 from Figs. 12 to 13 and Table 5, it is found that the A3 scheme has smaller deformation and higher energy absorption. Whether the degree of radome deformation or the energy absorption capacity, A3 is better than A2. One can predict that:

(1) Inserting part of the sub-layers with other angles in the pure orthogonal plies can improve the material properties of the honeycomb sandwich structure.

(2) In the case of a total of 5 layers of laminates, the improvement effect of inserting another orthogonal layer of ±45° in the orthogonal layer of 0°/90° is better than the angle of ±30°. It is predicted that it may be better than any other angle ±θ°.

Therefore, when designing the composite honeycomb sandwich structure, the influence of the layup angle should be considered.

5. Conclusions

From the finite element model and simulation results, some properties of the impact resistance of composite honeycomb sandwich structures can be found. The increase in the thickness distribution of the upper panel is a positive gain for the impact resistance of the radome. A thicker upper panel not only has a better energy absorption effect, but also has better resistance to deformation. Allocating more thickness to the upper panel can improve its impact resistance and energy absorption effect. The overall thickness can be reduced, and the distribution thickness of the upper panel can be increased for lightweight design. That will give people a new sight to design lightweight structures, while maintaining the impact resistance.

Inserting some other angled sub-layers in the orthogonal plies can improve the material properties of the honeycomb sandwich structure. In the case of a total of 5 layers of laminates, the effect of inserting another orthogonal layer ±45° into the orthogonal layer 0°/90° is better than ±30°, and it is predicted that it may be better than any other angle ±θ°. Therefore, design of laying angle has important effect on the impact resistance of composite structures.

References

[1] Long S, Mu X, Yuan H, et al. Failure modeling of composite wing leading edge under bird strike[J]. Composite Structures, 2021, 255: 113005.

[2] Qiu J, Wang D, Liu C, et al. Dynamic response of bird strike on honeycomb-based sandwich panels of composite leading edge[J]. International Journal of Crashworthiness, 2020, 9: 1-14.

[3] Giglio M, Giliol A. Investigations on sandwich core properties through an experimental–numerical approach[J]. Composites Part B: Engineering, 2012, 43(2): 361-374.

[4] Tolga T, Yavuz S M. Investigation of low velocity impact behaviors of honeycomb sandwich composites[J]. Journal of Mechanical Science & Technology, 2018, 32(7): 3161-3167.

[5] Akatay A, Mustafa B, Onur O, et al. The influence of low velocity repeated impacts on residual compressive properties of honeycomb sandwich structures[J]. Composite Structures, 2015, 125: 425-433.

[6] Nishikawa M, Takeda H. Finite-element simulation for modeling composite plates subjected to soft-body high-velocity impact for application to bird-strike problem of composite fan blades[J]. Composite Structures, 2010, 93(5): 1416-1423.

[7] Arachchige B, Ghasemnejad H, Yasaee M. Effect of bird-strike on sandwich composite aircraft wing leading edge[J]. Advances in Engineering Software, 2020, 148: 102839.

A Model for the T-shaped Beams with Straight Haunches for Uniformly Distributed Load

Arnulfo Luévanos-Rojas*, Sandra López-Chavarría and Manuel Medina-Elizondo

Institute of Multidisciplinary Research, Autonomous University of Coahuila, México

*Corresponding author email: arnulfoluevanos@uadec.edu.mx

Abstract. This paper shows a model for the T-shaped beams with straight haunches under uniformly distributed load that considers shear and bending deformations to obtain the fixed-end moments, carry-over and stiffness factors, which is the main part of this investigation. The methodology used is the conjugate beam method to obtain the rotations in supports and by the superposition method is solved this problems type. The current model considers only the bending deformations, and some authors consider bending and shear deformations for some proportions that are shown in tables. A comparison among the proposed approach that considers shear and bending deformations against the current model that considers bending deformations only are presented in the tables and graphics. A significant advantage of this paper over any other document is that is not limited to certain dimensions or proportions, and it can also be used for TT-shaped or TTT-shaped beams, which is commonly applied in highway bridges.
Keywords: T-shaped beams; Carry-over factors; Fixed-end moments; Stiffness factors; Straight haunches; Uniformly distributed load; Bending and shear deformations.

1. Introduction

The reinforced concrete beams with haunches at their ends differ from prismatic beams because the height of the beam presents gradual variations in all or part of its length. The main application is in buildings of medium height, and highway and railway bridges.

The main problems in the analysis of structures with moment of inertia variable along of its length are to obtain the fixed-end moments, carry-over and stiffness factors.

In the last century, the PCA (Portland Cement Association) have been published several aids and tables for the design of non-prismatic beams by that presents the stiffness factors and the fixed-end moments [1]. In these tables the bending deformations are considered only and the shear deformations are neglected, and also the length-height relationship of the beam is not considered in definition of the various stiffness factors, simplifications that can lead to significant errors in the determination of the stiffness factors.

There are important contributions in the non-prismatic beams analysis that are based on the theory of Euler-Bernoulli beams as Just [2], Schreyer [3], Medwadowski [4], Brown [5], Banerjee and Williams [6], Eisenberger [7, 8], Tena-Colunga [9], Failla and Santini [10], Failla and Impollonia [11].

The new design aids are found in appendix B of the book of "Analysis of structures with matrix methods" that replace to the old PCA tables, and the tables provided the fixed-end moments, carry-over, and stiffness factors for beams of sections in shape of "I" and "T" [12]. But these are restricted to some relations, and the height of the haunches is the same in the two ends.

The topic of non-prismatic beams in the last 13 years has aroused great interest among researchers, and the main researchers with different contributions are: Yuksel [13, 14], Luévanos-Rojas and Montoya-Ramírez [15], Luévanos-Rojas *et al.* [16], Raju *et al.* [17], Albegmprli *et al.* [18], Luévanos-Rojas [19], Beltempo *et al.* [20], Auricchio *et al.* [21], Luévanos-Rojas *et al.* [22, 23], Balduzzi *et al.* [24], Luévanos-Soto and Luévanos-Rojas [25], Balduzzi *et al.* [26], Balduzzi *et al.* [27], Qissab and Salman [28], Rezaiee-Pajand *et al.* [29], Velázquez-Santillán *et al.* [30], Sandoval-Rivas *et al.* [31].
The main objective of this research is to show a model for the T-shaped beams with straight haunches under a uniformly distributed load considering the shear and bending deformations to find the fixed-end moments, carry-over and stiffness factors. The methodology used is the conjugate beam method to obtain the rotations in supports and by the superposition method is solved this problems type. The current model only considers the bending deformations, and some authors consider bending and shear deformations for some proportions that are shown in tables. A comparison among the proposed approach that considers shear and bending deformations against the current model that considers bending deformations only are presented in the tables and graphics to observe differences.

2. Methodology

2.1. Theoretical Foundations

The conjugate beam method is a fictitious beam of length equal to that of the real beam and whose load is the reduced moment diagram (*M/EI*) applied on the compression side. The conjugated beam is always a beam statically determined. The method of the conjugate beam consists in finding the moment in the real beam and loading it to the conjugate beam. Then, by cutting and isolating one of the parts of the best convenience, it is obtained the shear in the conjugate beam that will be the rotation of the real beam and the moment in the conjugate beam that will be the displacement of the real beam.
This method consists in changing the problem of finding, the slopes and deflections caused in a beam by a system of applied loads. It has the advantage that it does not need to know a zero tangent point beforehand, so that the slope and deflection can be ascertained directly at any point of the elastic.
The conjugate beam method by the Timoshenko theory (shear and bending deformations) is used to obtain the rotations in the support "A" (left support), and in the support "B" (right support) of a simply supported beam.
Now, the sum of the moments is developed in support "A" to obtain the rotation in support "B" and is presented as follows:

$$\sum M_A = 0 \leftrightarrow \theta_B L - \left[\int_0^L \frac{V}{GA}dx + \int_0^L \frac{Mx}{EI_x}dx\right] = 0 \tag{1}$$

$$\theta_B = \frac{1}{L}\int_0^L \frac{V}{GA}dx + \frac{1}{L}\int_0^L \frac{Mx}{EI_x}dx \tag{2}$$

where: E is the modulus of elasticity, I is the moment of inertia, G is the shear modulus, A_s is the shear area, M is the moment, V is the shear force.
Now, the sum of the forces is considered to obtain the rotation in support "A" and is presented below:

$$\sum F = 0 \leftrightarrow \theta_A + \theta_B - \int_0^L \frac{M}{EI_x}dx = 0 \tag{3}$$

$$\theta_A = \int_0^L \frac{M}{EI_x}dx - \theta_B \tag{4}$$

2.2. Properties of the T-shaped Beam

Figure 1 presents in detail the asymmetric T-shaped beam under a uniformly distributed load where the flange width "b_f", flange thickness "t_f", and web thickness "b_w" are constant, and the web depth "d_x" varies with the linear shape at the ends, and the beam in the center has a constant section.
The geometrical properties of the cross section of the beam in general form are:

$$y_g = \frac{b_f t_f^{\,2} + b_w d_x(d_x + 2t_f)}{2(b_f t_f + b_w d_x)} \tag{5}$$

$$I_x = \frac{b_f t_f^3 + 3b_f t_f (2y_g - t_f)^2 + b_w d_x^3 + 3b_w d_x (d_x + 2t_f - 2y_g)^2}{12} \tag{6}$$

$$A_{sx} = b_w (d_x + t_f) \tag{7}$$

where: y_g is the center of gravity of the cross section, I_x is the moment of inertia at a distance "x" in function of "d_x", A_{sx} is the shear area at a distance "x" in function of "d_x".

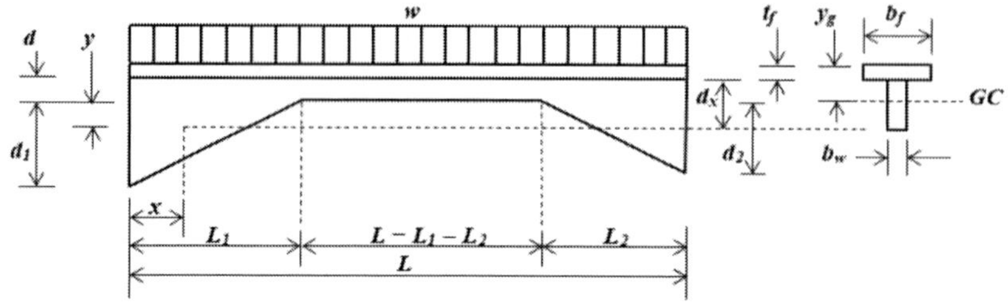

Figure 1. T-shaped beam under a uniformly distributed load.

The geometrical properties for the first cross section of the beam are:

$$d_{x1} = \frac{dL_1 + d_1(L_1 - x)}{L_1} \tag{8}$$

$$I_{x1} = \frac{b_f t_f^3 L_1^3 + b_w [dL_1 - d_1(x - L_1)]^3}{12L_1^3} + \frac{b_w b_f t_f [dL_1 - d_1(x - L_1)][t_f L_1 + dL_1 - d_1(x - L_1)]^2}{4L_1^2 \{b_f t_f L_1 + b_w [dL_1 - d_1(x - L_1)]\}} \tag{9}$$

$$A_{sx1} = \frac{b_w [t_f L_1 + dL_1 - d_1(x - L_1)]}{L_1} \tag{10}$$

The geometrical properties for the second cross section of the beam are:

$$d_{x2} = d \tag{11}$$

$$I_{x2} = \frac{(b_f - b_w)(b_f t_f^4 - b_w d^4) + b_f b_w (d + t_f)^4}{12(b_f t_f + b_w d)} \tag{12}$$

$$A_{sx2} = b_w (d + t_f) \tag{13}$$

The geometrical properties for the third cross section of the beam are:

$$d_{x3} = \frac{dL_3 + d_2(x - L + L_2)}{L_1} \tag{14}$$

$$I_{x3} = \frac{b_f t_f^3 L_2^3 + b_w [dL_2 - d_1(L - L_2 - x)]^3}{12L_1^3} + \frac{b_w b_f t_f [dL_2 - d_1(L - L_2 - x)][t_f L_2 + dL_2 - d_1(L - L_2 - x)]^2}{4L_2^2 \{b_f t_f L_2 + b_w [dL_2 - d_1(L - L_2 - x)]\}} \tag{15}$$

$$A_{sx3} = \frac{b_w [t_f L_1 + dL_1 - d_1(L - L_2 - x)]}{L_1} \tag{16}$$

2.3. Formulation for the Fixed-end Moments

Figure 2(a) shows a beam with fixed ends under a uniformly distributed load. The fixed-end moments are obtained by the principle of superposition (sum of the individual rotations in each support "A" and "B"). The individual rotations are: "β_{A1}" and "β_{B1}" due to the load applied with the simply supported beam, "β_{A2}" and "β_{B2}" due to a moment "M_{AB}" applied in the support "A", and "β_{A3}" and "β_{B3}" due to a moment "M_{BA}" applied in support "B". The moments are positive in counterclockwise direction, and negative in clockwise direction. Figure 2(b) presents the same simply supported beam in its two supports under the same load applied to obtain the rotations "β_{A1}" and "β_{B1}". Figure 2(c) shows the

moment "M_{AB}" applied in the support "A" that generates the rotations "β_{A2}" and "β_{B2}". Figure 2(d) presents the moment "M_{BA}" applied in the support "B" that generates the rotations "β_{A3}" and "β_{B3}".
Now, by the principle of superposition and using the beam compatibility equations for the rotations in each support are obtained the following equations:

$$\beta_{A1} - \beta_{A2} + \beta_{A3} = 0 \tag{17}$$

$$\beta_{B1} - \beta_{B2} + \beta_{B3} = 0 \tag{18}$$

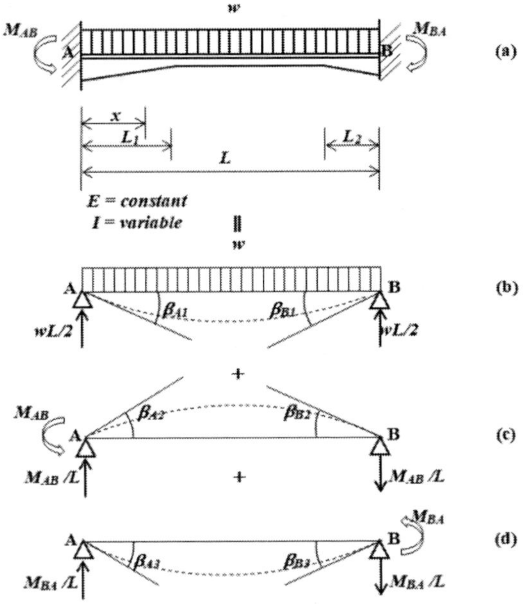

Figure 2. T-shaped beam with fixed supports.

Figure 2 is analyzed to obtain the rotations (β_{A1} and β_{B1}, β_{A2} and β_{B2}, β_{A3} and β_{B3}) and the flexural and shear deformations are taken into account due to the loads applied on beam.
By the equations (2) and (4) the rotations in non-prismatic members (β_{Ai} and β_{Bi}) for general case are obtained in the following way:

$$\beta_{Bi} = \frac{1}{L}\int_0^L \frac{V_x}{GA_{sx}}dx + \frac{1}{L}\int_0^L \frac{M_x x}{EI_x}dx \tag{19}$$

$$\beta_{Ai} = \int_0^L \frac{M_x}{EI_x}dx - \beta_{Bi} \tag{20}$$

where i = 1 (Due to the uniformly distributed load), 2 (Due to the moment "M_{AB}" applied in the support "A"), 3 (Due to the moment "M_{BA}" applied in the support "B"), V_x is the shear force at a distance x, G is the shear modulus, M_x is the moment at a distance x and E is the modulus of elasticity.
The shear forces and moments located in anywhere on the beam according to Figure 2(b) are:

$$V_x = \frac{w(L-2x)}{2} \tag{21}$$

$$M_x = \frac{w(L-x)x}{2} \tag{22}$$

Substituting the equations (21) and (22) into the equations (19) and (20) and the rotations "β_{A1}" and "β_{B1}" are obtained:

$$\begin{aligned}\beta_{A1} = {} & \frac{w}{EL}\left[\int_0^{L_1}\frac{(L-x)^2x}{2I_{x1}}dx + \int_{L_1}^{L-L_2}\frac{(L-x)^2x}{2I_{x2}}dx + \int_{L-L_2}^{L}\frac{(L-x)^2x}{2I_{x3}}dx\right] \\ & - \frac{w}{GL}\left[\int_0^{L_1}\frac{(L-2x)}{2A_{sx1}}dx + \int_{L_1}^{L-L_2}\frac{(L-2x)}{2A_{sx2}}dx + \int_{L-L_2}^{L}\frac{(L-2x)}{2A_{sx3}}dx\right]\end{aligned} \tag{23}$$

$$\beta_{B1} = \frac{w}{EL}\left[\int_0^{L_1}\frac{(L-x)x^2}{2I_{x1}}dx + \int_{L_1}^{L-L_2}\frac{(L-x)x^2}{2I_{x2}}dx + \int_{L-L_2}^{L}\frac{(L-x)x^2}{2I_{x3}}dx\right] + \frac{w}{GL}\left[\int_0^{L_1}\frac{(L-2x)}{2A_{sx1}}dx + \int_{L_1}^{L-L_2}\frac{(L-2x)}{2A_{sx2}}dx + \int_{L-L_2}^{L}\frac{(L-2x)}{2A_{sx3}}dx\right] \quad (24)$$

The shear forces and moments located in anywhere on the beam according to Figure 2(c) are:

$$V_x = \frac{M_{AB}}{L} \quad (25)$$

$$M_x = -\frac{M_{AB}(L-x)}{L} \quad (26)$$

Substituting the equations (25) and (26) into the equations (19) and (20) and the rotations "β_{A2}" and "β_{B2}" are obtained:

$$\beta_{A2} = \frac{M_{AB}}{EL^2}\left[\int_0^{L_1}\frac{(L-x)^2}{I_{x1}}dx + \int_{L_1}^{L-L_2}\frac{(L-x)^2}{I_{x2}}dx + \int_{L-L_2}^{L}\frac{(L-x)^2}{I_{x3}}dx\right] + \frac{M_{AB}}{GL^2}\left[\int_0^{L_1}\frac{1}{A_{sx1}}dx + \int_{L_1}^{L-L_2}\frac{1}{A_{sx2}}dx + \int_{L-L_2}^{L}\frac{1}{A_{sx3}}dx\right] \quad (27)$$

$$\beta_{B2} = \frac{M_{AB}}{EL^2}\left[\int_0^{L_1}\frac{(L-x)x}{I_{x1}}dx + \int_{L_1}^{L-L_2}\frac{(L-x)x}{I_{x2}}dx + \int_{L-L_2}^{L}\frac{(L-x)x}{I_{x3}}dx\right] - \frac{M_{AB}}{GL^2}\left[\int_0^{L_1}\frac{1}{A_{sx1}}dx + \int_{L_1}^{L-L_2}\frac{1}{A_{sx2}}dx + \int_{L-L_2}^{L}\frac{1}{A_{sx3}}dx\right] \quad (28)$$

The shear forces and moments located in anywhere on the beam according to Figure 2(d) are:

$$V_x = \frac{M_{BA}}{L} \quad (29)$$

$$M_x = \frac{M_{BA}x}{L} \quad (30)$$

Substituting the equations (29) and (30) into the equations (19) and (20) and the rotations "β_{A3}" and "β_{B3}" are obtained:

$$\beta_{A3} = \frac{M_{BA}}{EL^2}\left[\int_0^{L_1}\frac{(L-x)x}{I_{x1}}dx + \int_{L_1}^{L-L_2}\frac{(L-x)x}{I_{x2}}dx + \int_{L-L_2}^{L}\frac{(L-x)x}{I_{x3}}dx\right] - \frac{M_{BA}}{GL^2}\left[\int_0^{L_1}\frac{1}{A_{sx1}}dx + \int_{L_1}^{L-L_2}\frac{1}{A_{sx2}}dx + \int_{L-L_2}^{L}\frac{1}{A_{sx3}}dx\right] \quad (31)$$

$$\beta_{B3} = \frac{M_{BA}}{EL^2}\left[\int_0^{L_1}\frac{x^2}{I_{x1}}dx + \int_{L_1}^{L-L_2}\frac{x^2}{I_{x2}}dx + \int_{L-L_2}^{L}\frac{x^2}{I_{x3}}dx\right] + \frac{M_{BA}}{GL^2}\left[\int_0^{L_1}\frac{1}{A_{sx1}}dx + \int_{L_1}^{L-L_2}\frac{1}{A_{sx2}}dx + \int_{L-L_2}^{L}\frac{1}{A_{sx3}}dx\right] \quad (32)$$

Now, the equations (23), (27) and (31) that correspond to support "A" are substituted into equation (17), and the equations (24), (28) and (32) that correspond to support "B" are substituted into equation (18). Subsequently, the two generated equations are solved to obtain the values of "M_{AB}" and "M_{BA}". The equations for moments are presented as follows:

$$M_{AB} = \frac{wL^2}{m_{AB}} \quad (33)$$

$$M_{BA} = \frac{wL^2}{m_{BA}} \quad (34)$$

where m_{AB} and m_{BA} are the fixed-end moment factors.

2.4. Formulation for the Carry-over and Stiffness Factors

The principle of superposition and the beam compatibility equations are used to obtain the carry-over and stiffness factors. The procedure is the following: when the moment "M_{AB}" is applied clockwise at support "A" of a simply supported straight member of non-constant cross-section at one end (support

"A") and fixed at the other end (support "B") to obtain the rotation "β_{A1}" in the support "A" and the fixed-end moment "M_{BA}" in the support "B" (see Figure 3(a)). Rotations "β_{A1}" and "$\beta_{B1} = 0$" in the supports are caused by the moments "M_{AB}" and "M_{BA}", respectively. Figure 3(b) shows that the rotations "β_{A2}" and "β_{B2}" are caused by "M_{AB}", and Figure 3(c) presents the rotations "β_{A3}" and "β_{B3}" caused by "M_{BA}".

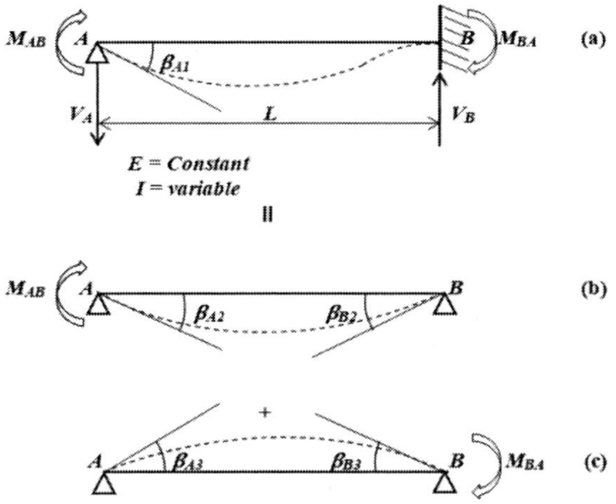

Figure 3. Simply supported beam at one end and fixed at the other.

Now, by the principle of superposition and using the beam compatibility equations for the rotations in each support are obtained the following equations:

$$\beta_{A1} = \beta_{A2} - \beta_{A3} \quad (35)$$

$$0 = \beta_{B2} - \beta_{B3} \quad (36)$$

The beam is analyzed to obtain "β_{A2}" and "β_{B2}" in the function of "M_{AB}" (see Figure 3(b)); these appear in equations (27) and (28). The beam is analyzed to obtain "β_{A3}" and "β_{B3}" in the function of "M_{BA}" (see Figure 3(c)); these appear in equations (31) and (32).

Now, substituting equations (28) and (32) into equation (36), and "M_{BA}" is obtained in the function of "M_{AB}". Therefore, the carry-over factor of "A" to "B" (C_{AB}) that is the percentage of the moment produced in the support "B" due to a moment applied in the support "A"; the procedure to find the carry-over factor of "B" to "A" (C_{BA}) is the same used above. Therefore, equations are shown as follows:

$$M_{BA} = C_{AB}M_{AB} \quad (37)$$

$$M_{AB} = C_{BA}M_{BA} \quad (38)$$

where: "C_{AB}" and "C_{BA}" are the carry-over factors.

Now, substituting equations (27) and (31) into equation (35) is presented in the function of "M_{AB}", "M_{BA}" and "β_{A1}", and subsequently equation (37) is substituted in this same equation to find "M_{AB}" in function of "β_{A1}". Therefore, the stiffness "K_{AB}" is the moment applied in the support "A" to produce a rotation of 1 radian in the support "A"; the procedure to find the stiffness "K_{BA}" is the same used above. Therefore, equations are presented as follows:

$$M_{AB} = K_{AB}\beta_{A1} = \frac{k_{AB}EI_{x2}}{L}\beta_{A1} \quad (39)$$

$$M_{BA} = K_{BA}\beta_{B1} = \frac{k_{BA}EI_{x2}}{L}\beta_{B1} \quad (40)$$

where: "k_{AB}" and "k_{BA}" are the stiffness factors.

3. Results

Figures 4, 5, 6, 7, 8 and 9 show the comparison of the proposed model against the current model. The proposed model (PM) considers the shear and bending deformations, and the current model (CM) only takes into account the bending deformations without considering the shear deformations.

Figure 4 shows the fixed-end moment factors (m_{AB} and m_{BA}) for d = 0.1L of the two models. Figure 5 presents the fixed-end moment factors (m_{AB} and m_{BA}) for d = 0.2L of both models. Figure 6 shows the carry-over factors (C_{AB} and C_{BA}) for d = 0.1L of the two models. Figure 7 presents the carry-over factors (C_{AB} and C_{BA}) for d = 0.2L of both models. Figure 8 shows the stiffness factors (k_{AB} and k_{BA}) for d = 0.1L of the two models. Figure 9 presents the stiffness factors (k_{AB} and k_{BA}) for d = 0.2L of both models.

Figures 4, 6 and 8 show the fixed-end moments factors, carry-over factors, and stiffness factors of the members between the two models for L = 10d → d = 0.10L, $b_f = b_w + 16t_f = 0.475L$, $b_w = 0.75d = 0.075L$, $t_f = 0.25d = 0.025L$, $d_1 = d_2$, $\nu = 0.20$ of concrete. Figures 5, 7 and 9 show the fixed-end moments factors, carry-over factors, and stiffness factors of the members between the two models for L = 5d → d = 0.20L, $b_f = b_w + 16t_f = 0.95L$, $b_w = 0.75d = 0.15L$, $t_f = 0.25d = 0.05L$, $d_1 = d_2$, $\nu = 0.20$ of concrete. The values shown in Figures 4, 6 and 8 of the proposed model are identical to those in the tables presented in appendix B by Tena-Colunga [12].

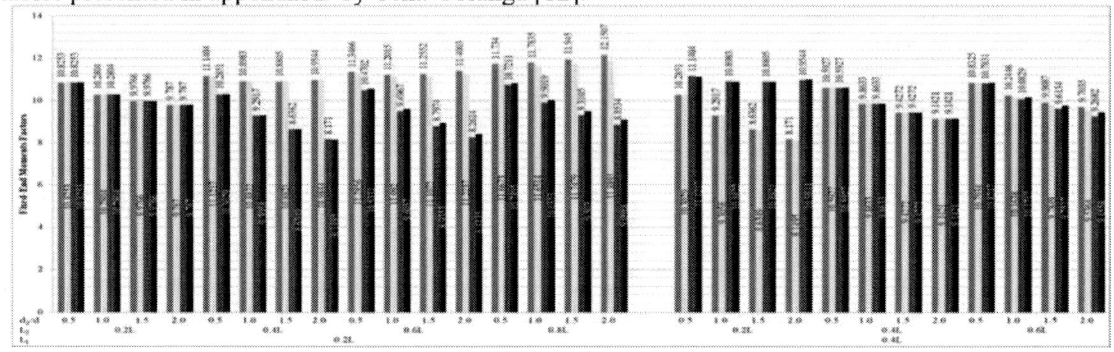

PM-m_{AB} CM-m_{AB} PM-m_{BA} CM-m_{BA}

Figure 4. Fixed-End Moments Factors (m_{AB} and m_{BA}) for d = 0.1L.

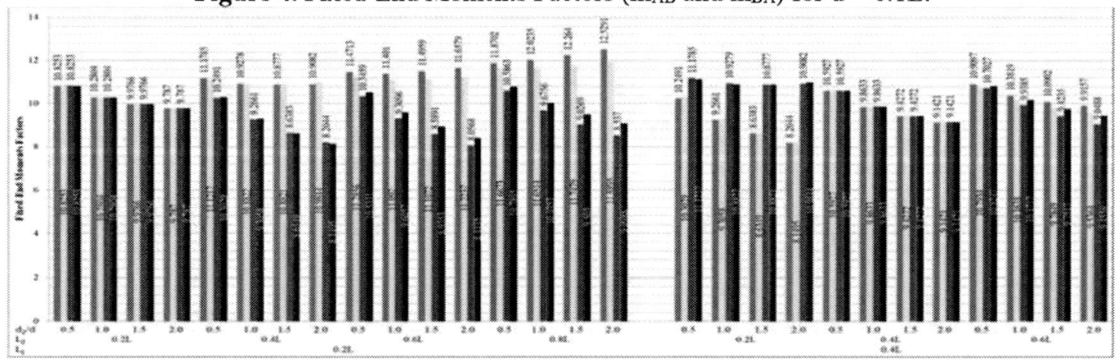

PM-m_{AB} CM-m_{AB} PM-m_{BA} CM-m_{BA}

Figure 5. Fixed-End Moments Factors (m_{AB} and m_{BA}) for d = 0.2L.

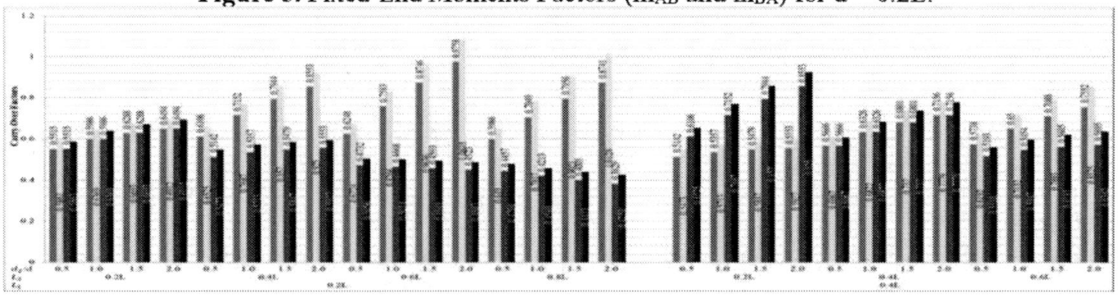

PM-C_{AB} CM-C_{AB} PM-C_{BA} CM-C_{BA}

Figure 6. Carry-Over Factors (C_{AB} and C_{BA}) for d = 0.1L.

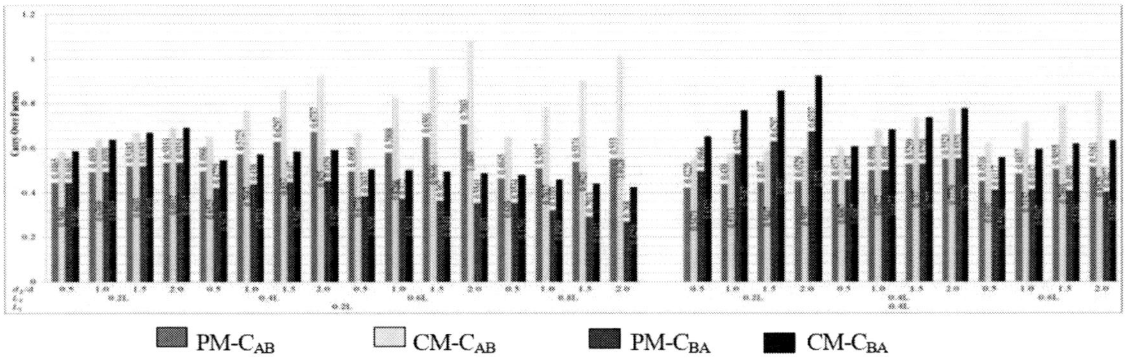

Figure 7. Carry-Over Factors (C_{AB} and C_{BA}) for d = 0.2L.

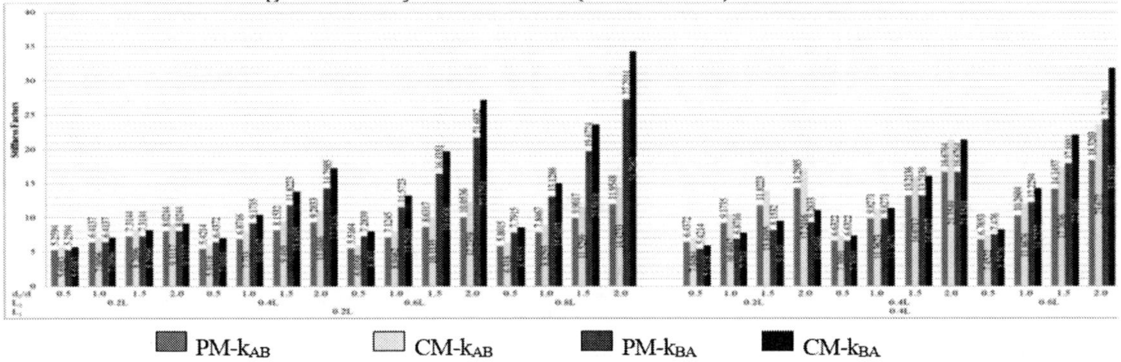

Figure 8. Stiffness Factors (k_{AB} and k_{BA}) for d = 0.1L.

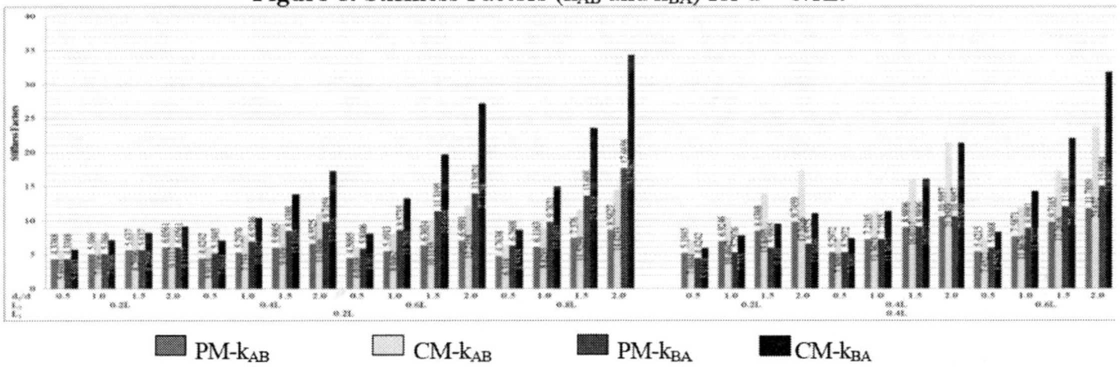

Figure 9. Stiffness Factors (k_{AB} and k_{BA}) for d = 0.2L.

Table 1 shows the ratio of CM/PM for the factors of the mechanical elements for d = 0.1L, and Table 2 shows the ratio of CM/PM for the factors of the mechanical elements for d = 0.2L.

Another way to verify the proposed model for constant cross sections is as follows:

1.- If "$L_1 = L_2 = 0L$" or "$d_1 = d_2 = 0$" is substituted into equations (33) and (34) and also shear deformations are neglected, the fixed-end moments factors are obtained "$m_{AB} = m_{BA} = 12$. Thus, fixed-end moments are "$M_{AB} = M_{BA} = wL^2/12$".

2.- If "$L_1 = L_2 = 0L$" or "$d_1 = d_2 = 0$" is substituted into equations (37) and (38) and also shear deformations are neglected, the carry-over factors are obtained "$C_{AB} = C_{BA} = 0.5$".

3.- If "$L_1 = L_2 = 0L$" or "$d_1 = d_2 = 0$" is substituted into equations (39) and (40) and also shear deformations are neglected, the stiffness factors are obtained "$k_{AB} = k_{BA} = 4$". Therefore, the stiffness are obtained "$K_{AB} = K_{BA} = 4EI_{x2}/L$".

Thus, the mechanical elements are verified for constant or uniform cross sections.

Table 1. Ratio of CM/PM for the factors of mechanical elements for d = 0.1L.

L_1	L_2	d_2/d	m_{AB}	m_{BA}	C_{AB}	C_{BA}	k_{AB}	k_{BA}
0.2L	0.2L	0.5	1.0000	1.0000	1.0627	1.0627	1.0835	1.0835
		1.0	1.0000	1.0000	1.0640	1.0640	1.1054	1.1054
		1.5	1.0000	1.0000	1.0644	1.0644	1.1224	1.1224
		2.0	1.0000	1.0000	1.0644	1.0644	1.1355	1.1355
	0.4L	0.5	0.9983	1.0017	1.0686	1.0644	1.0905	1.0949
		1.0	0.9985	1.0015	1.0748	1.0665	1.1254	1.1342
		1.5	1.0001	0.9998	1.0788	1.0672	1.1099	1.1715
		2.0	1.0027	0.9974	1.0815	1.0674	1.1902	1.2058
	0.6L	0.5	0.9946	1.0060	1.0762	1.0685	1.0907	1.0985
		1.0	0.9902	1.0115	1.0901	1.0741	1.1294	1.1463
		1.5	0.9869	1.0157	1.1018	1.0775	1.1717	1.1981
		2.0	0.9845	1.0184	1.1114	1.0796	1.2171	1.2530
	0.8L	0.5	0.9943	1.0064	1.0842	1.0756	1.0890	1.0978
		1.0	0.9888	1.0135	1.1090	1.0892	1.1252	1.1456
		1.5	0.9830	1.0207	1.1339	1.1014	1.1644	1.1987
		2.0	0.9785	1.0277	1.1587	1.1123	1.2065	1.2567
0.4L	0.2L	0.5	1.0017	0.9983	1.0644	1.0686	1.0949	1.0905
		1.0	1.0015	0.9985	1.0665	1.0748	1.1342	1.1254
		1.5	0.9998	1.0001	1.0672	1.0788	1.1715	1.1099
		2.0	0.9974	1.0027	1.0674	1.0815	1.2058	1.1902
	0.4L	0.5	1.0000	1.0000	1.0708	1.0708	1.1021	1.1021
		1.0	1.0000	1.0000	1.0784	1.0784	1.1562	1.1562
		1.5	1.0000	1.0000	1.0837	1.0837	1.2163	1.2163
		2.0	1.0000	1.0000	1.0871	1.0871	1.2805	1.2805
	0.6L	0.5	0.9964	1.0039	1.0791	1.0761	1.1015	1.1045
		1.0	0.9921	1.0088	1.0972	1.0904	1.1566	1.1638
		1.5	0.9875	1.0144	1.1136	1.1028	1.2204	1.2324
		2.0	0.9828	1.0202	1.1288	1.1136	1.2921	1.3097

Tables 1 and 2 show that when the beam is symmetric, the fixed-end moments are the same in the current model and the proposed model, i.e., the shear deformations not affect to the fixed-end moments for symmetrical beams.

Regarding the fixed-end moment factors, if the moment is greater in one support than the opposite end, the current model is greater in one support, and the proposed model is greater in the opposite end, with the difference being up to 2.77% in Table 1 (m_{BA}: $L_1 = 0.2L$, $L_2 = 0.8L$, $d_2 = 2d$) and 6.58% in Table 2 (m_{BA}: $L_1 = 0.2L$, $L_2 = 0.8L$, $d_2 = 2d$). Regarding the carry-over factors, the current model is greater in all cases, and the difference reaches up to 15.87% in Table 1 (C_{AB}: $L_1 = 0.2L$, $L_2 = 0.8L$, $d_2 = 2d$) and 83.15% in Table 2 (C_{AB}: $L_1 = 0.2L$, $L_2 = 0.8L$, $d_2 = 2d$). Regarding the stiffness factors, also the current model is greater in all cases, with a difference of 30.97% in Table 1 (k_{BA}: $L_1 = 0.4L$, $L_2 = 0.6L$, $d_2 = 2d$) and 110.73% in Table 2 (k_{BA}: $L_1 = 0.4L$, $L_2 = 0.6L$, $d_2 = 2d$).

Therefore, the results show that the major differences are presented in Table 2 with the parameter d = 0.20L.

Table 2. Ratio of CM/PM for the factors of mechanical elements for d = 0.2L.

L_1	L_2	d_2/d	m_{AB}	m_{BA}	C_{AB}	C_{BA}	k_{AB}	k_{BA}
0.2L	0.2L	0.5	1.0000	1.0000	1.3127	1.3127	1.3133	1.3133
		1.0	1.0000	1.0000	1.2895	1.2895	1.3939	1.3939
		1.5	1.0000	1.0000	1.2913	1.2913	1.4564	1.4564
		2.0	1.0000	1.0000	1.2910	1.2910	1.5046	1.5046
	0.4L	0.5	0.9949	1.0052	1.3139	1.2942	1.3362	1.3569
		1.0	0.9958	1.1122	1.3427	1.3043	1.4597	1.5025
		1.5	1.0004	0.9984	1.3610	1.3081	1.5773	1.6412
		2.0	1.0070	0.9933	1.3730	1.3087	1.6862	1.7691
	0.6L	0.5	0.9838	1.0181	1.3532	1.3177	1.3342	1.3699
		1.0	0.9729	1.0318	1.4232	1.3478	1.4653	1.5472
		1.5	0.9662	1.0403	1.4822	1.3671	1.6045	1.7395
		2.0	0.9628	1.0443	1.5311	1.3790	1.7507	1.9435
	0.8L	0.5	0.9829	1.0192	1.3972	1.3565	1.3263	1.3662
		1.0	0.9691	1.0372	1.5336	1.4372	1.4436	1.5407
		1.5	0.9579	1.0527	1.6786	1.5142	1.5632	1.7330
		2.0	0.9490	1.0658	1.8315	1.5892	1.6844	1.9414
0.4L	0.2L	0.5	1.0052	0.9949	1.2942	1.3139	1.3569	1.3362
		1.0	1.0043	0.9958	1.3043	1.3427	1.5025	1.4597
		1.5	0.9996	1.0004	1.3081	1.3610	1.6412	1.5773
		2.0	0.9933	1.0070	1.3087	1.3730	1.7691	1.6862
	0.4L	0.5	1.0000	1.0000	1.3264	1.3264	1.3798	1.3798
		1.0	1.0000	1.0000	1.3649	1.3649	1.5741	1.5741
		1.5	1.0000	1.0000	1.3908	1.3908	1.7878	1.7878
		2.0	1.0000	1.0000	1.4085	1.4085	2.0154	2.0154
	0.6L	0.5	0.9893	1.0114	1.3711	1.3573	1.3746	1.3886
		1.0	0.9790	1.0235	1.4684	1.4358	1.5642	1.5995
		1.5	0.9697	1.0348	1.5614	1.5079	1.7767	1.8401
		2.0	0.9617	1.0450	1.6518	1.5749	2.0094	2.1073

4. Conclusions

The main findings of this investigation are the following:

1) For the fixed-end moment factors: When the fixed-end moment factors are greater in one support than the opposite end, the current model is greater in one support, and the proposed model is greater in the opposite end. When the straight haunches of beams are symmetrically, the fixed-end moment factors are the same for the current model and the proposed model, and the fixed-end moment factors are influenced by the web depth in straight haunches; i.e., if the straight haunches volume is increased, the factor is greater in the same support.

2) For the carry-over factors: The proposed model is smaller compared to the current model in all cases, because the lower stiffness is presented in the proposed model and shear deformations are considered.

3) For the stiffness factors: The proposed model is smaller than the current model in all cases, because the beams have higher rotations and displacements, and the stiffness is obtained from the rotations and displacements, and the proposed model considers the bending and shear deformations.

4) If the parameter "d/L" is higher, the differences are greater in the current model compared to the proposed model for the fixed-end moments factors, carry-over factors, and stiffness factors, because the factors for the proposed model are reduced.

5) If the parameter "d/L" for the current model is modified, the fixed-end moments factors, carry-over factors, and stiffness factors do not change their values.

The benefits of non-prismatic beams are as follows:
1) The T-shaped beams with straight haunches have been used in various structures including buildings and bridges.
2) With the beams being tapered, the architects would be able to create and implement novel aesthetic architectural designs.
3) Structural engineers could seek for the optimum low weight-high strength systems through a redistribution of materials and beams shape.
The fixed-end moments, the carry-over factors, and the stiffness factors of a member will be helpful for the matrix methods of structural analysis (see application example) and also for the method of moment distribution (Hardy Cross method).
The main advantages of the proposed model in this paper is not limited to certain dimensions or proportions, as tables presented by Tena-Colunga [12] for the T-shaped beams consider the flange width "b_f" is equal to the web thickness "b_w" more 16 times flange thickness "t_f", the relationship between the beam length and the effective depth is equal to 10, the relationship between the flange thickness "t_f" and the effective depth is equal to 1/4, the relationship between the web thickness "b_w" and the effective depth is equal to 3/4, and the height of the straight haunches is the same in the two ends.
Another advantage of this model presented in this document is that it can be used for TT-shaped or TTT-shaped beams, which is commonly applied in highway bridges.
The proposed model is more suitable for structural engineering analysis and is more adapted to real conditions, since bending moments and shear forces act in any type of structures, and therefore, the shear and bending deformations are present.
Therefore, the proposed model is valid and is not restricted to certain dimensions or proportions as shown in some books, and also the shear and bending deformations are considered.
Suggestions for future research may be as follows:
1) Mathematical model for another type of non-constant cross section.
2) Mathematical model for other types of haunches as may be parabolic haunches.
3) Mathematical model for another type of load such as concentrated load located anywhere of the beam.

References

[1] Portland Cement Associations (PCA) (1958). *Beam Factor and Moment Coefficients for Members of Variable Section*, Handbook of Frame Constants, Chicago, IL, USA.
[2] Just, D.J. (1977). Plane frameworks of tapering box and I-section. *J. Struct. Eng.-ASCE*, 103(1), pp 71-86.
[3] Schreyer, H.L. (1978). Elementary theory for linearly tapered beams. *J. Struct. Eng.-ASCE*, 104(3), pp 515-527.
[4] Medwadowski, S.J. (1984). Nonprismatic shear beams. *J. Struct. Eng.-ASCE*, 110(5), pp 1067-1082.
[5] Brown, C.J. (1984). Approximate stiffness matrix for tapered beams. *J. Struct. Eng.-ASCE*, 110(12), pp 3050-3055.
[6] Banerjee, J.R. and Williams, F.W. (1986). Exact Bernoulli-Euler static stiffness matrix for a range of tapered beam-columns. *Int. J. Numer. Meth. Eng.*, 23, pp 1615-1628.
[7] Eisenberger, M. (1991). Stiffness matrices for non-prismatic members including transverse shear. *Comput. Struct.*, 40(4), pp 831-835.
[8] Eisenberger, M. (1991). Exact solution for general variable cross-section members. *Comput. Struct.*, 41(4), pp 765-772.
[9] Tena-Colunga, A. (1996). Stiffness formulation for nonprismatic beam elements. *J. Struct. Eng.-ASCE*, 122(12), pp 1484-1489.
[10] Failla, G. and Santini, A. (2007). On Euler-Bernoulli discontinuous beam solutions via uniform-beam Green's functions. *Int. J. Solids Struct.*, 44(22-23), pp 7666-7687.
[11] Failla, G. and Impollonia, N. (2012). General finite element description for non-uniform and discontinuous beam elements. *Arch. Appl. Mech.*, 82, pp 43-67.

[12] Tena-Colunga, A. (2007). *Análisis de estructuras con métodos matriciales*. Limusa, México.
[13] Yuksel, S.B. (2009). Behaviour of symmetrically haunched non-prismatic members subjected to temperature changes. *Struct. Eng. Mech.*, 31(3), pp 297-314.
[14] Yuksel, S.B. (2012). Assessment of non-prismatic beams having symmetrical parabolic haunches with constant haunch length ratio of 0.5. *Struct. Eng. Mech.*, 42(6), pp 849-866.
[15] Luévanos-Rojas, A. and Montoya-Ramírez, J. (2014). Mathematical model for rectangular beams of variable cross section of symmetrical linear shape for uniformly distributed load. *Int. J. Innov. Comput. I.*, 10(2), pp 545-564.
[16] Luévanos-Rojas, A., Luévanos-Rojas, R., Luévanos-Soto, I., Luévanos-Vazquez, R.G. and Ramírez-Luévanos, O.A. (2014). Mathematical model for rectangular beams of variable cross section of symmetrical linear shape for concentrated load. *Int. J. Innov. Comput. I.*, 10(3), pp 851-881.
[17] Raju, M., Rajsekhar, K. and Raghuram, T. (2014). Performance of non-prismatic simply supported prestressed concrete beams. *Struct. Eng. Mech.*, 52(4), pp 723-738.
[18] Albegmprli, H.M., Çevik, A., Gülşan, M.E. and Kurtoglu, A.E. (2015). Reliability analysis of reinforced concrete haunched beams shear capacity based on stochastic nonlinear FE analysis. *Comput. Concrete*, 15(2), pp 259-277.
[19] Luévanos-Rojas, A. (2015). Modelado para vigas de sección transversal "I" sometidas a una carga uniformemente distribuida con cartelas rectas. *Ingeniería Mecánica Tecnología y Desarrollo*, 5(2), pp 281-292.
[20] Beltempo, A., Balduzzi, G., Alfano, G. and Auricchio, F. (2015). Analytical derivation of a general 2D non-prismatic beam model based on the Hellinger-Reissner principle. *Eng. Struct.*, 101, pp 88-98.
[21] Auricchio, F., Balduzzi, G. and Lovadina, C. (2015). The dimensional reduction approach for 2D non-prismatic beam modelling: a solution based on Hellinger-Reissner principle. *Int. J. Solids Struct.*, 63, pp 264-276.
[22] Luévanos-Rojas, A., López-Chavarría, S. and Medina-Elizondo, M. (2016). Modeling for mechanical elements of rectangular members with straight haunches using software: part 1. *Int. J. Innov. Comput. I.*, 12(3), pp 973-985.
[23] Luévanos-Rojas, A., López-Chavarría, S. and Medina-Elizondo, M. (2016). Modeling for mechanical elements of rectangular members with straight haunches using software: part 2. *Int. J. Innov. Comput. I.*, 12(4), pp 1027-1041.
[24] Balduzzi, G., Aminbaghai, M., Sacco, E., Füssl, J. and Eberhardsteiner, J. (2016). Non-prismatic beams: A simple and effective Timoshenko-like model. *Int. J. Solids Struct.*, 90, pp 236-250.
[25] Luévanos-Soto, I. and Luévanos-Rojas, A. (2017). Modeling for fixed-end moments of I-sections with straight haunches under concentrated load. *Steel Compos. Struct.*, 23(5), pp 597-610.
[26] Balduzzi, G., Morganti, S., Auricchio, F. and Reali, A. (2017). Non-prismatic Timoshenko-like beam model: Numerical solution via isogeometric collocation. *Comput. Math. Appl.*, 74(7), pp 1531-1541.
[27] Balduzzi, G., Aminbaghai, M., Auricchio, F. and Füssl, J. (2018). Planar Timoshenko-like model for multilayer non-prismatic beams. *Int. J. Mech. Mater. Des.*, 14(1), pp 51-70.
[28] Qissab, M.A. and Salman, M.M. (2018). Shear strength of non-prismatic steel fiber reinforced concrete beams without stirrups. *Struct. Eng. Mech.*, 67(4), pp 347-358.
[29] Rezaiee-Pajand, M., Masoodi, A.R. and Mokhtari, M. (2018). Static analysis of functionally graded non-prismatic sandwich beams. *Adv. Comput. Des.*, 3(2), pp 165-190.
[30] Velázquez Santillán, F., Luévanos-Rojas, A., López-Chavarría, S. and Medina-Elizondo, M. (2019). Modeling for Beams of Rectangular Cross Section with Parabolic Haunches: Part 1. *Computación y Sistemas*, 23(2), pp 547-556.
[31] Sandoval-Rivas, R., Luévanos-Rojas, A., López-Chavarría, S. and Medina-Elizondo, M. (2019). Modeling for Beams of Rectangular Cross Section with Parabolic Haunches: Part 2. *Computación y Sistemas*, 23(3), pp 1115-1124.

Analytical Modeling for Magnetic Field Calculation of SPMSM Based on Equivalent Magnetic Network Method

Jin Wang, Jialin Zhang, Zhanyang Yu, Jianmin Du and Yan Li*

National Engineering Research Center for Rare Earth Permanent Magnet Machines, Shenyang University of Technology, China.

*Corresponding author email: eeliyan@126.com

Abstract. In order to quickly calculate electromagnetic parameters and reduce total harmonics distortions (THD) of air-gap flux density for surface-mounted permanent magnet synchronous motor (SPMSM), the equivalent magnetic network method (EMNM) is used. Compared with traditional magnetic circuit method, the calculation result is more accurate and the calculation speed is faster than finite element method FEM). The EMNM considering the stator groove shape and core saturation is established by piecewide fitting of annular equivalence and magnetic ring curve formulas of the groove fan. The accuracy of EMNM is verified by FEM.
Keywords: Analytical modeling; Equivalent magnetic network; Air-gap flux density; Surface-mounted permanent magnet synchronous motor (SPMSM).

1. Introduction

In recent years, SPMSM is widely used in various fields, because of its advantages of high torque density, high power density, high power factor and high efficiency [1]. Therefore, electromagnetic analysis of surface-mounted permanent magnet synchronous motor (SPMSM) quickly and accurately becomes the key of motor design. At present, the analytical calculation of motor magnetic circuit is mainly divided into traditional magnetic circuit method, finite element method (FEM) and equivalent magnetic network method (EMNM). Based on ampere law, the traditional magnetic circuit method equates the air-gap to a magnetoresistance, and the stator and rotor to a parallel model of magnetoresistance and magnetomotive force respectively. This method has simple structure and fast operation speed, but its calculation accuracy is low due to excessive use of empirical formulas and conversion coefficients, so it can only be applied to general engineering requirements [2].

The FEM mainly uses software to solve nonlinear problems such as silicon steel sheet. Compared with traditional magnetic circuit method, its calculation accuracy is higher and its application scope is wider. However, the modeling process is complex and the calculation time is long, so it is not suitable for analytical calculation of a large number of schemes.

According to the flux-tube principle, the EMNM divides the motor into a large number of unit grids, and different medium correspond to different magnetic permeance or magnetic potential. Then, the unit grid is formed into an overall magnetic network model by matrix operation, and the magnetic potential of each node is solved by the node voltage method[3]. The calculation accuracy of EMNM is higher than traditional magnetic circuit method, and the calculation time is much less than FEM, which is very suitable for preliminary motor design [4].

Based on EMNM, an analytical model of external SPMSM is carried out. By equivalent treatment of stator slot structure, the workload of motor splitting is reduced, and the equivalent magnetic permeance and equivalent magnetic potential of different motor structures are calculated. The nonlinearity influence of silicon steel sheet is considered. According to motor motion equation, the EMNM is established, and the distribution of radial air-gap flux density on motor circumference at different times could be calculated. Finally, the accuracy of proposed model is verified by FEM.

2. Princiole of Equivalent Magnetic Network Method

2.1. The Resemblance of a Magnetic Circuit to An Electric Circuit

Table 1. Comparison of physical quantities corresponding to magnetic field and electric field.

parameter	magnetic field	electric field
dielectric constant	magnetic conductivity μ	electrical conductivity σ
density	magnetic flux density B	electric current density J
intensity	magnetic field intensity H	electric field intensity E
impedance	$R_m = \int_0^l \frac{dx}{\mu(x)\cdot S(x)}$	$R_c = \int_0^l \frac{dx}{\sigma(x)\cdot S(x)}$
pressure drop	$F = \int_0^l H dl$	$V = \int_0^l E dl$
flux	$\Phi = \int B dS = \frac{F}{R_m}$	$I = \int J dS = \frac{U}{R_c}$
stored energy	$\int F d\Phi$	$\int I dt$

The calculation method can be used for reference in the calculation of magnetic circuit. The correspondence of common physical quantities in Table 1 can be obtained through the similarity between magnetic circuit and circuit.

2.2. Calculation of Equivalent Magnetic Permeance

Material properties of different structures should be considered to calculate the equivalent magnetic conductivity. For example, the magnetic permeance of air and PM is linear, while the magnetic permeance of silicon steel sheet is nonlinear. The value of linear permeability is not affected by magnetic field, and the value of nonlinear permeability changes with the intensity of magnetic field.

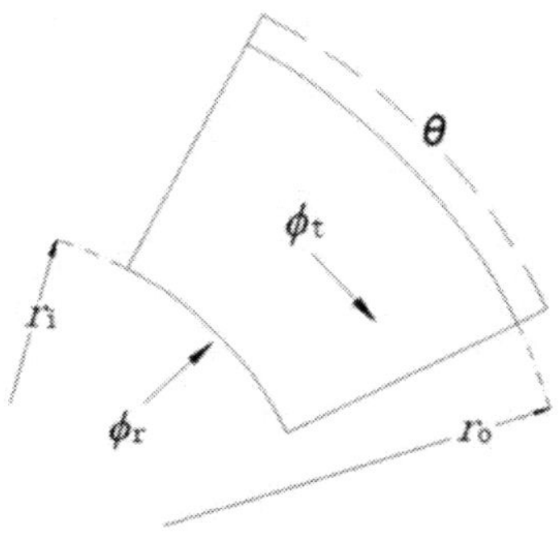

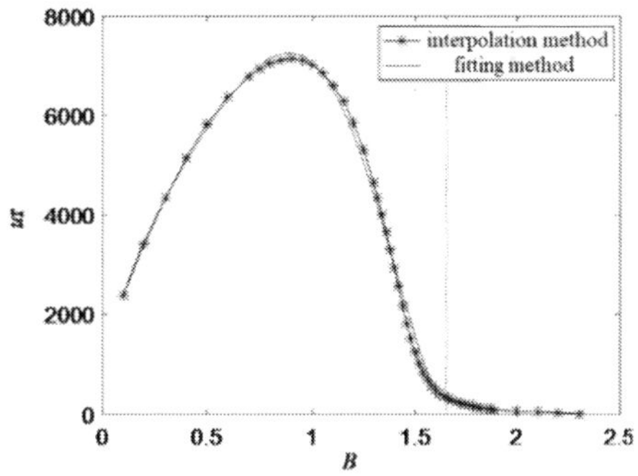

Figure 1. (a) Fan annular unit magnetic permeability of the SPMSM. (b) Curves of relative conductivity and magnetic density.

As shown in Figure 1(a), θ is the angle of fan annular medium, G_r is the radial magnetic permeance generated by radial magnetic flux ϕ_r, and G_t is the tangential magnetic permeance generated by radial magnetic flux ϕ_t. ϕ_r is the tangential permeance generated by the tangential magnetic flux ϕ_t [5]. The radial magnetic permeance G_r and the tangential magnetic permeance G_t can be obtained as follows.

$$G_r = \frac{\mu l \theta}{\ln \frac{r_o}{r_i}} \tag{1}$$

$$G_t = \mu \frac{\ln \frac{r_o}{r_i}}{\theta} l \tag{2}$$

where, r_i and r_o are the inner and outer radius of fan annular medium respectively. For different media, μ has different values. When the medium is air, the permeability is $\mu_0=4\pi\times10^{-7}$T·m/A. When the medium is silicon steel, the *B-H* curve is needed to obtain the corresponding μ_r under different flux density, which is shown in Figure 1(b).

2.3. Calculation of Equivalent Magnetic Potential of PM

In no-load operation of SPMSM, its current can be regarded as 0A, and only PMs in the magnetic field provide magnetic potential for the motor. According to the classical electromagnetic field theory [6], the equivalent magnetic potential of PM *F* can be calculated by (3).

$$F = hH_c \tag{3}$$

$$\mu = \frac{B_r}{H_c} \tag{4}$$

where, *h* is magnetization length of PM, H_c is coercivity of PM, B_r is remanence induction intensity of PM.

Since only the midpoint of the PM is radial magnetizing, and the PMs on both sides have a certain offset angle with the magnetization direction, the equivalent magnetic potential of the fan-annular PM is shown in Figure 2(a). And the radial and tangential magnetic potentials of the unit magnet can be caculated by (5) and (6).

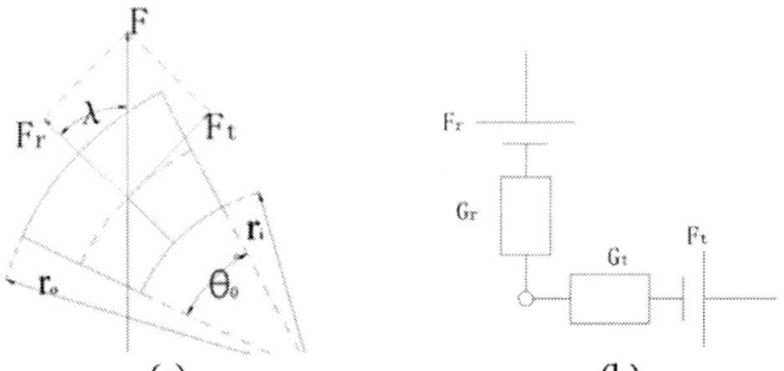

Figure 2. (a) Magnetic potential vector decomposition diagram of unit PM. (b) Magnetic network model of single node.

$$F_{r1} = hH_c \cos\lambda = (r_o - r_i) H_c \cos\lambda \tag{5}$$

$$F_{t1} = hH_c \sin\lambda = (r_o + r_i) H_c \sin\lambda \tag{6}$$

where, *F* is the synthetic magnetic potential of PM, λ is the included angle between the magnetization direction of the PM and the normal direction of the unit PM.

For the whole unit fan annular PM, the magnetic network model of a single node can be constructed according to its equivalent magnetic permeance and magnetic potential, as shown in Figure 2(b). And each node is connected with two branches, one of which is composed of radial magnetic conductivity and radial magnetic potential, and the other is composed of tangential magnetic conductivity and tangential magnetic.

3. Equivalent Magnetic Network of SPMSM

3.1. The Equivalent of Slot

This article takes a 24slots 20 poles SPMSM as an example. In order to facilitate the EMNM and ensure the impact of equivalent slot shape is minimized, the stator slot shape should be equivalent to a three-layer fan annular, and the equivalent structure is shown in Figure 3. And the equivalent parameters are shown in the following formulas.

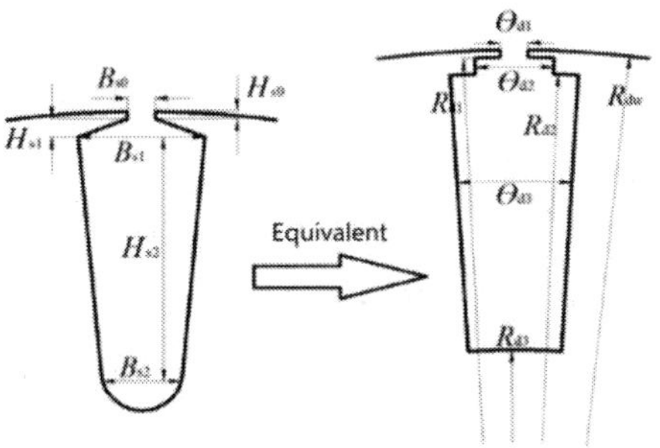

Figure 3. Equivalent structure of SPMSM stator slot.

$$R_{d1} = R_{dw} - H_{s0} \tag{7}$$

$$R_{d2} = R_{d1} - H_{s1} \tag{8}$$

$$R_{d3} = R_{d2} - H_{s2} - \frac{B_{s2}}{2} \tag{9}$$

$$\theta_{d1} = \frac{360 \cdot B_{s0}}{\pi \left(R_{dw} + R_{d1} \right)} \tag{10}$$

$$\theta_{d2} = \frac{180 \cdot \left(B_{s0} + B_{s1} \right)}{\pi \left(R_{d1} + R_{d2} \right)} \tag{11}$$

$$\theta_{d3} = \frac{180 \cdot \left(B_{s1} + B_{s2} \right)}{\pi \left(R_{d2} + R_{d3} \right)} \tag{12}$$

where, B_{s0}, B_{s1}, B_{s2}, H_{s0}, H_{s1} and H_{s2} are the structural parameters of slot respectively, R_{dw} is the outer radius of stator, R_{d1}, R_{d2} and R_{d3} are the inner radius of fan annular from the first to third layer respectively, θ_{d1}, θ_{d2} and θ_{d3} are the annular angles of the first to third layer fan, respectively.

3.2. Equation of Motion of EMNM

In the analysis of magnetic network, the node voltage method is firstly used to solve the magnetic potential value of each node in the matrix, and the other variables of motor are analyzed. According to Kirchhoff's flux law, the flux flowing into node is the same as that flowing out of node.

$$\boldsymbol{A} \cdot \varphi = 0 \tag{13}$$

where, φ is the branch magnetic flux, $\boldsymbol{A}$ is the association matrix between nodes and branches, where rows represent nodes and columns represent branches corresponding to nodes. The direction from the center of circle to the circle is radial positive direction, and the clockwise direction is tangential positive direction. According to the law of magnetic pressure drop, equation (14) can be derived.

$$\varphi = \boldsymbol{G} \cdot \boldsymbol{U} = \boldsymbol{P} \cdot \boldsymbol{A} \cdot \boldsymbol{U} \tag{14}$$

where, $\boldsymbol{G}$ is the node permeance matrix, $\boldsymbol{P}$ is the relative permeance matrix of each node. $\boldsymbol{\Lambda}$ is the node permeance matrix without considering relative conductivity. $\boldsymbol{\Lambda}$ is the branch magnetic pressure drop vector. According to the node voltage method we could know.

$$\boldsymbol{U} = \boldsymbol{A}^{\mathrm{T}} \cdot \boldsymbol{U}_{\mathrm{n}} + \boldsymbol{U}_{\mathrm{s}} \tag{15}$$

where, $\boldsymbol{U}_{\mathrm{s}}$ is the magnetic potential source vector of branch PM, and $\boldsymbol{U}_{\mathrm{n}}$ is the node magnetic potential vector. Equation (16) can be obtained from equation (13-15).

$$\boldsymbol{A} \cdot \boldsymbol{P} \cdot \boldsymbol{\Lambda} \cdot \left(\boldsymbol{A}^{\mathrm{T}} \cdot \boldsymbol{U}_{\mathrm{n}} + \boldsymbol{U}_{\mathrm{s}} \right) = 0 \tag{16}$$

Magnetic saturation exists in the magnetic field to silicon steel sheet, and the μ values of each node in $\boldsymbol{P}$ should be determined according to magnetization curve. As well known, *B-H* curve is composed of discontinuous points. In order to accurately determine the magnetization curve and avoid the error caused by interpolation function, the magnetization curve is divided into two sections with B=1.65T as boundary, and the formula fitting of the two regions is carried out respectively. Then, μ_0 a initial value is given, and calculate the B value of each node by equation (17), μ_1. μ_1 is iterated according to equation (18), and the calculated μ_2 is used as new relative conductivity of silicon steel sheet to continue cycle iteration.

$$B = \frac{\phi}{S} = \frac{P \cdot \Lambda \cdot U}{S} \tag{17}$$

$$\mu_{k+1} = \mu_k^p \cdot \mu_{k-1}^q \tag{18}$$

where, k is the number of iterations, p=0.95, q=0.05. When the absolute values of the difference between two relative permeability values before and after the iteration result is less than the given error ε, the iteration is terminated [7].

$$\left| \mu_k^p \text{-} \mu_{k-1}^q \right| \leq \varepsilon \tag{19}$$

Because the SPMSM network model built is divided into fine sections and has a large number of nodes, vector and matrix are needed for calculation. In the process of matrix construction, sparse matrix should be used to reduce the storage space occupied by the matrix and improve the calculation speed of the matrix.

4. Calculation Results by EMNM

When calculating no-load radial air-gap flux density, the difference between the magnetic potential φ_{m} of air-gap node and magnetic potential φ_{n} of next node is taken as magnetic pressure drop at both ends of flux.

$$U = \varphi_{\mathrm{m}} - \varphi_{\mathrm{n}} \tag{20}$$

By substituting equation (20) into equation (17), the distribution of no-load radial air-gap flux density on the circumference of motor can be obtained. Since the unit node is small enough, there is no need to post-process motor air-gap results.

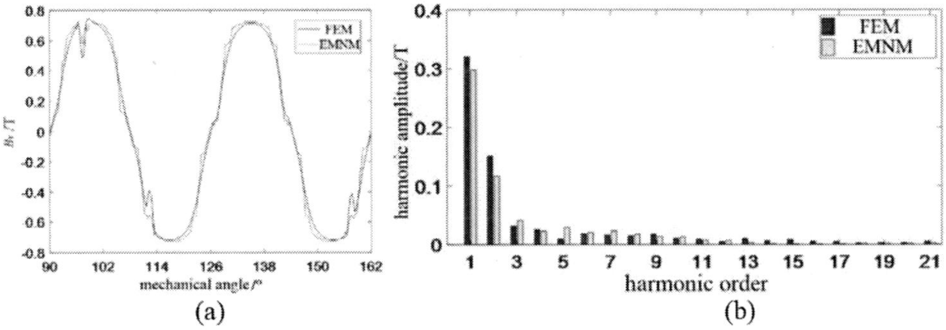

Figure .4 (a) Comparisons for no-load radial air-gap flux density (b) Harmonic analysis.

It can be seen that the calculation results of EMNM are very close to FEM in both variation trend and amplitude from Figure 4(a). As Table 2 shows, the error between EMNM and FEM in an effective value, maximum value, distortion rate and fundamental wave amplitude of air-gap flux density is very small, which meets the requirements of motor simulation.

Table 2. Harmonic comparison of no-load radial air-gap flux density between FEM and EMNM.

Air-gap	FEM	EMNM	Error
Effective value/T	0.55	0.53	3.6%
Maximum value/T	0.753	0.747	0.8%
Waveform distortion rate/%	51.4	45.7	11.1%
Base amplitude/T	0.32	0.30	6.2%

The changes of EMNM and FEM in local area under iteration and non-iteration conditions are compared in Figure 5. It can be seen that the analytical results of EMNM after iteration are closer to the results of FEM.

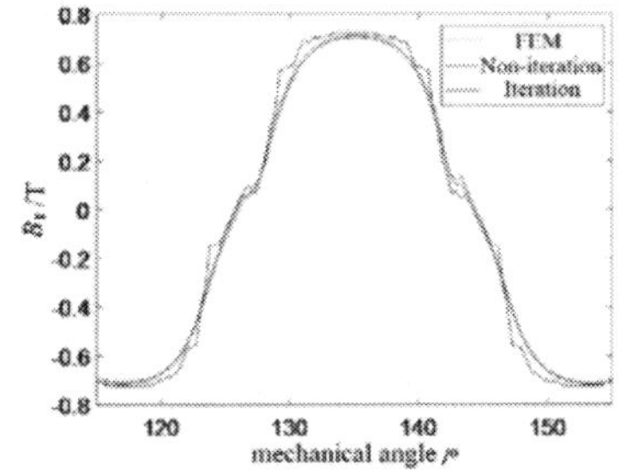

Figure 5. Comparison diagrams of magnetic network iteration and non-iteration.

As motor rotating, the air-gap flux density changes with the relative position between stator and rotor. When the motor rotation time is t, the structure of $\boldsymbol{A}$ does not change, the stator region in $\boldsymbol{G}$ remains unchanged, the rotor and PM region need to rotate m nodes, and $\boldsymbol{U}_s$ rotates m nodes.

$$m = \frac{M\pi nt}{360\times 30} \tag{21}$$

$$G = \begin{bmatrix} G_{z(t,M-m+1)}\cdots G_{z(t,M)}\, G_{z(t,1)}\cdots G_{z(t,M-m)} \\ G_{q(t,1)} \quad \cdots \quad \cdots \quad \cdots \quad \cdots \quad G_{q(t,M)} \\ G_{d(t,1)} \quad \cdots \quad \cdots \quad \cdots \quad \cdots \quad G_{d(t,M)} \end{bmatrix} \tag{22}$$

$$\boldsymbol{U}_s = \left[\boldsymbol{U}_{(1,\ M-m+1)}\cdots \boldsymbol{U}_{(1,\ M)}\boldsymbol{U}_{(1,1)}\cdots \boldsymbol{U}_{(1,\ M-m)}\right] \tag{23}$$

Figure 6 shows the distribution of air-gap flux density on circumference at electrical angles of 0°, 90° and 180° respectively. Compared with FEM, the established EMNM model can be applied to the magnetic field analysis of SPMSM with high accuracy.

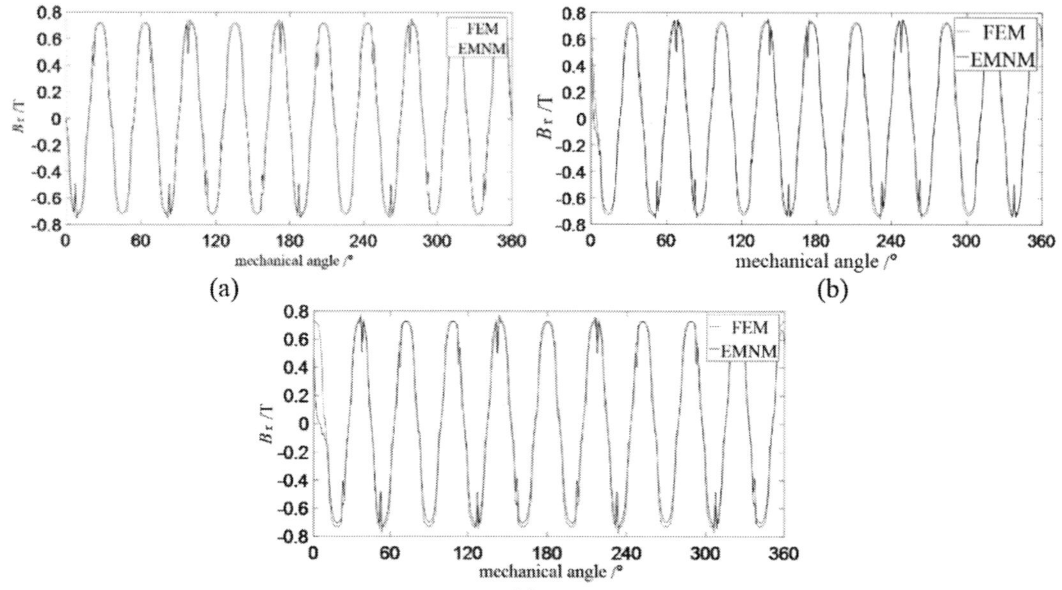

Figure 6. Air-gap flux density at different electrical angles. (a) The electric point is 0°. (b) The electric point is 90°. (c) The electric point is 180°.

5. Conculusion

The air gap flux density of SPMSM can be calculated by the magnetic network method. Compared with the finite element method, the calculated results are within the allowable error range, which can be used to accurately calculate the flux density distribution. At the same time, the calculation results of the air gap flux density under different mechanical angles show that the method has high accuracy. And the stator slot shape and core saturation were considered by piecewide fitting of annular equivalence and magnetic ring curve formulas of the groove fan.

References

[1] A Yousefi-Talouki, P Pescetto and G Pellegrino. (2017). Sensorless direct flux vector control of synchronous reluctance motors including standstill, MTPA and flux weakening. *IEEE Transactions on Industry Applications*, 53(4), 3598-3608.

[2] W. Su, Dong. W and Y. J. Guo. (2018). Analytical calculation of no-load air-gap magnetic field of non-salient pole synchronous generator by applying the distributed magnetic circuit approach. *Proceedings of the CSEE,* 38(9), 2736-2744.

[3] Z. H. Xu. (2016). Permanent magnet synchronous machine modeling and rotor eddy current loss analysis based on magnetic network. *Micromotors*, 49(3), 1-3.

[4] M. Zhang, A Macdonald and K Tseng. (2013). Magnetic equivalent circuit modeling for interior permanent magnet synchronous machine under eccentricity fault. *Power Engineering Conference (UPEC), 2013 48th International Universities*. Dublin, Ireland.

[5] J. Hu, X. Y Huang. (2016). Evaluation of PMSM Using Dynamic Reluctance Mesh Model. *Micromotors*, 49(1), 5-8.

[6] J. Zhang, J. Wang, Z. Yu, & Y. Li. (2020). Analytical Calculation for Two Layers PMSRM based on Equivalent Magnetic Network Method. *2020 IEEE International Conference on Applied Superconductivity and Electromagnetic Devices (ASEMD)*. Tianjin, China.

[7] W. Tong, S. Wang and S. Dai. (2018). A quasi-three-dimensional magnetic equivalent circuit model of a double-sided axial flux permanent magnet machine considering local saturation. *IEEE Transactions on Energy Conversion*, 33(4), 2163-2173.

Simulation and Analysis on the Influence of Diamond Blade Cutting Parameters of Grinding Wheel Dicing Saw

Yan Ma*, Xingqiao Jiang, Zhengdi Cao, Huang Liu and Dongxu Li

Shenyang Academy of Instrumentation Science Co., LTD. Shen yang, China

*Corresponding author email: MY9079cn@163.com

Abstract. In this paper, the cutting model of the dicing saw based on ANSYS simulation was established. Through coupling analysis, the force distribution and deformation of the cutting chip of the diamond grinding wheel cutter were simulated. The full-parameter solid modeling of the aerostatic spindle was carried out to analyze the influence of fixed cutter diameter, machining parameters and condition changes on spindle vibration. The blade thickness, whether to add coolant or not, and the influence of different spindle speed on spindle vibration were studied. The influence of different blade thickness on spindle vibration when the tool diameter is fixed is analyzed. The effect of different tool diameters matching different tool thicknesses on spindle vibration is discussed. Analyze the vibration of the spindle under different diameters. In the experiment, the spindle vibration was measured by changing the cutting parameters, and the root mean square value was used as the evaluation index to analyze the influence of the cutting parameters of the grinding wheel's diamond blade. The vibration of different diameters and different spindle directions was analyzed. The experimental results show that the above analysis method is feasible within the allowable error range, which provides a powerful reference for the future theoretical research.
Keywords: Simulation; Dicing saw; Diamond blade;The spindle vibration.

1. Introduction

Grinding wheel slicer is a machine for cutting silicon integrated circuits, light-emitting diodes, solar cells and other materials by using the cutting ability of thin slice grinding wheel [1,2] at high speed rotation. The grinding wheel blade is one of the important parts of the chip machine [3]. In the working process, its rotation will produce centrifugal stress, and excessive centrifugal stress may cause the fracture of the grinding wheel. The shape of the cutting tool has a direct effect on the scratch width and slice effect of hard and brittle materials. There are optimization methods to study cutting process parameters through vibration and research on diamond grinding wheel blade[4,5], but there is no influence on cutting parameters of diamond grinding wheel cutting machine blade through simulation study.Therefore, to study the performance of grinding wheel blade and ensure that the grinding wheel blade has enough strength is one of the important problems of grinding wheel scribing machine.

In this paper, the influence of diamond blade cutting parameters of diamond grinding wheel is analyzed by means of the experiment of diamond grinding wheel scribing machine. The influence of fixed tool diameter and variation of machining parameters and conditions on spindle vibration [6] was analyzed. According to the cutting test, the vibration curves of x, Y and Z directions under different tool diameters were obtained. Meanwhile, the vibration curves of the three directions with the same tool diameters were compared in the same graph. Analyze the influence of cutting parameters change

when the tool diameter is 54mm, 52mm and 50.8mm. The influence of machining parameters and conditions on spindle vibration was analyzed when the tool diameter was fixed. The position of the vibration peak and the situation of the minimum vibration point were analyzed. The change of spindle vibration with the increase of spindle speed; The influence on vibration after adding coolant; Blade thickness increases, vibration changes, etc. The influence analysis of diamond blade cutting parameters of diamond grinding wheel scribing machine is obtained.
Based on the robust design method, this paper makes a quantitative study on a variety of controllable factors through orthogonal experiments. The results were analyzed by range analysis, analysis of variance and factor index relationship trend chart analysis. Carry out the test of cutting width and the test of maximum cutting width, find out the factor with the greatest contribution, select the appropriate level, and obtain the optimal processing parameters. The main factors affecting the width and maximum width of the cut are described.

2. Experimental Analysis

A precision grinding wheel scribing machine was selected as the research object to carry out the scribing experiment. The speed is 18000r/min and 3000r /min. The coolant flow rate is 1.5L/min, which is divided into cooling effect and no cooling effect. The tool diameter is 54 mm, 52 mm, 50.8mm; Tool thickness 0.150 mm, 0.10 mm, 0.08 mm, 0.05mm; Perform cutting tests under different diameters and conditions. The spindle vibration was tested by changing the cutting parameters, and the root mean square value of spindle vibration was used as the evaluation index of the cutting. The results were analyzed as shown in Figure 1 to Figure 3. Figure 1 is the vibration curve in the X direction under different tool diameters, Figure 2 is the vibration curve in the Y direction under different tool diameters, and Figure 3 is the vibration curve in the Z direction under different tool diameters.

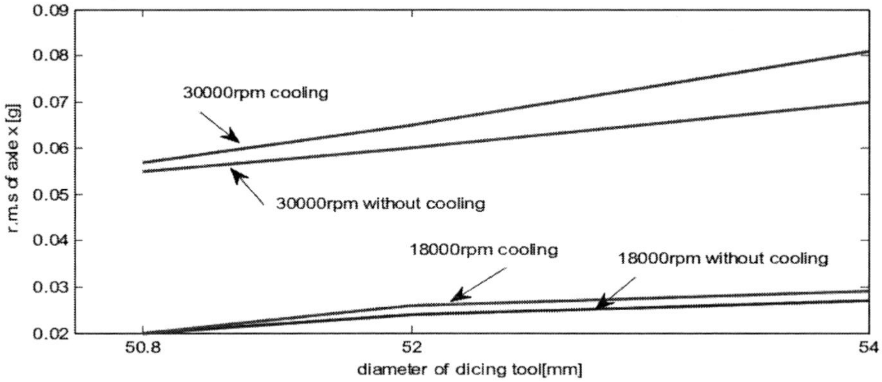

Figure 1. Vibration in X direction with different tool diameters.

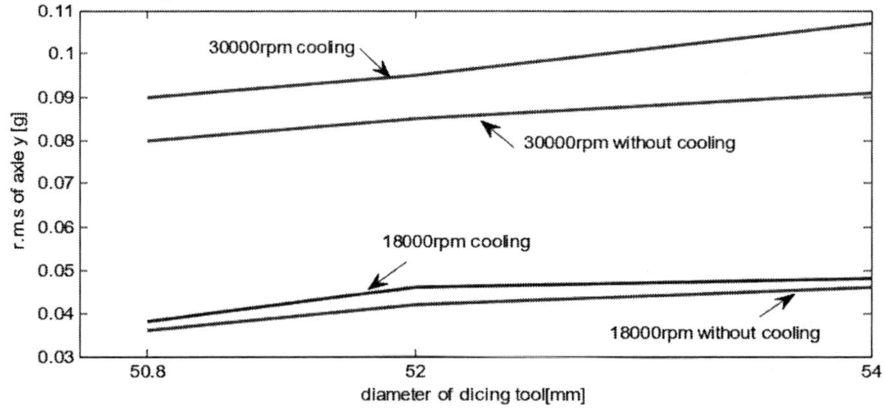

Figure 2. Y-direction vibration of different tool diameters.

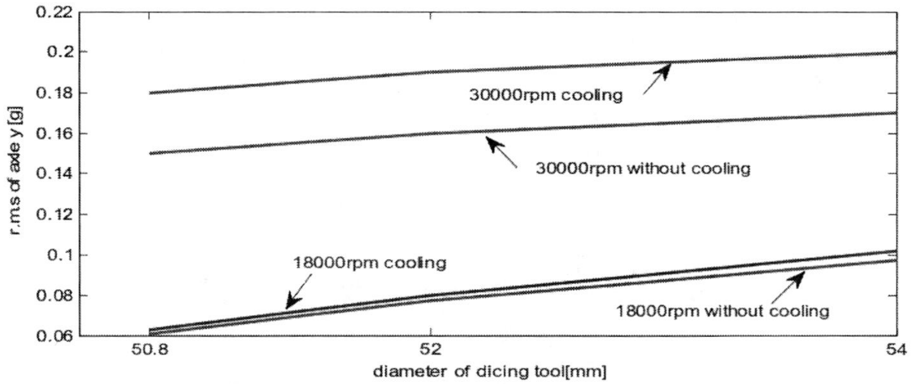

Figure 3. Z - direction vibration with different tool diameters.

Figure 1 to Figure 3 show the comparison of cutting conditions under different cutter diameters. It can be seen from the results that the vibration degree of spindle is enhanced after the cooling effect is added. The vibration degree of spindle increases with the increase of tool diameter and speed. With the increase of tool thickness, the increasing speed of vibration slows down.

3. Influence Analysis of Diamond Blade Cutting Parameters

Simulation Study on the influence of diamond Blade cutting parameters of Grinding wheel dicing SAW was carried out with the help of the matlab Simulink.

3.1. Change of Cutting Parameters when the Tool Diameter is 54mm

Figure 4 shows the influence of machining parameters and conditions on X-axis spindle vibration when the tool diameter is fixed at 54mm. Among them, blade thickness, whether to add coolant or not, different spindle speed have an impact on spindle vibration. The blade thickness of tool with different diameter has different influence on spindle vibration. According to the analysis in FIG. 4, when the tool diameter is 54mm, the vibration peak in the X direction occurs when the blade thickness is the thickest, the spindle speed is the largest and coolant is added. Spindle vibration increases with the increase of spindle speed. After adding coolant, the vibration intensifies, which is verified by experiments. The blade thickness increases and the vibration increases slightly. However, the location of the minimum vibration point in the X direction is affected by the processing parameters and conditions.

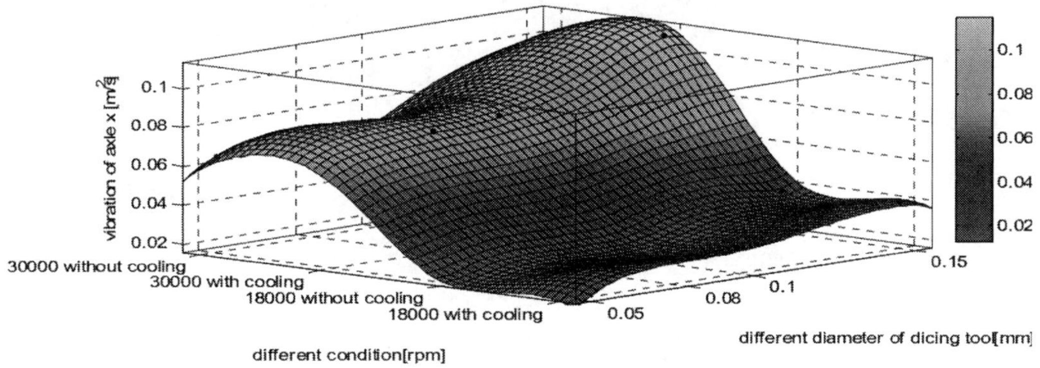

Figure 4. Influence of cutting parameters change on x - axis spindle vibration with a 54mm tool diameter.

Figure 5 shows the influence of machining parameters and conditions on Y-axis spindle vibration when the tool diameter is fixed at 54mm. Among them, blade thickness, whether to add coolant or not, different spindle speed have an impact on spindle vibration. The blade thickness of tool with different diameter has different influence on spindle vibration. According to the analysis in Figure 5, when the tool diameter is 54mm, the vibration peak in the Y direction appears when the blade thickness is the thickest and the spindle speed is the largest. Whether the coolant is added or not has little influence on the vibration peak at the high speed. Spindle vibration increases with the increase of spindle speed. The addition of coolant intensifies the vibration. The blade thickness increases and the vibration increases slightly. The minimum vibration point in y direction occurs when the spindle speed is not large and the blade thickness is appropriate.

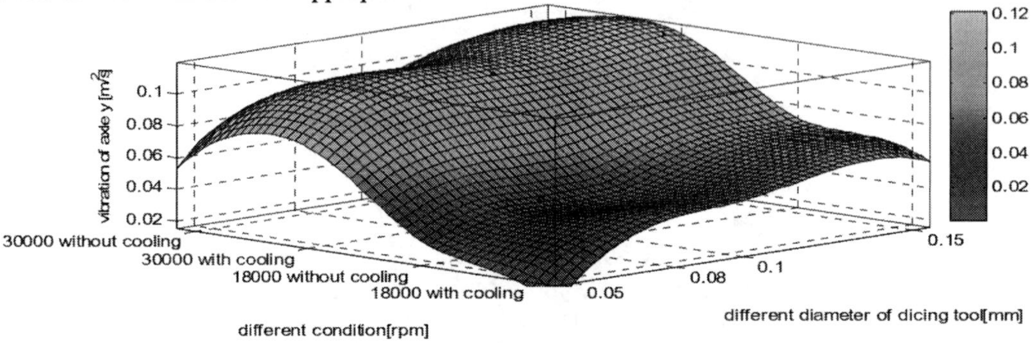

Figure 5. Influence of cutting parameter variation on Y-axis spindle vibration with a 54mm cutter diameter.

Figure 6 shows the influence of machining parameters and conditions on z-axis spindle vibration with fixed tool diameter. Among them, blade thickness, whether to add coolant or not, different spindle speed have an impact on spindle vibration. The blade thickness of tool with different diameter has different influence on spindle vibration. According to the analysis in Figure 6, when the tool diameter is 54mm, the vibration peak in the Z direction appears when the blade thickness is the thickest, the spindle speed is the largest, and coolant is added. Spindle vibration increases with the increase of spindle speed. The addition of coolant intensifies the vibration. The blade thickness increases and the vibration increases slightly. The minimum vibration point in z direction occurs when the spindle speed is not large and the blade thickness is appropriate.

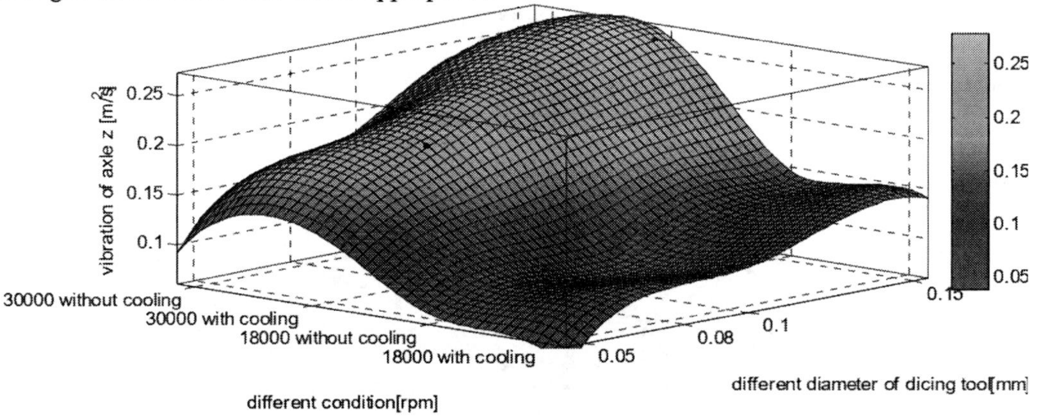

Figure 6. Influence of cutting parameter variation on z-axis spindle vibration with a 54mm cutter diameter.

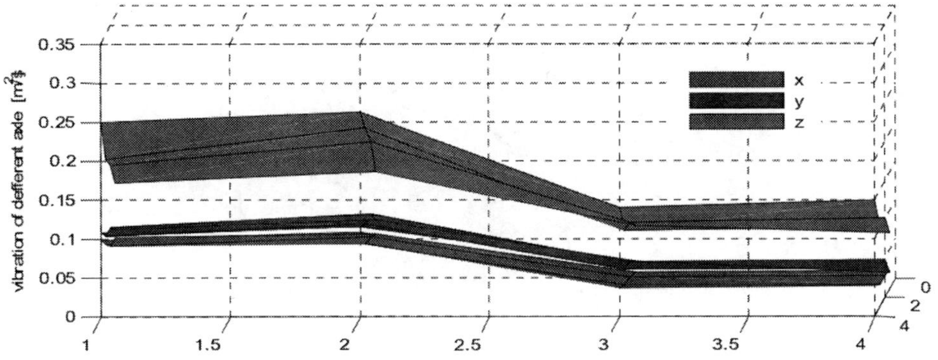

Figure 7. Comparison of vibration in three directions with tool diameter of 54mm.

Meanwhile, the vibration in the three directions with the tool diameter of 54mm was compared in the same figure, as shown in Figure 7. It can be seen that for this type of tool, vibration in the Z direction is maximum at a diameter of 54mm.

3.2. Change of Cutting Parameters when the Tool Diameter is 52mm

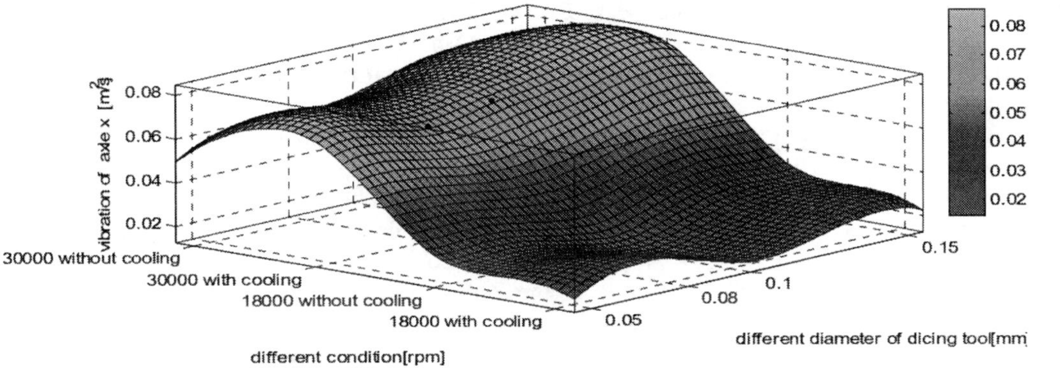

Figure 8. Shows the influence of machining parameters and conditions on X-axis spindle vibration when the tool diameter is fixed at 52mm.

Figure 8 shows the influence of machining parameters and conditions on X-axis spindle vibration when the tool diameter is fixed at 52mm. Among them, blade thickness, whether to add coolant or not, different spindle speed have an impact on spindle vibration. The blade thickness of tool with different diameter has different influence on spindle vibration. According to the analysis in Figure 8, when the tool diameter is 52mm, the vibration peak in the X direction occurs when the spindle speed is the maximum and coolant is added, and the blade thickness has little influence. Spindle vibration increases with the increase of spindle speed. After adding coolant, the vibration intensifies, which is verified by experiments. The blade thickness increases and the vibration increases slightly. However, the location of the minimum vibration point in the X direction is affected by the processing parameters and conditions.

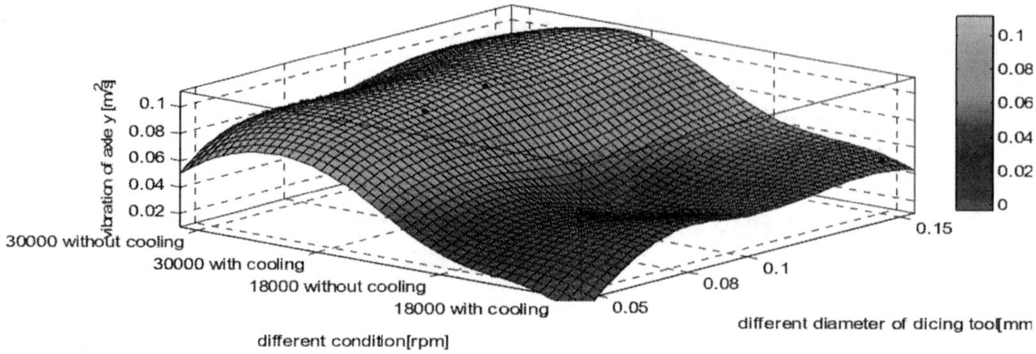

Figure 9. Influence of cutting parameter variation on Y-axis spindle vibration with a tool diameter of 52mm.

Figure 9 shows the influence of machining parameters and conditions on Y-axis spindle vibration when the tool diameter is fixed at 52mm. Among them, blade thickness, whether to add coolant or not, different spindle speed have an impact on spindle vibration. The blade thickness of tool with different diameter has different influence on spindle vibration. According to the analysis in FIG. 9, when the tool diameter is 52mm, the vibration peak in the y direction occurs when the spindle speed is the maximum and coolant is added, and the vibration peak in the case of high blade thickness and high speed has little influence. Spindle vibration increases with the increase of spindle speed. The addition of coolant intensifies the vibration. The blade thickness increases and the vibration increases slightly. The minimum vibration point in y direction occurs when the spindle speed is not large and the blade

Figure 10 shows the influence of machining parameters and conditions on z-axis spindle vibration when the tool diameter is fixed at 52mm. Among them, blade thickness, whether to add coolant or not, different spindle speed have an impact on spindle vibration. The blade thickness of tool with different diameter has different influence on spindle vibration. According to the analysis in Figure 10, when the tool diameter is 52mm, the vibration peak in the Z direction occurs when the spindle speed is the maximum, the blade thickness is the thickest and relatively thin, and coolant is added at the same time. Spindle vibration increases with the increase of spindle speed. The addition of coolant intensifies the vibration. The blade thickness increases and the vibration increases slightly. The minimum vibration point in z direction occurs when the spindle speed is not large and the blade thickness is appropriate.

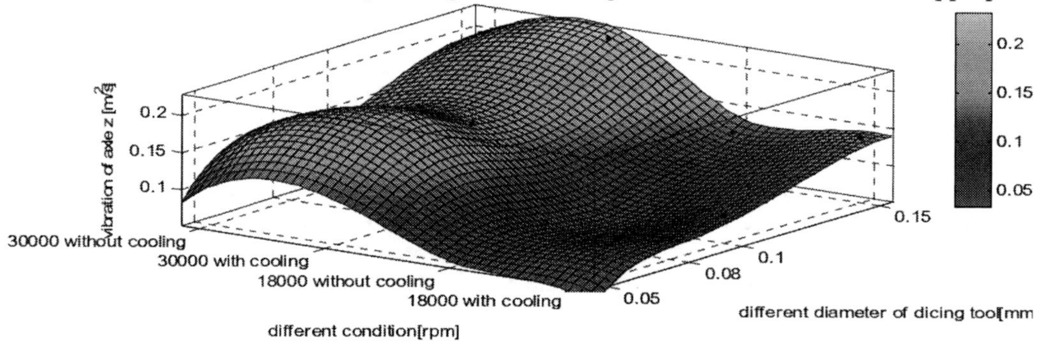

Figure 10. Influence of cutting parameter variation on z-axis spindle vibration with a tool diameter of 52mm.

Meanwhile, the vibration in three directions with a tool diameter of 52mm was compared in the same figure, as shown in Figure 11. It can be seen that for this type of tool, the vibration in the z direction is maximum when the diameter is 52mm. There is no significant difference in vibration values between x and Y directions.

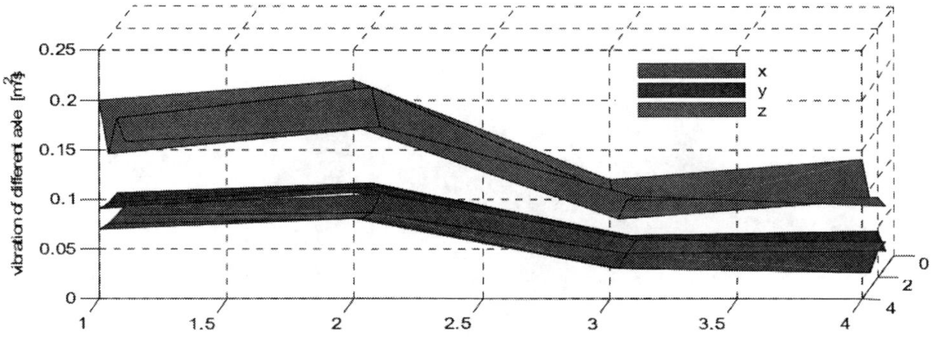

Figure 11. Comparison of vibration in three directions with a tool diameter of 52mm.

3.3. Change of Cutting Parameters when the Tool Diameter is 50.8mm

Figure 12 shows the influence of machining parameters and conditions on X-axis spindle vibration when the tool diameter is fixed at 50.8mm. Among them, blade thickness, whether to add coolant or not, different spindle speed have an impact on spindle vibration. The blade thickness of tool with different diameter has different influence on spindle vibration. According to the analysis in Figure 12, when the tool diameter is 50.8mm, the vibration peak in the X direction occurs when the blade thickness is the thickest, the spindle speed is the largest and coolant is added. Spindle vibration increases with the increase of spindle speed. After adding coolant, the vibration intensifies, which is verified by experiments. The blade thickness increases and the vibration increases slightly. The location of the minimum vibration point in the X direction is affected by the processing parameters and conditions.

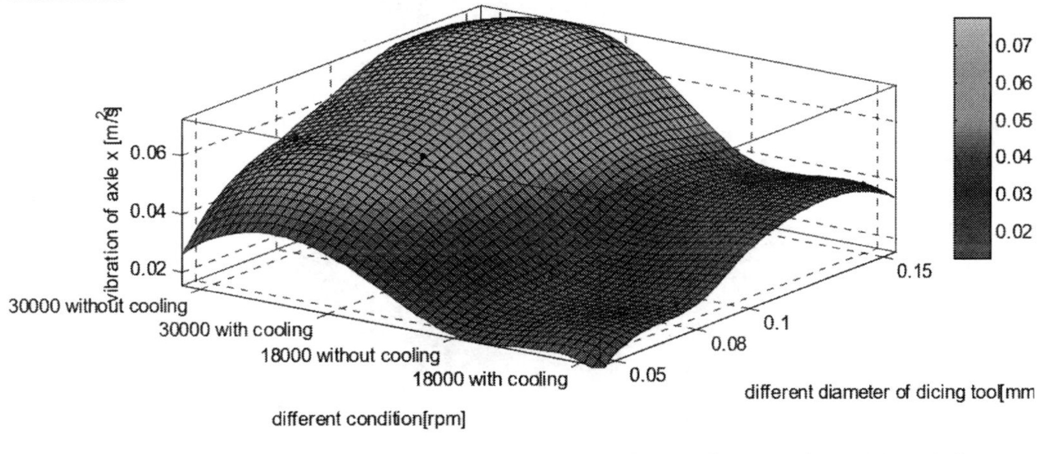

Figure 12. Influence of cutting parameter variation on X-axis spindle vibration with tool diameter of 50.8mm.

Figure 13 shows the influence of machining parameters and conditions on Y-axis spindle vibration when the tool diameter is fixed at 50.8mm. Among them, blade thickness, whether to add coolant or not, different spindle speed have an impact on spindle vibration. The blade thickness of tool with different diameter has different influence on spindle vibration. According to the analysis in Figure 13, when the tool diameter is 50.8mm, the vibration peak in the Y direction appears when the blade thickness is the thickest and the spindle speed is the largest. Whether the coolant is added or not has little influence on the vibration peak at the high speed. Spindle vibration increases with the increase of spindle speed. The addition of coolant intensifies the vibration. The blade thickness increases and the vibration increases slightly. The minimum vibration point in y direction occurs when the spindle speed is small and the blade thickness is small or large.

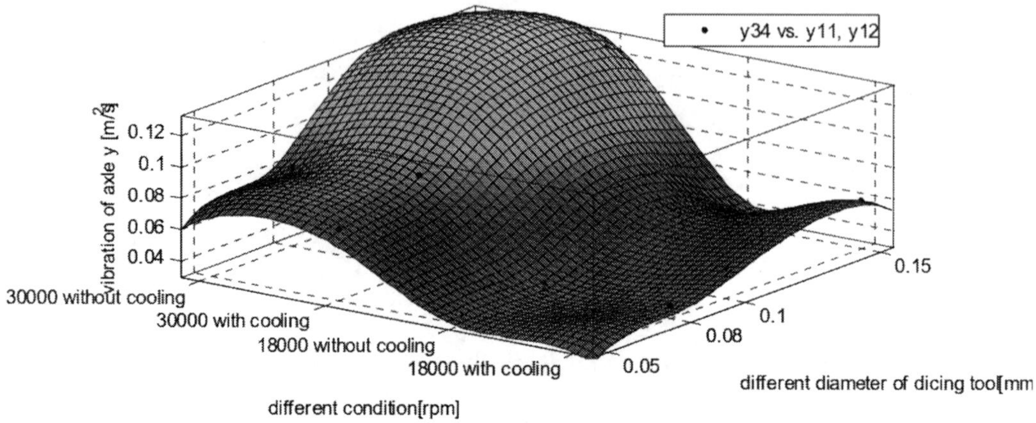

Figure 13. Influence of cutting parameter variation on Y-axis spindle vibration with tool diameter of 50.8mm.

Figure 14 shows the influence of machining parameters and conditions on z-axis spindle vibration when the tool diameter is fixed at 50.8mm. Among them, blade thickness, whether to add coolant or not, different spindle speed have an impact on spindle vibration. The blade thickness of tool with different diameter has different influence on spindle vibration. According to the analysis in Figure 14, when the tool diameter is 50.8mm, the vibration peak in the Z direction appears when the blade thickness is the thickest, the spindle speed is the largest, and coolant is added. Spindle vibration increases with the increase of spindle speed. The addition of coolant intensifies the vibration. The blade thickness increases and the vibration increases slightly. The minimum vibration point in z direction occurs when the spindle speed is not large and the blade thickness is appropriate.

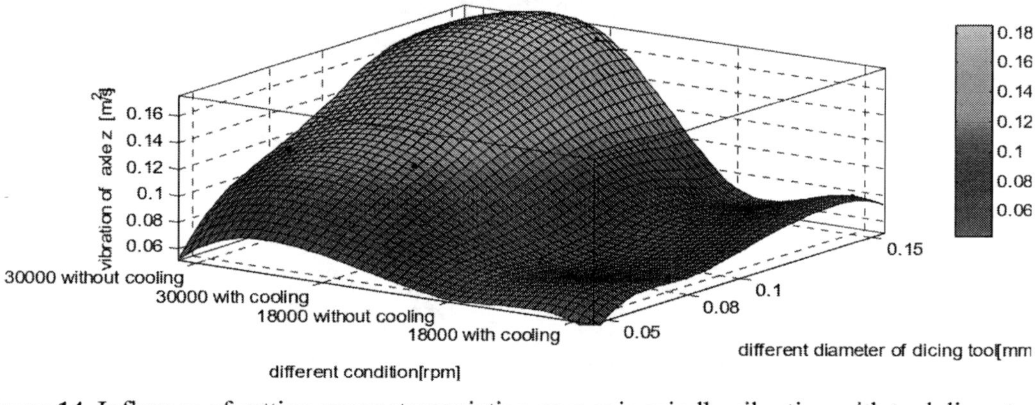

Figure 14. Influence of cutting parameter variation on z-axis spindle vibration with tool diameter of 50.8mm.

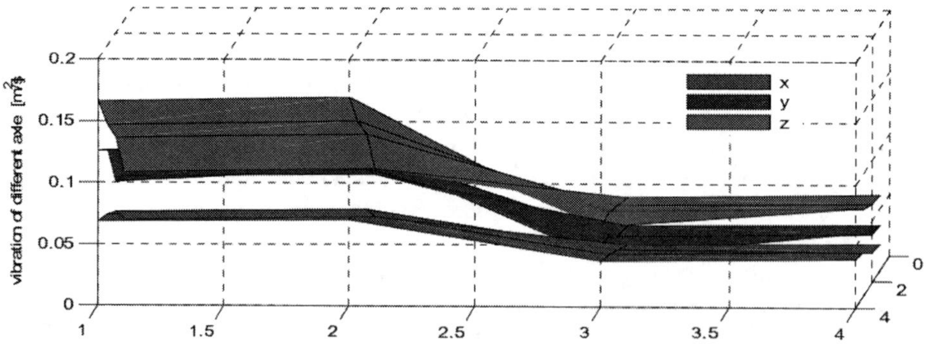

Figure 15. Comparison of vibration in three directions with tool diameter of 50.8mm.

Meanwhile, the vibration in the three directions with the tool diameter of 54mm was compared in the same figure, as shown in Figure 15. It can be seen that for this type of tool, the vibration in the Z direction is maximum when the diameter is 50.8mm. When the blade thickness is small, the vibration of z direction and Y direction is larger.

4. Conclusion

According to the result curve, for the tool studied in this paper, when the diameter is 54mm, the vibration in the Z direction is the largest; When the diameter is 52mm, the vibration in the Z direction is the largest, and there is little difference between the vibration in the X direction and the vibration in the Y direction. When the diameter is 50.8mm, the vibration in the Z direction is the largest. When the blade thickness is small, the vibration of z direction and Y direction is larger.

Analyze the influence of cutting parameters change when the tool diameter is 54mm, 52mm and 50.8mm. The influence of machining parameters and conditions on spindle vibration was analyzed when the tool diameter was fixed. Among them, blade thickness, whether to add coolant or not, different spindle speed have an impact on spindle vibration. The blade thickness of tool with different diameter has different influence on spindle vibration.

It can be seen from the analysis results that the vibration degree of the spindle is enhanced after the cooling effect is added. The vibration degree of spindle increases with the increase of tool diameter and speed. The increasing speed of vibration decreases with the increase of tool thickness.

References

[1] Li Bomin, Zhao Bo. Modern Grinding Technology. Beijing: China Machine Press, 2003.
[2] GUO Dongming, KANG Renke, Jin Zhuji, et al. Journal of manufacturing science and technology, 2007,5(4):1-32.
[3] Ma Yan, YUAN Huizhu, ZHANG Mingming, Sun Jiaquan. Study on performance of diamond Grinding Wheel blade of scribing machine. Special Equipment for Electronic Industry. 2008, 166:30-32-37.
[4] Hong-chun,SUN Hong-bao WANG,Yong XU,Li-yang XIE, Optimization Method of Cutting Process Parameters for Dicing Saw .Journal of Northeastern University(Natural Science), 2017, 38(4)
[5] Junming Cao. Design and Error Compensation research of Grinding Wheel Blade Wear Detection System for Dicing Saw [D]. Hunan University,2019.
[6] Zhang Ming, Zhu Yu, DUAN Guanghong. Microvibration and Its Suppression of ultra-precision Air-floating workpiece table, Manufacturing Technology & Machine Tool, 2005, (11) : 47-49
[7] WANG J. Water jet machining of Engineering Materials. Switzerland: Trans. Tech. Publications Ltd., 2003.

Simulation and Analysis of Characteristics on Type of Variable Cross-section Free Film Air Spring

Chuanyin Tang[1,*], Tao Shen[1], Lv Pan[1] and Shi Fan[2]

[1]School of Mechanical Engineering and Automation, Northeastern University, Shenyang, 110819, China
[2]School of Science, Northeastern University, Shenyang, 110819, China

[*]Corresponding author email: monsigor@aliyun.com

Abstract. The stiffness formula of rolled ear variable cross-section air spring is obtained. Due to the strong nonlinear characteristics of the free-film air spring with variable cross section, the possible chaotic phenomena of the air spring are analyzed, and its behavior under deterministic excitation is explored, and simulation of the phase trajectory diagram and time history curve of the system under periodic motion and chaotic motion are obtained. The vehicle analysis model of variable section free film air suspension with rolled lug is established. According to the requirements of national standards, the simulation analysis of manipulation stability experiment is carried out, and the evaluation value of simulation results is obtained.
Keywords: Simulation; Manipulation stability evaluation; Rolled ear variable cross-section air spring.

1. Introduction

With the rapid development of vehicle technology, vehicle users pay more attention to the comfort of vehicles on the basis of paying attention to driving safety. It is more and more important to obtain excellent ride comfort and handling stability during driving. Air suspension [1-2] can effectively improve vehicle ride comfort and handling stability.

Air spring is a complex system with compressed air and rubber capsule as elastic medium and gas as transmission medium. Compared with ordinary coil spring, the advantage of air spring is that the speed changes relatively slowly and the change of dynamic force is controlled within a certain range. Due to the complexity of air spring, the combination of computer simulation and test has become the main means for scholars to study the characteristics of air spring. Professor Yang Shaopu studied the chaotic phenomenon of automobile suspension system [3], and literature [4] established three kinds of air spring stiffness characteristic models based on van der Waals equation. Reference [5] proposed a method to study the stiffness of air spring by nonlinear fitting method. Many scholars have made very important attempts in the analysis and modeling of air spring.

This paper takes the membrane air spring as the research object. Based on the chaotic vibration theory [6], the nonlinear characteristics of the ear variable cross-section free film air spring system are analyzed. The dynamic model of rolling ear variable cross-section free film air spring suspension is established. According to the national standard [7], the handling and stability experiment simulation is carried out, including steady-state rotation simulation, steering wheel angle pulse input simulation test and hunting simulation test. And obtain the result evaluation value.

2. Establishment of Air Spring Model

The ear variable cross-section free film air spring is used as the research object. The spring has small bearing capacity and large dynamic stiffness under the same atmospheric pressure and size. The change rate of area and volume of air spring can be realized by controlling the shape of piston. The stiffness of the air spring is determined by the initial pressure, the volume of the initial design position, the effective area, the change rate of the effective area, the initial atmospheric pressure, the relative volume, the volume and the change rate of the volume. It is assumed that the outer diameter of the airbag and the length of the meridian in the airbag are fixed values. The calculation model of ear variable cross-section free film air spring is shown in Figure 1.

Where: spring height is defined as H ; The piston tip radius is defined as R_a ; Indicates H_i the height of the piston; 2α is the top angle of piston cone; The radius of the inner surface of the airbag is defined as R_r . $H_{design} = 400\text{mm}$ .Poly tropic index of gas motion $m = 1.3$; Atmospheric pressure $p_a = 0.1\text{MPa}$ and Polytropic index of gas movement process $m = 1.3$; Airbag initial pressure $p_{ro} = 0.3\text{MPa}$.

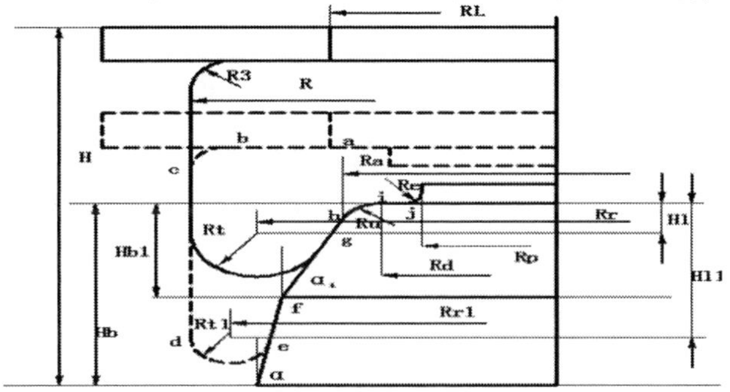

Figure 1. Calculation model of rolled ear variable cross-section free film air spring.

The piston is divided into two parts from the design height, and one part only contains H_{b1} , the other part contains only greater than H_{b1} .

The area change rate is:

$$R_r = R_a + \frac{R - R_a}{1 + \cos\alpha_1} + H_l \tan\frac{\alpha_1}{2} \tag{1}$$

The volume change rate is:

$$\frac{\mathrm{d}V'}{\mathrm{d}h} = \frac{\mathrm{d}V_1}{\mathrm{d}h} + \frac{\mathrm{d}V_2}{\mathrm{d}h} - \frac{\mathrm{d}V_7}{\mathrm{d}h} - \frac{\mathrm{d}V_8}{\mathrm{d}h} \tag{2}$$

Effective area change rate:

$$\frac{\mathrm{d}A_{e1}}{\mathrm{d}h} = 2\pi R_{r1} \frac{\mathrm{d}R_{r1}}{\mathrm{d}h} \tag{3}$$

Volume change rate:

$$\frac{\mathrm{d}V''}{\mathrm{d}h} = \frac{\mathrm{d}V_1}{\mathrm{d}h} + \frac{\mathrm{d}V_2}{\mathrm{d}h} - \frac{\mathrm{d}V_7}{\mathrm{d}h} - \frac{\mathrm{d}V_8}{\mathrm{d}h} \tag{4}$$

The formula of effective area and volume change rate is:

$$\frac{\mathrm{d}A}{\mathrm{d}h} = \begin{cases} \frac{\mathrm{d}A_e}{\mathrm{d}h}(H_b < H_{b1}) \\ \frac{\mathrm{d}A_{e1}}{\mathrm{d}h}(H_b > H_{b1}) \end{cases} ; \frac{\mathrm{d}V}{\mathrm{d}h} = \begin{cases} \frac{\mathrm{d}V'}{\mathrm{d}h}(H_b < H_{b1}) \\ \frac{\mathrm{d}V''}{\mathrm{d}h}(H_b > H_{b1}) \end{cases} \tag{5}$$

By substituting the stiffness formula (6), the air spring stiffness of the piston stroke within H_{b1} and H_{b2} can be obtained.

$$C=\frac{dp}{dh}=\left[(p_{r0}+p_a)\left(\frac{V_0}{V}\right)^m-p_a\right]\frac{dA_e}{dh}-A_e m(p_{r0}+p_a)\frac{V_0^m}{V^{m+1}}\frac{dV}{dh} \tag{6}$$

3. Nonlinear Characteristic Analysis of Variable Cross-section Free Film Air Spring with Rolling Lug

When the excitation amplitude is in a certain range, chaos may occur. Taking 1 / 4 vehicle as the research object, based on the nonlinear theory, the excitation is $x_0 = F_1 \sin \omega t$, and the motion equation of the system is as follows:

$$M\ddot{x}+c(\dot{x}-\dot{x}_0)+k_1(x-x_0)+k_f(x-x_0)^2+k_{ff}(x-x_0)^3=0 \tag{7}$$

Let $y = x - x_0$ and $k = -k_1$, the system state equation is as follows:

$$\dot{Z}=f(Z)+\varepsilon g(Z);\ f(Z)=\begin{pmatrix} f_1(Z) \\ f_2(Z) \end{pmatrix};Z=\begin{pmatrix} z_1 \\ z_2 \end{pmatrix}\ g(Z,t)=\begin{pmatrix} g_1(Z,t) \\ g_2(Z,t) \end{pmatrix}=\begin{pmatrix} 0 \\ -c_1 z_2-k_2 z_1^2+h\sin\frac{\omega}{\omega_g}t \end{pmatrix}$$

Where: $z=\frac{y}{F_1}$, $T=\sqrt{\frac{M}{k_1}}$, $\tau=\frac{t}{T}$, $T=\frac{1}{\omega_g}$, $\varepsilon=T$, $c_1=\frac{c}{M}$, $k_2=\frac{k_f F_1 T}{M}$, $k_3=\frac{k_{ff}F_1^2 T}{M}$, $h=\frac{\omega^2}{\omega_g}$

When $\varepsilon = 0$, simplified formula (4) is a two-dimensional conservative system [8].

$$\begin{cases} \dot{z}_1 = z_2 \\ \dot{z}_2 = z_1 - k_3 z_1^3 \end{cases} \tag{8}$$

Let: $\dot{z}_1 = \dot{z}_2 = 0$ .When $k_3 < 0$, there is only one fixed point $O(0,0)$, the system is a stable periodic motion without chaos.; When $k_3 > 0$, there are three fixed points: $O(0,0)$, $A\left(\sqrt{1/k_3},0\right)$, $B\left(-\sqrt{1/k_3},0\right)$, the system is a Hamilton plane system. Solving Hamilton function:

$$\begin{cases} \frac{\partial H}{\partial z_2} = f_1(Z) = z_2 \\ -\frac{\partial H}{\partial z_1} = f_2(Z) = z_1 - k_3 z_1^3 \end{cases} \tag{9}$$

Melnikov function $M(t_0)=\int_{-\infty}^{+\infty} f\left(q_\pm^0(t)\right)\wedge g\left(q_\pm^0(t),t+t_0\right)dt$

If there is a simple zero point for k = 1, it needs to meet: $\frac{\sqrt{2}h\pi\omega}{\omega_g}\sqrt{\frac{1}{k_3}}\cos\left(\frac{\omega t_0}{\omega_g}\right)\mathrm{sech}\left(\frac{\pi\omega}{2\omega_g}\right)>\frac{4}{3}\frac{c_1}{k_3}$

When chaos occurs in the system, the threshold value of pavement displacement excitation amplitude is shown as :

$$F_1>\frac{2\sqrt{2}c\omega_g^2\cosh\left(\frac{\pi\omega}{2\omega_g}\right)\sec\left(\frac{\omega t_0}{\omega_g}\right)}{3\pi\omega^3\sqrt{k_{ff}MT}}\mathrm{def}R(\omega) \tag{10}$$

If the excitation amplitude of the system falls below the curve, the system does not appear chaos. If the excitation amplitude is above the curve, chaos may appear in the system. Set the zero point to the design height of the air spring. The air spring travel displacement is positive and the air spring return displacement is negative. For the research object of this paper, the critical condition for chaos is $F_1 = 0.005m$. When $F_1 < 0.005m$ does not occur, chaos may occur when $F_1 > 0.005m$.

In the simulation experiment, different excitations are taken for testing. Two typical excitations, 0.004m and 0.05m respectively, are taken in the interval where chaos occurs and does not occur in the system. Based on the fourth-order Runge -Kutta method, the phase trajectory and time history curve are obtained, as shown in Fig. 2-Fig.3.

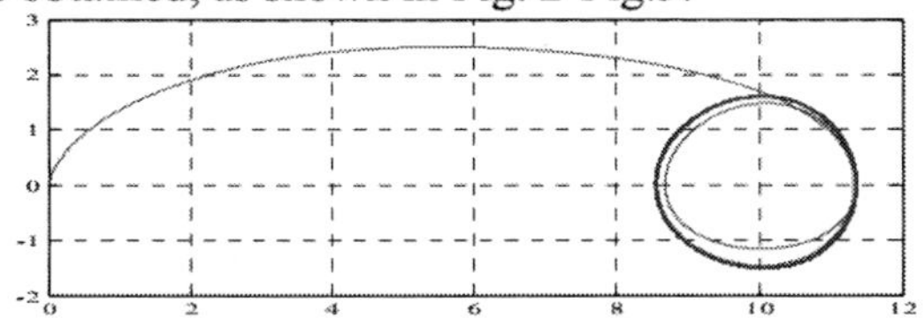

Figure 2. Phase diagram with 0.004m excitation.

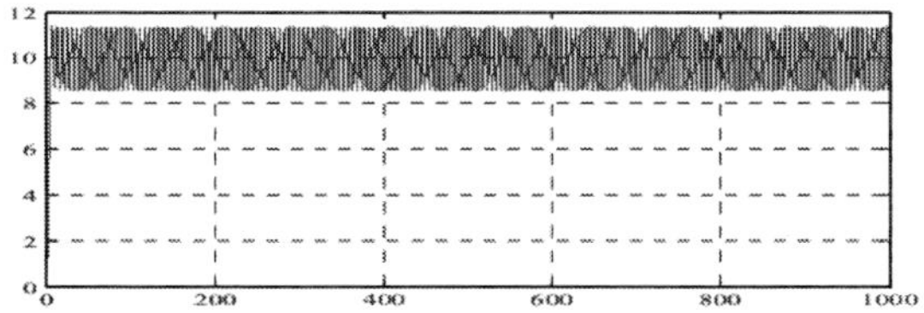

Figure 3. Time curve with 0.004m excitation.

When the excitation is 0.004m, based on Melnikov method, there is a limit cycle in the phase plane diagram, and the time history curve shows periodicity, which is judged as periodic motion.

When the excitation is 0.05m, the phase plane diagram is disordered, and the time history curve has no periodicity, which can be judged as chaotic motion, as shown in Fig. 4-Fig.5.

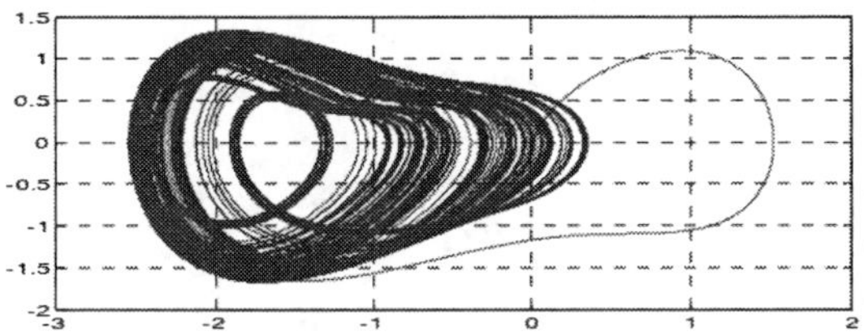

Figure 4. Phase diagram with 0.05m excitation.

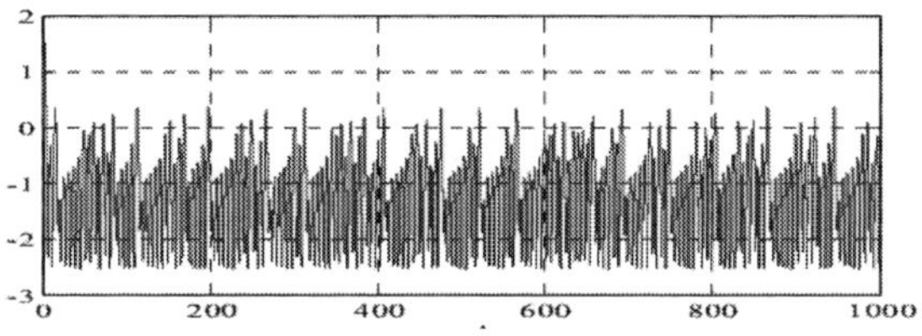

Figure 5. Time curve with 0.05m excitation.

Study a class B pavement, if the wavelength is10m, and the excitation frequency is f, $v = 10m/s$, then $f = 1Hz$. When the excitation amplitude is 0.05m, the frequency range of chaos is: $9.7-11.9Hz$, its speed is: $15.4-18.9m/s$. Excitation amplitude, frequency, nonlinear stiffness and other factors may affect the occurrence of chaos. Chaos may occur when the vehicle speed is at medium speed.

4. Dynamic Analysis of Air Suspension Vehicle

Using multi-body dynamics analysis tool, the dynamic model of rolling ear variable cross-section free film air suspension is established. According to the requirements of national standards, the handling and stability simulation analysis is carried out, and the simulation experimental evaluation value of the air suspension vehicle is obtained.

Take the steady-state rotation simulation experiment as an example to illustrate the analysis process. The steady-state rotation simulation test curves are shown in Fig. 6(a)-Fig. 6(f) respectively represent the yaw rate, body roll angle, centroid roll angle with lateral acceleration, roll angle with lateral acceleration, the difference of front and rear axle roll angle with lateral acceleration, and the motion trajectory of the whole vehicle.

4.1. Steady State Rotation

Evaluation of steady-state rotation simulation results: evaluate according to the lateral acceleration value a_n of neutral steering point, insufficient steering degree U and car roll degree K_ϕ, and the formula is as follows:

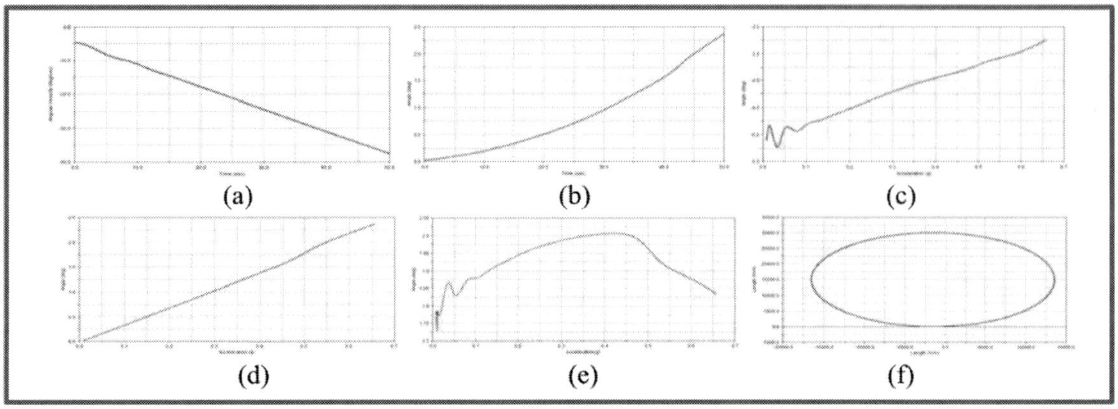

Figure 6. Curve of constant radius steering.

$$N_{a_n} = 60 + \frac{40}{a_{n100} - a_{n60}} \cdot (a_n - a_{n60}) \tag{11}$$

Where: N_{a_n} -Evaluation score of lateral acceleration value; a_n -Test value of lateral acceleration; a_{n60} -Lower limit of lateral acceleration value; a_{n100} Upper limit of lateral acceleration value.

$$N_U = 60 + \frac{U(U_{60} - U)(\lambda - U)}{U_{100}(U_{60} - U_{100})(\lambda - U_{100})} \cdot 40 \tag{12}$$

Where: N_U -Evaluation score of under steer; U -Test value of under steer; $\lambda = \frac{2 \cdot U_{60} / U_{100}}{U_{60} / U_{100} - 2} \cdot U_{100}$;

U_{60} -Lower limit of under steer; U_{100} -Upper limit of under steer.

$$N_\phi = 60 + \frac{40}{K_{\phi 60} - K_{\phi 100}} (K_{\phi 60} - K_\phi) \tag{13}$$

Where: N_ϕ -Evaluation score value of car roll; $K_{\phi 60}$ Upper limit value of body roll; $K_{\phi 100}$ -Lower limit of body roll; K_ϕ -Test value of vehicle body roll.

$$N_w = \frac{N_{a_n} + N_U + N_\phi}{3} \tag{14}$$

Where: N_w -Comprehensive evaluation score of steady-state rotation test.

The results and evaluation values of steady-state rotation simulation experiment are shown in Fig. 7 and Fig. 8. In addition, the results and evaluation values including steering wheel angle pulse input and serpentine test are shown in Fig. 9- Fig. 12 respectively.

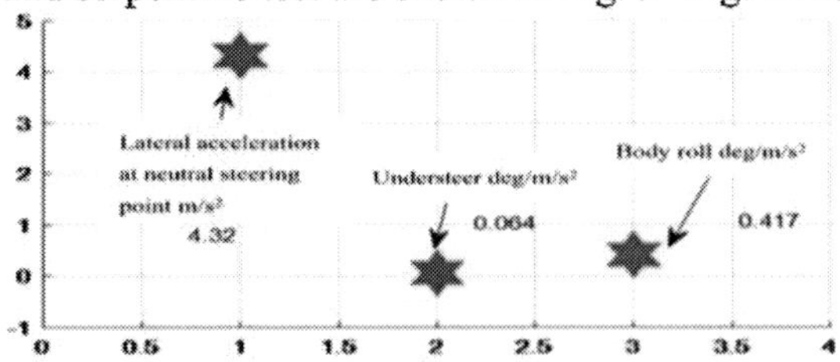

Figure 7. Experiment results of constant radius steering.

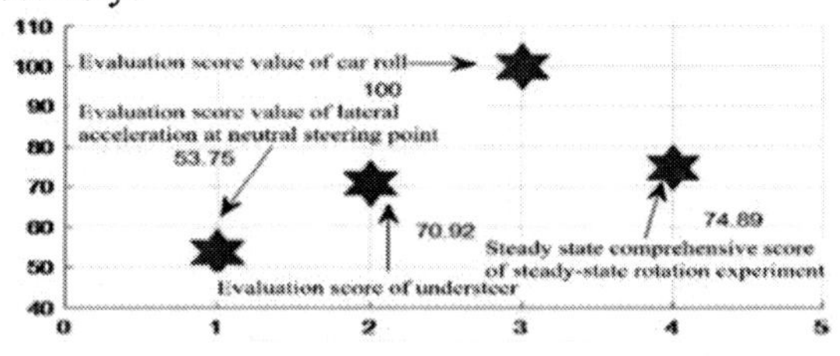

Figure 8. Evaluation results of constant radius steering.

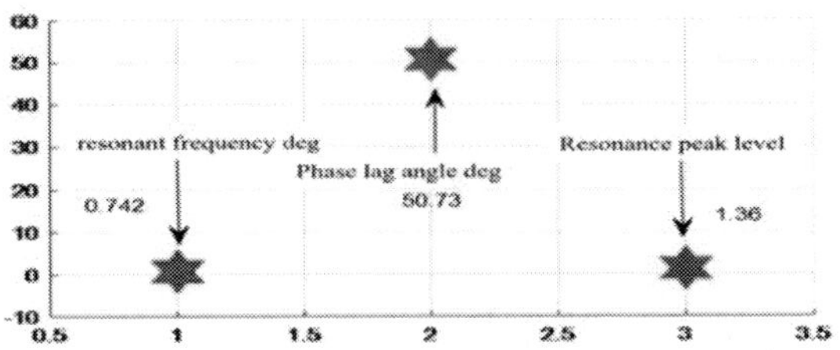

Figure 9. Experiment results of impulse response input.

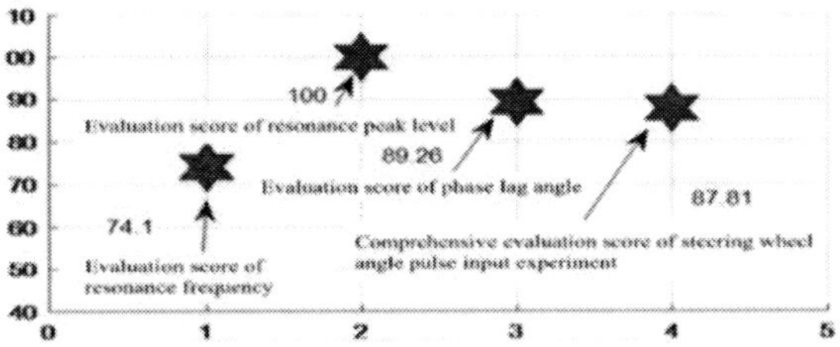

Figure 10. Evaluation results of impulse response input.

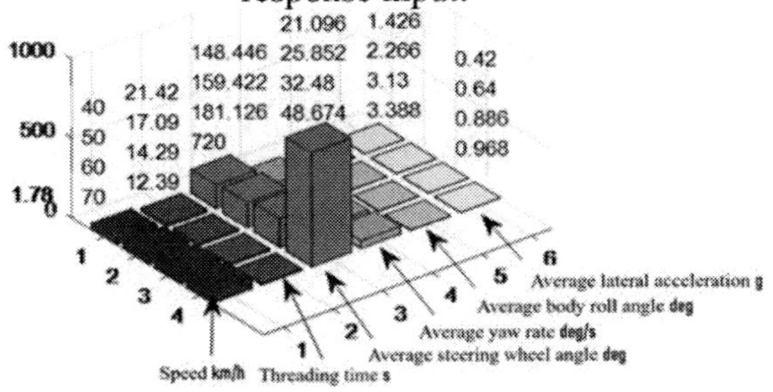

Figure 11. Experiment results of sine swept.

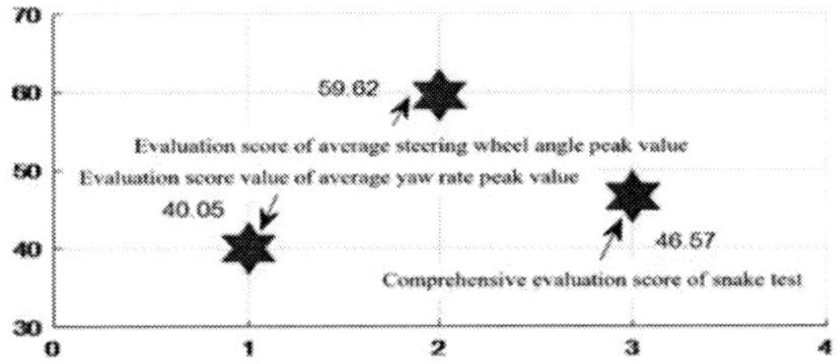

Figure 12. Evaluation results of sine swept.

5. Conclusion

In this paper, the ear variable cross-section free film air spring is taken as the research object. Through theoretical analysis, the stiffness formula of the ear variable cross-section free film air spring is obtained; The possible chaotic phenomena of the ear variable cross-section free film air spring are analyzed, and the dynamic behavior under deterministic excitation is analyzed; The dynamic model of air spring suspension system is established. The handling and stability simulation analysis is carried out, and the experimental evaluation value is obtained according to the national standard QC / t480-1999.

Acknowledgments

Thanks for the financial support from National Key Research and Development Program of China, grant number N2103028.

References

[1] YU Fan, LIN Yi. Vehicle System Dynamics[M]. Machinery Industry Press, 2005.
[2] Xing X, Kongkang Z, Nannan Z I, et al. Hierarchical Control of Ride Height System for Electronically Controlled Air Suspension Based on Variable Structure and Fuzzy Control Theory, Chinese Journal of Mechanical Engineering, 2015, 28(5): 945-953.
[3] YANG Shaopu, SHEN Yongjun. Bifurcation and Singularity of Nonlinear Systems with Hysteresis[M]. Beijing: Science Press, 2003.
[4] CHEN Long, ZHAO Huawei, JIANG Haobin, et.al. Study on Air Spring Stiffness Characteristic Model and Amendment to the Non-ideal Gas[J]. Automobile Technology, 2010, 4: 29-32.
[5] CHEN Liao, ZHOU Kongkang, LI Zhongxing. Dynamic Characteristics Fitting of Air Springs and Numerical Analysis of Air Suspension with Variant Stiffness[J]. Journal of Mechanical Engineering, 2010, 46(4): 93-98.
[6] Guekenheimer J, Homles P. Nonlinear oscillations: Dynamical Systems and Bifurcations of Vector Fields [M]. Springer-Verlag, New York, 1997.
[7] Criterion thresholds and evaluation of control ability and stability for automobiles.
[8] Yin Wanjian. Investigation on Nonlinear Dynamical in Automobile Air Spring Suspension System[D]. Beijing Jiaotong University, 2007.

Research on Underground Space Application of Multi-storey Complex Structure Based on Agent-Base Continuous Model

Desheng Xu[1], Xiangling Duanmu[2], Yanfeng Li[1,*], Peng Zhang[3] and Junmei Li[1]

[1] Beijing Key Laboratory of Green Built Environment and Energy Efficient, Beijing University of Technology Beijing, China
[2] China Academy of Building Research Limited Company, China
[3] China National Machinery Import & Export Corporation, China

*Corresponding author email: liyanfeng@bjut.edu.cn

Abstract. The fire emergency evacuation in a transfer station, with complicate structure and large number of passengers, is more difficult than that in a single subway station. A three-storey subway transfer station with cross type and island-side transfer form was selected as the research object. Through a field investigation, it was determined that there were 368 occupants should be evacuated in B3 platform and 212 people in B2 platform. And the proportion of adults attached 84%. An agent-based egress simulator was adopted to investigate the effect of different evacuation schemes. Results show that the optimal scheme is to change the arrival field for stair 1 and the stair 4 at the both ends of B3 platform to the concourse in B1 layer directly. In addition, fire curtains on B2 should land to an occupant height in the early stage of evacuation and then fall down to the floor after 240 s. It could improve the evacuation efficiency of the B2 and the B3 train fire, and the evacuation time is decreased by 110 s and 20 s respectively. This study would be helpful for safety evacuation design in a transfer station.

1. Introduction

Subway transfer station, as the node of metro line network, has characteristics such as complicated structure, personnel with high density. When a fire occurs, it is more difficult for passenger to evacuate from transfer station than that in a single subway station. Therefore, how to ensure the safe evacuation at a transfer station with fire scenario is an important issue in fire safety engineering.

Many scholars have made investigation on the fire scenario and occupant evacuation in a single subway station by numerical simulation method due to its efficiency and convenience. Tsukahara et al. [1] analysed the influence of smoke on personnel evacuation by using FDS (Fire Dynamics Simulator) numerical simulation method for a subway station. Zhou et al. [2] found the waiting pedestrians have different influences on the delay time, walking speed and turn angle of passing pedestrians, especially under high density level. Song et al. [3] changed the choice of pedestrian exits constantly to obtain the optimized pedestrian evacuation partition map. Results showed that the evacuation strategy of pedestrian partition could shorten the evacuation time of stadiums. In addition, some scholars have carried out a large number of studies on methodological approaches for evacuation, in order to make a better reflection of people behaviour in fire emergency. Yang et al. [4] introduced a modified model which has a definition of navigational force based on social force model. The simulation results verified that guide would play an essential role on efficient evacuation during an emergency. Guo et al. [5] proposed a modified social force model includes spatial and temporal separation rules to study the pedestrian counter flow through a bottleneck. Lo et al. [6] introduced an agent-based pedestrian movement model—city flow, and simulated a metro station case. Kallianiotis et al. [7] studied the evacuation process, users load and concentration, travel speed and maximum allowable time reach a point of safety.

Current researches are mainly focused on the occupant evacuation in a single subway station. The transfer station evacuation scheme is more complicate and important than that in a single station for fire safety. Furthermore, in the transfer stations, the evacuation in the station with island–side form is much more difficult than that of transfer station with island-island form. However, few studies have been conducted on the evacuation of subways in an island–side form transfer station.

The aim of this paper is to investigate the passenger safety evacuation in a three-storey underground transfer station with island–side transfer form. A field investigation was made in the island-side transfer subway station to determine the number of people should evacuate during the peak passenger flow. An agent-based egress simulator was adopted to investigate the effect of different evacuation schemes. The evacuation time of people in different zones were obtained and analysed. Optimization of evacuation scheme was conducted by comparing schemes with different settings of elevators and landing strategies of fire curtains.

2. Methods

2.1. Physical Model

Figure 1 shows a three-story island-side subway interchange station, which is the intersection of line 9 and line 10. The exits and transfer mode of the station could be clearly seen from the picture. There are four stairs from B3 to B2 layer. The stair 1 and stair 4 have the same width of 4 m while the remaining two stairs are 2.2 m wide. There are six stairs connecting the B1 and the B2. The middle staircases (staircases 5 and 6) are 4 m wide and the stairs on both sides are 2.6 m wide.

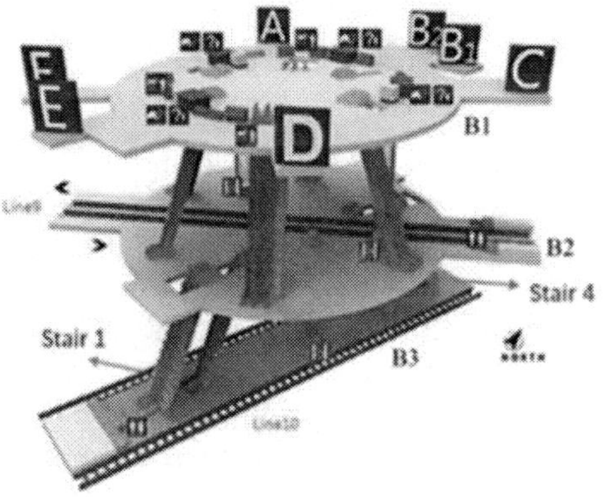

Figure 1. The structure diagram of a three-story island-side subway interchange station.

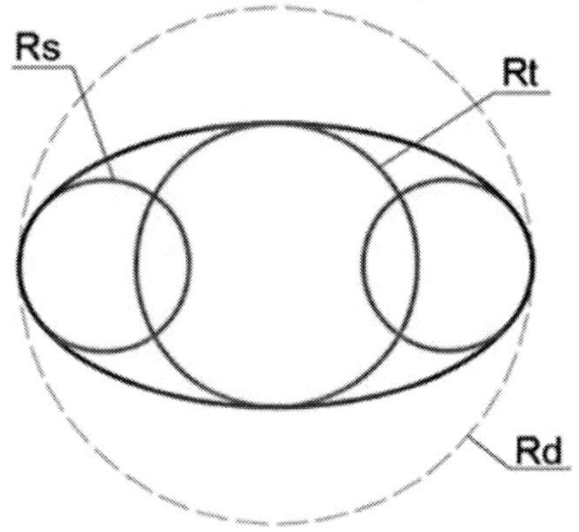

Figure 2. The shape of occupant size.

2.2. Mathematical Model

In this study, an agent model was used to show the movement circumstances of people. People are considered as agents [8], and moved in a plane which represents the floor of above ground or underground building. Occupant body can be simplified as a three-round combination shape in terms of size, as shown in figure 2.

The model of the agent movement follows the trajectory of the agent in the process of calculating the agent motion model through the laws of mechanics. And each agent should obey its motion equation:

$$m_a \frac{d^2 \vec{x}_a(t)}{dt^2} = \vec{f}_a(t) + \vec{\varepsilon}_a(t) \tag{1}$$

In formula (1), $\vec{x_a}(t)$ is the location of agent a at the moment of t, $\vec{f_a}(t)$ is the force of surrounding environment acting on agent a, m_a is agent mass, $\vec{\varepsilon_a}(t)$ is a random weak fluctuation force.

In this model, people are considered as individual agents, and controlled by their specific governing equation. This model can simulate high density crowds and many other situations. In many validation cases, the simulation data of this model has a correct agreement with the experimental data.

2.3. Numerical Simulation Method
An island-side transfer station simulation model was established by Pathfinder. The platform model of the B3 layer is shown in figure 3. The B2 layer of the station is a side platform. There are fire curtains between the station waiting area and the transfer area, as shown in figure 4. Once the station fire occurs in platform or rail area, fire curtains on both sides of the platform would drop to ground to form fire partition area.

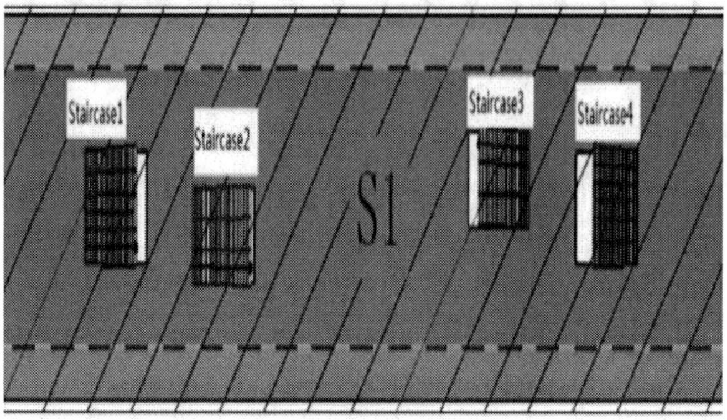

Figure 3. The diagram of B3 layer platform model.

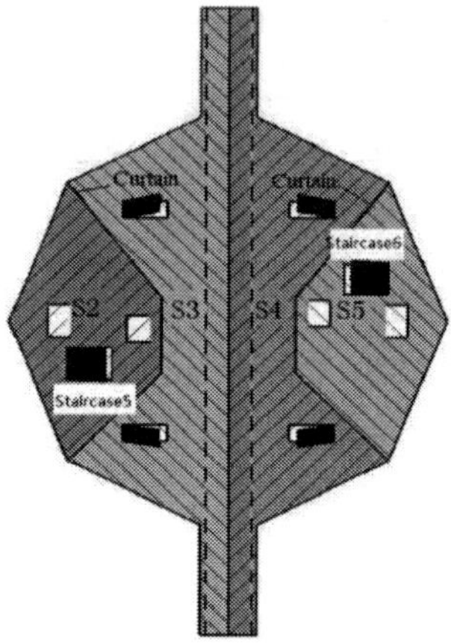

Figure 4. The diagram of B2 layer platform model.

2.4. Evacuation Time for Metro Station
From the perspective of personnel safety, the Chinese subway design code requires all personnel on the platform to stay away from the platform level within 6 minutes in the event of a fire. It is called the 6-minute principle. In NFPA130 [9], the export capacity of the platform is required to ensure that the load of the platform occupant is evacuated within 4 minutes or less. In comparison, it is slightly shorter than Chinese code. However, the prescribed evacuation time for the safe area is the same, both standards are 6 minutes. In CFDOM, the upper level of the platform can be a safe area because the exhaust ventilation system should control the smoke in the fire floor. Therefore, it could be stipulated that the time for the occupant to reach the safe area cannot exceed 6 minutes. And requirements of CFDOM was selected for judging safety evacuation in this study.

3. Results and Discussions

3.1. The Designed Evacuation Schemes
Due to the inner complicated structure, three evacuation schemes including the fire curtain operation and route selection are designed and listed in table 1. For each scheme, fire scenarios [10] have been included as shown in table 2. Type power of 2MW was selected for platform fire accident and 5MW was used for train fire accident [11]. The number of people evacuate at the subway transfer station during the peak passenger flow was determined by a field investigation in the island-side transfer subway station. The survey time was selected in the rush period from 7:45am to 8:15am. It was found that there are 15456 people on the Line 10 platform within 30 minutes and the train interval of Line 10 was 1.4 min. It was determined that during the rush period, there were 368 persons on Line 10 platform. Therefore, there were 368 occupants should be evacuated during this time. The Line 9 personnel load was determined in the same way as Line 10. There were 212 people waiting for the platform on one side during the peak period, and the number of passengers in the train was 1380. Besides, it can be concluded that the proportion of adults attach 84% through calculation.

For this three-storey transfer subway station, it is necessary to divide the station into different zones for safety evacuation. For example, when the fire accident occurs on the B3 platform, the areas shown in figure 3 and figure 4 can be marked as follows: S1 is the danger zone, S3 and S4 are the primary safety zones then S2 and S5 are the secondary safety zones, the concourse in B1 is the ultimate safety zone for evacuation. The primary safety zone is the initial position for passenger departing from fire source. Passengers need to arrive at secondary safety zones and research the ultimate safety zone. The waiting passengers in the concourse are considered in the safe area and they would no longer be taken into account as emergency evacuation person.

Table 1. Description of three evacuation schemes.

		Scheme No.1	Scheme No. 2	Scheme No.3
Fire curtains	Landing directly	+		
	Landing in stages		+	+
Stairs	Initial state	+	+	
	Improved state			+

Note: '+' means select this item

Table 2. The corresponding fire scenarios for subway transfer station evacuation schemes.

	Scenario 1	Scenario 2	Scenario 3	Scenario 4	Scenario 5
Fire position	Concourse	B2 train	B2 platform	B3 train	B3 platform
Evacuation number	0	2172	792	2172	792

3.2. Evacuation Time of No.1 Scheme

In the No.1 scheme, fire curtains in B2 were designed to drop to ground level directly when an emergency occurs. At the same time, the route of B3 to B1 was simple and lack of selectivity. Figure 5 shows the comparison of evacuation time in each zone when the fire source is located in the B2 floor transfer hall. The evacuation time of S1 and the primary safety zone is the longest, all of which are greater than 300 s. In addition, considering of the reaction time, the required safety evacuation time for the station is greater than 360 s, which cannot satisfy the safety evacuation criteria. This may be due to that the B3 floor occupant evacuate to the station hall must first evacuate to the B2 floor transfer hall. Therefore, once a fire breaks in the B2 floor transfer hall, the occupant in B3 would evacuate from the other side of the transfer hall stairs, which greatly increases the evacuation load of the transfer stairs in the opposite side. This would result in a serious evacuation bottleneck near the stairs, further affecting the station's evacuation efficiency.

3.3. Evacuation Time of No.2 Scheme

The program is mainly aimed at solving the problem that B2 floor transfer hall cannot meet the occupant evacuation of the B3 floor in a short period of time. In order to solve this kind of problem, the scheme proposes B2 floor fire curtains to be landed in stages. That is to say, when a fire occurs, fire curtains firstly drop to a height of 2 m from the ground. At this time, S2 and S3, S4 and S5 are still connected, and people are free to choose the stairs to evacuate to the station hall. At 240 s, the fire curtains are completely landed, S2 and S3, S4 and S5 are separated by fire curtains at this time, and people in each area could only be evacuated by their respective stairs.

It can be seen from figure 6 that when train fire occurs in B2 floor fire and B3 floor, the evacuation time is more than 350 s, and the reaction time of the occupant is 60 s, the required safety evacuation time is greater than 410 s, which does not meet the safety evacuation criteria. That is because when a fire occurs on a B2 train, the fire curtains are only dropped to the height of the occupant at the beginning of the fire. The passengers in the train can be evacuated from the B2 floor platform or the stairs of the transfer hall to the station hall during this time. However, passengers on the B3 floor also need to be evacuated to the transfer hall at the beginning of evacuation. Which results in the transfer hall stairs would bear both the evacuation load of the B3 floor and B2 train. The evacuation load of the stairs is increased and severe congestion occurs at the stairs compared to the initial situation, as shown in figure 7.

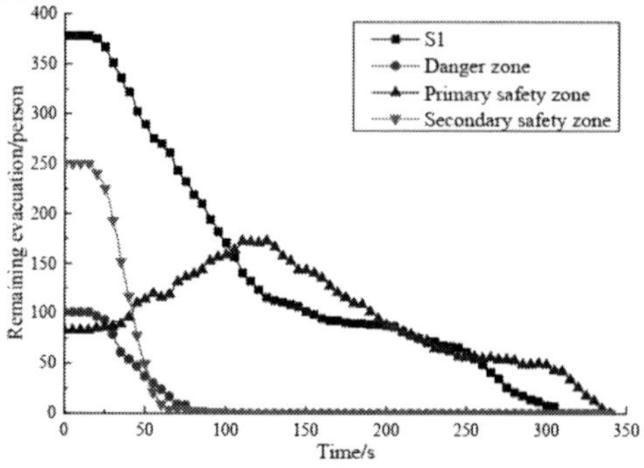

Figure 5. Comparison of evacuation time in each area for platform fire in transfer hall of B2 layer.

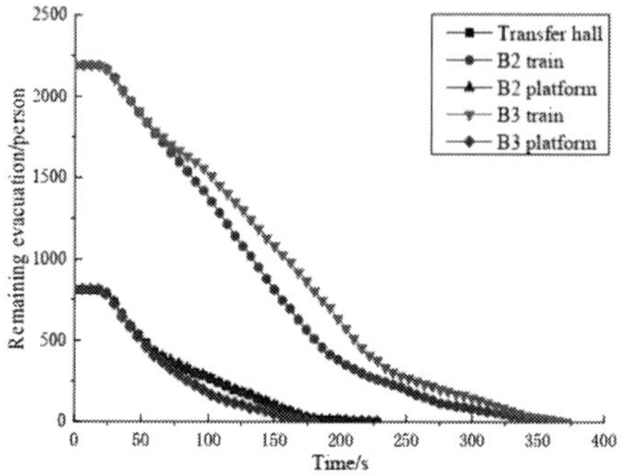

Figure 6. Comparison of evacuation time for No.2 scheme under different fire conditions.

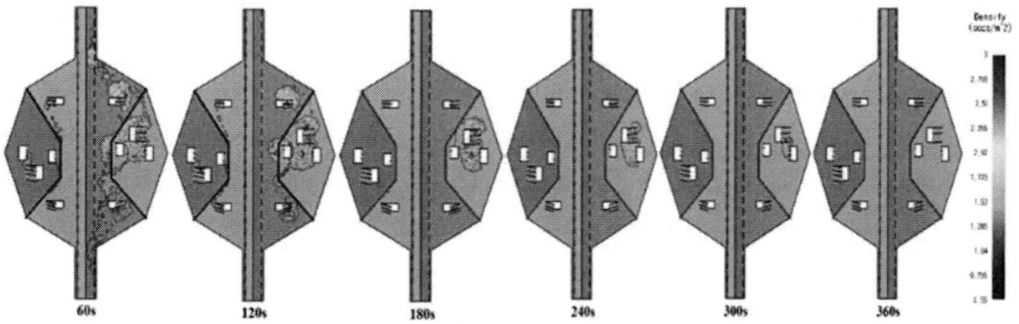

Figure 7. B2 floor evacuation process under B2 train fire.

3.4. Evacuation Time of No.3 Scheme

In this scheme, the fire curtains are lowered to the height of the occupant at the beginning of the evacuation and fall to the ground after 240 s. At the same time the arrival level of the stairs 1 and 4 of the B3 floor change to B1 directly, mainly to solve the problem that the B3 floor stairs throughput capacity cannot meet the full occupant evacuation of the B3 floor in a short time. The evacuation time of personnel under different fire conditions using this scheme is shown in figure 8. The scheme mainly improves the evacuation efficiency of the B2 and the B3 train fire, the evacuation time is decreased by 110 s and 20 s respectively.

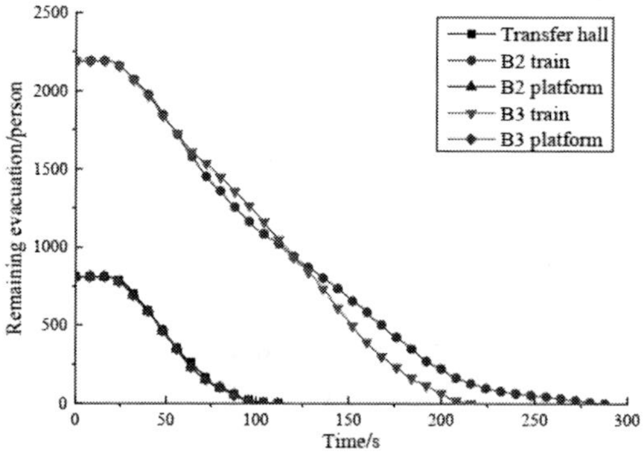

Figure 8. Comparison of evacuation time for No.3 scheme under different fire conditions.

4. Conclusions

The evacuation in an existing three-layer underground transfer station has been simulated by an agent-based egress simulator. Three schemes of evacuation were designed and the optimal scheme for this side-island transfer station was decided. The conclusions are drawn as follows:

1) In this three-storey transfer station with side-island transfer form, the disadvantage is that passenger in B3 layer has to move to B2 layer then to concourse in B1 layer. In the No.1 scheme, when fire occurred in B2 side platform, only limited region can be used for passenger in B3 and B2 layer if fire curtains dropped directly to the ground level. The safety evacuation criteria cannot be achieved.

2) One improvement issue in No. 2 scheme was to change the stage of fire curtains dropping. This method had no significant effect on improve the evacuation efficient in B2 layer when fire occurd at side platform in B2 layer. B2 layer passenger include train passenger and await passenger both on B3 and B2 would gather on the stairs from B2 to B1. The safety evacuation criteria cannot also be achieved.

3) In No 3 scheme, both setting two stairs directly from B3 to B1 and changing the dropping stage of fire curtain in B2 layer were considered. Since, the passenger moving to B2 from B3 layer decreased and the stairs could be used for passengers from B2 to B1 were kept for a certain time period. The safety evacuation criteria could be achieved. Therefore, it was determined as the optimal scheme for designing the safety evacuation in this three-layer underground transfer station with island–side transfer form.

In the future work, we will study the density of people flow and speed in subway transfer stations.

Acknowledgments

This work has been supported from the Beijing Natural Science Foundation under Grant No. 18H10009.

References

[1] Tsukahara M, Koshiba Y, Ohtani H (2011) Effectiveness of downward evacuation in a large-scale subway fire using Fire Dynamics Simulator[J]. *Tunnelling & Underground Space Technology*. 26(4):573-581.

[2] Zhou M, Dong H, Wang F Y, Zhao Y B, Gao S G, Ning B (2019) Field observations and modeling of waiting pedestrian at subway platform. *Information Sciences*. 504:136-160.

[3] Y Song, Cheng H, Wang Q, Lyu W (2021) Study on optimization strategy and simulation of pedestrian partition evacuation in large stadiums[J]. *Journal of Safety Science and Technology*. 17 (3):19-25.

[4] Yang X, Dong H, Wang Q (2014) Guided crowd dynamics via modified social force model[J]. *Physica A: Statistical Mechanics & Its Applications*. 411(10):63-73.

[5] Guo R Y (2014) Simulation of spatial and temporal separation of pedestrian counter flow through a bottleneck[J]. *Physica A Statistical Mechanics & Its Applications*. 415(415):428-439.

[6] Lo S M, Wang W L, Liu S B (2014) Using Agent-based Simulation Model for Studying Fire Escape Process in Metro Stations[J]. *Procedia Computer Science*. 32:388-396.

[7] Kallianiotis A, Papakonstantinou D, Arvelaki V, Benardos A (2018) Evaluation of evacuation methods in underground metro stations[J]. *International Journal of Disaster Risk Reduction*. 31:526-534.

[8] Thompson P A, Marchant E W (1995) A computer model for the evacuation of large building populations[J]. *Fire Safety Journal*. 24:131–148.

[9] Anastasios K, Despina P, Vasiliki A (2018) Evacuation Method Evaluation in Underground Metro Station[J]. *International Journal of Disaster Risk Reduction*. S2212420918302152.

[10] Zhang P, Li Y F (2018) Experimental Study on Shrinking Dimensions of Subway Transfer Station Based on Similar Analysis[J]. *Fire Science and Technology*. 1009-0029.

[11] Li Y F (2015) Fire safety technology for large subway interchange stations. [M]. *Science press*. China, 69-75.

Numerical Simulation Study of Effect of Tunnel Slope on Smoke Propagation

Dandan Wan[1], Yanfeng Li[1,*], Yunfei Li[2] and Huiqiang Liu[3]

[1] Beijing Key Laboratory of Green Built Environment and Energy Efficient Technology, Beijing University of Technology, Beijing, China
[2] China Railway Construction Real Estate Group, Beijing, China
[3] Beijing Engineering Corporation Limited, Beijing, China

*Corresponding author email: liyanfeng@bjut.edu.cn

Abstract. Tunnel slope is an important factor affecting smoke spreading for a stopped subway train on fire. In this paper, a multi-scale method where one-dimensional (1D) network is coupled with three-dimensional (3D) field simulation is used to calculate smoke and temperature distributions for a fire occurred in the middle part of the subway train in tunnel. The tunnel slopes of 0‰, 12‰, 21‰ and 30‰ are considered. By comparing the temperature distribution in a condition with natural ventilation and with 2 m/s longitudinal ventilation, we comprehensively analyzed temperature, smoke spread characteristics. Results show that tunnel slope has a great influence on temperature and smoke profile when slope is higher than 12‰. The temperature at 1.8m would exceed the safe limit of 60℃ at 60 s after fire ignition. The larger the slope, the higher temperature would be in the train carriage. For a train fire scenario with natural ventilation, within initial 120 s period the train compartment could keep a safe evacuation condition. This research would provide reference for making right evacuation plan of subway tunnel fire.

1. Introduction

With the development of urbanization, rail transport has attracted more and more attention because of its large passenger capacity, comfort and safety [1]. The metro system has been developed rapidly due to its ability to ease the city traffic jam. Due to the long and confined structure character of subway tunnel, passengers would be greatly threatened by smoke in fire scenario [2]. Therefore, it is of great importance to study smoke propagation effect in a subway tunnel fire for safety evacuation [3].

Many scholars have studied the effect of slope on smoke propagation in road tunnel. Zhao et al. [4] conducted numerical simulation and theoretical analysis on the influence of slope on the fire of subway tunnel. It revealed that the effect of tunnel slope on smoke spread increases with the increasing slope. Based the 1/20 reduced-scale model tunnel, G.H. k et al. [5] adopted experiment to study the effect of tunnel slope on the critical velocity in tunnel fires. The results indicated that critical velocity increases due to the stack effect caused by slope. Chen et al. [6] carried out a series of experiments in a model bifurcated tunnel with different ramp slopes under natural ventilation. The results showed that the maximum ceiling temperature in the main tunnel increases with the slope increasing. Wang et al. [7] used the full-scale Highway Slope Tunnel to study the influence of tunnel width and slope on ceiling gas temperature distribution by numerical simulations. It proposed characteristic temperature distribution in slopping tunnel. Li et al. [8] proposed that the temperature of the underground inclined tunnel changes with the exponential law, and introduced the simplified method of the temperature and pressure of the underground inclined tunnel. Chow et al. [9] found that the smoke temperature characteristic and smoke propagation in tunnel fire when slope varied. The results indicated that the high temperature area extends to the downstream area with the increasing tunnel slope.

For the research mentioned above, previous researches mainly focused on road tunnel fire while the subway tunnel was rarely investigated. However, there are some differences between subway tunnel and road tunnel. For example, compared with road tunnel, the subway tunnel has a greater blockage ratio than 0.5 and the tunnel block length by train is much longer than a car. Furthermore, cross-sectional geometries are different. More importantly, the slope of subway tunnel is relatively small compared with road tunnel.

The aim of the paper is to investigate the temperature distribution and smoke spread law in a subway tunnel where the train on fire is stopped in the middle of tunnel. A multi-scale simulation method combining field and network simulation was proposed to analyze smoke and temperature profile under the different slopes. It could establish reasonable time and endurable environment for safe evacuation of personnel according to simulate the smoke propagation.

2. Methods

A multi-scale method where one-dimensional network is coupled with three-dimensional field simulation was selected. This paper adopted multi-scale method by combining the one-dimensional simulation software RTV (road tunnel ventilation) with three-dimensional simulation software (FDS).

Fire Dynamic Simulator (FDS) is used to solve a form of the Navier–Stokes equations for thermally-driven flow, which was proposed by the National Institute of Standard Technology (NIST) Building Fire Research Laboratory. In addition, DNS model and LES model also can be found in FDS.

IDA Road Tunnel Ventilation (RTV) is a modular tunnel ventilation simulation software. It uses IDA solver to solve modular differential algebraic equations.

2.1. Physical Model

The tunnel of a Beijing subway line is the object of the research, which is a standard shield tunnel with a length of 2.2 km and an inner diameter of 5.4 m. The dimension of the train is 120m(L) × 3m(W) × 3.8 m(H). (excluding the roof air conditioner and pantograph), as shown in figure 1.

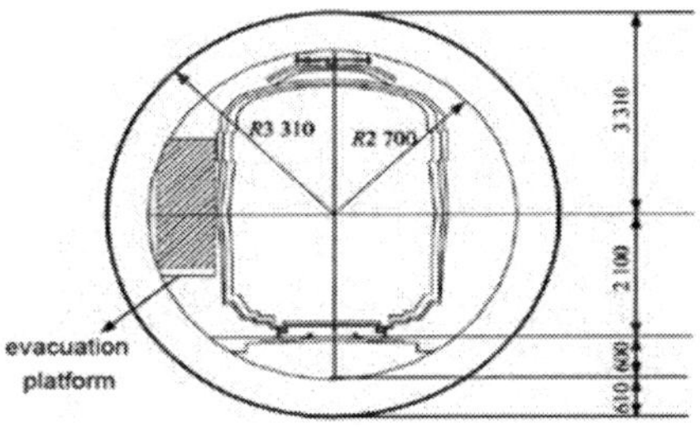

Figure 1. Section size and train size of interval tunnel.

2.2. Gird Independence Test

According to the previous research, the heat release rate selected 7.5 kW and the reference point was at 10 m downstream of the fire source. Grid sizes of 0.3 m, 0.25 m, 0.2 m and 0.15 m were selected to simulate vertical temperature distribution 10 m downstream of the fire. After the flow field was fully developed for 10 minutes, the simulation result was shown in figure 2.

As shown in figure 2, the vertical temperature distribution curve was relatively rough when the mesh grid equaled to 0.25 m and 0.3 m. When the mesh size was refined to 0.20 m and 0.15 m, the vertical temperature distribution curve was relatively smooth. Considering the calculation resources and time cost comprehensively, all the grid sizes were 0.2 m × 0.2 m × 0.2 m.

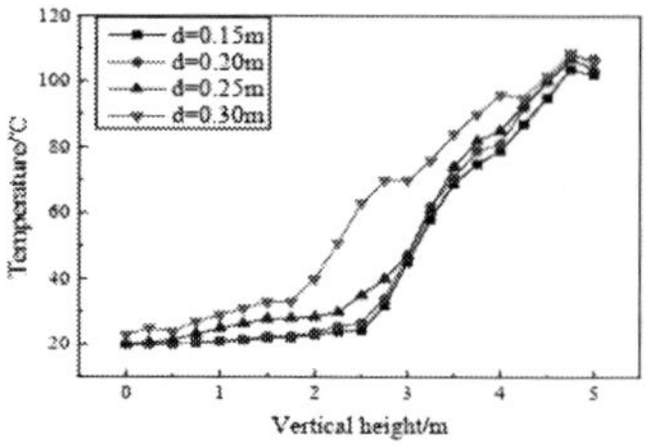

Figure 2. Vertical temperature distribution at 10 m downstream of fire source under different grid sizes.

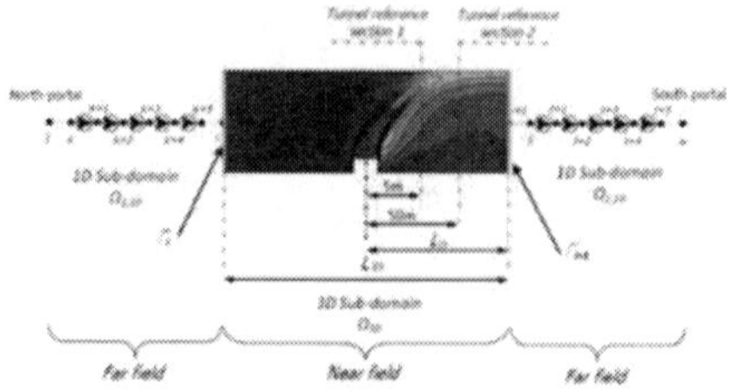

Figure 3. Multi-scale model.

2.3. *Multi-Scale Method*

This study used a multi-scale simulation method combining field and network and introduced NF (Near Field) and FF (Far Field). There is a huge temperature and velocity change rate near the fire source and obvious 1D flow in the far fire source region, so the behavior of near field regions is modelled using a FDS model and the far field regions using the 1D model. Figure 3 shows the model of the multi-scale simulation.

2.3.1. Establishment of 1D numerical model. RTV software was used to simulate the situation that the train stops in the middle of the long interval tunnel. In the 2200 m long tunnel, we set the middle smoke exhaust shaft at 1050 m away and set the fire source at 250 m away from the left platform. The simulated ambient temperature was 293 K. The 1D numerical model at standard atmospheric pressure was shown in figure 4. The elevation difference of model at different transverse positions in the figure did not represent the actual tunnel slope. On the contrary, the different slopes of the tunnel should be determined by internal parameters.

Based on different reference surfaces, the results calculated by 1D model as boundary conditions were brought into FDS model. The results of 1D model were compared with 3D model to obtain different error rates. According to the simulation results, the coverage area of the field simulation should be expanded by 150 m at the reference section to improve the accuracy of simulation results.

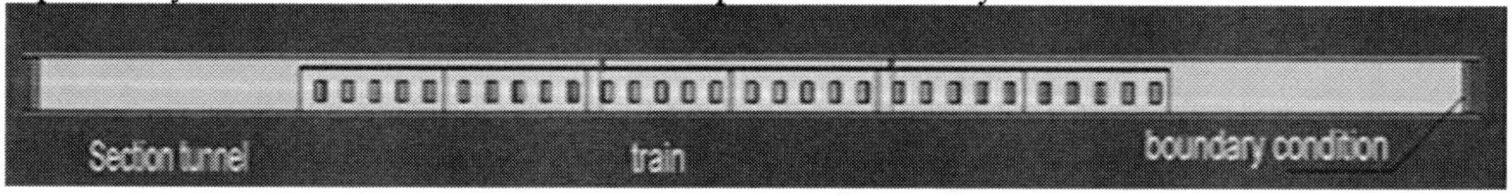

Figure 4. 1D numerical model

2.3.2. Establishment of 3D numerical model. In order to satisfy the accuracy of the simulation results, the interface location was expanded by 150 m on the basis of research points. Therefore, the simulation length of 3D field model was approximately 420 m, as shown in figure 5.

The boundary conditions of the local 3D field model were provided by the simulation results of the 1D network model. Among them, the velocity was set to be 0.02 m/s. Moreover, the temperature was set to be 295 K when the wind blows from left to right and was set to be 294 K when the wind directions is from right to left.

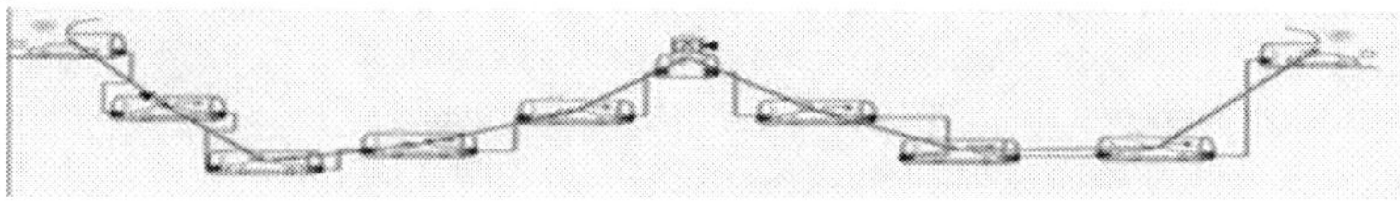

Figure 5. 3D field model.

2.4. *Simulated Conditions*

There are many factors that have a significant impact on smoke spreading in a subway tunnel fire. The influence of slope was mainly studied in this paper. According to the Metro Design Code (GB50157-2013), the tunnel slope should not exceed 35‰ and the critical velocity is usually 2 m/s. Under the natural velocity, the slopes of 0‰, 12‰, 21‰ and 30‰ were simulated when the fire occurred in 60 s and 360 s. The temperature distribution with the slope of 30‰ at the height of human eye was simulated under two conditions: natural velocity and longitudinal velocity of 2 m/s in uphill tunnel.

3. Results and Discussion

Due to the particular structure of subway long interval tunnel, the safety of people will be greatly threatened when the fire occurs. This study aims to provide a reference for safe evacuation in tunnel fire by studying a series of parameter changes during the smoke propagation.

3.1. Smoke Propagation under Natural Ventilation in Tunnel with Difference Slopes

It could be presented in figure 6 and figure 7 that the smoke on both sides of the fire source in the horizontal interval tunnel was basically symmetrically diffuse under natural ventilation conditions with time when the fire was in the middle of the train. The smoke distributions in the tunnels with slopes of 12‰, 21‰, and 30‰ show few significant differences after 60 s. After 360 s, the smoke diffused and filled the entire train compartment. For the tunnel with slopes of 12‰ and 21‰, the thickness of the smoke layer increased in the uphill direction, and the smoke flew back in the downhill direction. The smoke accumulated and began to gradually settle down in the uphill direction under the slope of 30‰, meanwhile, the smoke diffused to the end train in the downhill direction, where the smoke back-flow is obvious. The results showed that as the tunnel slope increased, the smoke spreading distance increased in the uphill direction and decreased in the downhill direction.

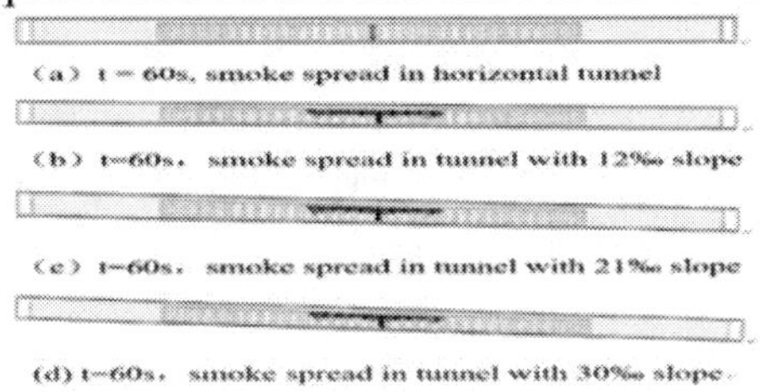

Figure 6. t=60 s, the smoke distribution under natural ventilation in interval tunnels with different slopes.

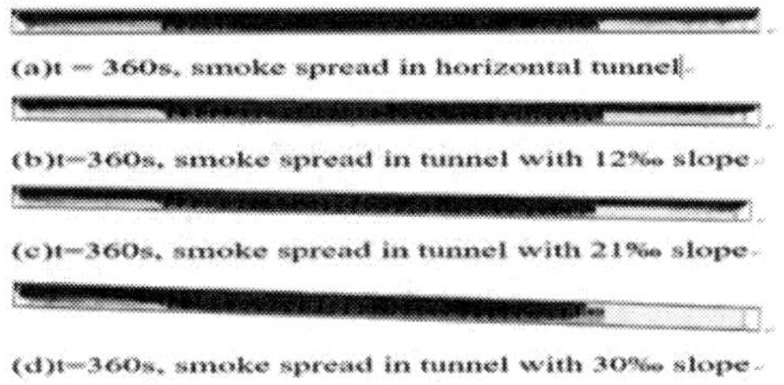

Figure 7. t=360 s, the smoke distribution under natural ventilation in interval tunnels with different slopes.

3.2. Smoke Temperature Distributions

Personnel height refers to the vertical distance from the ground to the head when the human body is upright, whose research range is generally from 1.6 m to 2 m. This study selected the value of the personnel height as 1.8 m. According to the recommended value of NFAPA130, the temperature at personnel height should not exceed 60°C, which is the limit temperature of the safe evacuation. Based on this standard, the temperature distributions with different slopes under natural ventilation were investigated, as shown in figure 8 to figure 11.

These results showed that the temperature distribution in the horizontal tunnel was basically symmetrical when the fire was in the middle of the train. The smoke almost filled the whole train in h the tunnel with 12‰ slope, and the temperature in the train was in the range of 65-90°C after 120 s. The smoke spreading range expanded and the smoke temperature increased with time. After 360 s, the smoke spreading range continued to increase, but the smoke temperature no longer showed an obvious upward trend. When the slope is 21‰, the one-sided distribution of temperature appeared. Affected by stack effect, the temperature along the uphill direction was generally higher than that in the downhill direction.

After 120 s, the temperature was above the personnel safety limit of 60°C during the smoke propagation, but the temperature in the downstream tunnel decreased faster, moreover, the safety range was expanding. When the slope was 30‰, the temperature difference between the upstream direction and the downstream direction was remarkable. After 120 s, the smoke spread out of the train about 10 m upstream of the tunnel. During the smoke coverage, the temperature was in the range of 80-105°C. Compared with the tunnel with the slope of 21‰, the overall temperature in the tunnel increased to a certain extent. In addition, the high temperature range has further spread with time. The smoke caused by stack effect appeared back-laying. Compared with the temperature distribution after 240 s, the safe area was expanded.

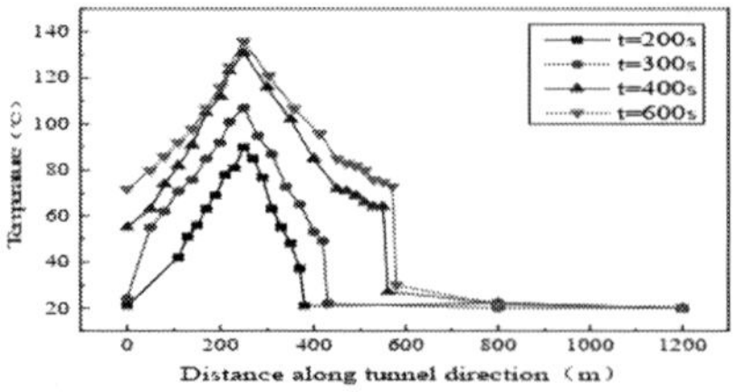

Figure 8. Temperature profile in horizontal interval tunnel under natural ventilation with time.

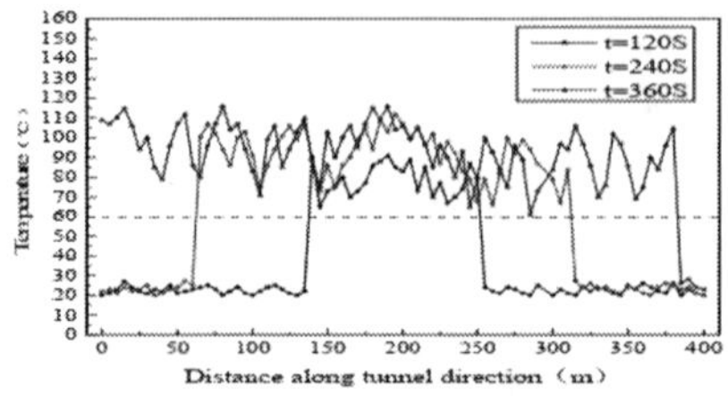

Figure 9. Temperature profile in 12‰ slopping tunnel under natural ventilation with time.

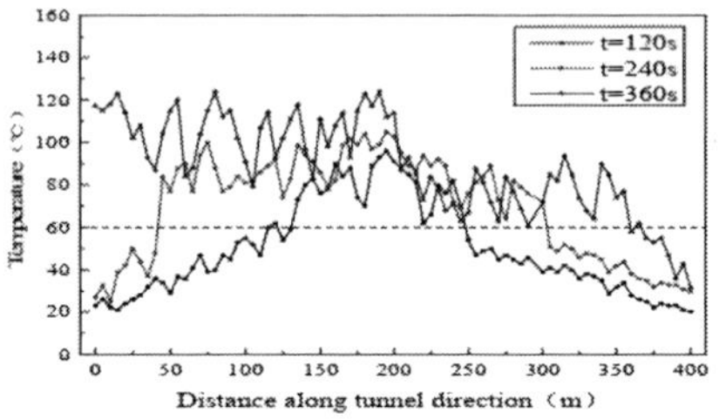

Figure 10. Temperature profile in 21‰ slopping tunnel under natural ventilation with time.

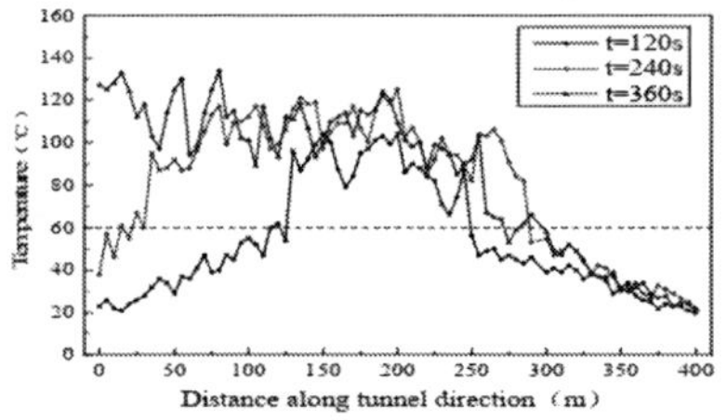

Figure 11. Temperature profile in 30‰ slopping tunnel under natural ventilation with time.

3.3. Comparison of Temperature Distribution between Natural Ventilation and Longitudinal Ventilation

In previous studies, many smoke exhaust modes were proposed to control the smoke spread in the long interval tunnel. Longitudinal ventilation was one of the common methods. Taking the 30‰ slopping tunnel as an example, the comparison of the temperature distribution with and without longitudinal ventilation was studied to take a reference for safe evacuation of personnel. Figure 12 and figure 13 presented the simulation results. It can be found that temperature distribution was basically symmetrical under natural velocity after 120 s. With the passage of time, the asymmetry of temperature distribution caused by stack effect appeared and the asymmetry was more obvious under longitudinal velocity. During the smoke spreading process, the temperature under longitudinal velocity was lower than under natural velocity, which is due to that longitudinal velocity will bring cool air to cool part of the high temperature flue gas. After 120 s, compared with no longitudinal velocity, the high-temperature area on the right side of the fire source expanded. It was due to that the longitudinal wind energy tilted the flame slightly. It could also be seen that after 120 s, the flue gas temperature under the longitudinal velocity was controlled within the safe limit of 60°C.

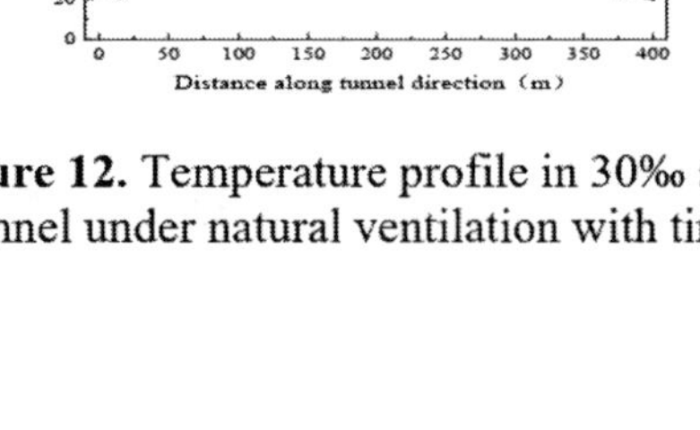

Figure 12. Temperature profile in 30‰ slope tunnel under natural ventilation with time.

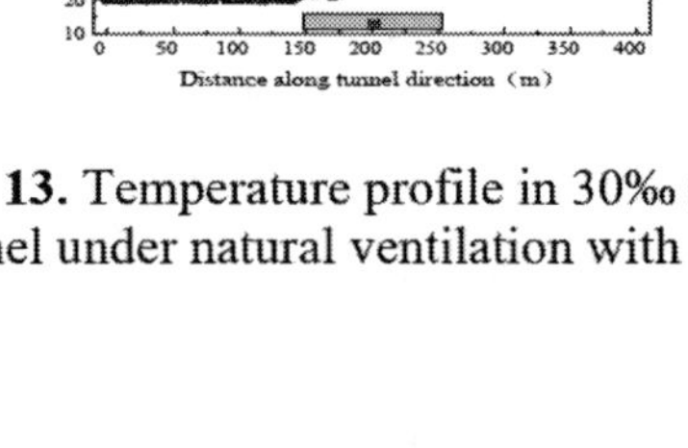

Figure 13. Temperature profile in 30‰ slopping tunnel under natural ventilation with time.

4. Conclusion

This paper investigated the influence of slope on smoke and temperature distribution in subway tunnel by using multi-scale simulation. By analyzing the simulated results, main conclusions can be drawn as follows:

Different slopes have great effects on smoke spread and safe evacuation. Smoke spreading distance in upstream direction increased and decreased in downhill direction with increasing slope. When the slope was higher than 12‰, the smoke back-laying would appear after 360 s. When the slope equal to 30‰, the back-laying phenomenon was obvious. After 120 s, the smoke almost filled the entire train under the tunnel slope of 12‰. The temperature in train was in the range of 65℃ to 90℃.

For tunnel with a slope of 30‰, the temperature under natural ventilation can be maintained below 60℃ in the initial period of 120 s. It is necessary to turn on mechanical exhaust system to control smoke after 120 s of train fire. In the future study, we would conduct related experiments to verify the simulation results.

Acknowledgments

This work has been supported from the Beijing Natural Science Foundation under Grant No.8172006.

References

[1] Yang Y X, Liu C, Qiu P Y, Long Z, Zhong M H (2020) Full-scale fire experiment under vehicle blockage in a tunnel with slope[J]. *Journal of Tsinghua University (Natural Science Edition)*. 60(12):1030-1038.

[2] Zhang S G, Yao Y Z, Zhu K, Li K Y, Zhang R F, Lu S, Cheng X D (2016) Prediction of smoke back-layering length under different longitudinal ventilations in the subway tunnel with metro train[J]. *Tunnelling and Underground Space Technology incorporating Trenchless Technology Research*. 53:13-21.

[3] Lei F X (2015) *Study on fire ventilation of subway tunnels with long slopes*, M.A. Thesis. Cheng Du: Southwest Jiao tong University.

[4] Zhao W D, Li H (2009) Mechanism analysis of the influence of slope on railway tunnel fire[J]. *Transportation Science and Engineering*. 25(01): 53-57.

[5] Gwon Hyun K, Seung Ryul K, Sun R H (2009) An experimental study on the effect of slope on the critical velocity in tunnel fires[J]. *Fire Sci*. 28:27–47.

[6] Chen C K, Nie Y L, Zhang Y L, Lei P, Fan C G, Wang Z Y (2021) Experimental investigation on the influence of ramp slope on fire behaviors n a bifurcated tunnel[J]. *Tunnelling and Underground Space Technology*. 104.

[7] Wang Z Y, Ding L, Wan H X, Ji J, Gao Z H, Yu L X (2020) Numerical investigation on the effect of tunnel width and slope on ceiling gas temperature in inclined tunnels[J]. *International Journal of Thermal Sciences*. 152.

[8] Li A G, Gao X P, Ren T (2017) Study on thermal pressure in a sloping underground tunnel under natural ventilation[J]. *Energy & Buildings*. 147.

[9] Chow W K, Gao Y, Zou Z F, Liu Q K, Chow C L, Miao L (2018) Numerical studies on thermally-induced air flow in sloping tunnels with experimental scale modelling justifications[J]. *Fire Technol*. 54(4):867-892.

Modeling of Film Formation in Airless Spray

Guichun Yang, Yan Chen*, Shiming Chen and Fei Zhang

Department of Petroleum, Army Logistics University, Chongqing, China

*Corresponding author e-mail: yansohucom@sohu.com

Abstract. Aiming at the spraying process of airless atomization spraying, a coating film forming model based on Euler-Lagrangian method is proposed. A Rosin-Rammler atomization model was established. In the spraying model, for the discrete phase, a force balance equation is established. At the same time, a continuity equation and a momentum equation are established to describe the motion of the continuous phase. The momentum equation of the continuous phase is closed by the standard *k-ε* model. The continuity equation and momentum equation are established in the adhesion deposition model to solve the coating thickness. PC-SIMPLE algorithm was used to solve the model. Finally, the dynamic spraying simulation is verified through experiments, and the numerical simulation results are in good agreement with the experiments as a whole. It is proved that the film forming model can be used to predict the film forming results of airless spraying.
Keywords: Airless spray; Film model; Numerical simulation.

1. Introduction

For large-scale equipment in the oil and gas industry, airless spraying to form a protective layer on the metal surface is the most effective method of anti-corrosion [1]. The rationality of the airless spraying film-forming model directly affects the high utilization rate of the paint, the smoothness, and the uniformity of the thickness of the coating. Due to the lack of a reasonable film forming model, there are problems such as unsatisfactory spraying trajectory planning, inaccurate numerical simulation of spraying thickness, and difficulty in controlling the quality of the coating film. Therefore, the establishment of a reasonable airless spray coating film forming model is of great significance to improve the quality of the coating, reduce the waste of the coating, protect the environment, and improve the working environment of workers.

This article mainly establishes the airless spraying film formation model to predict the airless spraying film formation results. The second part introduces the research progress of airless spray modeling. In the third part, a film-forming model of airless spraying was established according to the spraying process. The fourth part verifies the rationality of the model through numerical simulation. The fifth part verifies the correctness of the model through experiments.

2. Progress in Modeling Film Formation by Airless Spraying

Early scholars used experimental methods to summarize empirical formulas in the research of spray coating modeling to establish the film formation model. There are mainly two models: Infinite range model and limited range model. In the infinite range model, the infinite distance from the spray point is defined as the point where the film thickness is zero, and the spray flow field has no boundaries. In the limited range model, the spray flow field has a boundary, and the point of zero film thickness in the flow field lies outside the boundary of the flow field. Most of these models use the controlled

variable method to simply study the influence of various factors on the quality and thickness of the film, which is of little significance to actual engineering guidance.
With the rapid development of computer technology, computational fluid dynamics (CFD) can be used to accurately simulate complex turbulence results at a relatively low cost. According to different paint drop treatment methods, there are two main CFD methods for establishing models at this stage: Euler-Euler method [2] and Euler-Lagrangian method [3]. The Euler-Euler method regards paint and air as continuous phases for description, so it cannot accurately describe the movement of paint droplets. The Euler-Lagrangian coordinate method considers the mass, momentum, energy and component exchange between the discrete and continuous phases through the physical properties of a single droplet, and can describe each droplet in the flow field environment better and accurately The overall state of motion.

3. Spray Model

The spraying process of airless spraying can be divided into the atomization process and the coating film forming process. Since the focus of this paper is the film formation process, only the atomization process is used as the boundary condition and initial condition of the paint incidence [4]. The coating film formation process includes the movement process of the coating droplets and the air phase and the droplet collision adhesion deposition process. The movement process of the paint droplet and the air phase is the process in which the paint is accelerated from the nozzle by high pressure, collides with the air many times, and finally shoots to the surface of the workpiece.

3.1. Atomization Model

The paint distribution in the atomization field conforms to the Rosin-Rammler distribution:

$$V_d = 1 - \exp\left[-\left(\frac{d}{\bar{d}}\right)^n\right] \tag{1}$$

where V_d is the ratio of the volume of all paint droplets with a diameter below d to the total paint droplet volume; d is the average diameter of all paint droplets (m); n is uniformity index, the larger the value of n, the more uniform the diameter distribution of paint droplets.

3.2. Coating Film Forming Model

The movement process of airless spray paint is the transfer process of momentum and quality between the paint phase and the air phase, and the heat transfer phenomenon in the flow can be ignored. Using the Euler-Lagrangian method, the discrete phase is the atomized paint droplets, and the continuous phase is the air in the flow field, and the momentum and mass conservation equations are established.

3.2.1. Discrete phase model. The force balance equation of paint droplets is:

$$\frac{du_s}{dt} = \frac{g_x}{\rho}(\rho - \rho_s) + F_D u - F_D u_s + F_x \tag{2}$$

where u is the velocity of the outside air phase (m/s); u_s is the velocity of the paint droplet phase (m/s); ρ is the density of the outside air phase (kg/m^3); ρ_s is the paint droplet phase Density (kg/m^3); g_x is the local gravity (kg·m/s^2); F_D is the drag force of air relative to paint droplets (N); F_x is other forces (N).
The differential equation of the trajectory of the paint droplet can be obtained by gradually performing the integral operation on the discrete time and step length:

$$\frac{dx}{dt} = u_s \tag{3}$$

If in each time step, the force on the droplet can be maintained constant, the trajectory equation can be written as:

$$\frac{du_s}{dt} = \frac{1}{t_s}(u - u_s) \tag{4}$$

where t_s is the time of the droplet relaxation(s).

3.2.2. Continuous phase model. In the airless spraying process, the air in the environment is used as the continuous phase. Because the heat transfer phenomenon in the process is ignored, the governing physical equations of the continuous phase air mainly have two basic equations: mass equation and momentum equation.
Quality equation:

$$\frac{\partial \rho}{\partial t} = -\frac{\partial(\rho u_i)}{\partial x_i} \tag{5}$$

where ρ is the air phase density (kg/m^3); u_i is the i direction speed (m/s).
Momentum equation:

$$\frac{\partial(\rho u_i)}{\partial t} + \frac{\partial(\rho u_i u_j)}{\partial x_j} + \frac{\partial p}{\partial x_i} = \frac{\partial}{\partial x_j}\left(\mu \frac{\partial u_i}{\partial x_j}\right) - \frac{\partial\left(\rho u_i' u_j'\right)}{\partial x_j} \tag{6}$$

where u_j is the speed in j direction (m/s); p is the pressure (Pa).
Since in the continuous phase air flow field, the air flow belongs to turbulent motion, it is necessary to establish a corresponding turbulence model to close the momentum equation of the air phase. The standard k-ε model is introduced to close the momentum equation. The transport equation representing the turbulent kinetic energy (m^2/s^2), namely the k equation, is:

$$\rho \frac{\partial k}{\partial t} + \rho \frac{\partial(k u_i)}{\partial x_i} = \frac{\partial}{\partial x_j}\left[\left(\mu + \frac{\mu_t}{\sigma_k}\right)\frac{\partial k}{\partial x_j}\right] + G_k + G_b + S_k - Y_M - \rho\varepsilon \tag{7}$$

The transport equation that expresses the turbulent dissipation rate(m^2/s^3), namely the ε equation, is:

$$\rho \frac{\partial \varepsilon}{\partial t} + \rho \frac{\partial(\varepsilon u_i)}{\partial x_i} = \frac{\partial}{\partial x_j}\left[\left(\mu + \frac{\mu_t}{\sigma_\varepsilon}\right)\frac{\partial \varepsilon}{\partial x_j}\right] + \frac{\varepsilon}{k} G_{1\varepsilon}\left(G_b + G_{3\varepsilon} G_b\right) + \frac{\varepsilon^2}{k}\rho C_{2\varepsilon} + S_\varepsilon \tag{8}$$

3.2.3. Adhesion deposition model. Scholars such as Mirko [5] have shown that most of the droplets that hit the wall will not splash and rebound. Therefore, it can be considered that as long as the droplets hit the wall, they will deposit and form a liquid film:

$$\frac{\partial h}{\partial t} + \nabla_s \cdot \left[h \vec{V}_l\right] = \frac{\dot{m}_s}{\rho_l} \tag{9}$$

where ρ_l is the liquid film density (kg/m^3); h is the liquid film thickness (m); $\vec{V}_l$ is the average velocity of liquid film flow (m/s); $\dot{m}_s$ is the mass source term of liquid film on the wall (kg/(m^2·s)).
Momentum equation:

$$\frac{\partial h \vec{V}_l}{\partial t} + \nabla_s \cdot \left[h \vec{V}_l \vec{V}_l\right] + \frac{3}{h}\vec{V}_l v_l = \frac{h \nabla_s P_L}{\rho_l} + \vec{g}_\tau h + \frac{3}{2}\frac{\vec{\tau}_{fs}}{\rho_l} \tag{10}$$

In this equation, the three terms on the left represent instantaneous terms, convection terms, and viscous pressure in the mucosa; the first term on the right represents the sum of air flow pressure, liquid film pressure, and liquid film surface tension, and the second term represents the direction of gravity along the wall The third term represents the effect of viscous shear stress at the liquid film interface.

4. Numerical Simulation

4.1. Nozzle Model

The standard olive-type 521 nozzle is used for calculation. The simplified model of the nozzle designed by SolidWorks is shown in Figure 1(a). Taking the center of the central atomization hole as the coordinate origin, the established coordinate system is shown in Figure 1(b) and 3(c). The X axis of the coordinate system is along the direction of the groove, the Y axis is perpendicular to the direction of the groove, and the Z axis is vertically downward along the central atomization hole.

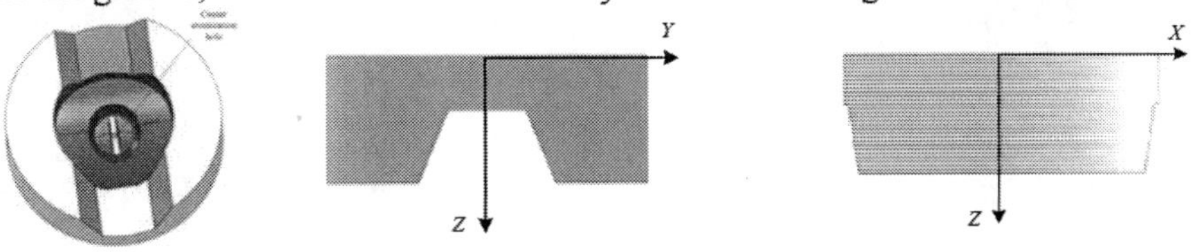

(a) Nozzle model **(b)***YZ* plane (long axis direction) ©*XZ* Plane (short axis direction)

Figure 1. Nozzle model and coordinate system.

4.2. Spraying Method

Static spraying and dynamic spraying are used for airless spraying on the plane. The static spraying is shown in Figure 2(a). During the spraying process, the nozzle and the plane to be sprayed remain static. Dynamic spraying is shown in Figure 2(b). During the spraying process, the nozzle moves in a uniform linear motion along the Y axis at a speed of 0.5m/s, and the spray plane remains stationary.

(a) Static spraying **(b)** Dynamic spraying

Figure 2. Plane spraying method.

4.3. Parameter Settings

Using hypermesh to mesh the spray flow field through a hybrid grid composed of tetrahedrons and hexahedrons, as shown in Figure 3. The calculation domain of the flow field near the nozzle is 220mm×220mm×50mm, which is encrypted with a tetrahedral mesh. The calculation domain of the spray flow field near the plane to be sprayed is a hexahedral grid of 2000mm×650mm×300mm. The same mesh model is used for static spraying and dynamic spraying. The distance between the origin of the central atomization hole and the plane to be sprayed is 300mm.

The gravitational acceleration is set to $9.8 m/s^2$, the particle velocity is set to 100m/s in the cone nozzle, the diameter is 2×10^{-5}m, and the mass flow rate is 0.03kg/s, the nozzle outlet diameter is 0.00053m. The pressure at the entrance of the paint is 15MPa, the density of the paint is $1320 kg/m^3$, the dynamic viscosity is 0.275kg/m·s, the surface tension coefficient is 0.036N/m, and the spraying distance is 300mm. The PC-SIMPLE algorithm is used to solve the equations, and the time step is set to 2×10^{-5}s, the number of time steps is 5×10^4, the maximum number of iterations is 10, and the residual is 1×10^{-6} when the calculation is completed.

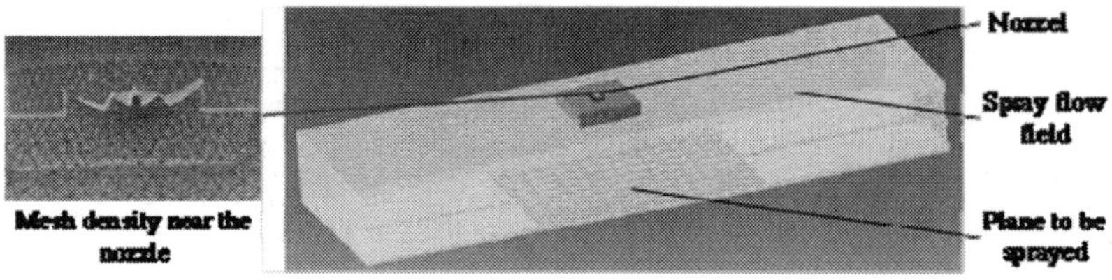

Figure 3. Grid division of spray flow field.

4.4. Fog Shape and Film Thickness

As shown in Figure 4, the paint is sprayed from the central atomization hole after being pressurized. Under the obstacle of the cut walls on both sides, the shape changes rapidly to form a flat fan-shaped spray.

(a) *XY* plane shape **(b)** *YZ* plane shape

Figure 4. Spray flow field shape.

(a) Static spraying film formation **(b)** Dynamic spray coating

Figure 5. Film formation.

The obtained static sprayed coating thickness growth rate cloud chart is shown in Figure 5(a). It can be seen that the shape of the coating film of the airless spraying corresponds to the cross-sectional shape of the spray flow field, which is a round rectangle. When the dynamic spraying numerical simulation is performed, the spraying time is set to 1s, and the spraying distance is 0.5m. The resulting flat coating is shown in Figure 5(b).

The coating growth rate on the long axis and short axis of the rectangular coating film is shown in Figure 6. It can be seen from the figure that there are more obvious double peaks in the figure 6(a), that is, the maximum thickness growth rate is on both sides of the origin, and the figure 6(b) also appears, but the performance is not obvious. The main reason is that after the paint of the origin attachment contacts the wall surface, it changes the direction of movement and has a sub-velocity along the tangential direction, which facilitates the accumulation of paint on both sides of the origin.

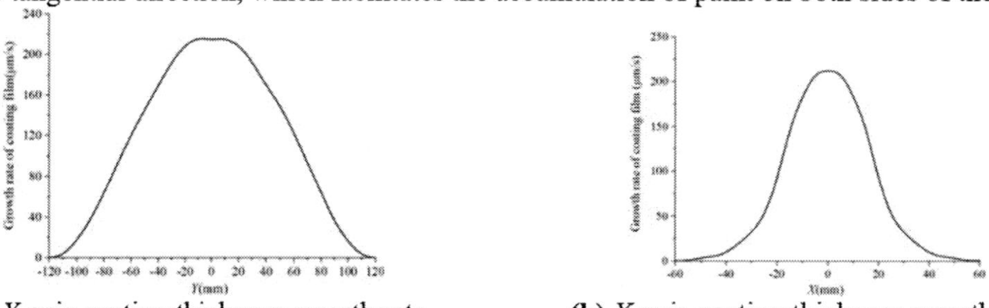

(a) *Y*-axis coating thickness growth rate **(b)** *X*-axis coating thickness growth rate

Figure 6. Coating thickness growth rate.

5. Experimental Verification

In order to further verify the simulation results, the dynamic spraying experiment was carried out according to the simulation content, and the coating situation obtained is shown in Figure 7. The coating thickness distribution of the long axis obtained by the dynamic spraying simulation was compared with the thickness distribution of the long axis obtained by the experiment. The result is shown in Figure 8. It can be seen from the results that the overall error between the numerical simulation results and the experimental results is relatively small. However, there are still errors in the thickness results of simulation and experiment. With the volatilization of the liquid components, the paint gradually flows from the place with high thickness to the place with low thickness, so the film thickness distribution of numerical simulation is slightly higher in the middle part than in the experiment. Due to the influence of surface tension and surface tension, there is a certain weak motion, and the thickness simulation at the edge is lower than that in the experiment.

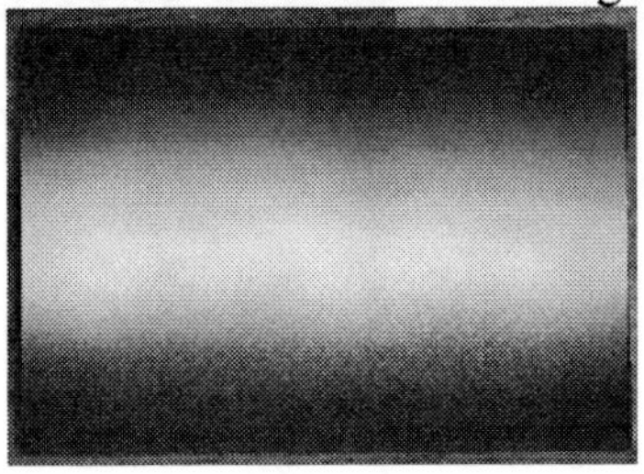

Figure 7. Dynamic spraying experimental coating.

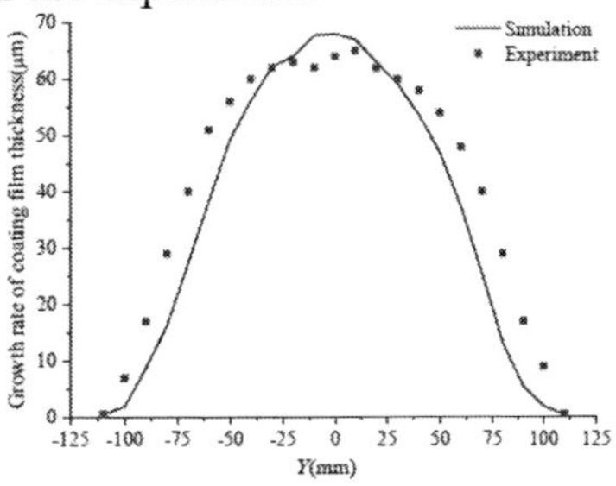

Figure 8. Simulation and experiment coating thickness.

6. Conclusion

- Aiming at the spraying process of airless spraying, the atomization model and the coating film formation model were established respectively, and the film formation model was divided into discrete phase model, continuous phase model and adhesion deposition model;
- The spray shape of airless spraying is a flat fan shape. The main reason is that it is affected by the wall of the outlet. The narrow wall in the short axis direction makes the coating leave a flat jet after spraying. During static spraying, the coating film width and the maximum thickness growth rate in the long axis direction are both greater than the short axis;
- Finally, a dynamic spraying experiment was carried out to compare with simulation. The thickness of the coating film obtained by the experiment simulation in the middle part is larger, and the thickness of the experiment in the edge part is larger. During the construction process, the thickness of the coating film in the middle part of the simulation can be slightly shifted to the edge part.

Acknowledgments

This project is supported by the National Natural Science Foundation of China (Grant No. 51475469) and the Science and Technology Research Program of Chongqing Municipal Education Commission (Grant No. KJZD-M201912901 and KJQN202012906).

References

[1] Yangsong Wang, Yingdan Wang, Shuai Yu, et al. Research on metal anticorrosion and its anticorrosion measures[J]. Liaoning Chemical Industry, 2020, 49(3): 315-318.

[2] Zvonimir Petranović, Wilfried Edelbauer, Milan Vujanović, et al. Modelling of spray and combustion processes by using the Eulerian multiphase approach and detailed chemical kinetics[J].Fuel, 2017, 191: 25-35.

[3] Josef Tausendschön, Jari Kolehmainen, Sankaran Sundaresan, et al. Coarse graining Euler-Lagrange simulations of cohesive particle fluidization[J]. Powder Technology, 2020, 364: 167-182.

[4] [3] Xiangyu Li, Jianjun Du, Licheng Wang, et al. Effects of different nozzle materials on atomization results via CFD simulation[J]. Chinese Journal of Chemical Engineering, 2020, 28(2): 362-368..
[5] Mirko Garbero, Marco Vanni, Giancarlo Baldi. CFD modelling of a spray deposition process of paint[J]. Macromolecular Symposia, 2002, 187(1): 719-730.

Design and Development of Wind Turbine Performance Analysis System Modeling

Debin Chen[1], Binbin Cheng[1], Dexin Li[1], Zhouyu Li[1], Tianye Zhao[1] and Jiasheng Zhang[2,*]

[1]New Energy Branch of Huaneng Jilin Power Generation Co. LTD, Changchun, jilin,130000
[2]School of Energy and Power Engineering, Huazhong University of Science & Technology, 430074 Wuhan, China

*Corresponding author email: 919729665@qq.com

Abstract. Wind turbines have been subjected to alternating loads for a long time, and the working environment is harsh, and their performance gradually deteriorates over time. Therefore, it is very necessary to understand the operating status of each component of the wind turbine unit through detailed evaluation of its performance through the monitoring data of the wind turbine unit.This paper designs and develops a wind turbine performance analysis software system based on the related business processes and functional requirements of wind farms. The system can calculate all the indicators related to wind farm and generating capacity, lost electricity and performance of wind farm, helping operators and analysts to achieve more comprehensive, in-depth and professional performance analysis of wind farm, so as to achieve the purpose of intelligent operation and maintenance of enterprises.
Keywords: Wind turbine; Generating capacity; Performance analysis; Software development; Intelligent operation.

1. Introduction

As a kind of renewable clean energy, wind energy has abundant storage and wide distribution, and its commercial development value is attracting worldwide attention. Different from the traditional way of generating, wind turbines environment complex, under high temperature, high humidity and salt fog corrosion, sand and dust, typhoons, storms and other severe weather conditions, as the growth of the runtime asked, makes the unit running process performance degradation, capacity reduction, reduces the efficiency of wind turbines, even has some influence to the safety of [1].Therefore, it is necessary to carry out wind turbine performance evaluation and understand the operating state of the turbine. Relevant researchers at home and abroad have carried out studies on wind turbine performance evaluation, aiming at extending the service life of wind turbines, improving the utilization efficiency of the output electric energy of wind farms and reducing costs[2]. At present, most wind turbine performance analysis methods are offline, which are modeled or simulated by various software, so it is difficult to realize the online real-time analysis of wind turbine performance.

Based on this, based on the performance analysis requirements of the wind farm group of a certain wind power regional company, this paper conducts special and in-depth design for the performance benchmarking of wind turbines and low-efficiency wind turbine excavation. Through the system, the monitoring of wind turbines Real-time data is effectively analyzed to help the company increase the power generation of wind farms and reduce the corresponding power loss. The entire system covers

related business processes, functional requirements, user interfaces, etc., using HTML+JAVA language to develop a wind turbine performance analysis software system. The overall design architecture of the system includes a data acquisition module, a data analysis module, and a data display module.
Compared with the traditional offline analysis method, the data acquisition module of the system can collect real-time operating status data of wind turbine components online and real-time, and then through the real-time analysis of the data analysis module, it can early warning of faults and early maintenance, etc., so as to achieve Extend the service life of wind turbines, improve effective power generation time and output power utilization.

2. System Business Design

Performance analysis system access of the unit operation of the center for data, alarm, data report, wind power and other data, wash and filter the data, calculate the wind farm and the unit capacity, power loss and all performance indicators, help operators and analysts to more comprehensive, in-depth and professional performance analysis of the implementation of the wind farm.
The system is divided into three layers: platform construction layer, data analysis layer and performance analysis layer.
Platform construction layer: access the operation data, alarm data and other data, the data state identification, generate the actual wind power curve of the wind turbine, and then calculate a variety of power loss and performance indicators of the unit, used for the performance evaluation of a single unit, benchmarking analysis, and data cross analysis; Access electricity metering data, downtime information, etc., calculate the production statistics of wind farms, wind turbine models, fan manufacturers and other statistical dimensions.
Data analysis layer: Perform quantitative analysis of wind turbine performance by means of wind power curve of wind turbine based on the operating data of wind turbine after state recognition, and support customized analysis functions such as BIN curve query and cross analysis. In addition, according to the fault information, relying on the fault maintenance knowledge base, the production management system to provide cause analysis, maintenance guidance and other data support.
Performance analysis layer: based on the production statistics, from the dimensions of wind farm, wind turbine model and wind turbine manufacturer, the statistical display of power generation, power loss, failure, downtime and other core performance indicators; And provide low efficiency unit mining, fault analysis and other analysis modules to help managers find the problem wind field.

3. Software Function Design

The performance analysis software system consists of data acquisition module, data calculation module and interface display module, and its data flow is shown in Figure 1.

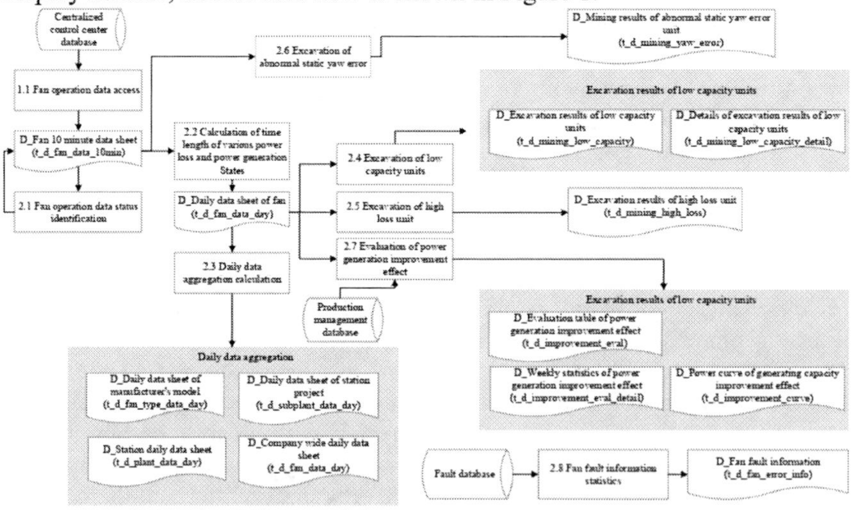

Figure 1. Data processing flow

It can be seen from Figure 1 that the data sources of the software system come from wind turbine operating data, production management data and fault data.The above data are stored in the historical database of the centralized control center. The main function of the data acquisition module is to obtain the second-level data of the access points by processing them one by one according to the API provided by the big data platform. The data acquisition module has also developed a data preprocessing function to help improve the reliability of sampled data and filter interference data through preprocessing.The purpose of data pretreatment is to distinguish valid data from invalid data, eliminate unreasonable data generated in the process of data collection or recording, and improve the accuracy of the calculation of the following key indicators.

Data calculation module is the core module of performance analysis software, which is responsible for the calculation and statistical analysis of all indicators. Generally speaking, the data calculation module includes the function modules of unit operating data status identification, daily loss of electricity, power status duration calculation, daily data aggregation calculation, low generating unit mining, high loss generating unit mining, static yaw error abnormal generating unit mining, power generation improvement effect evaluation and unit fault information statistics, etc. The unit operating data status identification module identifies whether the unit is in normal or abnormal power generation state. Then, according to IEC61400-12-1 standard, more accurate actual power curve is obtained by using the data points of normal power generation. The calculation module of daily lost electric quantity and power generation state duration is executed regularly to calculate the daily lost electric quantity and power generation state duration of yesterday's month, and calculate the actual power curve of lost electric quantity.On this basis, the daily data aggregation calculation module aggregates the number of days of a single unit according to the model, wind field project, wind farm and the whole company, so as to facilitate the query and display on the page. The low-generating unit mining module automatically mines low-generating units of various types within a given time range according to the set mining conditions and time Settings. The high-loss unit mining module automatically excavates the high-loss unit of the whole company within a given time range according to the set mining conditions and time Settings. The static yaw error abnormal unit mining module automatically mines the static yaw error abnormal unit of the whole company within a given time range according to the fixed mining conditions and time Settings. According to the interface provided by the production management system, the evaluation module of electricity generation improvement effect obtains the rectification work order information of inefficient units processed last week, conducts statistical analysis on their electricity generation, and generates the evaluation result of electricity generation improvement effect. The unit fault information statistics module performs the wind turbine fault information statistics processing, calculates the fault start time and the loss of electricity within the fault calculation end time.

The interface display module displays the calculation results of the data calculation module according to the operation characteristics of personnel and the wind farm level and regional company level. Specifically, it can be divided into: Power generation overview interface; Wind turbine performance list includes two sub-interfaces: wind farm performance list and unit performance list. Performance analysis interface includes time availability analysis, energy availability analysis and lost electric quantity analysis. The three analysis interfaces can be displayed by wind farm and unit respectively. Inefficient wind turbines mining display interface, including the inefficient units summary, inefficient units ordered list interface, power generation units display interface, deviate from the design curve units mining interface, interface, high performance degradation unit mining damage system with mining interface, static drift error exception mining interface, power promotion effect evaluation and alarm limit load query interface unit; Fault analysis interface, including fault shutdown statistics, fault cause analysis and fault sequence diagram three sub-interfaces; Power curve analysis interface, including power curve analysis and crossover analysis sub-interface display; The system Settings screen allows you to manage and configure system data and functions.

4. Software System Implementation

To ensure the applicability of the system, support both Windows and Linux during system development. The development language of the system is JAVA and HTML. Data acquisition module and data calculation module are implemented by JAVA. Java language is an excellent programming language,

with platform-independent, its language design is completely object-oriented, can well achieve the system required data acquisition and computing functions. The front - end interface display is realized by HTML. HTML is powerful, supports file insertion in different data formats, is simple, extensible, and platform independent. The interface developed with it can be used on a wide range of platforms.
After the software system is designed and developed, it is arranged in the centralized control center of jilin Branch of a certain company. Its main operating interfaces are shown in Figure 2-Figure 4. Figure 2 shows power generation across the company in selected years. FIG. 3 shows the monthly mining results of low-efficiency units within the selected time interval, and marks low-generating units, units deviating from the design curve, units with degraded performance and units with high loss respectively. Figure 4 shows the power generation improvement effect of the fan with work order processing in the selected week, and the power generation improvement of the selected fan in the low-efficiency fan processing list, as well as the comparison of the actual power curve and the designed power curve before and after the selected fan in the low-efficiency fan processing list.

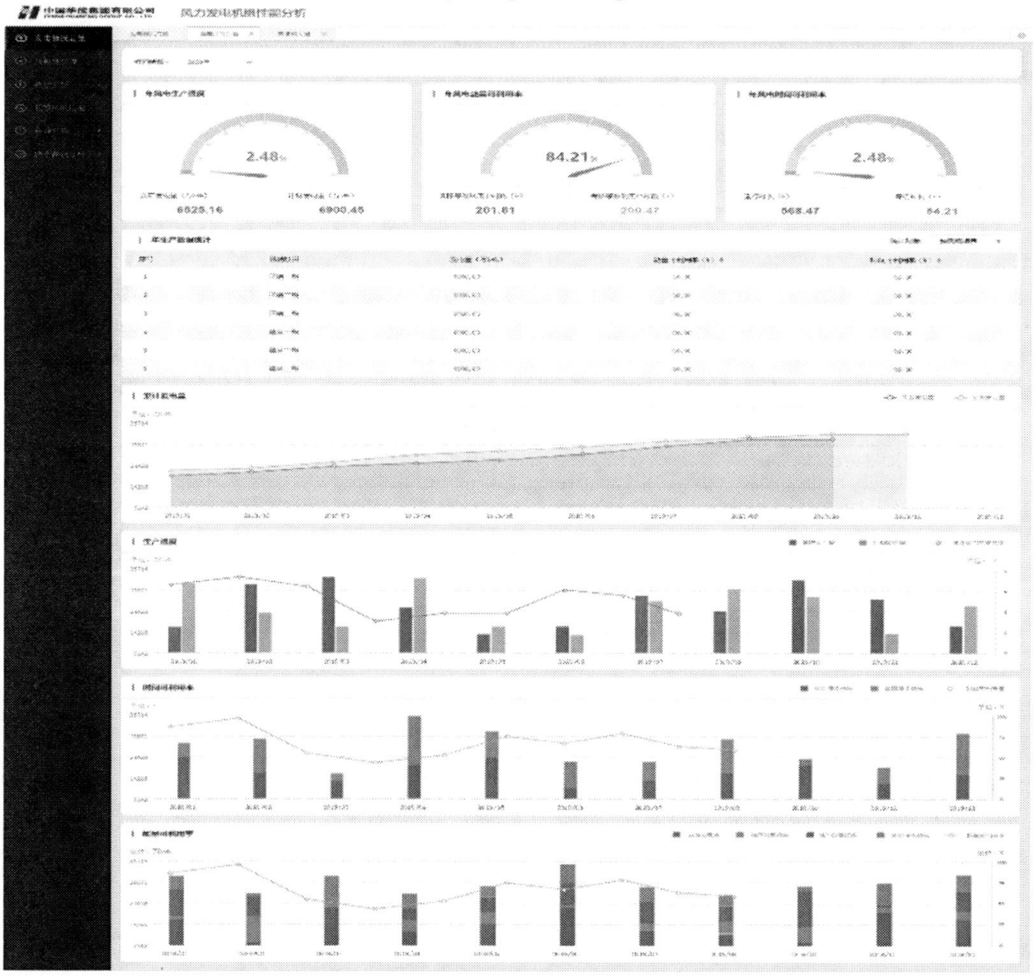

Figure 2. Power generation overview interface

Journal of Physics: Conference Series **2179** (2022) 012025 doi:10.1088/1742-6596/2179/1/012025

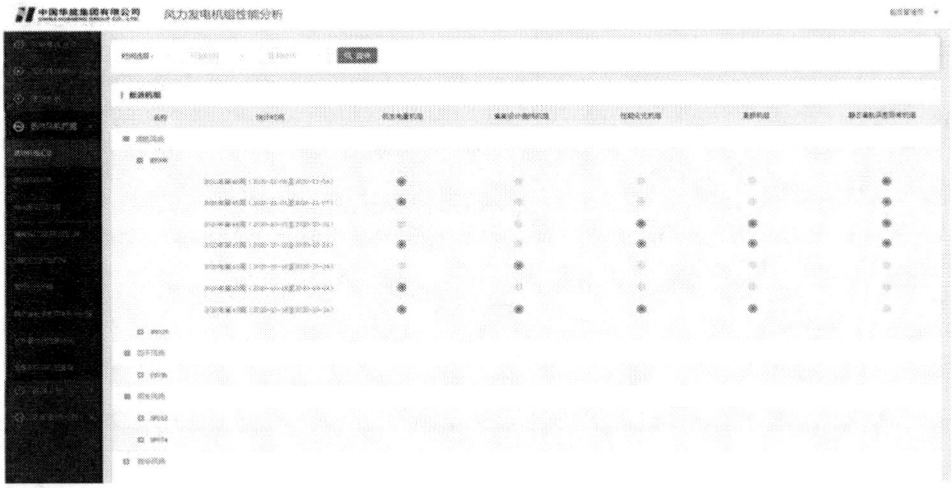

Figure 3. Interface for mining low-efficiency units

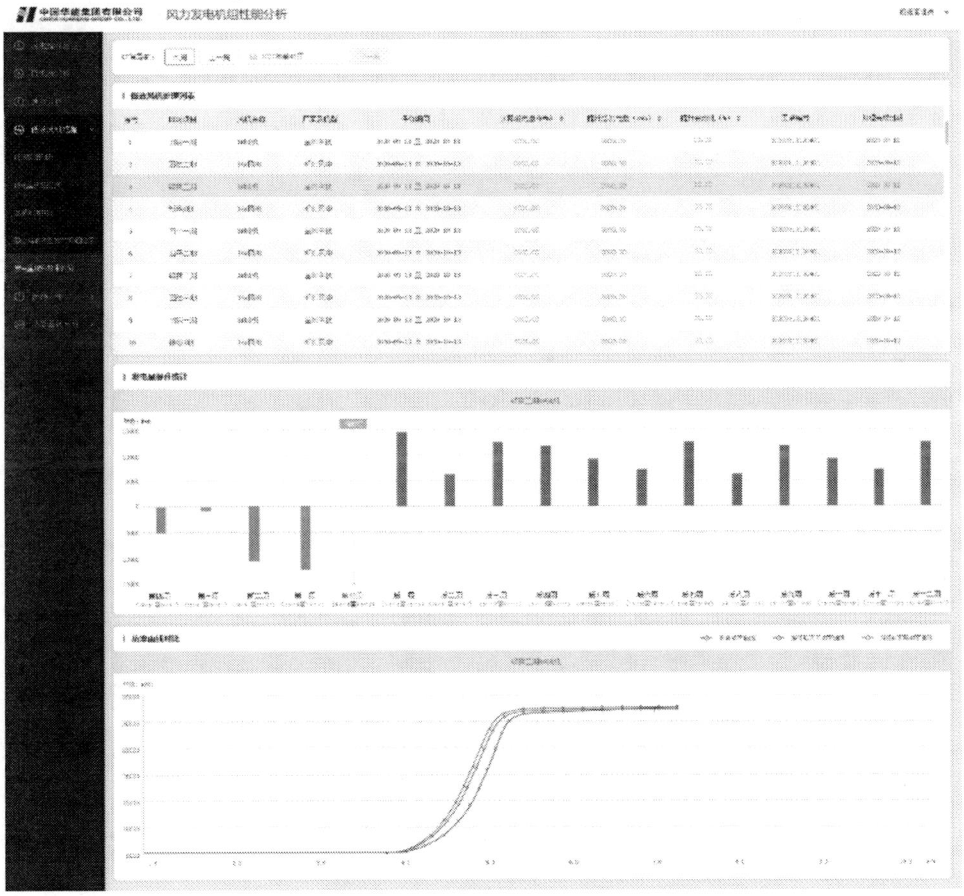

Figure 4. Evaluation interface of power generation improvement Effect

5. Conclusion

In this paper, the performance of wind turbines is subject to alternating loads for a long time, and its performance gradually deteriorates over time, which leads to a decrease in the service life of the components of the unit, and an increase in failure rate leads to a decrease in the utilization rate of the output power of the wind farm. The JAVA+HTML language is used to develop the wind turbine. The performance analysis software system can collect and monitor the operating status data of the wind turbine in real time, and analyze the performance indicators of the data to realize the early warning of the wind turbine failure and the early maintenance, so as to improve the power generation of the wind farm. The purpose of improving the service life of the unit and reducing the failure rate. This paper compares and analyzes the data based on the actual operating conditions of the system in a wind power company's centralized control center. The results show that the wind power performance analysis software developed in this paper can realize real-time status monitoring and performance evaluation of wind turbines, and has complete functions and software. Self-friendliness, easy operation, accurate data processing results and other characteristics.

References

[1] Bangalore P, Patriksson M. Analysis of SCADA data for early fault detection, with application to the maintenance management of wind turbines. Renewable Energy, 2017, 115: 521-532.

[2] Du Ming-hui. Performance research and software development of variable speed wind Turbine[D], 2002.

[3] Li Qing. Research on power characteristics and power quality test of wind turbine and wind farm[D], 2006.

[4] Liu Hao. Research on comprehensive performance evaluation method of large wind turbine[D],2006.

[5] Yang Tianshi. Research on comprehensive performance evaluation system of large wind turbine[D], 2011.

Correlation and Analysis of Operating Temperature Data for Direct Olefin Conversion from Heavy Crude

Ahmad M. Al-Otaibi and Meshal Al-Samhan*

Petroleum Research Center. Kuwait Institute for Scientific Research (KISR), P.O. Box 24885, 13109-Safat-Kuwait

*Corresponding author email: msamhan@kisr.edu.kw

Abstract. Data analytics is a complex process of examining available data to reveal information such as hidden arrangements and correlations that can help enhance production or cost cutting decisions. The world's demand and production of olefins are now higher than any other chemicals especially light olefins. This demand is supported by the fact that currently, Crude Oil-to-Chemicals (COTC) is an influential driver and a strong trend of high interest to all integrated refineries and chemical producers. Zeolite-based catalyst is a major cracking catalyst used in the industry, where Y-zeolites have progressively attracted attention as good adsorbents and stable acid catalysts, and are characterized by large, essentially spherical and internal cavities. This work focuses on analyzing operating data and correlating reaction temperatures for heavy crude to enhance the production of olefin using zeolite based catalyst. The prepared Y-zeolite was characterized for Specific surface area (SSA), total pore volume (TPV), average pore diameter (APD), and pore area (PA) according to the ASTM methods. The prepared catalyst showed a large pore diameter 211 Å and a high specific area of 425.6 m2/g, then the prepared catalyst was tested under varying conditions in a 250-mL autoclave-type reactor equipped with a stirrer, online Gas Chromatography (GC) and a gas sampling port. Different samples of crude with catalyst were tested at 380-450 °C and analysed via the online GC, where high peaks of ethylene and propylene were observed mainly at 430 °C reaching 3000 ppm wt. Moreover, the accumulative data of ethylene and propylene production from the experiments were recorded.
Keywords: Correlation; Data analysis; Operating temperature; Olefin; Zeolite catalyst.

1. Introduction

Products and materials derived from the petrochemical industry are fundamental to many aspects of modern society nowadays, among these are chemical plastics and fertilisers, which are considered the two largest groups of chemical products widely used and considered essentials in our everyday lives (1,2). Olefins feedstock is used to manufacture several key raw materials of plastics, polymers and resins such as Polyethylene, Polypropylene (3). Recently several works focus on assessing the possibility of producing olefins directly from the heavy oil and evaluate the performance of zeolite as the cracking catalyst for the heavy crude oil. Catalyst selection for such a process is grounded on coke resistance, a complex association of particle size, activity, and pore size grading (4, 5, 6). A good study targets 10–15% conversion from heavy oil to usable olefins. The Zeolite- based catalyst is a major cracking catalyst used in the industry (7). The Sodium –Y zeolite support composition can be optimized with the combination of alumina, which will dilute zeolite acid sites and contribute to the larger pores. The accessibility of zeolite pore becomes crucial in heavy crude or residue feedstock; thus, the larger pore may enhance feed diffusion and enhance stability toward deactivation (8, 9). The large pore catalyst is expected to allow adequate diffusion of heavy oil and to achieve a catalytically active ratio between the

external and internal surfaces. Also, it is known that operating conditions such as temperature has a direct effect on the yield of the products (10, 11).
In this work large pore aluminium oxide based on zeolite (Al-Z) catalyst was prepared in the laboratory and subjected to enhance the reactivity of heavy crude for direct production of olefins using a batch reactor at different temperatures, to evaluate the performance of the prepared catalyst. Such investigation will enhance the understanding of a new technique of heavy oil processing, help in improving operations, optimization, and modifications of existing facilities in continuous modernization efforts of refineries upgrading and integrating with a petrochemical plant. As well, results will help in increasing the profitability of final products.

2. Materials and Experimental Procedure

In this work, the reactivity of heavy crude under the influence of zeolite based catalyst was carried out to evaluate the catalyst performance under different operating temperatures. In the following description of the materials used and the testing steps.
Crude: The lower fares feedstock planned to be used in the autoclave experiment was collected from a local production company and was analysed and prepared (desalted) for an autoclave test. Crude properties are shown in table 1.

Table 1. Feed Properties.

Parameter	Unit	LF
Density @ 15°c	g/ml	0.981-1.000
API	API	13.2
Total sulfur	wt %	5.3-6.6
CCR	wt%	15.0-17.0
Asphaltene	wt%	9-10

Catalyst: The catalysts were prepared using a dry extrusion technique for mixing various components of the catalysts. The catalyst component such as Ultra Stable Y zeolite (USY zeolite) was used as an acidic component, PEG was used as a pore enlarging agent, while alumina (Versal Boehmite) was used as a binder, which was peptized with water and 2.5 % (vol/vol) HNO3. For catalyst specification see table 2.

Table 2. Prepared catalyst properties.

Sample ID	*Support Composition*	*BET*			*Porsimetry*		
		SSA, m^2/g	*TPV, cc/g*	*APD, Å*	*TPV, cc/g*	*APD, Å*	*PA, m^2/g*
Al-Z-50	CBV-300 (50%) +Versal alumina (50%)	425.60	0.4109	72	0.7018	211	132.74

*Specific Surface Area(SSA); Total Pore Volume (TPV); Average Pore Diameter (APD); Pore Area (PA), Polyethylene glycol (PEG)for all experiments = 25%

Testing catalysts in a batch Reactor: The prepared feed (125 ml heavy oil) was placed inside an autoclave (250 ml) to evaluate the cracking under varying temperatures. Each experiment was conducted for a period of 2 hours starting from 25 °C (room temperature) up to the selected temperature and pressures of 90-100 pounds per square inch (psi) and was maintained throughout the experiment. The experiment was carried out with a condenser installed in the outlet of the gas section, see figure 1. The same experimental steps were followed when the prepared catalyst (1.25 g) was added to the 125ml crude, where The speed of the agitator was kept 750rpm to ensure uniform mixing. After collecting the gas samples, the reactor was cooled down to room temperature.

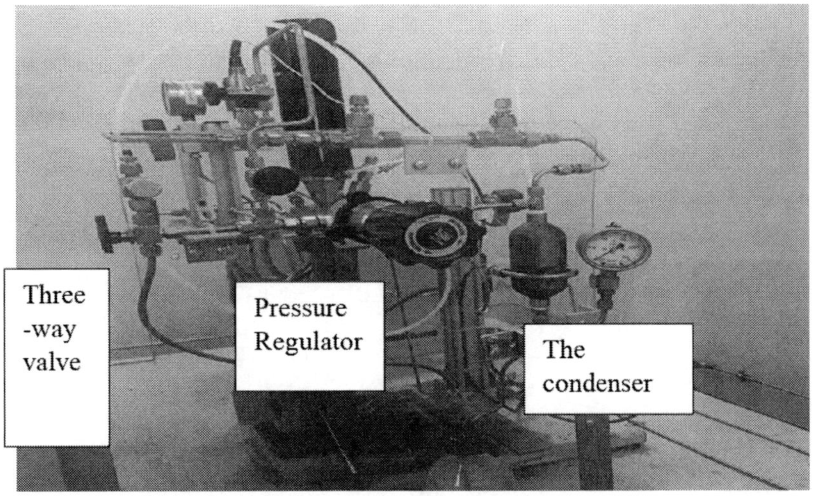

Figure 1. Autoclave arrangements.

Online GC: The single channel GC -(FID) is equipped with a Split-splitless injector capable of diluting samples in any ratio needed to avoid overloading the FID detector and heating to reduce condensation in the liner. The GC was calibrated using a standard gas mixture of all possible classes of hydrocarbons ranging from methane to pentane. The standard gas mixture is composed of saturates, olefins, and isomers, the calibration test results are shown in table 3. Later, the gas outlet of the autoclave was connected to a GC system for online gas analysis, the gas outlet line was passed through cooling water to a condenser pot and then connected with a back pressure regulator for controlling the pressure.

Table 3. Batch reactor online gas analysis calibration.

Compound	RT
Methane	1.17
Ethane	1.23
Ethylene	1.3
propane	1.46
propylene	1.8
Iso-Butane	2.12
Butane	2.25
Propadiene	3.25
1 butene	3.68
1,3 butadiene	6.6

3. Results and Discussion

The heavy feedstock processes have two definite roles: (i) difficult to react on catalyst surface (diffusion problem) and (ii) deactivate the catalyst faster. Considering the complex nature of feedstock, the main goal in this study is to find an adequate initial conversion and information about the catalyst response to the operating temperature and whether it is a linear relation, as well as if it can handle the high sulfur, metal, and asphaltene contents from the heavy feedstock.

Figure 2 shows the ethylene and propylene yield of the selected crude at different temperatures, it can be noticed that the highest yield of both olefins is reached between 400-430 °C with a noticeable accumulation of propylene over ethylene as shown in figure 3. This demonstrates that around this range of temperatures olefin gases are in a dynamic status to leave the surface of the crude. This conversion in the lighter molecule will increase the asphaltene and micro residue content in the liquid products. Where in figure 4 the favourable light olefin gases (ethylene and propylene) started the dynamic status after 430C indicating that the designed zeolite catalyst AlZ-50 contains 50 wt% aluminas and 50 wt% zeolites, which have a balance between acid sites and textural properties. Hence it is a stoichiometric ratio that will be generated the presence of zeolite (acid site) and the alumina as a binder and provide larger pores in the catalyst. These results can indicate that the higher the acid the larger the initial gas formation, particularly at lower temperatures. These results indicated that catalyst compositions have a relatively high impact at lower temperatures, but operating at a financial gain may not be feasible for refineries, particularly using these compositions. However, the accumulative amount of propylene remains higher than ethylene as shown in figure 5.

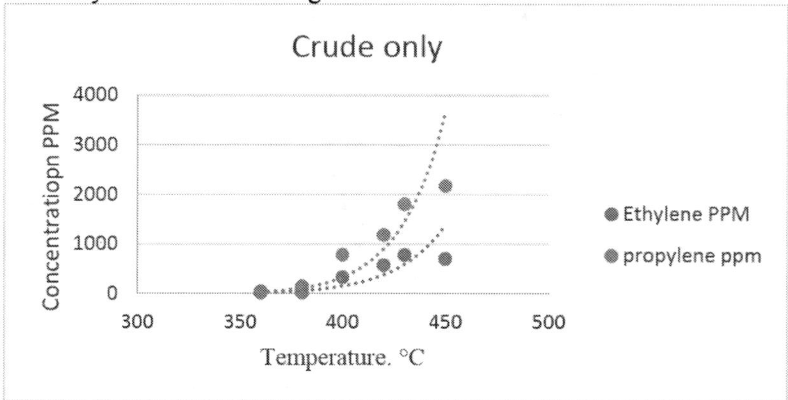

Figure 2. Ethylene and propylene yield for LF at different temperatures.

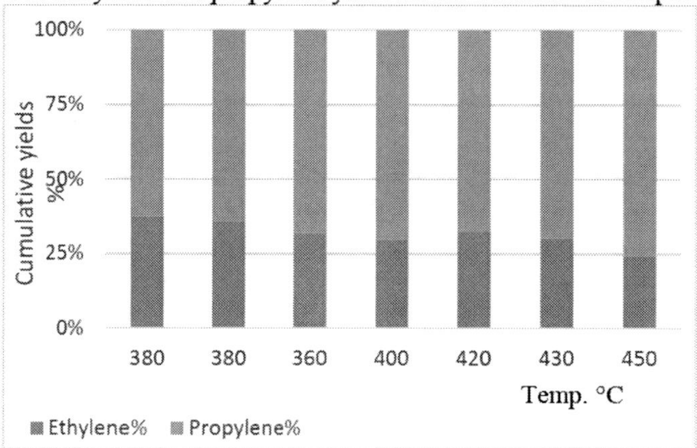

Figure 3. Cumulative products yields percentage for LF only.

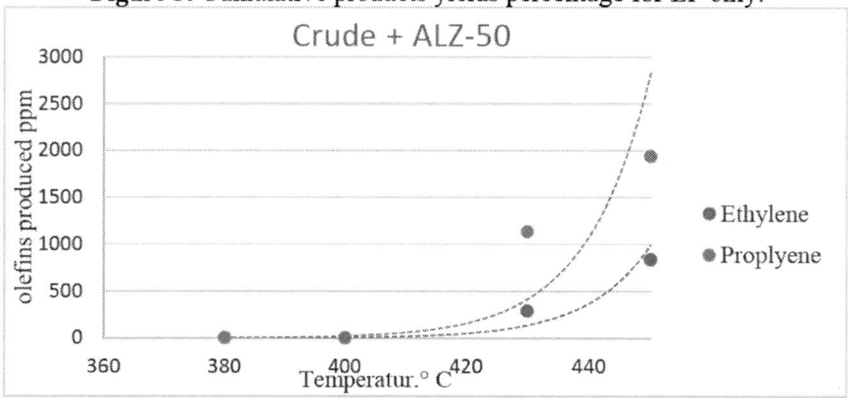

Figure 4. Ethylene and propylene yield for LF with ALZ-50 at different temperatures.

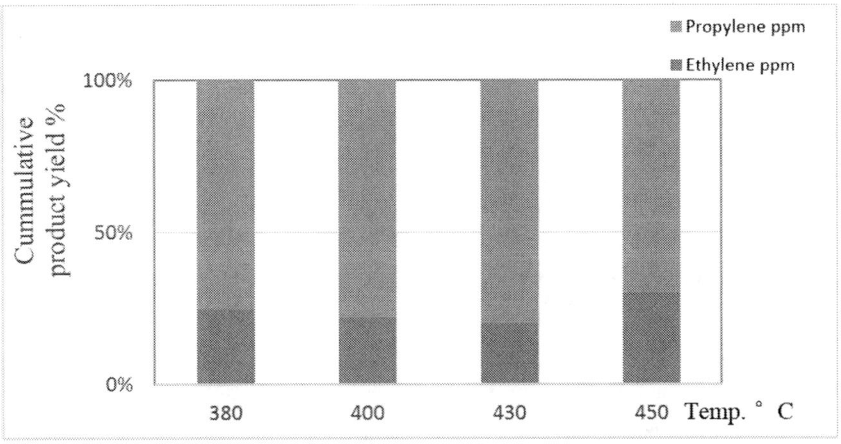

Figure 5. Cumulative products yields percentage for LF with ALZ-50.

4. Conclusion

Data analysis and correlation techniques were used to help improve the operational efficiency of direct olefin production from heavy oil. Upgrading or maximizing the benefits of heavy crude oil was an objective of this study as well as assessing the catalyst texture impact on reactivity towards olefin yield, which allows us to better understand a new technique for maximizing petrochemical feedstock. The prepared zeolite catalyst is optimized with alumina, which will dilute zeolite acid sites and contribute to the larger pores, and the accessibility of zeolite pores becomes crucial in heavy crude. The data collected indicated that there is no linear relation between the change in the operating temperature and the olefins yield, rather it follows exponential patterns. This reactivity trend will open the need for evaluating the residence time and durability of the catalyst in phase two of the study. Also, the catalyst acid has an impact on the initial gas formation, particularly at lower temperatures. In general, the data analysis for this study revealed that the reaction at high temperatures 420-450 °C, and propylene dominate the selectivity.

References

[1] Cetinkaya E, Liu N, Simons TJ, and Wallach J. Petrochemicals 2030: Reinventing the way to win in a changing industry, McKinsey on Chemicals Report, February **(2018)**.

[2] Siirola JJ. The Impact of Shale Gas in the Chemical Industry. AIChE J. **(2014)**; 60: 810–819.

[3] Murphy JJ, and Payen CF. Oil-to-chemical: new approach, PTQ, **(2020)**; Q2: 49-53.

[4] Corma A, Corresa E, Mathieu Y, Sauvanaud L, Al-Bogami S, Al-Ghrami MS, Bourane A. Crude oil to chemicals: light olefins from crude oil. Catal. Sci. Technol. **(2017)**; 7(1): 12–46.

[5] Blazek, J. J., Zeolitic catalyst proved on wax stock. Oil Gas J. **(1971)**;69 (45):66-68.

[6] Holland BT et al., Utilizing Colloidal Silica and Aluminum-Doped Colloidal Silica as a Binder in FCC Catalysts: Effects on Porosity, Acidity, and Microactivity. Industrial & Engineering Chemistry Research, **(2007)**; 46(13): 4486–4496.

[7] Scherzer J. Octane-enhancing, zeolitic FCC catalysts: scientific and technical aspects. Catal. ReV.sSci. Eng. **(1989)** a;31 (3): 215-354.

[8] Scherzer J. Octane-Enhancing, Zeolitic FCC Catalysts: Scientific and Technical Aspects. Catalysis Reviews, **(1989)** b; 31(3): 215–354.

[9] Zhang, Y.; Li, Z.; Wang, Z.;Jin, Q. "Optimization Study on Increasing Yield and Capacity of Fluid Catalytic Cracking (FCC) Units. Processes". **(2021)**; 9: 1497.

[10] Ren,T; Patel,M.; Blok,K."Olefins from conventional and heavy feedstock: Energy use in steam cracking and alternative processes, Energy **(2006)**; 31: 4, 425-451.

[11] Meshal Al-Samhan et al., Prospects of refinery switching from conventional to integrated: An opportunity for sustainable investment in the petrochemical industry, Fuel, **(2022)**; 310, Part B, 122-161.

Research on Passenger Throughput Forecast of Civil Aviation Airport Based on Multi-source Data

Daoyue Peng[1], Meng Zhang[2], Yun Xiao[2,*] and Yidan Wang[2]

[1]Anhui Road Transportation Management Service Center, Hefei, Anhui, China
[2]School of Urban Construction and Transportation, Hefei University, Hefei, Anhui, China

*Corresponding author email: xiaoyun@hfuu.edu.cn

Abstract. In order to scientifically predict the passenger throughput of civil aviation airports. The paper taked Hefei Xinqiao International Airport as the research object, comprehensively considers multi-source data such as GDP, permanent population, airport competition, and spatial distance, used multiple linear regression, gravity-regression and other methods to construct a prediction model, and the results were obtained by different methods. The prediction result errors were compared and analyzed. The results of the study show GDP and population are important factors affecting the passenger throughput of civil aviation airports. The multiple linear regression model that considers the factors affecting airport competition can improve the accuracy of prediction, and the gravity-regression model can effectively predict the passenger throughput of civil aviation airports.
Keywords: Civil aviation airport; Data analysis; Passenger throughput; Multiple linear regression; Gravity-regression model.

1. Introduction

Air traffic and regional development have a direct correlation. Zhang Lei et al. [1] took Nanjing Lukou International Airport as an example, used Granger causality test method to analyze the economic growth relationship between Nanjing Lukou International Airport and different spaces in the metropolitan area, and concluded that the urban economy has driven the increase of airport passenger and freight. Liu Lanjuan et al. [2] selected civil aviation transportation indicators and GDP data from 1978 to 2008 to establish a Vector Auto Regression Model(VAR), conducted Granger Causality Testing on the long-term equilibrium relationship between indicator variables, and verified that civil aviation transportation has a long-term equilibrium relationship with the national economy. Transportation has played a role in promoting the development of the national economy. Liu Yuezhong [3] analyzed the relationship between the airports of different regions and different scales in my country and the regional economy from 2003 to 2013 through the use of neo-classical endogenous growth theory, agglomeration theory, accessibility theory, and trade theory, and panel analysis. The conclusion is that airport operations and urban economic growth promote each other.

Airport passenger throughput is an important indicator reflecting the level of airport operations and is a key measurement basis for the formulation of the development plan for the air transport industry. Domestic scholars have studied more airport passenger throughput forecasts and have achieved corresponding results. Cai, Wenting et al. used a multiple regression model prediction method and found that the relative error between the predicted value of the multiple regression model and the historical data is small [4-6]. Wang Cui [7] used the time series forecasting method, RBF neural network, and grey forecasting model to carry out a detailed study on the airport passenger throughput forecasting method, and confirmed that the forecasting accuracy of the grey RBF neural network forecasting model is better. Yao Yanbin [8] used the Grey System Theory Model(GM) to predict passenger throughput, and the results showed that the advantages of the gray prediction method are simple, efficient, accurate, with good credibility and practical value, and the prediction results are more reasonable.

Existing studies generally use the overall data of administrative divisions to predict the throughput of civil aviation airports. For example, the overall data of the administrative region of Anhui Province is used to predict the throughput of civil aviation at Hefei Xinqiao International Airport. In actual operation, although Wuhu City, Ma'anshan City, and other regions belong to the same province as Hefei Xinqiao International Airport, they are closer to airports in neighboring provinces, and there is not much demand for throughput of the Hefei Xinqiao International Airport. Perform split processing. In addition, the different spatial distances of various regions have significant differences in the throughput requirements of civil aviation airports. The spatial distances of different regions need to be considered when constructing the prediction model, but There is little literature to study this content.

This paper comprehensively considers the impact data of gross product (GDP), permanent population, airport competition, and spatial distance, and constructs a prediction model based on multiple linear regression analysis, gravity-regression analysis, and other methods to predict the passenger throughput of the airport. Operation management, traffic planning, etc., provide an important basis.

2. Research Methods

This article takes GDP and permanent population as basic data, and exploratoryly adds factors such as airport competition and spatial distance to the forecast model to improve forecast accuracy and reliability.

2.1. Multiple Linear Regression

Multiple linear regression models are usually used to study the relationship between a dependent variable and multiple independent variables. A multiple linear model can be established for analysis if the relationship between the two can be described in a linear form.

The multiple linear regression model is usually used to describe the random linear relationship between the variables y and x. The general form of the multiple linear regression model is:

$$Y_i = \beta_0 + \beta_1 X_1 + \beta_2 X_2 + \cdots + \beta_i X_k + \mu_i \tag{1}$$

Among them: Y_i is the dependent variable; $X_{i,}\cdots,X_k$ is the independent variable; μ_i is the error term; i = 1, 2…n; k is the number of explanatory variables; β is called the regression coefficient, which means that when the other explanatory variables remain unchanged, X changes by 1 unit each time, The mean value of Y is the change of $E(Y)$.

The model uses the province's gross product (GDP) and permanent population as the explanatory variables of airport passenger throughput, and a multiple linear regression model can be constructed:

$$Q_1 = \beta_0 + \beta_1 G + \beta_2 P \tag{2}$$

Among them: Q_1 is the airport passenger throughput; G is the province's GDP value; P is the province's permanent population; β_0, β_1 and β_2 are the parameters that can be obtained through specific calculations.

2.2. Multiple Linear Regression Considering Airport Competition

Consider the competing airports around some cities. These airports are more appealing to the region, and residents have more travel options.

Regarding the main factors affecting the passenger throughput of central city airports, the model uses the province's gross product (GDP) and permanent population after excluding regions with competing airports as the explanatory variables of airport passenger throughput, which can construct a multivariate linearity considering airport competition. Regression model:

$$Q_2 = \beta_0' + \beta_1' G' + \beta_2' P' \tag{3}$$

Among them: Q_2 is the airport passenger throughput; G' is the GDP value after considering the airport competition; P' is the permanent population after considering the airport competition; β_0', β_1' and β_2' are the parameters that can be obtained through specific calculations.

2.3. Gravity-regression model analysis

The gravity model originally originated from the application of the law of universal gravitation in physics in the study of space interaction theory. The law of universal gravitation states that there is an attraction between any two objects, and the magnitude of gravity is proportional to the mass of the two and inversely proportional to the square of the distance between the two. The general form of the gravity model is:

$$G_{ij} = \frac{m_i m_j}{d_{ij}^2} \tag{4}$$

After considering the competitiveness of airports, it is determined that the province's gross product (GDP), permanent population, and distance from the city center to the airport, after removing the regions with competing airports, are the influencing factors for the passenger throughput of central city airports. On the basis of the gravity model, linearization is performed; that is, the sum of the ratios of the GDP of each city and the distance d from the city center to the airport; the sum of the ratios of the permanent population of each city and the distance d from the city center to the airport are used as explanatory variables. The basic form of the model for:

$$Q_3 = \beta_0'' + \beta_1''\left(\sum \frac{G_i}{d_i}\right) + \beta_2''\left(\sum \frac{P_i}{d_i}\right) \tag{5}$$

Among them: Q_3 is the airport passenger throughput; G_i is the GDP of each city after considering the airport competition, P_i is the permanent population of each city after considering the airport competition, d_i is the distance from the city center to the airport after considering the airport competition; β_0'', β_1'' and β_2'' are the parameters that can be calculated through specific calculations. Take advantage of this offer.

2.4. Predictive Effect Evaluation Index

In order to evaluate the prediction effect of the model, the average absolute error (MAE), mean square error (MSE), average absolute percentage error (MAPE), and mean square percentage error (MSPE) are used as the model's performance evaluation indicators. The calculation formula is:

$$MAE = \frac{1}{n}\sum_{i=1}^{n}\left|\hat{y}_i - y_i\right| \qquad MSE = \sqrt{\frac{1}{n}\sum_{i=1}^{n}(\hat{y}_i - y_i)^2}$$

$$MAPE = \frac{1}{n}\sum_{i=1}^{n}\left|\frac{\hat{y}_i - y_i}{y_i}\right| \qquad MSPE = \sqrt{\frac{1}{n}\sum_{i=1}^{n}\left(\frac{\hat{y}_i - y_i}{y_i}\right)^2} \tag{6}$$

Among them: $\hat{y}_i$ is the predicted value; y_i is the actual value; n is the number of predicted values; n =1, 2, 3,..., n.

3. Research Area and Data Source

Take Anhui Province, as an example. The airport passenger throughput data comes from the Anhui Provincial Statistical Yearbook, the official website of the people's government of each city, and the official website of the Civil Aviation Administration of China. The data is authoritative and reliable. The passenger throughput data comes from the statistics published on the website of the Civil Aviation Administration of China. The GDP and permanent population data of Anhui Province and its cities come from the Anhui Provincial Statistical Yearbook and the official websites of the people's governments of each city. On Baidu's map.

4. Empirical Analysis

According to the passenger throughput of Hefei Xinqiao Airport from 2014 to 2019, the GDP and permanent population data of 16 cities in Anhui Province, the above three models were used to predict and analyze the passenger throughput of the airport.

4.1. Multiple Linear Regression Analysis

4.1.1. Passenger throughput prediction model. According to the principle of multiple linear regression model, Anhui Province's GDP and permanent population can be used as explanatory variables of airport passenger throughput to construct a multiple linear regression model. The independent variables and dependent variable values in the model are shown in Table 1.

Table 1. Multivariate linear model independent variable and dependent variable value.

Year	G (100 million yuan)	P (Ten thousand people)	Q_1 (ten thousand people * times)
2014	21249.31	6083	597.4599
2015	22220.86	6143.64	661.3111
2016	24662.78	6195.6	739.1998
2017	27423.21	6254.8	914.7128
2018	33722.78	6323.34	1111.0596
2019	37083.61	6367.35	1228.2384

According to historical data, the parameters β_0, β_1, β_2 of the model can be obtained using SPSS software. According to the output of the software, the specific form of the model is as follows:

$$Q_1 = 0.026G + 0.805P - 4867.066 \tag{7}$$

Among them: Q_1 is the passenger throughput of Hefei Xinqiao Airport; G is the GDP value of Anhui Province; P is the permanent population of Anhui Province.

4.1.2. Testing of multiple linear regression models. According to the output of SPSS software, the coefficient of determination R^2 of the model is 0.9915, which is close to 1, and the reliability is very high. At the same time, the model is tested with f and the result of f test is 0.0003646, which is less than the significance level of 0.05. The regression equation has a significant regression effect and can be used for prediction and control.

4.1.3. Passenger throughput forecast results and error analysis. According to the multiple linear regression model, the passenger throughput of Hefei Airport can be predicted from 2014 to 2019. According to the prediction effect evaluation index, the error of the prediction value can be calculated. From the predicted and actual values, the average absolute error (MAE) is 12.0537, the mean square error (MSE) is 16.5809, the average absolute percentage error (MAPE) is 0.0152, and the mean square percentage error (MSPE) is 0.0210. From the perspective of the prediction error results, the results obtained by the four prediction error methods are all small, and the average absolute percentage error is controlled within 2%, indicating that the multiple linear regression model is used to predict passenger throughput with high accuracy. The specific calculation is shown in Table 2.

Table 2. Forecast value absolute percentage error calculation table.

Year	Actual value	Predictive value	$\lvert\hat{y}_i - y_i\rvert$	MAE_1	$\hat{y}_i - y_i$	MSE_1	$\left\lvert\frac{\hat{y}_i - y_i}{y_i}\right\rvert$	$MAPE_1$	$\frac{\hat{y}_i - y_i}{y_i}$	$MSPE_1$
2014	597.4599	588.6588	8.8011	12.0537	-8.8011	16.5809	0.0147	0.0152	-0.0147	0.0210
2015	661.3111	662.9774	1.6663		1.6663		0.0025		0.0025	
2016	739.1998	768.8827	29.6829		29.6829		0.0402		0.0402	
2017	914.7128	888.9742	25.7386		-25.7386		0.0281		-0.0281	
2018	1111.0596	1109.4382	1.6213		-1.6213		0.0015		-0.0015	
2019	1228.2384	1233.0501	4.8117		4.8117		0.0039		0.0039	

4.2. Multiple Linear Regression Considering Airport Competition

4.2.1. Passenger throughput prediction model. According to the principle of the multiple linear regression model and considering the factors of airport competition, there are Nanjing Lukou International Airport, Hangzhou Xiaoshan International Airport, and other airports around the four cities of Wuhu, Maanshan, Xuancheng, and Huangshan, and they are far away from Hefei Xinqiao Airport. Therefore, the gross product (GDP) and permanent population after excluding the four cities of Wuhu, Maanshan, Xuancheng, and Huangshan are used as the explanatory variables of the airport passenger throughput, and a multiple linear regression model is constructed. The independent variables and dependent variable values in the model are shown in Table 3.

Table 3. Multivariate linear model independent variable and dependent variable value.

Year	G'(100 million yuan)	P'(Ten thousand people)	Q_2 (ten thousand people * times)
2014	16181.84	5104.7	597.4599
2015	16895.88	5155.36	661.3111
2016	18831.28	5203	739.1998
2017	20952.98	5255.2	914.7128
2018	26358.32	5309.34	1111.0596
2019	28975.05	5345.25	1228.2384

According to historical data, the parameters β_0', β_1', β_2' of the model can be obtained using SPSS software. According to the output of the software, the specific form of the model is as follows:

$$Q_2 = 0.031G' + 1.017P' - 5106.795 \tag{8}$$

Among them: Q_2 is the passenger throughput of Hefei Xinqiao Airport; G' is the GDP value after considering the airport competition; P' is the permanent population after considering the airport competition.

4.2.2. Testing of multiple linear regression models. According to the output of SPSS software, the coefficient of determination R^2 of the model is 0.9912, which is close to 1, and the reliability is very high. At the same time, the model was tested with f and the result of f test was 0.000384, which was less than the significance level of 0.05. This regression equation has a significant regression effect and can be used for prediction and control.

4.2.3. Passenger throughput forecast results and error analysis. According to the multiple linear regression model, the passenger throughput of Hefei Airport can be predicted from 2014 to 2019. According to the prediction effect evaluation index, the error of the prediction value can be calculated. From the predicted and actual values, the average absolute error (MAE) is 11.6025, the mean square error (MSE) is 16.8681, the average absolute percentage error (MAPE) is 0.0148, and the mean square percentage error (MSPE) is 0.0216. Judging from the prediction error results, the results obtained by the four prediction error methods are all small, and the average absolute error (MAE) and average absolute percentage error (MAPE) of the model are smaller than the prediction error of Model 1, indicating that the accuracy of the multiple linear regression model prediction under consideration of airport competition factors is better than that of Model 1. The specific calculation is shown in Table 4.

Table 4. Forecast value error calculation table.

Year	Actual value	Predictive value Q_2'	$\lvert\hat{y}_i - y_i\rvert$	MAE_2	$\hat{y}_i - y_i$	MSE_2	$\left\lvert\frac{\hat{y}_i - y_i}{y_i}\right\rvert$	$MAPE_2$	$\frac{\hat{y}_i - y_i}{y_i}$	$MSPE_2$
2014	597.4599	588.0216	9.4383		-9.4383		0.0158		-0.0158	
2015	661.3111	661.7346	0.4235		0.4235		0.0006		0.0006	
2016	739.1998	770.3257	31.1259	11.6025	31.1259	16.8681	0.0421	0.0148	0.0421	0.0216
2017	914.7128	889.3435	25.3693		-25.3693		0.0277		-0.0277	
2018	1111.0596	1112.3628	1.3032		1.3032		0.0012		0.0012	
2019	1228.2384	1230.1933	1.9549		1.9549		0.0016		0.0016	

4.3. Gravity-regression Model Analysis

4.3.1. Passenger throughput prediction model. It is known from Model 2 that considering airport competition can reduce the prediction error, so Model 3 further considers the spatial distance between the airport and each city on the basis of considering airport competition, so as to predict the passenger throughput of Hefei Xinqiao Airport.

According to the principle of the gravity model and the principle of the multiple linear regression model, the sum of the ratios of the GDP of each city in Anhui Province, and the distance d from the city center of Anhui Province to the airport, the permanent population of each city in Anhui Province, and the distance d from the city center of Anhui Province to the airport A multiple linear regression model was constructed as an explanatory variable. The values of the independent variables and dependent variables in the model are shown in Table 5.

Table 5. Multivariate linear model independent variable and dependent variable value.

Year	$\sum \frac{G_i}{d_i}$	$\sum \frac{P_i}{d_i}$	Q_3 (ten thousand people * times)
2014	209.0299	48.6183	597.4599
2015	222.3176	48.7320	661.3111
2016	247.4529	49.1800	739.1998
2017	275.1775	49.6892	914.7128
2018	341.2044	50.2512	1111.0596
2019	374.0022	50.6866	1228.2384

According to historical data, the parameters β_0'', β_1'', β_2'' of the model can be obtained using SPSS software. According to the output of the software, the specific form of the model is as follows:

$$Q_3 = 1.464\sum\frac{G_i}{d_i} + 188.501\sum\frac{P_i}{d_i} - 8867.765 \tag{9}$$

Among them are: Q_3 is the passenger throughput of Hefei Xinqiao Airport; G_i is the GDP of each city in Anhui Province after considering the airport competition, P_i is the permanent population of each city in Anhui Province after considering the airport competition, and d_i is the distance from the city center of Anhui Province to the airport after considering the airport competition .

4.3.2. Testing of multiple linear regression models. According to the output of SPSS software, the coefficient of determination R^2 of the model is 0.9935, which is close to 1, and the reliability is very high. At the same time, the model is tested with f and the result of f test is 0.000243, which is less than the significance level of 0.05. The regression equation has a significant regression effect and can be used for prediction and control.

4.3.3. Passenger throughput forecast results and error analysis. According to the gravity-regression model, the passenger throughput of Hefei Airport from 2014 to 2019 can be predicted. According to the prediction effect evaluation index, the error of the prediction value can be calculated. From the predicted and actual values, the average absolute error (MAE) is calculated to be 12.4721, the mean square error (MSE) is 14.4923, the average absolute percentage error (MAPE) is 0.0160, and the mean square percentage error (MSPE) is 0.0195. Judging from the prediction error results, although the average absolute error (MAE) and average absolute percentage error (MAPE) of the model are slightly larger than the prediction error values of model 1 and model 2, the results obtained by each prediction error method show that Both are too small, indicating that the prediction accuracy of the multiple linear regression model under consideration of the spatial distance is high, and it can be used to predict the airport passenger throughput. The specific calculation is shown in Table 6.

Table 6. Forecast value error calculation table.

Year	Actual value	Predictive value	$\lvert\hat{y}_i - y_i\rvert$	MAE_3	$\hat{y}_i - y_i$	MSE_3	$\left\lvert\frac{\hat{y}_i - y_i}{y_i}\right\rvert$	$MAPE_3$	$\frac{\hat{y}_i - y_i}{y_i}$	$MSPE_3$
2014	597.4599	602.9151	5.4552	12.4721	5.4552	14.4923	0.0091	0.0160	0.0091	0.0195
2015	661.3111	643.8018	17.5093		-17.5093		0.0265		-0.0265	
2016	739.1998	765.0623	25.8625		25.8625		0.0350		0.0350	
2017	914.7128	901.6265	13.0863		-13.0863		0.0143		-0.0143	
2018	1111.0596	1104.2390	6.8206		-6.8206		0.0061		-0.0061	
2019	1228.2384	1234.3369	6.0985		6.0985		0.0050		0.0050	

4.4. Comparison and Analysis of Three Forecasting Models

The error results obtained by the three prediction error methods of the three models are all small, and the average absolute percentage error is controlled within 2%, indicating that the three models have high accuracy in predicting passenger throughput. Among them, the average absolute error (MAE) and average absolute percentage error (MAPE) of model 2 are smaller than the prediction error of model 1, indicating that the accuracy of the multiple linear regression model prediction is improved compared with model 1. The average absolute error (MAE) and average absolute percentage error (MAPE) of model 3 are slightly larger than the prediction error values of models 1 and 2, but the result values obtained by each prediction error method are all smaller, indicating the gravity-regression model is stable. Can effectively predict airport passenger throughput.

Table 7. Three model prediction error values.

	MAE	MSE	$MAPE$	$MSPE$
Multiple linear regression model (Model 1)	12.0537	16.5809	0.0152	0.0210
Multiple linear regression model considering the competitive relationship (Model 2)	11.6025	16.8618	0.0148	0.0216
Gravity-regression model (Model 3)	12.4721	14.4923	0.0160	0.0195

By comparing the difference between the predicted value of the passenger throughput of the three models and the actual value, it is concluded that the predicted residual fluctuation range of the three prediction models is small, and the predicted value of the gravity-regression model is the closest to the actual value.

5. Conclusion

Based on the existing multi-source data, this paper comprehensively considers the airport competition relationship, spatial distance, and other data, studied three civil aviation airport passenger throughput prediction models, Compared the prediction accuracy and reliability of different models, and predicted the passenger throughput of Hefei Xinqiao International Airport. The data shows:

(1) Compared with the general multiple linear regression model, the prediction model that comprehensively considers airport competition can improve the accuracy of airport passenger throughput prediction.

(2) Considering the spatial distance of different regions, constructing a gravity-regression prediction model can effectively predict the passenger throughput of the airport.

(3) The residual fluctuation of the airport passenger throughput predicted by the gravity-regression prediction method is smaller than that of the single prediction method, and its predicted value is the closest to the actual value. This model can significantly reduce the prediction error and improve the accuracy and reliability of the prediction.

6. Practical Implications and Directions for Further Research

There are many factors influencing airport passenger throughput. This article selects GDP, population, competitive relationships, and distance as the analysis factors. However, different GDP structures and population structures have different effects on airport passenger throughput. Therefore, how to further divide the structure of influencing factors and improve the prediction model is the direction to be studied.

References

[1] Zhang Lei, Chen Wen, Song Zhengna and Xue Junfei 2010 The correlation between airport operations and regional economic growth: Taking Nanjing Lukou International Airport as an example [J]. *Advances in Geographical Sciences*, 29(12): 1570-1576.

[2] Liu Lanjuan and Dong Wanhao 2009 Empirical analysis of the relationship between civil aviation transportation industry and GDP based on VAR model [J]. *Financial Research,* 35(8):69-78.

[3] Liu Yuezhong 2016 Airport Operation and Regional Economic Development-Analysis of the Coupling and Cooperative Development of Multi-airport Area [D]. Guangzhou: Jinan University.

[4] Cai Wenting, Cai Wenting, Peng Yi and Chen Qiuji 2019 Forecast of air transportation passenger volume based on multiple regression model [J]. *Aviation Computing Technology*, 49(04): 50-53.

[5] Huang Bangju, Lin Junsong, Zheng Xiaoyu and Fang Xuedong 2013 Passenger throughput prediction of civil transportation airport based on multiple linear regression analysis [J]. *Mathematical Practice and Knowledge*, 43(4): 172-178.

[6] Yue Hui and An Ran 2010 Research on the Application of Multiple Regression Model in the Passenger Traffic Forecast of Newly-built Airports——Based on Dynamic Cluster Analysis [J]. *Modern Business and Trade Industry*, 1(20): 13-15.

[7] Wang Cui 2008 Research on Civil Aviation Passenger Traffic Forecast Method Based on Grey Theory and RBF Neural Network [D]. Beijing Jiaotong University.

[8] Yao Yanbin 2006 Diversion prediction of large-capacity airport rail transit to landside traffic [D]. Civil Aviation University of China.

Power Data Integration Method Based on Database-table Metadata Semantic

Lei Zheng[1, *], Jianhong Pan[2] and Kai Zhang[1]

[1]Information and Communication Company, State Grid Jilin Electric Power Co., Ltd., Jilin Changchun, China
[2]State Grid Jilin Electric Power Co., Ltd., Jilin Changchun, China

*Corresponding author email: zhenglei@jl.sgcc.com.cn

Abstract. Aiming at the characteristic of power data with multiple sources, wide distribution and complex association, this paper proposes a power data integration method based on Database-Table Metadata Semantic (DTMS), which can realize the integration of grid data at the metadata layer. Firstly, based on the structure of the database, the semantic expression is established between database and table; Secondly, this paper combines the Jaccard distance similarity method and the edit distance similarity method to achieve semantic matching between metadata and technical terms, which can solve the shortcomings of the single similarity method. Finally, according to the characteristics of power data assets, a power data integration scheme based on DTMS is proposed for multi-source power data integration. Through experiments, the accuracy of the algorithm in this paper has reached 95.8%.
Keywords: Metadata; Database-table semantic; Data asset; Jaccard similarity; Edit distance.

1. Introduction

Currently, data asset has become a very important basic strategic resource in all walks of life. In the process of power information construction, the maturity of data resource construction is directly related to the advanced level of power enterprises. Data asset management has changed the perception of power data in the past: The old cognition believed that data is only an accessory product of power companies; while in the new concept, data can be managed as assets [1-3]. In the process of data asset management, the first problem to be solved is data integration [4-6]. Without considering the specific application scenarios of data integration, current research is mainly divided into two levels: One is the data integration method at the framework level [7-11], and the second is the data integration method at the specific technical level [12-17].

In terms of data integration architecture, Liwei Kuang[8] proposed a data integration framework based on cloud computing, which consists of five components: data representation, data dimensionality reduction, data relationship establishment, data sorting and data retrieval. Chuanzhao Tian [9] proposed an integration framework of different Internet open data sources, including how to identify data entities, how to load data, and how to resolve data conflicts. Kalaycı Tahir [10] proposed a data integration framework based on knowledge graphs, which simplifies the process of data extraction and management, and improves the efficiency of data integration greatly. Liao Rongtao[11] proposed a visual data integration framework in order to solve the problem of massive data resources and information islands, which includes a system integration layer, an intelligent information engine layer and The visual display layer is used to enhance the operator's ability to control the real-time dynamics of the power grid.

In terms of the realization of data integration technology, Berlin J [12] introduced a bayesian model in the data integration process to evaluate the correlation between data of different data sources. Nottelmann H [13] proposed an automatic data integration method for heterogeneous data integration scenarios. Chen Z [14] proposed a method to integrate spreadsheet data. However, this method requires some manual operations to ensure the accuracy of data acquisition. Chanial C Dziri R [15] proposed a data integration method for the digital multimedia environment. This method can search heterogeneous data sources on the Internet through the ConnectionLens system and perform keyword search on the important French journal Le Monde. In reference to the basic characteristics of government big data sources, ZHAO Jie [16] proposed a method based on knowledge graphs to realize data integration. Halevy A [17] elaborated on the implementation of Google's data integration technology Google Dataset Search (GOODS). GOODS collects metadata of datasets in a non-intrusive way in the background, and then empowers some services for the convenience of Google engineers to Find suitable data set.

2. Traditional Metadata-based Integration Scheme

The so-called metadata refers to data about data. Using metadata can help data analysts to easily grasp the basic information of the data corresponding to the metadata, and play a very important role in the unified standardization of data, the integration of multiple data sources, and data quality control[18,20]. In electric power industry, faced with a large amount of intelligent power distribution data, data integration often uses based on metadata extraction methods. Metadata contains important information in data assets, and its extraction and management are important links in the grid data asset inventory. With the construction of smart grid and informatization, the company has accumulated massive data resources, which have the basic characteristics of big data in terms of data volume, diversity, speed and value. However, most business systems do not achieve efficient integration, which also brings many problems to metadata extraction, including: the diversification of data sources, the diversification of data types, and the inconsistent metadata standards. In the use of metadata-based data integration methods, the technical metadata of different data sources are basically directly extracted and applied [18-20]. WAN Chan [18] proposed a power data resource management scheme based on standardized metadata. Firstly, the program obtains metadata of various power data sources through preprocessing processes such as data screening, data transformation, and normalization. Secondly, it uses middleware technology to uniformly access data, and finally uses Hadoop technology to build a relational database. Centralized management and storage. The experimental results show that there are certain advantages in the query time of data integration. Jiecai Zheng [19] proposed a metadata modeling program, which uses XML-structured document form to model business data and uses MDA to reduce the difficulty of data manipulation. In a university information system, the integration and sharing of various heterogeneous data on the platform has been achieved, and good results have been achieved relatively. In [20], ZHANG Xin-yang proposed a metadata management method based on big data in the electric power industry, sorts out the source and data flow of each data source, achieves the purpose of unlocking the "black box" of the big data platform, and guides the data management of electric power enterprises.
Summarizing the above documents, the basic solutions based on metadata in power industry applications are shown in figure 1:

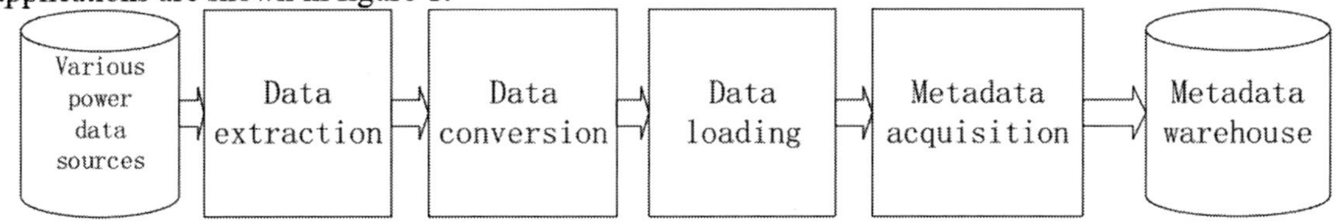

Figure 1. Basic scheme of metadata extraction in traditional power industry.

(1) Various power data sources: mainly refer to various power data sources that need to be integrated for a unified platform, such as financial data sources, infrastructure data sources, PMS2.0 etc.
(2) Data extraction: The access to various power data sources is realized through data plug-ins, and the main access is relational database data.

(3) Data conversion: Perform data format conversion operations for data sources read from different databases, including operations such as unified data metrics, data desensitization, and data de-redundancy.
(4) Data loading: load the converted data, and load the data into the designated data storage device.
(5) Metadata acquisition: Through the data structure analysis of the loaded data, the description of these data is obtained to form metadata, which generally includes technical metadata, business metadata, etc.
(6) Metadata warehouse: It is mainly used to store different metadata and manage the metadata. Including metadata extraction, deletion, update and other operations.
In traditional metadata-based data integration solution, the storage and management of metadata sets can be realized with the help of a metadata management platform. However, the methods for capturing metadata from different types of data sources lack uniformity, completeness, especially for the relationship semantic information of metadata tables in relational databases. So it is difficult to realize the metadata association analysis.

3. Data Integration Scheme Based on DTMS

Aiming at the practical problems encountered in the above metadata integration scheme, this paper realizes the associated analysis of metadata by constructing DTMS to solve the problems caused by the inconsistent data standards of electric power enterprises.
The data integration scheme based on DTMS is shown in Fig. 2:

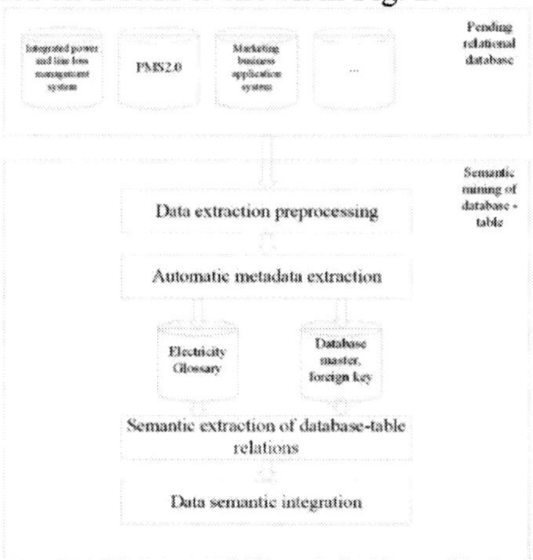

Figure 2. Data integration scheme based on DTMS.

After pre-processing the core business database data of the power grid, the metadata is automatically extracted by the metadata extraction algorithm, and the database table relationship semantic discovery is carried out by using the main foreign key of the database. The metadata of a database table includes: tables, fields (attributes), data types, table constraints, primary keys, foreign keys, comments, etc. However, these metadata cannot complete the semantic description of the relationship between the database tables, and use primary and foreign keys to semantically mark the relationship between the database tables to generate semantic information that can express the relationship.
(1) Data extraction pre-processing. It can realize the integration problems of data preloading, data standardization, data de-redundancy, and data conflict resolution of different data sources.
(2) Automatic extraction of metadata. By using metadata extraction tools to extract metadata, in order to accurately mine the semantics of subsequent database-tables.
(3) Extraction of semantic relations between database and tables. As mentioned in [21], metadata is a description of the structure of data tables and fields, and contains corresponding semantic information. Therefore, it is necessary to dig out the implicit semantic information between tables and fields in the database.
Currently, there are many standards for similarity measurement. In actual application, there are two commonly used similarity measures: edit distance similarity [21] and Jaccard distance similarity [22].

Although the accuracy rate is high based on the edit distance, the recall rate is low. In some power application business scenarios, some semantic information with specific meanings will be missed. Based on Jaccard distance similarity is very sensitive to character order,

This paper uses the method based on the edit distance similarity and the Jaccard distance similarity to measure the similarity to solve the shortcomings of a single method. The specific steps are shown in figure 3:

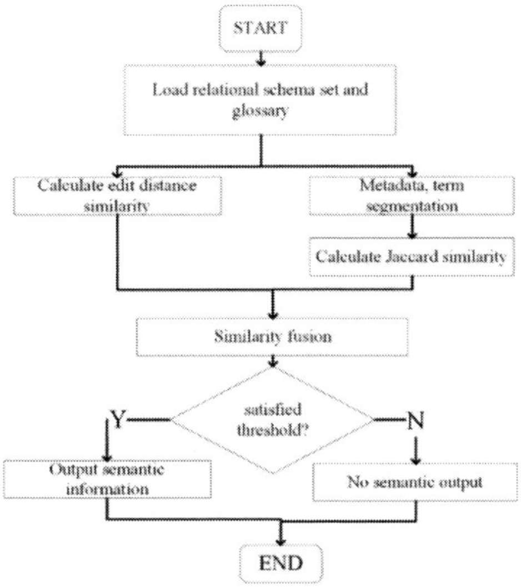

Figure 3. The execution steps of the two similarity measures fusion method.

Key step 1: Load relational schema and glossary. Suppose the metadata relational pattern set of the electric power database is $K = \{k_i, i = 1,2, \ldots, M\}$, M represents the total number of database metadata relational patterns. Suppose the electric power glossary (or entity table) is denoted as $S = \{s_j, j = 1,2, \ldots, N\}$, N represents the number of glossaries. Since there is a certain semantic relationship between the power database data relationship model K and the power glossary S, the semantic relationship can be measured by the similarity of the two strings.

Key step 2: Metadata, terminology segmentation. In order to use the measurement method based on Jaccard distance similarity to analyses the semantics of the database tables, it is necessary to perform word segmentation processing on the power database data relation pattern K and the power glossary S. $\forall k_i \in K$, the word segmentation set obtained after word segmentation is shown in formula (1):

$$k_i = \{k_{i1}, k_{i2}, \ldots, k_{im}\} \quad (1)$$

Among them, $k_{it}$is the t-th vocabulary obtained by segmentingk_i, and t satisfies: $t \in (0, m)$.
$\forall s_j \in S$, the word segmentation set obtained after word segmentation is shown in formula (2):

$$s_j = \{s_{j1}, s_{j2}, \ldots, s_{jn}\} \quad (2)$$

Among them, s_{ju} is the u-th vocabulary obtained by segmenting s_j, and u satisfies: $u \in (0, n)$.

Key step 3: Calculate Jaccard similarity. Calculate the Jaccard distance similarity between $k_i$and s_j using formula (3):

$$J(k_i, s_j) = \frac{|k_i \cap s_j|}{|k_i \cup s_j|} \quad (3)$$

Key Step 4: Calculate the edit distance similarity. Calculate the edit distance similarity between k_i and s_j using formula (4) :

$$B(k_i, s_j) = 1 - \frac{E(k_i, s_j)}{M(k_i, s_j)} \quad (4)$$

Where $E(k_i, s_j)$ represents the edit distance betweenk_i *and* s_j, and $M(k_i, s_j)$represents the maximum length of k_i, s_j.

Key Step 5: Similarity Fusion. In order to solve the problems of a single similarity measure, this paper adopts the formula (5) strategy for fusion:

$$R(k_i, s_j) = w_1 J(k_i, s_j) + w_2 B(k_i, s_j) \tag{5}$$

In the above formula, $w_1 \in [0,1], w_2 \in [0,1]$ and $w_1 + w_2 = 1$.

4. Experimental Analysis

In this paper, the 100 tables have been selected to verify the effectiveness of the algorithm. These tables have 500 metadata items and 1500 related glossary. The similarity threshold is set to 0.5. We conducted a comparison of algorithms.

(1) $\boldsymbol{w_1}, \boldsymbol{w_2}$parameter selection experiment

Since w_1, w_2 satisfies $w_1 + w_2 = 1$, this paper only conducts experiments on the value of w_1, and takes the value at an interval of 0.1, and carries out the library table semantic extraction experiment respectively. The experimental results of w_1 are shown in figure 4:

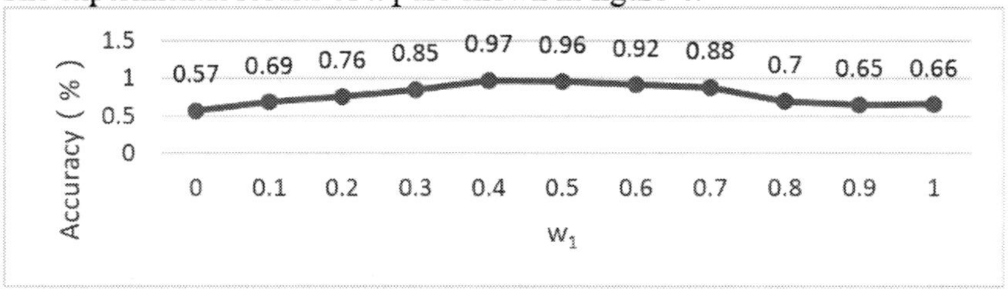

Figure 4. w_1 selection experiment.

It is found from experiments that when$w_1 = 0.4$, the accuracy rate is the highest, reaching 97%. Therefore, it is determined that $w_1 = 0.4$, $w_2 = 0.6$. If you want a more accurate value of w, you can perform a least-square fitting of the experimental results.

(2) Semantic generation accuracy rate

After determining the value of w, this paper compares the semantic labelling methods of database-tables using different similarity measurement algorithms, and verifies their accuracy. The experimental results are shown in Table 1:

Table 1. Comparison of accuracy of different algorithms.

Methods	Accuracy
Jaccard	93.5%
Edit distance	91.7%
Our Method	95.8%

5. Conclusion

Aiming at the characteristics of grid data sources with multiple sources, wide distribution and complex data association, this paper proposes a power data integration method based on DTMS, which can realize the integration of grid data at the metadata layer. When extracting the semantics of the database table, this paper combines the Jaccard distance and the edit distance to solve the shortcomings of the single method. The accuracy of semantic annotation of database-tables to a certain extent in this paper under the experiments. In future, we will consider better expression of semantic features of database tables to improve the accuracy of data fusion.

Acknowledgements

This project is funded by the Science and Technology Project of State Grid Jilin Electric Power Co., Ltd. [Project Title: Research and Application of Key Technologies of Provincial Power Grid Data Asset Inventory. Grant Number:2020-32]

Journal of Physics: Conference Series **2179** (2022) 012028 doi:10.1088/1742-6596/2179/1/012028

References

[1] YANG Yurong XU Yongjun DU Dongwei and HAN Chao 2019 Application of Distribution Network Security Authentication Scheme Based on DNP 3.0 *ZHEJIANG ELECTRIC POWER* pp 48-53
[2] GUO Yi YU Lei ZHANG Hanwen YAN Yong and SUN Yi 2018 Trading Platform of Power Data Asset Based on Consortium Blockchain *Frontiers of Data & Computing* pp 48-59
[3] Dai Linjie 2018 Research on the Guarantee of Power Data Assets *Science and Technology Innovation Herald* pp 177-178
[4] Wang S Peng YW Lan H Luo QW and Peng ZY 2020 Survey and prospect: Data integration methodologies *Journal of Software* pp 893-908
[5] Du XY Lu W and Zhang F 2019 History, present, and future of big data management systems *Journal of Software* pp 127-141
[6] Meng XF and Du ZJ Research on the big data fusion: Issues and challenges *Journal of Computer Research and Development* pp 231-246
[7] Yang XD Peng ZY and Liu JQ 2016 An overview of information integration *Computer Science* pp 55-59
[8] Liwei Kuang Laurence T Yang and Yang Liao 2019 An Integration Framework on Cloud for Cyber-Physical-Social Systems Big Data *IEEE Transactions on Cloud Computing* pp 363-374
[9] Chuanzhao Tian and Guoqing Li 2019 A Framework for the Data Integration of Earthquake Events. *IEEE Access* pp 172628-172637
[10] Kalaycı Tahir Emre Bricelj Bor Lah Marko Pichler Franz Scharrer Matthias K. and RubešaZrim Jelena 2021 A Knowledge Graph-Based Data Integration Framework Applied to Battery Data Management *Sustainability* pp 1583-1583
[11] Liao Rongtao and Yu Zheng, Feng Hao 2016 Design of Visual Display Framework for Power System Information Integration *SCIENTIST* pp 66-68
[12] Berlin J and Motro A 2002 Database schema matching using machine learning with feature selection *In: Proc. of the Int'l Conf. On Advanced Information Systems Engineering* pp 452–466
[13] Nottelmann H and Straccia U 2007 Information retrieval and machine learning for probabilistic schema matching *Information Processing & Management* pp552-576
[14] Chen Z and Cafarella M 2014 Integrating spreadsheet data via accurate and low-effort extraction. *Proc. of the 20th ACM SIGKDD Int'l Conf. on Knowledge Discovery and Data Mining. ACM* pp 1126–1135
[15] Chanial C Dziri R and Galhardas H 2018 Connectionlens: Finding connections across heterogeneous data sources *Proc. of the VLDB Endowment* pp 2030-2033
[16] ZHAO Jie GONG Zheng and LI Shengfei 2020 Research on Institutional Big Data Integration Platform Based on Knowledge Map *STANDARD SCIENCE* pp 74-78
[17] Halevy A and Korn F Noy NF 2016 Goods: Organizing Google's datasets. *Proc. of the 2016 Int'l Conf. on Management of Data. ACM* pp 795-806
[18] WAN Chan WEI Lihao YANG Qiuyong YANG Zhaoyi and SU Huaquan 2021 Research on Data Storage Stratagy for Metadata Integration in Power Grid Industry *Microcomputer Applications* pp 26-32
[19] Jiecai Zheng and Xueqing Li 2016 Research and Application of Data Modeling and Integration Based on Metadata. *2015 7th International Conference on Information Technology in Medicine and Education (ITME)* pp 13-15
[20] ZHANG Xin-yang ZHANG Mei and MA Wen 2018 Metadata Management Method in the Era of Big Data Based on Power Industry *Computer Knowledge and Technology* pp 258-260
[21] DONG Guoqing and TONG Weiqin 2012 Automatic Semantic Annotation for Metadata in Databases *Computer Science* pp 159-162
[22] Jiang Jirong 2021 Interval number multi-attribute decision-making method based on modified Jaccard similarity *Educational Standardization* pp 125-127

Grid Data Inventory Model Based on Improved Redundancy Elimination and Label Extraction

Ming Chen[1,*], Xue Ding[2] and Limin Yuan[1]

[1]Information and Communication Company, State Grid Jilin Electric Power Co., Ltd., Jilin Changchun, China
[2]State Grid Jilin Electric Power Co., Ltd., Jilin Changchun, China

*Corresponding author email: chenming@jl.sgcc.com.cn

Abstract. In view of the difficulty of data inventory caused by the large amount of accumulated data of the current grid business data, multiple data sources, and complex data, this paper proposes a data inventory scheme for structured data and unstructured data. Firstly, this paper uses the YOLOv3 deep learning network to label unstructured data to convert unstructured data into structured data; Secondly, the Semantic Hypergraph Clustering (SHC) method is used to realize the identification of the same entity of the power grid data and the elimination of redundancy. This method introduces semantic information in the block; Finally, in the comparative experiments of AR and RR, 96.88% and 95.03% respectively verify the effectiveness of the scheme.
Keywords: Data inventory model; Redundancy elimination; Label extraction; Semantic hypergraph clustering.

1. Introduction

With the continuous development of power science and technology and the continuous improvement of informatization, more and more power data have been accumulated, and "power big data" has basically formed. However, it is not difficult to find in the development process of the power information system. Since all provinces and cities have established information systems that are convenient for their own operations when they first stepped into the information age, many systems are established independently, and there is little consideration of inter-system The data interaction caused by data sources are scattered and stored, and it is not easy to share. In order to deal with above problems, the power industry data center is gradually established. The data center is located between the foreground business application platform layer and the background data storage layer, which can provide data sharing services on the basis of ensuring data security [1,2]. The data center can fully integrate current multi-source structured power data sources, turning data into a reliable platform for assets. However, the current data center can better realize the integration of structured data sources, and it is difficult to integrate unstructured data sources. In addition, there is no reasonable strategy to deal with the redundancy problem in multi-source data integration and the problem of the same data entity.

In redundancy processing, many scholars at home and abroad have put forward fruitful results, among which the redundancy removal method based on data entity recognition has attracted much attention [3-9]. In 2015, Guo XiYue proposed the information extraction technology, the main purpose of which is to extract key semantic information in the case of big data and facilitate data management [3]. QIN Songtao studied the multi-source data fusion method of power grid, summarized the data fusion level of pixel level, feature level and decision level, and proposed a data fusion method based on mapping [4]. SHA Ren pointed out that entity recognition is an important means of data cleaning, and proposed a method of entity recognition process oriented to big data. This method uses the data set block to simplify the calculation steps, and then uses the hypergraph clustering algorithm to analyze the identity of the entities [5]. Isozaki H uses SVM to train data entities and their tags, generates data entity acquisition models, and uses the entities acquired by this model to perform redundancy checks [6]. Saiqin Long proposed a data redundancy elimination strategy based on similar clustering algorithm. This strategy consists of two stages: the preprocessing stage is responsible for classifying similar data together; the redundancy deletion stage is responsible for detecting and eliminating duplicate data using similarity clustering [7]. HONG De-hua proposed a data missing compensation method based on knowledge graph technology, which can solve the problem of data missing in multi-source data [8]. Ma X uses neural network to realize the knowledge extraction and entity recognition of the big data corpus, and has achieved good recognition results on the corpus and entity recognition library CoNLL 2003corpus of Penn Treebank WSJ [9].

In label extraction for unstructured data, there are more methods for labeling structured data, while labeling for unstructured data is relatively rare [10-15]. ZHANG Tao proposed a data sharing traceability method based on data tags. This method mainly uses special secret key technology to process data, generates a label system for each type of data, and uses the PADDING method to attach the label to the data entity [11]. Guo Min proposed a labeling method for power data. This method is designed to implement five data label systems, namely: customer label, equipment label, organization object label, supplier label and employee label [12]. YE Kang determined the data label generation method, including four steps: data extraction, data conversion, data identification and computing services, and gave the application of data labels in search services, application display, etc. [13]. Lin Sen implements customer labeling, including basic information tags, electricity usage behavior tags, payment behavior tags, and customer interaction tags, and verifies the importance of tag data from customer electricity bill risk analysis applications [14]. Cheng Rui proposed a labeling framework for power users, which implements structured management of multi-dimensional, multi-level labels and standardized management [15].

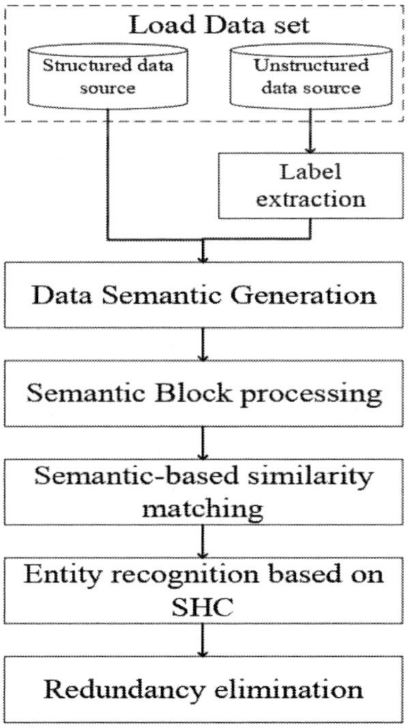

Figure 1. Overall framework of data inventory.

In view of the current problems encountered in grid data source inventory, the overall framework of grid data inventory based on redundancy elimination and label extraction proposed in this paper is shown in figure 1:

In the framework shown in figure 1, we propose a labeling method for unstructured data, which converts unstructured data into structured data for data inventory using YOLOv3. Regarding the redundancy problem of the same entity that may exist in structured data, this paper proposes the problem of identifying the same entity based on the semantic hypergraph clustering method to eliminate the redundancy of data entities.

2. Entity Recognition Based on SHC

(1) Hypergraph clustering entity recognition

In [5], the basic principle of the hypergraph clustering algorithm and the overall steps of entity recognition are given to solve the problem of entity analysis with multiple associations in practical applications. The concept of hypergraph is a basic concept proposed by Dan AS, which was later applied in many scenarios [16]. The related concepts are as follows:

Super edge: If an edge E can connect two or more nodes, then E is called a super edge. When the number of nodes of the super edge E is 2, it degenerates into an ordinary edge.

Hypergraph: Set up a graph structure G=(V,E), where V is a finite set of nodes, E is a set of hyperedges, then G is called a hypergraph.

Hypergraph clustering: The clustering method implemented on the hypergraph G is called hypergraph clustering.

According to the rough steps of hypergraph clustering entity recognition given in [5], as shown in figure 2:

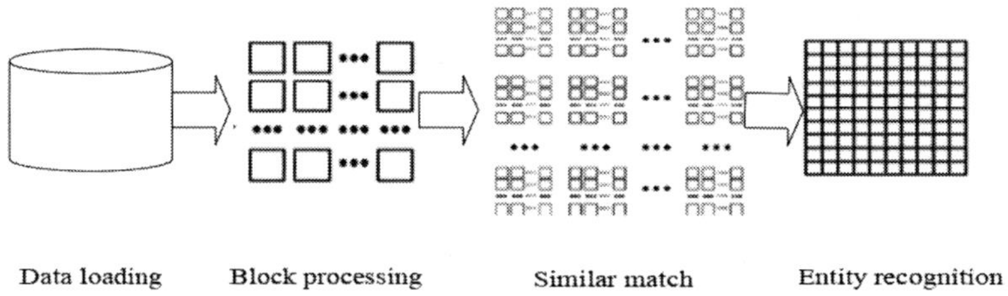

Figure2. Basic steps of data entity recognition.

Key step 1: Data block processing

This step is mainly to improve the data post-processing speed. The data is divided into blocks according to the block technique, only the data in the block is subsequently matched, and no operations are performed between the blocks [17]. According to the different block ideas, it can be divided into four categories: fixed block method, adjacent row method, tuple clustering method and mapping block method [5].

Key step 2: similarity matching to identify entities

For each block in the result of segmentation, data entity matching is performed. If the matching is successful, the two entities are classified as the same entity, and for the unsuccessful matching, the same entity cannot be divided. According to different matching methods, it can be divided into: threshold method, probability method, machine learning method, clustering method, etc. [5].

(2) Data entity recognition based on SHC

Although the hypergraph clustering method can make redundant judgments through entity recognition, the data block in the algorithm steps will greatly affect the final recognition result. Therefore, this paper proposes a data entity recognition method based on semantic hypergraph clustering, which can be applied to the entity recognition of structured data and unstructured data.

The main idea of the data entity recognition method based on semantic hypergraph clustering is to add semantic information of the data to the idea of blocking based on the basic idea of analyzing the block to improve the accuracy of entity recognition. The algorithm flow of this method is as follows:

1) Data Semantic Generation

For the loaded core business system data source after preprocessing, set the following symbols:

Business system data collection D: $D = \{d_1, d_2, \dots, d_{|D|}\}$
Record set R in data set D: $R = \{r_1, r_2, \dots, r_{|R|}\}$
The attribute set A contained in R: $A = \{a_1, a, \dots, a_{|A|}\}$
The corresponding attribute value in A is V: $V = \{v_1, v_2, \dots, v_{|V|}\}$
According to the above notation, the entity identification problem in the core business system can be transformed into: clustering the record set R in the core business system data set D into RL, and each set in the RL corresponds to the same entity.
2) Semantic Block processing
For all records in D, create a semantic triple (primary key, data column, column value). For all tuples, based on the primary key value, first perform the reverse sorting according to the method of literature [5], and then calculate the sorted semantic grouping according to the cosine similarity. Perform semantic similarity analysis on the sorted semantic groups, that is, for arbitrary semantic groups A and B, the cosine similarity is calculated according to formula (1):

$$\mathrm{S(A,B)} = \frac{A \cdot B}{\|A\|\|B\|} = \frac{\sum_{i=1}^{n} A_i B_i}{\left(\sum_{i=1}^{n} {A_i}^2\right)^{1/2} \left(\sum_{i=1}^{n} {B_i}^2\right)^{1/2}} \tag{1}$$

In formula (1), it is necessary to perform word segmentation processing on A and B first, and then convert the semantic characters involved into a vector form expressed by a number, and n represents the number of sets after vector segmentation.
In order to achieve segmentation, the similarity threshold μ needs to be set, and whether it is classified as a semantic segment $f(\mathrm{A,B})$ needs to satisfy the formula (2):

$$f(\mathrm{A,B}) = \begin{cases} 1 & \mathrm{S(A,B)} > \mu \\ 0 & \mathrm{S(A,B)} \le \mu \end{cases} \tag{2}$$

In the above formula, $f(\mathrm{A,B}) = 1$ means that A and B belong to the same semantic group.
This paper only improves the block part, and the rest of the operation steps are the same as those in reference [5].

3. Unstructured Data Label Extraction Based on YOLOv3
For unstructured data, it needs to be converted into structured data before processing, that is, the process of labeling unstructured data. This paper mainly takes picture data as an example, and uses the YOLO v3 deep learning network proposed by me in the literature [18] to identify entities in the picture data. The network architecture of YOLO v3 is shown in figure 3:

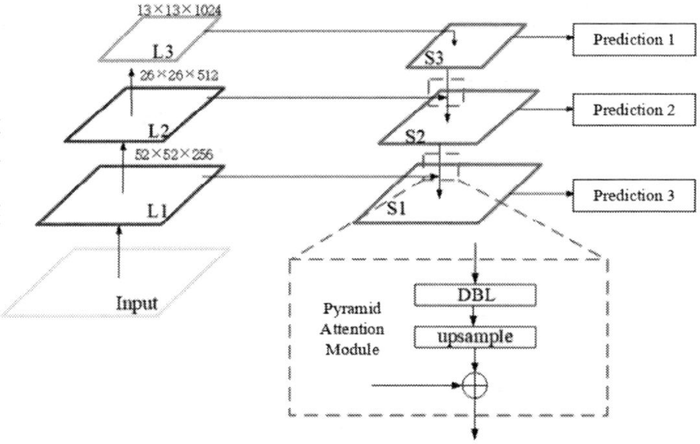

Figure 3. Network architecture of YOLOv3.
In figure 3, please refer to the description in Reference 18 for the structural meaning of the neural network. After the power picture passes through the above-mentioned neural network architecture, it can automatically identify the target in the picture and generate a structured label of the picture. The label information includes the name of the entity in the picture, the position x coordinate, the position

y coordinate, the height of the entity and the entity Width and other information. After that, perform a data inventory according to the structured data.

4. Results

In order to verify the effectiveness of our proposed in this paper, For the unstructured images, this paper select 1500 unstructured images of power were used for unstructured data in the experiment, and some common power labels (insulators, electric hammers, terminals, etc.) were labeled with LabelImg. These images are obtained during grid maintenance. For the structured data, Entity data includes 100 types of transformers, high-voltage capacitors, circuit breakers, power switches, etc. under the distribution network in a certain power supply company are used as an example to conduct experiments. The calculated indicators include accuracy rate and recall rate. The calculation formula is as follows:

$$AR = \frac{\sum_{i=1}^{m} \frac{p_i}{C_i}}{m} \tag{3}$$

Among them, m represents the total number of entities in the experiment m=100, $p_i$represents the number of records correctly identified in the i-th entity, and $C_i$represents the number of records identified by the algorithm in the i-th entity.

$$RR = \frac{\sum_{i=1}^{m} \frac{p_i}{k_i}}{m} \tag{4}$$

Among them, m and p_i have the same meaning as formula (3), and the only different k_i represents the actual number of records contained in the i-th entity.

The experimental results verify the effectiveness of the algorithm in this paper. The specific experimental results are shown in Table 1:

Table 1. Comparison experiment of AR and RR.

	AR	RR
Method in [5]	90.35%	92.18%
Our method	96.88%	95.03%

5. Conclusion

Aiming at the data inventory problem in the current grid big data scenario, this paper proposes a data inventory scheme for structured data and unstructured data. One is to use the YOLOv3 deep learning network to label unstructured data to convert unstructured data into structured data; the other is to use semantic hypergraph clustering to identify the same entity and eliminate redundancy in grid data.Technically, the integration of unstructured data sources and the redundancy of data entities are solved. In terms of economic benefits, more data are effectively utilized and the waste of data resources is reduced. In the future work, the algorithm needs to be optimized to improve the experimental results.

Acknowledgements

Our work is supported by the following fund: Science and Technology Project of State Grid Jilin Electric Power Co., Ltd. [Project Title: Research and Application of Key Technologies of Provincial Power Grid Data Asset Inventory. Grant Number:2020-32]

References

[1] LI Xinpeng LIU Wei YANG Zhiping and SHENG Pengfei 2020 Research on Data Mid-Platform Scheme of Power Grid Enterprises *ELECTRIC POWER ICT* pp 1-8

[2] XIA Hong-jun and AN Yan-na 2021 Construction of Data Asset Management Model for Power Supply Enterprises from the Perspective of Data Center *Information Scinece* pp 70-75

[3] Guo XiYue and He Ting-ting 2015 Survey about Research on information Extraction *Computer Science* pp 14-17

[4] QIN Songtao HUANG Chao TIAN Junyang YANG Yan and WEI Heng 2021 Research and Application of Multi-Source Big Data Fusion in Power Grid *Chinese Journal of Electron Devices* pp 480-484
[5] SHA Ren LIANG Qiong-fang LI Chang-ming and ZHANG Jia-xin 2020 Research on Related Technologies for Big Data Entity Identification *Software Guide* pp 125-127
[6] Isozaki H and Kazawa H 2002 Efficient Support Vector Classifiers for Named Entity Recognition *International Conference on Computational Linguistics-volume* pp 1-7
[7] Saiqin Long Zhetao Li Zihao Liu Qingyong Deng Sangyoon Oh and Nobuyoshi Komuro2020 A similarity clustering-based deduplication strategy in cloud storage systems *2020 IEEE 26th International Conference on Parallel and Distributed Systems (ICPADS)* pp 35-43
[8] HONG De-hua ZHANG Cui-cui XU Min and SUN Jia-li 2021 Research and Application of Data Arrangement Technology in Big Data Governance *Research and Development* pp 72-75
[9] Ma X and Hovy E 2016 End-to-end Sequence Labeling via Bi-directional LSTM-CNNs-CRF *In ACL* pp 1-12
[10] WANG Song PENG Yuwei and LAN Hai 2020 Survey and prospect: data integration methodologies *Journal of Software* pp 893-908
[11] ZHANG Tao 2020 Application of data labels in the traceability of shared data *Communications Technology* pp 221-224
[12] GUO Min LIN Xiaojing YIN Zenan and WAN Kai 2020 Research on construction and service capability of electric power data tag library *POWER SYSTEMS AND BIG DATA* pp 86-92
[13] YE Kang LENG Xiwu and XIAO Fei 2019 Intelligent analysis method of power grid monitoring based on big data label technology *Electrical Measurement & Instrumentation* pp 75-79
[14] Lin Sen Ouyang Liu 2016 Construction of power customer labeling system based on big data theory *Electrical Engineering* pp 98-101
[15] Cheng Rui Jia Jia Zhang Hao Chen Ying Duan Zhi Tian Wang Xia and Chen Qiuyu 2018 Framework Design of Electric Power Customer Labeling System Based on "Customer Insights" *The 3rd Smart Grid Conference* pp 63-69
[16] Dan A S and Djeraba C 1973 Graphs and Hypergraphs. Graphs and hypergraphs *North-Holland Pub. Co* pp 1307-1315
[17] FISHER J CHRISTEN P and WANG Q 2015 A clustering-based frame□work to control block sizes for entity resolution *21th ACM SIGKDD Int'l Conference on Knowledge Discovery and Data Mining* pp 279-288
[18] CHEN Ming ZHAO Lian-fei Yuan Li-min XU Feng HAN Mo and NI Hong-yin 2020 Insulator Detection Method based on Feature Selection YOLOv3 Network *INFRARED AND LASER ENGINEERING* pp 1-7

Fault Prediction of Elevator Door Lock Based on MPGA-BP Algorithm

Jun Qiu[1], Bei Qiu[2] and Leijing Yang[3,*]

[1]Department of Crane Inspection, Shanghai institute of special equipment inspection and technical research, Shanghai, China
[2]College of Marine Culture and law, Shanghai Ocean University, Shanghai, China
[3]Human Resources Department, Henan Special Equipment Safety Testing Institute, Zhengzhou, China

*Corresponding author email: 362823505@qq.com

Abstract. The door lock is one of the elevator fault prone parts, and the fault may further lead to personal injury accidents. In the process of elevator operation, the door lock acts frequently, resulting in a large amount of operation data. Using big data method, the fault prediction of elevator door lock could be realized, so as to prevent and reduce accidents. BP (Back Propagation) neural network algorithm and GA-BP (Genetic Algorithm) were used as door lock fault prediction algorithms, but there were some defects in performance. BP algorithm was easy to fall into local minima and the convergence speed was uncertain, while GA-BP algorithm would reduce the genetic diversity of the population. In this paper, we use MPGA (Multiple Population Genetic Algorithm) to improve the initial weight threshold of BP algorithm, adjust the neural network, and establish a door lock fault prediction model. The simulation results show that MPGA-BP model greatly improves the generalization ability and prediction accuracy of BP neural network, compared with traditional BP and GA-BP model.
Keywords: Elevator door lock; Fault prediction; MPGA optimized algorithm; BP neural network.

1. Introduction

The results of statistical analysis of elevator operation failure and accident data show that the door lock is one of the main components of elevator failure, and may further cause equipment damage or personal injury [1]. In recent years, based on the massive data generated in the elevator operation process, research on failure prediction of the main elevator components by using big data method is in full swing [2-6]. The main algorithms or models of the failure prediction have their own characteristics. Based on the typical elevator door lock structure (FIG.1) and main failure modes and cause analysis of the door lock [7], the associated operating parameters for fault monitoring of the door lock can be determined, including: door lock engagement depth, door lock circuit current and voltage (Table 1). Corresponding sensors were used to collect current and voltage signals, and sensors based on the principle of sliding resistance [8] were used to collect the meshing depth. The collected voltage signals were converted into the displacement value of the hook lock, so as to collect the accurate meshing depth of the door lock. The data sampling period was set at 0.1s. It is necessary to normalize the sampled data set to prevent the BP neural network from fitting badly due to excessive voltage data.

Figure 1. Typical elevator door lock structure (1- fixed hook, 2- movable hook, H-meshing depth).

Table 1. Common failures of elevator door locks, failure causes and monitoring parameters.

failure form	failure causes	monitoring parameters
abnormal of lock safety contact	• Safety contacts are short circuited • Adhesion of safety contacts • The grounding of the door lock loop is abnormal	• Door lock circuit current • Door lock circuit voltage
lock failure	• The magnetism of the permanent magnet decreases • The spring elasticity fails • The door lock is deformed and invalid	• Door lock circuit current • Door lock circuit voltage
Insufficient engagement depth of door lock	• The lock hook is damaged • Door locks are aging	• Door lock meshing depth

2. Comparison and Analysis of Algorithms for Door Lock Fault Prediction

2.1. Back Propagation (BP) Neural Network

BP (Back Propagation) neural network is a supervised learning algorithm. Its ultimate goal is to repeatedly correct weights and thresholds according to back propagation bias, so as to minimize the cost function of the output layer. Its algorithm structure is shown in FIG. 2.

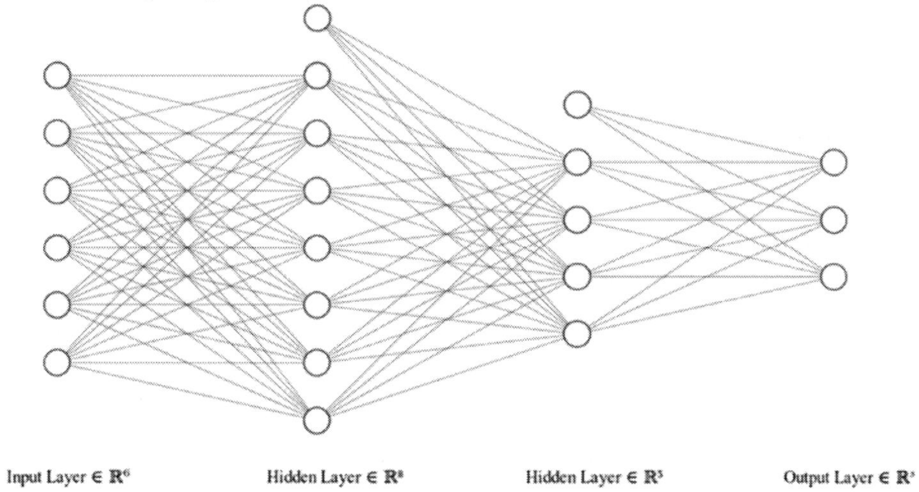

Figure 2. BP algorithm structure.

The BP algorithm itself has some deficiencies and defects [9-10], mainly including:

- It is not guaranteed that the final result of BP algorithm is globally optimal.
- The initial weight of BP neural network is set randomly, so once the initial weight is not appropriate, it will have a great impact on the training speed and training results of the subsequent algorithm, and may also bring the problem of local minimum value.
- The network convergence rate is slow.
- The memory of BP algorithm is unstable.

2.2. Genetic Algorithm (GA)

GA (Genetic Algorithm)is to imitate the operation process of individuals under natural selection conditions. With the help of genetic mechanism, the problem parameters are encoded by coding method, and then the more adaptable individuals are screened out through continuous iterative cyclic selection, crossover, heredity, variation and other operations, and finally the individual parameters that meet the improvement index are obtained.

GA has the characteristics of good global search, simple algorithm principle but significant effect, strong robustness and global parallel search. However, due to its single population, the value of crossover probability p_c and mutation probability p_m will affect the results of the algorithm, and under the action of selection operator, the algorithm will repeatedly select Exceptional individuals. As a result, the good genes of the individual will control the evolutionary trend of the next generation population, resulting in the decrease of genetic diversity and the gradual simplification of the population. It is usually called the precocious phenomenon of GA algorithm, which means that all individuals in the population stop evolution after a certain genetic algebra, but the evolutionary results tend to be exactly the same state, making GA unable to give satisfactory solutions [11-13].

2.3. Multiple Population Genetic Algorithm (MPGA)

MPGA is an extension of the algorithm structure on the basis of GA, which simultaneously conduct collaborative search on the solution space, and improves the global search ability of the algorithm, because it introduces the concept of multi-population. It also greatly reduced the influence of improper setting of genetic parameters, and had obvious effect on inhibiting the precocious phenomenon.

The core purpose of MPGA is to improve the operation processes such as selection, crossover and mutation in GA. The cooperation between various groups is carried out through the genetic operation between excellent individuals. The optimal individual in each population will be selected and transformed into the base individual for the genetic operation of that population, and then in the next population, the genetic operation will be carried out between the individuals of that population and the optimal individual. The optimal individual does not change in the process of crossover and mutation operation, which ensures that the information transmission mode among various groups is through the communication between the optimal individual of each population, and then the excellent information conveyed into the population is used to ensure that the offspring of the population can carry out good genetic operation. Therefore, this method can be used to solve the problem of random weight initialization of BP neural network and precocity of single population GA.

3. Fault Prediction of Elevator Door Lock Based on MPGA-BP

MPGA is used to improve the initial weight of BP algorithm, and the neural network is adjusted, hoping to significantly accelerate the training speed of the network, and solve the problem that BP algorithm is easy to fall into the local minimum value. Furthermore, the MPGA-BP algorithm is compared with the traditional GA-BP algorithm and BP algorithm to verify its advantages.

3.1. Structure Design of BP Neural Network

- The number of neurons in the input layer m=3, which represent the current of the door lock loop, the voltage of the door lock loop and the meshing depth of the door lock respectively.
- The Number of neurons in the output layer n=3, the output data type is 01 binary.
- The combination of the two can characterize the situation of 5 types of door locks, and its corresponding representation forms of each situation are shown in Table 2.

Table 2. Comparison table of door lock state and characterization form.

Door lock state	characterization form
normal operation	0 0 0
short circuit of the door lock circuit	1 0 0
insufficient engagement	0 1 0
short circuit + insufficient engagement	1 1 0
short circuit + non-engagement	1 0 1

3.2. Flow Chart of MPGA-BP Algorithm

As shown in FIG. 3, flow chart of MPGA-BP algorithm for elevator door lock fault prediction includes two parts: BP and MPGA.

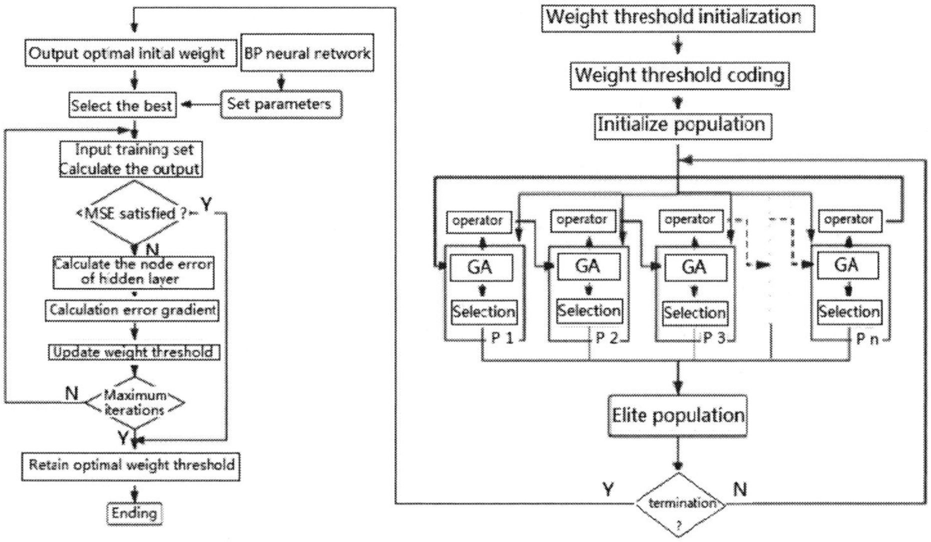

Figure 3. Flow chart of MPGA-BP.

- Firstly, the three-layer network structure of BP algorithm was defined, and the network structure was established as 3-3-7 according to the door lock data set and training target.
- The weight threshold of BP algorithm is initialized randomly. The 52 initial weight thresholds are encoded with 01, and the encoding length of the generated encoding string is 52.
- According to the rules of MPGA algorithm, each population was set to have 20 individuals, 10 populations were initialized, and 10 was set as the least maintaining algebra of the optimal individuals, 0.9 as the generation gap between each generation of the population, and the probability matrix of crossover and mutation was set.
- The genetic iteration of the population is carried out. After one iteration of each population, the individuals with high fitness are selected, which are allowed to replace the worst one of the next population through the migration operator, and then the individuals with the highest fitness are preserved in the elite population through the artificial selection operator.
- Repeat the above the operations of iterative loop, until the essence of population within the optimal individual fitness function value is greater than or equal to the termination conditions, the end of the cycle, the output of the highest fitness individual species encoded string, the optimal weights of BP neural network threshold vector encoded string translation can be, the BP neural network's initial weights threshold using the reverse translation vector to replace.
- After that, the model is trained according to the normalized training set of door lock data, and the iteration is repeated until the termination condition is reached (there are three termination conditions: the mean square deviation reaches the set value, the training reaches the maximum number of iterations, and the minimum gradient reaches the index). When the termination condition is reached, the weight threshold at this time is retained and the training ends.

3.3. Parameter Setting

According to the door lock failure sample data set, the algorithm test environment is Matlab2017A; In order to prevent the phenomenon of BP neural network over fitting, the data of 300 groups were divided into training set, test set and cross validation set according to the proportion of 60/20/20%. The purpose of the training set is to give the algorithm the right guidance to learn in the right direction. The function of cross validation set is to prevent over fitting. The principle is that when the output

MSE of the sample is no longer reduced for 6 consecutive times, then the neural network will appear over fitting. The parameter Settings of MPGA-BP are shown in Table 3.

Table 3. Parameters of MPGA-BP algorithm.

Parameters of MPGA-BP algorithm	Parameter selection
Number of nodes in the input layer: m	3
Number of nodes in the hidden layer: h	7
Number of nodes in the output layer: n	3
Number of weight thresholds: NumVar	52
Network layer: L	3
Weight threshold initialization range: D	(-0.5, 0.5)
The activation function σ	Sigmoid
Population size: Nind	20
Number of population: MP	10
Number of least maintained generations of the most optimal individual: Maxgen	10
Generation gap of population: GGAP	0.9
Cross probability matrix p_c	$p_c = 0.7 + 0.2 * rand(MP, 1)$
Mutation probability matrix p_m	$p_m = 0.001 + 0.05 * rand(MP, 1)$
Fitness function : Objfun	$Objfun(\omega, v) = \frac{1}{\sqrt{\frac{1}{R}\sum_{r=1}^{R} \sum_{t=1}^{n} \left(\delta_j^r - y_j^r\right)^2}}$

3.4. Training Results

As shown in FIG. 4, the evolution of genetic algorithm has reached the optimal population in the 13th generation, and the optimal solution of the population (i.e. 1/ fitness) is less than 0.02. The maximum genetic algebra of the optimal individual is set to be 10, but after the optimal solution is reached in the 13th generation, the iteration is continued for 10 generations until the optimal solution is no longer reduced for 10 consecutive generations, then the iteration is stopped and the optimal individual solution is output.

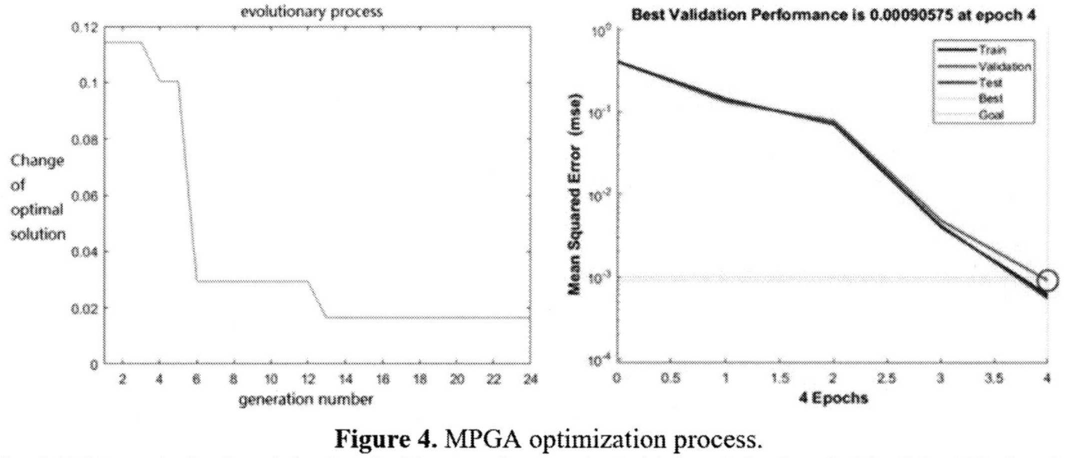

Figure 4. MPGA optimization process.

The MPGA optimized weight threshold was taken as the initial weight threshold of the BP algorithm, and the maximum iteration number epochs was set as 1000, the optimization goal was set as 0.001, and the learning rate l_r was set as 0.1. It can be seen from the MSE curve diagram obtained by learning and training that when the number of iterations reaches 4, the training has stopped. This is due to the limitation of cross validation, and if the training continues, the model is likely to over fit. The training curve and test curve in the figure are very close to the curve of the cross validation set, which may be due to the low dimension of the input feature vector of the data set and the insufficient difference between the samples. This problem can be solved by collecting more types of door lock elevator data, such as increasing the pressure sensor to test the pressure in the meshing part of the lock.

3.5. The Simulation Results

The comparison between the expected output of the test sample and the actual output is shown in Table 4, and it can be seen that the real output of BP algorithm after optimization is very close to the expected output, which indicating that the algorithm performs well and has a high degree of fitting.

Table 4. Actual output versus expected output.

Actual output	expected output	door lock state	**MSE**
0.0041 0.0032 0.0006	0 0 0	normal operation	0.0053
0.9932 0.0035 0.0011	1 0 0	short circuit of the door lock circuit	0.0077
0.0017 0.9817 0.0014	0 1 0	insufficient engagement	0.0131
0.9930 0.9912 0.0097	1 1 0	short circuit + insufficient engagement	0.0022
0.9971 0.0039 0.9882	1 0 1	short circuit + non-engagement	0.0124

3.6. Results Comparison of MPGA-BP, GA-BP and Traditional BP

The MSE of 60 groups of test samples shows that the mean square error of test samples is less than 0.014, which shows that the BP neural network using multi population genetic algorithm has good fitting. Through the initial weight threshold of BP neural network optimized by GA, it can be seen that in the case of 25 generations of genetic algebra, the final error value of MPGA is less than 0.02, while the final error value of GA is about 0.06. We can see that the optimization degree of MPGA is much higher than that of GA. As shown in FIG. 5, for the same set of training set and test set, the BP neural network optimized by MPGA is much less than the other two in terms of test error and training error, indicating that the generalization ability of MPGA-BP neural network is better than that of GA algorithm or random initial weight threshold method. The training error of the fourth MPGA-BP neural network is relatively large, which may be because the result error of the MPGA coevolution is relatively large at this time, which leads to the influence of the subsequent network training.

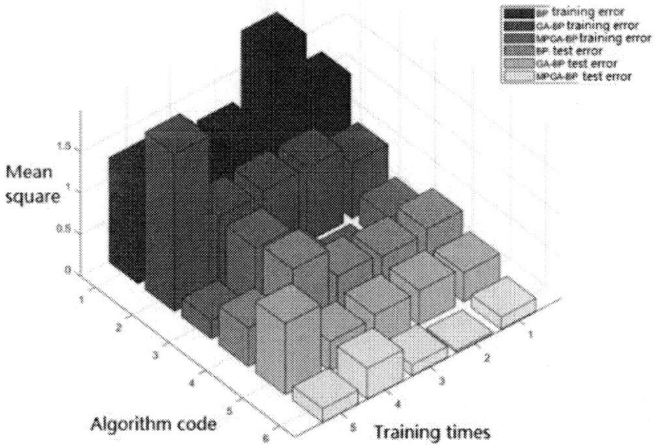

Figure 5. Error comparison of MPGA-BP, GA-BP and BP algorithm results.

4. Conclusion

- Aiming at the defects of BP neural network, GA optimization algorithm is introduced to improve, and then the defects of GA are made up by the characteristics of co evolution of MPGA. On the one hand, it can solve the weights of BP neural network random initialization of the problem of the great influence the result of the training, on the other hand it can restrain premature phenomenon of the single population genetic algorithm.
- The comparison results show that the performance of BP neural network optimized by MPGA algorithm is much better than GA algorithm or traditional BP neural network.

- Using MPGA to optimize the weight threshold of BP neural network will produce significant effects, but the optimization algorithm training time is long and the convergence speed is slow. If the predicted fault types are more complex, the training time of the optimization algorithm will be very long, and the speed of MPGA optimization algorithm needs to be improved.

Acknowledgments
This paper is support by The National Key Research and Development Program of China No.2018YFC0809005 and Science and Technology project of Shanghai Municipal Bureau of market supervision No.2021-21.

References
[1] Lan, S., Jiang, S., Qiu, J. Characteristics of typical elevator accidents in China from 2002 to 2019: A statistical investigation. Applied Mathematics and Nonlinear Sciences.
[2] Saimurugan, M.; Ramachandran, K.I.; Sugumaran, V.; Sakthivel, N.R. Multi component fault diagnosis of rotational mechanical system based on decision tree and support vector machine. Expert Systems with Applications. 2011, 38, 2923-4660.
[3] Zhang, Q.; Gao, J.W.; Zhang, B.; Mao, Y.L.; Dong, H.H. Fault Diagnosis of Rolling Bearing Based on Wavelet Packet and BBO-RBF. Industrial Control Computer.2017, 30, 83-84.
[4] Yaman, O.; Karakose, M. Auto Correlation Based Elevator Rope Monitoring and Fault Detection Approach with Image Processing. 2017 International Artificial Intelligence and Data Processing Symposium. 2017, 1-5.
[5] Mohammadi, R.; Fatemi Ghomi, S.M.T.; Zeinali, F. A new hybrid evolutionary based RBF networks method for forecasting time series: A case study of forecasting emergency supply demand time series. Eng. Appl. Artif. Intell., 2014, 36, 204-214.
[6] Lu, Z. Y.; Zhao, B.; Liu, T.; Song, Y. P.; Zhang, Y. Meng. Video detection system design of meshing depth of elevator's door lock. Mechanical Properties and Structural Materials, 2012, 580, 231-235.
[7] HE Jian. Several Cases of Failure of the Electrical Device for Proving the Elevator Landing Door Closed. China Elevator, 2018, 29(09):59-60+72.
[8] CHEN Jian-xun, LIN Xiao-ming, LIU Jian, TAI Sheng-lin.Correction of the Displacement Sensor for Measuring Engagement Length of Elevator Door Lock. Automation & Instrumentation, 2019, 34(10):81-84+94.
[9] Wen, P.G.; Zhi, M.; Zhang, G.Y.; Li, S.M. Fault Prediction of Elevator Door System Based on PSO-BP Neural Network. Engineering, 8, 761-766.
[10] Wang Haibo. Research on elevator real-time monitoring and online fault diagnosis based on fuzzy neural network [D]. Northeast University, 2013
[11] NIE Jun. Based on the Analysis and Prevent Premature Convergence of Genetic Algorithm. Software Engineer, 2013(11):42-44.
[12] ZHAO Jin-shuai, LU Rui-hua. An Improved Genetic Algorithm to Prevent Premature Convergence. JOURNAL OF SOUTHWEST UNIVERSITY(NATURAL SCIENCE EDITION), 2008(01): 156-159.
[13] QIN Guanghua, LI Zuoyong. Over-fitting of BP NN research and its application. ENGINEERING JOURNAL OF WUHAN UNIVERSITY, 2006(06):55-58.

Research on RCS Calculation for the Jet System of the Engine

Bingkun Jiang and Liangfa Xu*

Aviation Maintenance NCO Academy, AFEU, Xinyang, Henan, China

*Corresponding author email: ps20021119@qq.com

Abstract. In order to improve the stealth ability of the jet system of the engine, RCS calculation had been achieved through three steps, which were preprocess, core calculation and reprocessing. Some engine set as reference, the RCS of tail jet pipes with difference shapes had been calculated. The results show that the plug type tail pipe can advance stealth capability of the jet system. The numerical examples indicated the RCS calculation could play notable function in stealth design and promotion of engines.
Keywords: RCS features; Numerical calculation; Stealth capability; Engine.

1. Introduction

At present, many countries had been heavily equipped with long-range search and target indication radars. Therefore, radar stealth technology is also the main direction of the engine stealth technology research and application. At the same time, with the rapid development of electronic countermeasures, various radar detection technologies and radar guiding weapons, the backward stealth of aircraft has become an important factor to improve its survivability. The electromagnetic wave entering the engine through the tail nozzle of the jet system will produce strong echo, which will seriously reduce the stealth ability of the aircraft. Therefore, the research on radar stealth technology for the tail nozzle of engine jet system and necessary measures adopted to reduce the radar cross-section playing an important role in improving the overall stealth effect of engine, which has important significance.

This paper takes the tail nozzle of the jet system of a turbofan engine as a reference, nine types of plug jet nozzles had been modified and designed, and its radar cross-section characteristics were calculated and analyzed respectively. The influence of the nine plug nozzles on the stealth performance of the tail nozzle of the jet system had been analysed, which had highlighted the guiding significance of the radar cross-section calculation method in improving the stealth design of the jet engine.

2. Basic Concept of RCS

The physical quantity which quantitatively characterizing the scattering intensity of the target is called the effective scattering cross-sectional area of the target against the incident radar wave, which is usually referred to as the radar cross-section or RCS (Radar Cross Section). It is an imaginary area of the target, which comes from the research and design of the antenna. Obtaining the RCS of the target is the premise to study radar stealth technology of the target. RCS theoretical calculation by studying the scattering mechanism of the target, using various approximate methods and computer technology to quantitatively predict the radar cross-section characteristics of the target under various condition.

Generally, the radar transmitting antenna and receiving antenna are far away from the target, that is, the distance to the target is far greater than any meaningful size of the target. The radar wave incident on the target can be regarded as plane wave, while the target is basically a point scattering source. If we assume that the point scatter scatters energy isotropically, because the scattering field depends on the attitude of the target relative to the incident and scattering direction, the scattering intensity and

radar cross section of the hypothetical point scatter vary with the attitude angle of the target, that is, the RCS is not a constant, but a target characteristic closely related to the angle.
RCS is based on the theroy of isotropic scattering of targets illuminated by plane waves, and its definition formula is:

$$\sigma = \lim_{R\to\infty} 4\pi R^2 \frac{S^s}{S^i} = \lim_{R\to\infty} 4\pi R^2 \left|\frac{\vec{E^s}}{\vec{E^i}}\right|^2 = \lim_{R\to\infty} 4\pi R^2 \left|\frac{\vec{H^s}}{\vec{H^i}}\right|^2 \tag{1}$$

Where, S^s, S^i are the power density of scattered field and incident field respectively, $\vec{E^s}$, $\vec{E^i}$ are the intensity of scattered electric field and incident electric field respectively, $\vec{H^s}$, $\vec{H^i}$ are the intensity of scattered magnetic field and incident magnetic field respectively, R is the distance from the target to the radar. Because the incident wave is a plane wave and the target is assumed to be a point scatter, R tends to infinity, which is the most basic theoretical definition of radar cross section. In formula (1), $4\pi R^2 S^s$ represents the total scattering power of the scattered wave on the whole sphere with radius R. therefore, the meaning of RCS is the ratio of the total scattering power to the incident wave power per unit area when the target is isotropic scattering. This ratio has the dimension of area, in m^2. Its size indicates how much incident wave power the target intercepts and scatters it evenly in all directions.
RCS is a very complex physical quantity, which is not only related to the geometric parameters of the target, such as size, shape, material and structure, but also related to the parameters of the incident radar wave, such as frequency, polarization and waveform. At the same time, it is also necessary to consider the attitude angle of the target relative to the radar. The research of target radar scattering characteristics is to find out the relationship between these parameters and RCS, so as to provide basis for target radar stealth design.

3. RCS Characteristic Numerical Calculation

The numerical calculation process mainly includes three stages: pre-processing, core -computing and post-processing.

3.1. Pre-processing

The geometric modeling in pre-processing is completed by commercial CAD (UG, CATIA, SOLIDWORKS, etc.) software, making full use of its powerful geometric modeling function, so that the calculation is not limited by the geometry of jet system and has strong universality. After completing the modeling work, the geometric model is imported into the meshing software (ICEMCFD or GAMBIT) in the form of intermediate data format (IGES or PARASOLID) for mesh discretization. The density of mesh generation needs to be controlled according to the requirements of the incident wave length and IPO calculation, such as 9~16 triangular facets per square wavelength. In order to speed up the judgment process of occlusion relationship between panels, it is necessary to set a large enough growth factor in the wall grid of jet system cavity to generate a coarser internal connecting body grid. If the calculation domain contains a one-time reflection region that needs to be calculated by Po algorithm, such as an external adjusting plate, the grid division of the region can be in the form of Figure 1, that is, wrap the Po calculation domain with a transparent boundary region, and then generate a boundary grid as large as possible in the transparent boundary region, A coarser connecting mesh is generated between the Po calculation domain and the transparent boundary region to accelerate the judgment of the irradiation relationship of electromagnetic wave to Po panel (whether the panel is illuminated by electromagnetic wave). The judgment method is similar to the judgment method of occlusion relationship between panels. After the grid is generated, it is necessary to set the boundary conditions for different regions (the boundary conditions here are CFD type boundaries, which have nothing to do with electromagnetic calculation. The boundary conditions set here are only named for different regions to facilitate the subsequent definition of additional electromagnetic boundary conditions, and all regions can adopt the default settings), Then export the standard CFD unstructured grid data file. After importing the grid data into the program, set additional electromagnetic boundary conditions (define the boundary conditions related to electromagnetic calculation for the boundary grid of the original named area saved in the grid file). The types of electromagnetic boundary conditions include: solid boundary (non transparent) and transparent boundary. For the solid boundary, it is necessary to define whether it is a metal conductor wall or a coated medium wall. If it is a coated medium wall boundary, it is necessary to specify the relative dielectric constant, relative permeability and coating thickness of the medium. In addition, it is also necessary to define the calculation type to be adopted by the boundary panel, including IPO and PO. In order to search the contour line of the nozzle outlet edge of the jet system, it is necessary to define which boundary line of the two areas is the nozzle outlet edge calculated by EEC method. After the electromagnetic boundary conditions are set, the program needs to import calculation parameters to start calculation. The calculation parameters include incident wavelength, polarization mode, underrelaxation factor and calculation angle. The setting of the calculation angle should adapt to the coordinate system during geometric modeling, and the incident wave length should match the mesh generation density, otherwise the calculation accuracy will be reduced. After inputting the calculation parameters, the program will reorder the regional grid according to the nature of the boundary, and then judge the illumination relationship of Po panel and the occlusion relationship of IPO panel. Considering that the storage of occlusion relationship information between polygons requires a lot of memory, and for the same geometric model, the occlusion relationship between polygons does not change with the change of calculation angle and boundary condition properties, the principle of one-time calculation and multiple utilization is adopted, that is, the occlusion relationship information is compressed and stored in binary bits. This avoids the repeated calculation of occlusion relationship under the same geometric model and different calculation conditions, and improves the economy of calculation.

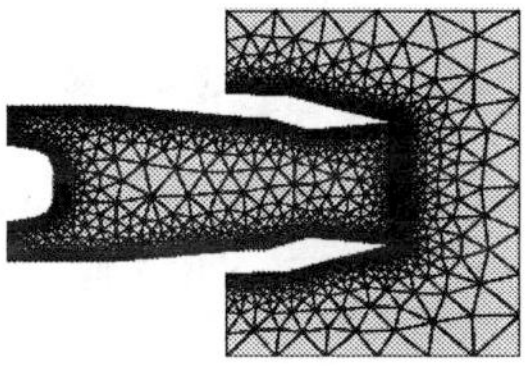

Figure 1. Schematic diagram of grid division of RCS calculation domain.

3.2. Core –computing

After completing the above pre-processing work, the program will use the corresponding algorithm to calculate it according to the different area types. For the IPO calculation area, firstly, the equivalent electromagnetic current on the aperture surface of the cavity is calculated according to the electromagnetic field equivalence principle, then the initial current on the inner wall of the cavity is calculated according to the Kirchhoff approximation formula, and then the forward and backward iterative calculation is carried out according to the occlusion relationship information between the panel elements; If the boundary region of the coating medium is encountered in the iterative process, the magnetic current term will be calculated according to the impedance boundary condition, and the iterative current will be corrected; When the iteration converges, the induced electromagnetic current on the cavity aperture surface is calculated according to the induced current and induced magnetic current on the cavity wall (there is a magnetic current term at the dielectric boundary), and then the far-field scattering field is calculated according to the far-field conditions. The RCS of the total field can be obtained by space vector superposition of the scattering field and diffraction field calculated by Po and EEC and the scattering field calculated by IPO.

3.3. Post-processing

The post-processing of the program includes the calculation of the induced current density vector and its modulus of the wall grid, and the calculation of the RCS of the scattered field of each component (different boundary regions). The wall induced current density distribution and streamline diagram under different calculation angles, as well as the RCS curves of various fields (IPO scattering field, Po scattering field, EEC diffraction field and total field) and components are output in Tecplot file format, which provides a reference for further analysis of RCS characteristics of jet system.

4. Simulation Analysis

This section takes the tail nozzle of a turbofan engine jet system as a reference, calculates the RCS of different shapes of tail nozzles through the methods described above, and finds the design scheme to reduce the RCS of this type engine.

According to the knowledge of geometry and physics, it is inferred that the existence of the plug cone in the plug nozzle is very likely to effectively block various specular echoes in the nozzle, so as to weaken the total electromagnetic scattering contribution of the nozzle to the engine, which is conducive to the electromagnetic stealth of the engine.

Table 1. Comparison of RCS mean values of different plug nozzles.

	P0	P1	P2	P3	P4	P5	P6	P7	P8	P9
Top angle of half cone of back cone of plug cone α	-	7.5	10	11.25	13	15	17	18.75	20	22.5
$\theta\theta$ polarization RCS mean	33.90	30.82	31.08	30.33	29.84	29.65	30.09	30.05	29.42	30.52
$\varphi\varphi$ polarization RCS mean	33.98	29.88	30.41	30.51	30.36	29.45	29.37	29.55	28.80	29.38
$\theta\theta$ polarization reduction percentage	0	9.1%	8.3%	10.5 %	12.0 %	12.5 %	11.2 %	11.4 %	13.2 %	10.0 %
$\varphi\varphi$ polarization reduction percentage	0	12.1 %	10.5 %	10.2 %	10.7 %	13.3 %	13.6 %	13.0 %	15.2 %	13.5 %

In order to compare with the reference nozzle, the throat and outlet flow area of the plug nozzle are consistent with the prototype nozzle, and the plug cone is formed by the combination of front cone and rear cone. The design parameters of the plug cone are: the top angle of the front cone and the half cone of the plug coneβFixed at 30°, the radius ratio of the plug cone is fixed at 0.5 (t Ratio of maximum radius of plug cone to inlet radius R of inner channel), The top angles of the half cone of the back cone of the plug cone are 7.5°, 10°, 11.25°, 13°, 15°, 17°, 18.75°, 20°and 22.5°respectively. Number all nozzles respectively. The axisymmetric nozzle without plug cone is coded as P0, which increases with. The nozzle number is p1-p9, as shown in Table 1. Figure 2 shows the comparison between prototype nozzle and plug nozzle, and the schematic diagram of plug cone section.

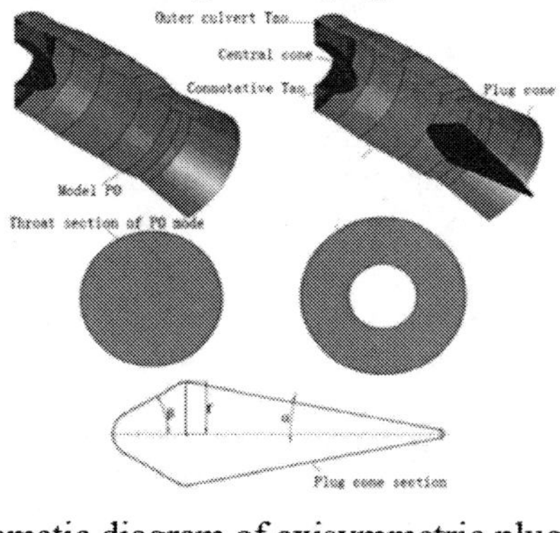

Figure 2. Schematic diagram of axisymmetric plug nozzle model.

Since there is a high-speed rotating turbine at the inlet end of the nozzle, only a small amount of electromagnetic wave can pass through the turbine, the inlet end (internal and external culvert section) can be regarded as the wall for short circuit treatment, that is, the nozzle can be calculated as a single end open metal cavity. The key pitch angle range of calculation is: θ=-40°~40°, Angular interval:Δθ=1°, A positive value indicates top view irradiation and a negative value indicates bottom view irradiation. The incident wave length is; Inλ(physical model feature size 10λ), The grid division density is 16 per square wavelength, and the total number of grids is about 20000 ~ 23000. The inner wall of the nozzle and the wall of the plug cone inside the nozzle cavity are calculated by IPO; the exposed part of the plug cone is calculated directly by Po without iteration; the diffraction field at the edge of the nozzle outlet is calculated by EEC; finally, the three fields are vector superimposed in space to obtain the RCS of the total field.

Table 1 shows the arithmetic mean of the total Radar Scattering Field RCS of the nozzle with different shape plug cones in the range of - 40°~ 40°pitch angle, and the reduction percentage of the RCS mean of each model nozzle relative to the original nozzle. It can be seen from the table that under the two polarization modes, the RCS value of all plugged nozzles is lower than that of the original nozzle. This is mainly because the plug cone effectively blocks the specular reflection echo of the simplified plane at the nozzle inlet in some pitch angle range, so that the mean RCS of the nozzle with plug cone is smaller than that without plug cone.

As can be seen from the mean table, the plug nozzle P8(α= 20°) the RCS is the smallest in both polarization modes. Therefore, P8 model is compared with the original nozzle. Figure 3 shows the variation curve of Total Field RCS of P0 and P8 models with elevation angle. Since the end of the plug cone of p8 model does not exceed the nozzle, the shape and size of the nozzle outlet edge of P0 and P8 model are consistent. The calculation results also show that the RCS values of the diffraction field at the outlet edge of the two models are the same. Therefore, the change in the total field is mainly due to the difference of the scattering field of the nozzle cavity. As can be seen from the figure, in the range of - 10 degrees to - 22 degrees and 19 degrees to 40 degrees of pitch angle, the reduction effect is obvious under the $\theta\theta$ polarization RCS mean mode, In the range of - 9 degrees to - 31 degrees and 1 degree to 37 degrees, the reduction effect is obvious under the $\varphi\varphi$ polarization RCS mean mode. This is mainly because in these angle ranges, the existence of the plug cone effectively blocks the specular reflection echo of the turbine end wall, thus reducing the RCS in these directions. However, there are also some angles where the echo is enhanced due to the coupling between the plug cone and the nozzle wall. For example, the RCS of nozzle P8 is greater than that of the original nozzle P0 at 0 degree to - 9 degrees as shown in Figure. 3 (a), and - 32 degrees to - 34 degrees as shown in Figure. 3 (b). However, in general, the RCS of the nozzle with plug cone is effectively reduced in most pitch angles.

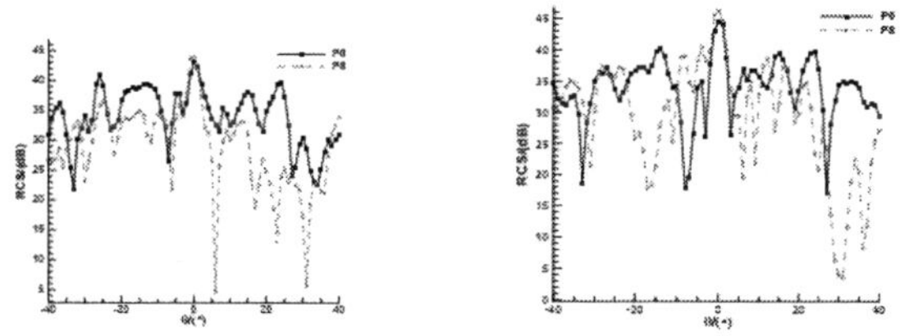

(a) (b)

Figure 3. RCS comparison between P8 nozzle model and original nozzle model.

It can be seen from this example that the RCS characteristics of engine jet system with different structures can be analyzed by numerical calculation method, and the results can be used to guide the RCS stealth design of the jet system.

5. Conclusion

RCS calculation has important guiding significance in stealth design. In this paper, RCS numerical calculation is realized through three steps: pre-processing, core calculation and post-processing, and RCS calculation of tail nozzle and its improved scheme of an aircraft. The example in this paper shows that the RCS numerical calculation can provide theoretical basis and improvement direction for the stealth improvement and upgrading of the engine, which has important practical significance for improving the survivability of the aircraft. Here, only a theoretical calculation method is provided, and several accurate jet system models will be made in the future to verify correctness of this method.

References

[1] Zhao Lei. High order finite element boundary integral method for RCS calculation of target [J]. Aerospace ECM, 2015,31 (6): 17-19
[2] Sun Jun. Field RCS measurement accuracy model and its engineering estimation [J]. Command control and simulation, 2015,37 (6): 118-121
[3] Zhang Xiaoyan, Liu Zhiwei, Liu Yingting. Fast calculation of target backscatter RCS Based on MBPE uniform sampling [J]. Journal of Beijing University of technology, 2015,35 (11): 1152-1157
[4] Huang Haoran, Wang Hongwei, Chang Xu. Analysis of vertical tail electromagnetic scattering characteristics and RCS Reduction Method [J]. Science and technology horizon, 2015, (15): 149-150
[5] Zhuang Yaqiang, Zhang Chenxin. Research on statistical modeling of meter band RCS of stealth aircraft [J]. Science, technology and engineering, 2014, (23): 219-223
[6] Yue Kuizhi, Jia Zhonghu. Numerical simulation of RCS characteristics of shipborne EW aircraft [J]. Systems engineering and electronic technology, 2014,36 (5): 852-858
[7] Zhuang Yaqiang, Zhang Chenxin. RCS statistical analysis of stealth targets based on Gaussian mixture density model [J]. Journal of Air Force Engineering University: Natural Science Edition, 2014,15 (2): 37-40
[8] Yang Xiaoshu, Zhang Yunfei, Bao Xiaoxiang, et al. Stealth design and experimental research of Military Helicopter Radar [J]. Helicopter technology, 2014, (1): 11-15

Research on a New Analysis Algorithm of Virtual Trajectory Cryptography Technology for Password Encryption and Decryption Through Base Expanding

Yuanyuan Chen[1,*] and Jing Chen[2]

[1]Haikou University of Economics, Haikou, China
[2]Qingdao City University, Qingdao, China

*Corresponding author email:252009674@qq.com

Abstract. In the rapidly developing Internet era, network security and data sharing coexist. With the progress of information reform, all walks of life need to ensure the security of data information. Password is a tool to protect their privacy, property, information and important data, and some criminals through technical means to crack the password, peeping and stealing other people's important information, data, meanwhile, the data caused information and property losses also happen. Information security has become a strategic issue that people must pay attention to, which is related to social stability, economic development and national security. In this paper, we study a new virtual trajectory password technology, analyzes the algorithm, complete the structure design of code disk and solve the problem of random generation of code base in code disk. The method of "base expanding" applied to enhance the strength coefficient of the password, and the random sequence generated in the chaotic encryption algorithm is used for encryption and decryption, and a random reference graph "code disk" is generated as an auxiliary memory method to help people remember some special lengthened passwords that cannot be remembered conventionally.
Keywords: Virtual trajectory; Password; Algorithm; Code disk; Code base

1. Introduction

With the continuous innovation and development of distributed network and heterogeneous network technology; digital, text, graphics, images, sound, video, animation and other information carriers in information technology are transmitted through computers and the Internet. Due to the openness, sharing and dynamics of the Internet, information security is threatened and disturbed.The problem of security vulnerabilities caused by network open sharing is becoming more and more serious. It is necessary to adopt data encryption technology to strengthen security protection. In the rapid development of the Internet era, it is quite common to use mobile phones for online shopping, information query, bank transfer, account recharge and other businesses. When users register and log in, they should register the account password for information security, that is, use the password for security verification. Password is a tool to protect their privacy, property, information and important data, and some criminals through technical means to crack the password, peeping and stealing other people's important information, data, meanwhile, the data vulnerability means that information and property losses also happen [1]. Information security has become a strategic issue that people must pay attention to, which is related to social stability, economic development and national security.
Based on the consideration of the above problems, people urgently seek a new password that can ensure the security of account and password, assist in memorizing password, prevent peeping and is

difficult to decipher. At the same time, when verifying through the mobile phone, it is often required to remember the passwords of various operations. There are many kinds of passwords, which is easy to cause confusion in memory. In view of this, people urgently need a password authentication method that can assist in memorizing passwords, preventing peeping, difficult to decipher and protect the security of password information.

2. Common Password Authentication Methods

As an important mechanism to ensure network information security, password authentication technology judges whether the authenticated person has secret information to prove his identity by some means. Those secret information can be the growth experience, biological characteristics or items recording identity information of the authenticated person [2]. As a means to realizing the identity authentication of both parties, identity authentication requires the participation of both parties and the design of a certificate recognized by both parties and strictly confidential to the third party. When one party presents the certificate, the legitimacy of the certificate can be judged according to the agreement at the beginning of the design. Therefore, key agreement protocol came into being. The commonly used registration verification passwords methods are as follows:

2.1. Memory Password

The recognizable information is converted into unrecognized information for use. During recovery, the unrecognized information is processed, recovered and cracked into recognizable information.

2.2. Biological Characterization Password

Using the characteristics of one or more parts of the human body for recognition.Fingerprint identification unlocking is to obtain fingerprint information by touching the unlocking area with a finger. As the fingerprint identification module is set in the system, it will transform the collected fingerprint information into characteristic data and store it for verification and identification. This part of data is encrypted and stored in a specific area. When the user needs to unlock the phone, it calls the data to identify it.

Face recognition unlock works through the acquisition device to obtain the user's face image, and then use the core algorithm to calculate and analyze the position of the facial features, face shape and Angle of the face, and then compare with the existing template in their database to determine the user's real identity[3].

Biometric authentication is universal, unique and permanent, but authentication information is often transmitted in plain text on the network, which is easy to be intercepted and affected by external factors, and errors are often made in feature extraction or matching.

2.3. User Name and Password Combination Authentication Method

Password is the most commonly used and easiest way to realize identity authentication. For example, 6-digit password technology is adopted for the login of commonly used systems, or online shopping and transfer, that is, the system establishes a combination of user name and password for each legitimate user. When the user logs in to the system, Confirm the user identity by checking whether the entered user name and password match the combination information saved by the system [4]. However, there are many problems in the security of this password authentication method, including the following.

2.3.1. Network data stream eavesdropping. In order to authenticate the password input, it needs to be transmitted over the network, commonly many systems transmit the unencrypted password directly over the network. It is easy for the attacker to eavesdrop on the data transmitted over the network through relevant software to identify the user name and password.

2.3.2. Snooping and social engineering. The attacker takes advantage of the opportunity to get close to the system to spy on the process of the user entering the password through the monitor, or impersonate a legitimate user to obtain the trust of the administrator to obtain the user password [5].

2.3.3. Authentication information interception playback. The attacker sends a packet that has been accepted by the server to cheat the authentication system, so as to obtain the login right to the system.

2.3.4. Dictionary attacks and exhaustive attempts. Users are used to using words that have some meaning for themselves, such as birthday, phone number, etc. as passwords. Attackers will use dictionaries or complete sets of strings containing most words to try to compare users' passwords [6]. Therefore, many systems suggest that users add special characters to the set password and limit the minimum length of password digits to enhance the security of password.

2.3.5. Spam search attacker. By searching the user's waste, such as written passwords on paper, from which to get the password information related to the system.
To sum up,with the core technology of information security,cryptography technology has many encryption methods. Although those methods are widely used in the management system of password verification at this stage,they have limitations and vulnerabilities to a certain extent, and are also facing many challenges. This paper studies a new "virtual trajectory password" technology, which breaks through the traditional method of memorizing characters, changes it into memory graphics, and changes the surface memory method from memorizing numbers and character,to memory track graphics, so as to make the traditional process of people memorizing passwords easier and safer.

3. The Research Methods

3.1. The "Virtual Trajectory"
Trajectory refers to starting from a starting point, and then drawing it according to the desired way to form the desired trajectory. In this paper, "virtual trajectory" refers to the image drawn by people's brain thinking, which is only stored in people's mind and filled in by inputting the code on the reference.

3.2. Virtual Trajectory Auxiliary Memory Method
Although the virtual trajectory is an individual's image thinking, it reflects the character information when inputed by reference code disk. Through our experiment, it is found that people's memory of the graphic information is better than that of a long string of numbers/characters. This design adopts the method of virtual trajectory forming graphics, which is convenient for people to remember. At the same time, due to the code disk with random transformation, the password is input into the password box according to the track in your memory every time you input the password. This method of obtaining the password seed is convenient for the system to convert the track remembered by people into character seed.

3.3. Virtual Trajectory Anti Peeping Method
When the user inputs the password at the input end, the system will pop up a code disk of virtual track password. They only need to input the corresponding password on the code disk according to their imagined track. Because the code disk popped up each time is random, the order of code bases is different, even if other peepers record the symbol password sequence entered by the user,they can't guess the input track, but they can log in successfully only by inputting the correct trajectory code. Therefore, the transformed code disk can prevent the input track password from being peeped.

3.4. Graphical Memory Password
Although there are more and more types of characteristic passwords and the technology is becoming more and more advanced, compared with memory passwords, the convenience, security and confidentiality of characteristic passwords are far less than memory passwords. Among many characteristic passwords, character password is the most widely used. In this design, "virtual trajectory password" completes the easy conversion from pattern password to character password, and provides a new security choice for the public.

4. The Specific Design

4.1. Design of Virtual Trajectory Cryptography Architecture

4.1.1. Code disk design. In order to facilitate the user to remember the password, the user can easily remember the simple "input seed code" by generating a constantly changing random reference figure - "code disk". Design a figure composed of 36 alphanumeric characters in 6 rows and 6 columns. We call it "code disk", and the user's virtual track password uses the code disk as a reference graph for auxiliary memory to memorize the password conveniently. The code disk is shown in Figure 1.

4.1.2. Code disk generation. The code disk is a 6-by-6 two-dimensional matrix, and a two-dimensional array is defined to describe it. Each array element can appear randomly in (0-36) code bases. The following is the instantaneous data of a random code disk, as shown in Figure 2:
Scd(1,1)=4,Scd(1,2)=t,Scd(1,3)=j,Scd(1,4)=0,Scd(1,5)=q,Scd(1,6)=6,Scd(2,1)=y,Scd(2,2)=b,Scd(2,3)=x,Scd(2,4)=m,Scd(2,5)=3,Scd(2,6)=1,Scd(3,1)=i,Scd(3,2)=n,Scd(3,3)=7,Scd(3,4)=h,Scd(3,5)=l,Scd(3,6)=a,Scd(4,1)=v,Scd(4,2)=c,Scd(4,3)=f,Scd(4,4)=u,Scd(4,5)=k,Scd(4,6)=g,Scd(5,1)=8,Scd(5,2)=9,Scd(52,3)=p,Scd(5,4)=z,Scd(5,5)=o,Scd(5,6)=5,Scd(6,1)=s,Scd(6,2)=w,Scd(6,3)=d,Scd(6,4)=2,Scd(6,5)=e,Scd(6,6)=r.

x	j	v	8	k	a
q	y	p	5	z	r
2	i	w	t	4	u
m	0	9	6	c	e
o	3	h	1	l	7
g	n	b	f	s	d

Figure 1. A random code salver diagram.

$$Scd = \begin{pmatrix} 4 & t & j & 0 & q & 6 \\ y & b & x & m & 3 & 1 \\ i & n & 7 & h & l & a \\ v & c & f & u & k & g \\ 8 & 9 & p & z & o & 5 \\ s & w & d & 2 & e & r \end{pmatrix}$$

Figure 2. Two dimensional matrix of code disk elements.

The code disk is generated by placing the continuous elements (code base) according to the randomly generated position, so as to ensure that all code base elements are placed on all code disks, and there are no repeated elements, so as to ensure that when the tracks of the selected password cross and coincide on the code disk, the system can still correctly distinguish the paths without ambiguity.

4.1.3. Generation of virtual trajectory password. The generation of virtual trajectory cipher is a cipher sequence obtained by using the chaotic sequence generated by the chaotic system and the reproducible random number generated by the user's seed trajectory code (initial value) through the transformation of the selected formula at the client. As we can see from Figure 3, each element on the array is the character displayed on the code disk. If the user selects an element and inputs that element, the system can get its position on the code disk. When continuous or discontinuous points are selected, there will be a certain position trajectory, the selected positions are grouped into a continuous (or discontinuous) sequence of positions and the path of the selected positions on the code disk, namely "virtual trajectory".

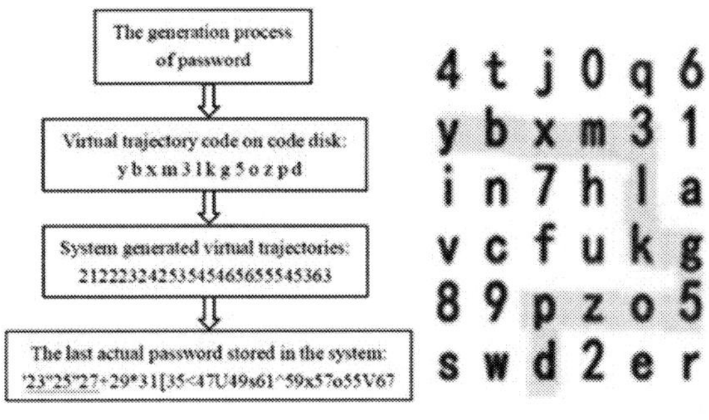

Figure 3. The schematic diagram of password generation process.

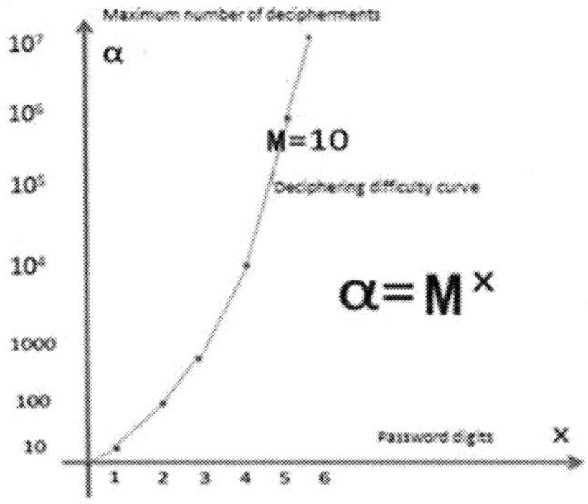

Figure 4. The relationship between the difficulty of deciphering password and the number of digits.

4.2. Design of Code Base

4.2.1. The code base of virtual trajectory password. Code base refers to the code range that can be changed by each matrix element on the random code disk. The code disk of this design is composed of 36 "code bases", that is, 10 digital characters 0-9 and 26 lowercase English characters A-Z. The code disk of the virtual trajectory password is composed of a 6-by-6 code base matrix, which will reduce the difficulty of using and remembering the password while forming the virtual trajectory password.

4.2.2. The influence of the choice of code base on the shape of code disk. The code base is the foundation of the code disk.Its selection will affect the resolution of the code disk and the appearance of the code disk matrix. For instance, if the code base of 0-9 is selected as the code disk, only a 3-by-3 code disk can be formed. At this time, the code base is less, the resolution is not enough, the seed password is too short, the security is not high, and it can only be repeated or crossed in a single cycle. If you choose to add the code base of lowercase a-z and uppercase A-Z as the code disk, it will form a code base matrix of 62 characters. The matrix composed of 62 characters is also asymmetric, which affects the appearance of the code disk and is not good for using the code disk as reference memory. If the 8-by-8 code disk is used,8*8=64, it is necessary to enable other characters to replace (such as: $, #, @, etc.), which will make the user more difficult to use, the efficiency is not high, the performance of the code disk is not very big. The resolution of the code disk of the 6-by-6 matrix is sufficient, and it can also form very complex trajectory graphics.

4.2.3. The relationship between password decoding and code base. The code base of "virtual trajectory password" is composed of 0-9, A-Z, 36 characters. The difficulty of deciphering the password conforms to exponential as in formula (1).
Decoding difficulty index:

$$\alpha = M^x \tag{1}$$

'M' is the number of code bases; 'x' is the number of digits in the password. For example, the bank password is 6 digits, the code base is composed of 0-9 numeric characters, and the decoding difficulty index is $\alpha = 10^6$, which means it can be deciphered up to 1 million times, and our code base is 36 characters. If the user inputs a password of 6 digits, the formula is $\alpha = 36^6 = 2.17$billion times, It can be seen that the difficulty of deciphering increases exponentially. Therefore, in the same number of digits, when the number of code bases is greater, the security of the password will increase, and the difficulty of deciphering will increase. It can also be seen that its security, operability will be higher than the original password, and easy to use. Thus,Figure 4 shows that for a given code base, the difficulty of deciphering a password increases exponentially with the number of digits.

5. Conclusion

This paper studies a new virtual trajectory cryptography technology, which breaks through the traditional method of remembering characters and changes it into memory graphics. In order to achieve the purpose of memorizing graphics, combined with the technical characteristics of Android system, the complete design of code disk structure and solving the problem of random generation of code base in the code disk was proposed. In order to prevent the password from being deciphered exhaustively, the password "base expansion" processing method is applied to enhance the strong coefficient of the password and solve the problem of generating a long 128 code coefficient to store and transmit the password after "base expansion". The random sequence of chaotic encryption algorithm is used to encrypt and decrypt, and a constantly changing random reference graph "code disk" is generated as an auxiliary memory method to help human remember some special lengthened passwords that cannot be remembered conventionally. The password verification is completed by matching the input virtual track password with the password stored in the computer after basic expansion and encryption.

Acknowledgments

This research was supported by Hainan Provincial Natural Science Foundation of China (620QN285).

References

[1] ALBADRI H, SULAIMAN R. A classification method for identifying confidential data to enhance efficiency of query processing over cloud[J].Journal of theoretical and Applied Information Technology,2016,93(2):412.
[2] Li Zhengtao. User Identity Authentication Technology for Mobile Social Network Applications, Nanjing University of Posts and Telecommunications, 2017.
[3] Cuiping Ma, Jiahui Liu." Design and Implementation of a Hash Algorithm Based on Chaotic Logistic Equation ". Harbin University of Science and Technology, 2018.
[4] Gao Fei, Application of Data Encryption Technology in Computer Network Security, Electronic test, pp.113-114, 2021.
[5] Decheng Si, Guangyi Wang, Xiang Wang, Rong Qiu, A sensitive data encryption algorithm based on Chaos, Journal of Hangzhou Dianzi University(Natural Sciences),2018.
[6] Research on security authentication method based on user behavior characteristics,The degree of master thesis Hangzhou Dianzi University, 2020.

Journal of Physics: Conference Series **2179** (2022) 012033 doi:10.1088/1742-6596/2179/1/012033

A Novel Method of Parallel Computing Based North-Finding of the Satcom-On-The-Move System and Its Simulation

Lei Han and Guangxia Li*

College of Communications Engineering, the Army Engineering University, Nanjing 210007, China

*Corresponding author email: satlab13905177686@163.com

Abstract. This paper presents a new north-finding method of the Satcom-On-The-Move (SOTM) system based on parallel computing. The characteristic of this method is that sets a range for the existence of north-pointing true value according to a piece of initial north-pointing information, selects several equally divided north-pointing value within this range. Take the value of each point as multiple initial north-pointing information, then calculate the position information accumulated over time, and compare the calculated position information with the position information obtained by the Global Navigation Satellite System (GNSS). Gradually reduce the value range of the true value of the north-pointing information until the position information obtained with the GNSS is less than a certain error range, then the initial north-pointing information value of the minimum error is obtained, which is approximately equal to the accurate value of the north-pointing information. The proposed algorithm is verified by a simulation case, and the simulation results show the effectiveness of the proposed algorithm.

1. Introduction

Since ancient times, location and bearing information has been very important for human production and living activities. North-finding is to find the north direction of the earth and is an important way to determine location and bearing information. North-finding technology has been widely applicated in mining, surveying and mapping, Satcom-On-The-Move (SOTM), navigation, astronomy, and other fields[1-6]. For the SOTM system, the pointing of the antenna is adjusted in real-time with the change of the carrier attitude and position information. The measurement of the attitude and position information of the carrier is the premise of the real-time adjustment of the SOTM antenna. For the measurement of carrier attitude, its pitch and roll can be obtained from the measurement data of the three-axis accelerometer, while for the measurement of carrier azimuth (i.e. north-pointing information), it is difficult.

Common north-finding methods include the north-finding method based on earth's magnetic field, the north-finding method based on GNSS, the astronomical observation method, and the inertial north-finding method. The north-finding method based on the earth's magnetic field may be impacted by some disturbance due to the possible local abnormal disturbance in the earth's magnetic field. In addition, the commonly used electronic compass often has poor effects because of its low accuracy. In complex environments, indoor environments, underground environments, and other application scenarios, north-finding methods based on GNSS and astronomical observation methods will be greatly impacted and even can not work.

The inertial north-finding method can work independently, completely free from the influence of the external environment[7][8], can work continuously, and has the advantage of high precision. However,

its application requires high-precision inertial navigation physical devices such as fiber optic gyroscope and laser gyroscope, which has a high economic cost and limits its large-scale application. Inertial north-finding methods mainly include the static multi-position method and dynamic continuous rotation method. Where the static multi-position method needs to use sensors to point to multiple directions and keep stationary in each direction for a period of time, and the dynamic continuous rotation method requires the measuring system to rotate continuously at the same speed in working engineering. These two inertial north-finding methods are impacted by the sensor error, installation error, and external disturbance. The commonly used methods to eliminate these errors include the Allan variance method, the least square method, the Kalman filter method, and so on. The static multi-position method is relatively simple, but it needs to be calibrated to eliminate the noise and drift error of the sensors. While the dynamic continuous rotation method, can eliminate the static error, but the dynamic error can not be eliminated, which will impact the system accuracy.
With the development of the technology, the MEMS inertial devices are more and more applicated in system design[9-13]. The MEMS devices have the advantages of convenient mass manufacturing, system-level integration, packaging integration, lightweight, small size, economy, and so on. But at the same time, taking the gyroscope as an example, compared with the fiber optic gyroscope and the laser gyroscope, the accuracy of the MEMS gyroscope is low, and it is difficult to be sensitive to the earth's rotation, so it cannot be directly applied to the design and application of the north-finding system.
In order to overcome the poor economy of the traditional inertial north-finding scheme, a north-finding scheme combining low-cost MEMS inertial physical devices, electronic compass, and GNSS is proposed in this paper. According to the initial measurement data and its error range, the scheme will equally divide the possible interval where the true value may locate, and then calculate the location information based on the north-pointing value at each point, and compare the calculated location information with the location information measured by the GNSS to obtain the new value interval closest to the location information given by the GNSS. Then, according to the above method, continue to divide the calculated new value interval, obtain a new basic value interval, and finally obtain a value that meets certain accuracy requirements.
The main contents of this paper are as follows: In Section 1, we introduced different north-finding technologies and analyzed their advantages and disadvantages. In Section 2, a MEMS inertial north-finding algorithm based on parallel computing is proposed. In Section 3, a parallel computing-based inertial north-finding algorithm combined with an electronic compass is proposed. In Section 4, the simulation experiment is carried out for the designed north-finding algorithm. Section 5 is the summary of the full paper.

2. Parallel Computing Based Inertial North-finding

Inertial north-finding has attracted the attention of researchers because it is not impacted by the external environment, meanwhile, it can work independently and continuously. However, due to the need for high-precision inertial physical devices, inertial north-finding has a poor economy, which limits its popularization and wide application. This section introduces a new inertial north-finding method based on multi-channel parallel computing for MEMS inertial physical devices.
The following settings are made in this section. The coordinate system adopts the East-North-Up (ENU). The true north direction is set to 0 degrees, the clockwise direction is positive and the counterclockwise direction is negative, the value range of north-finding true value is $(-180°, 180°]$.
Let the north-pointing value to be A_z, and its true value is $\hat{A}_z$。 The measured value of inertial physical device is matrix IN_c, which contains the pitch angle P_i, Roll angle R_o, as well as the three-axes accelerations a_e, a_n, a_u and other inertia information. The location value measured by GNSS is matrix GN_m, includes longitude and latitude values.
Since the north-pointing information cannot be obtained from a single gyroscope signal, the inertial navigation value IN can be expressed as:

$$IN = [A_z \quad IN_c]^T$$

The flow of north-finding algorithm based on parallel computing proposed in this section is as follows:

Step 1: Given the north-pointing information setting value A_z. Set an arbitrary value ε;
Step 2: Get the measured value of inertial physical device and obtain the matrix IN_c;
Step 3: According to the characteristics of inertial physical devices, the error ranges of pitch angle, roll angle and triaxial acceleration are obtained as follows: $\Delta P_i, \Delta R_o, \Delta a_e, \Delta a_n, \Delta a_u$;
Step 4: The values range of pitch angle, roll angle and three-axes acceleration are obtained as follows: $[P_i \pm \Delta P_i]$, $[R_o \pm \Delta R_o]$, $[a_e \pm \Delta a_e]$, $[a_n \pm \Delta a_n]$, $[a_u \pm \Delta a_u]$, which is called data extension;
Step 5: Since there is no physical device to measure north-pointing information, the possible value range of north-pointing information is: $(-180°, 180°]$;
Step 6: Divide the north-pointing information, pitch angle, roll angle and three-axes acceleration values into $N-1$ parts equally, then the north-pointing information, pitch angle, roll angle and three-axes acceleration values will get N values respectively, and then the inertial navigation matrix IN will get N^6 values;
Step 7: Make use of the conventional inertial navigation algorithm to calculate the carrier location at the following time steps as GN_c;
Step 8: Calculate the ζ values of different inertial navigation matrices IN, where $\zeta = \|GN_c - GN_m\|_2$;
Step 9: Find out a group of inertial navigation matrix IN, which could obtained ζ to be the smallest, then the north-pointing information value in this group of values A_z is the obtained optimal north-finding information.
The above algorithm scheme design is relatively simple, but it has the following disadvantages:
(1) Since the initial north-pointing information cannot be obtained, its value range is $(-180°, 180°]$, which covers all possible value of azimuth angles, and the value range is too large;
(2) The unprocessed inertial navigation data such as pitch angle, roll angle and three-axes accelerations are used at the initial time, and its accuracy is too low. Therefore, the above data are also extended when the initial data is extended. After the data is extended, the volume of the possible value of inertial navigation matrix is too large;
(3) In order to obtain the north-pointing information with sufficient accuracy, when dividing the inertial navigation data, the number of equal parts is high. Because the number of inertial navigation matrices is the product of the number of equal parts of each element, the number of parallel computing channels is too high.
Although in theory, the algorithm can obtain enough high-precision north-pointing information and other inertial data information. However, due to the above limitations, the amount of data and calculation of the algorithm is too large, and the real-time performance is difficult to meet. For the Satcom-On-The-Move system, the antenna adjusts the pointing in real-time with the change of the carrier attitude and position. Due to the continuous change of the attitude and position of the carrier, the Satcom-On-The-Move system has high requirements for real-time attitude acquisition. Otherwise, the antenna may deviate from the satellite and affect the communication effect, and even lead to the antenna of the Satcom-On-The-Move system being unable to receive the signal from the satellite, resulting in the interruption of the communication.

3. Algorithm Improvement

Aiming at the defects of the algorithm proposed in Section 2, this section proposes the following improvement scheme to overcome its shortcomings.
Firstly, the electronic compasses are added to the system, the initial north-pointing information is given by the electronic compass, and the initial value range of north-pointing information is obtained according to the measurement accuracy of the electronic compasses; Secondly, it is easy to obtain the high-precision values of pitch angle, roll angle and acceleration value of each axis in the inertial navigation matrix. Therefore, firstly, the inertial data of the above original measurements are processed to obtain its high-precision value. Then when the data extension task is executed, the above data will not be extended, only the north-pointing information will be extended. Finally, using the dynamic extension scheme, the system performs multiple iterations, and the north-pointing information data expansion range is reset in each iteration. At the same time, the equally divided parts in each iteration process are small. In this way, with the iteration, the data extension range becomes

smaller, so as to obtain the high-precision north-finding information. In the above process, only one-dimensional data in the inertial navigation matrix is extended, and the quantity of equal division of the extended data each time is small, which greatly reduces the number of parallel computing channels and greatly improves the realizability of the system.

The improved algorithm flow proposed in this section is as follows:

Step 1: Set the constant value ε;

Step 2: Obtain the values of pitch angle, roll angle and three-axes acceleration measured by inertial physical devices;

Step 3: The measured values of inertial physical devices are processed by conventional methods, and then the high-precision pitch angle, roll angle and three-axes accelerations are obtained as follows: $\breve{P}_i$, $\breve{R}_o$, $\breve{a}_e$, $\breve{a}_n$, $\breve{a}_u$;

Step 4: Obtain the north-pointing information measured by the electronic compass denoted by A_z. According to the measurement error range of the electronic compass which is denoted by ΔA_z, the north-pointing information is extended. Then the range of the initial north-pointing information is obtained as $[A_z \pm \Delta A_z]$;

Step 5: Divide the north-pointing information into $N-1$ parts equally, and N north-pointing information values can be obtained; Combine N divided north-pointing information values with the high-precision pitch angle, roll angle and three-axis acceleration values obtained in Step 2 to get N inertial navigation matrix IN values;

Step 6: According to the conventional inertial navigation algorithm, calculate the carrier location at the following time steps as GN_c;

Step 7: Calculate the ζ values of different inertial navigation matrices IN, where $\zeta = \|GN_c - GN_m\|_2$;

Step 8: Judgment of $\zeta < \varepsilon$?

If it is judged to be true, take the north-pointing information of value A_z which takes the minimum ζ as the north-pointing information to be calculated. Go to step 10; If the judgment is false, select the two north-pointing values A_σ, A_τ ($A_\sigma < A_\tau$) that minimizes ζ, Then go to Step 9;

Step 9: Set the north-pointing value range $[A_\sigma, A_\tau]$ as the new north-pointing information range, then go to step 5;

Step 10: The algorithm ends.

Compared with the algorithm proposed in Section 2, the amount of the calculation and data of the algorithm proposed in this section are greatly reduced. For the algorithm proposed in Section 2, N^6 parallel computing is required. For the algorithm proposed in this section, only N parallel computing is required, which greatly reduces the challenge of computing to hardware computing and storage performance. In addition, since there is no physical device providing north-pointing information in Section 2, its value covers the whole circle, which is $(-180°, 180°]$. For the algorithm in this section, because its hardware system contains an electronic compass and has initial north-pointing information, although its accuracy is not high, it can still greatly reduce the possible value range of its true value. For the equal component N of the two algorithms, due to the different value range, the size of their equal component N is also different. The equal component of the scheme in this section is smaller than that in the scheme in Section 2. In this way, the calculation difficulty and amount of the algorithm are further reduced, which is helpful for improving the real-time performance of calculation, it is significant for the SOTM system.

In this paper, parallel computing means that the above groups of inertial navigation data values IN are calculated respectively to obtain their latest position information values GN_c, and these calculations are carried out synchronously. This parallel computing does not mean that the hardware structure is parallel. It means that the process of calculating GN_c on the software operation is parallel, and there are no excessive requirements for the hardware structure.

4. Simulation Analysis

In order to verify the effectiveness of the algorithm proposed in this paper, the simulation analyses are carried out in this section. For the simulations in this section, the following assumptions are made: it is assumed that the pitch angle and roll obtained by inertial navigation physical devices have high enough accuracy; it is assumed that the precise value of azimuth angle is known in the initial state, so as to compare with the calculated value obtained by the algorithm proposed in this paper.

In this section, the East-North-Up coordinate system is adopted, and the following settings are made: It is 0 in the due east direction, positive in the counterclockwise direction, and negative in the clockwise direction. Then the due north direction is 90, the due west direction is 180, and the due south direction is - 90. Then the azimuth value range is $(-180°, 180°]$.

Due to the azimuth angle cannot being measured directly by the three-axes accelerometer, in this section, the electronic compass is used to collect the azimuth angle. At the same time, the measurement accuracy of the electronic compass can be obtained directly from the characteristics of the electronic compass. In the following simulation, in each step of the calculation of the algorithm, the possible value range of the true value is divided into 3 parts, and a total of 4 possible truth values are obtained.

Simulation 1: The precise north-pointing value of the initial state is 78.724° east by north (1.374 for radian value), and the measurement error of the electronic compass is 5.729° (0.1 for radian value). First, take the initial measurement north-pointing value of the electronic compass as 84.397° east by north (1.473 for radian value), that is, take the electronic compass as a large measurement error (the error is 5.672°, 0.099 for radian value, close to the maximum measurement error of the electronic compass of 5.729°, 0.1 for radian value), and it is larger than the true value. In this situation, the variation trend of the north-pointing value calculated at each step of the algorithm is shown in Figure 1.

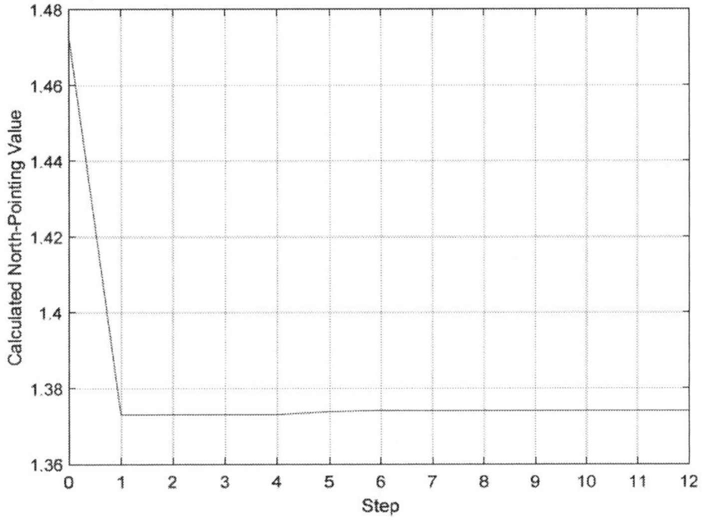

Figure 1. The calculated north-pointing value when the initial measured value of the electronic compass is 84.397° east by north (1.473 for radian value)

As shown in Figure 1, the initial north-pointing value, which is obtained from the electronic compass, has a large error. After processing by the algorithm proposed in this paper, with the algorithm steps go forward, the calculated north-pointing value converges to the true north-pointing value quickly. After a small amount of algorithm steps, the error between the calculated value and the true value decreases rapidly.

Simulation 2: The previous simulation is about the situation that the measured value of the electronic compass is larger than the true value. Here, take the initial measured north-pointing value of the electronic compass as 73.052° east by north, 1.275 for radian value, that is, take the electronic compass as a large measurement error (the error is 5.672°, 0.099 for radian value, close to the maximum measurement error of the electronic compass of 5.729°, 0.1 for radian value), and less than the true value. In this situation, the variation trend of the north-pointing value calculated at each step of the algorithm is shown in Figure 2.

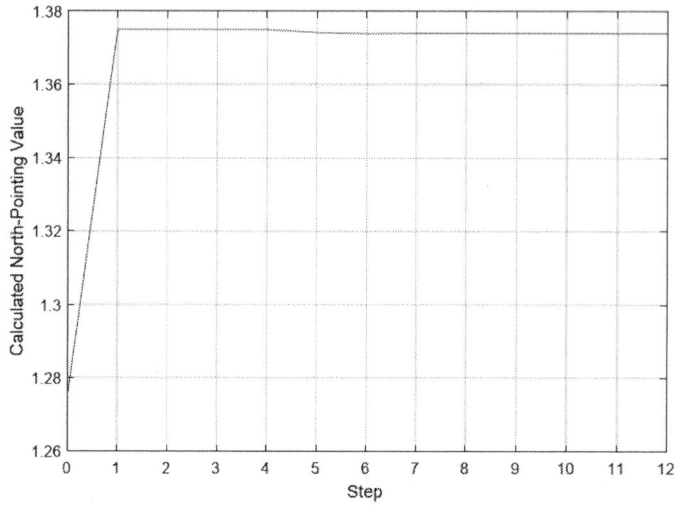

Figure 2. The calculated north-pointing value when the initial measured value of the electronic compass is 73.052° east by north (1.275 for radian value)

From Figure 2, the initial north-pointing value, which is obtained from the electronic compass, has a large error. After processing by the algorithm proposed in this paper, like the result in Figure 1, with the algorithm steps going forward, the calculated north-pointing value converges to the true north-pointing value quickly. After a small number of algorithm steps, the error between the calculated value and the true value decreases rapidly.

The above simulations verify the characteristics of the algorithm when there is a large error in the initial measured value of the electronic compass. The simulation results show that under the condition of large measurement error of electronic compass, whether the measured value is large or small than the true value, the algorithm proposed in this paper can converge to the true value rapidly. It shows that the algorithm proposed in this paper has the characteristics of good convergence and fast convergence speed. In addition, in each step of the calculation, the possible value range of truth value is divided into three parts, and a total of four possible truth values are obtained. For parallel computing, channels are needed to realize the algorithm. This means that the algorithm has low requirements for hardware and physical devices and is easy to be applied in engineering.

Compared with traditional methods, in the SOTM system, high-precision inertial navigation devices such as fiber optic gyroscope and laser gyroscope need to be used for carrier azimuth information measurement. Although these methods can obtain high accuracy and meet the needs of mobile communication systems, this scheme has great disadvantages because the price of high-precision devices such as fiber optic gyroscope and laser gyroscope is too high, The overall price of mobile communication systems has been improved. By the method in this paper, these high-precision devices with high prices can be avoided, and low-cost and low-precision devices such as MEMS gyroscope can be used to meet the system requirements.

5. Conclusion

This paper presents a novel north-finding method of the Satcom-On-The-Move system based on parallel computing. According to the initial measured data and its error range, the scheme will equally divide the possible interval of the true value, and then calculate the location information based on the value at each point, and compare the calculated location with the location measured by the GNSS to obtain the new possible value interval closest to the data collected by the GNSS. During the running of the algorithm, the possible value range of the true value is dynamically adjusted, and finally the value meeting the certain accuracy requirement is obtained. The simulation results show that the proposed algorithm has the characteristics of good convergence and fast convergence speed. In addition, the algorithm has low requirements for hardware and physical devices, and it has high real-time performance, which is suitable for the application on the Satcom-On-The-Move system.

Journal of Physics: Conference Series **2179** (2022) 012033 doi:10.1088/1742-6596/2179/1/012033

References

[1] Z Shi, ZQ Yang. The Key Technology of Seeking North's by GAT Magle Gyro Stationits and Application in Underground Engineering. *Advanced Materials Research*, vol. 694-697, p. 1051-1054. 2013.
[2] Y Liu, XC Xu, Q Liu, TH Ji. Autonomous Underground Azimuth Measuring Drive System Based on MEMS Gyroscope. *Adv Mater,* vol. 1042, p. 159-164. 2014.
[3] J.I. de Lang, G.D. N Twihaar, J.J. Pelgrom. Accurate Survey: An operator's Point of View. *IADC/SPE Drilling Conference*, Dallas, 1988.
[4] YY Lin. Applied Research on Electronic Compass in Vehicle-mounted Satellite Communication Mobile Station. *Coal Geology of China*. 2013, vol. 25(05), p, 65-67.
[5] M Sadiq, Z Zaman, A Jehanzeb, S Javed, N Qaiser. Design and Development of Precision Astronomical North-finding System: A Software and Hardware Perspective. *2017 14th International Bhurban Conference on Applied Sciences and Technology*, Islamabad, Pakistan, 2017.
[6] Z Zhou, Z Tan, X Wang, Z Wang. Modified Dynamic North-Finding Scheme with a Fiber Optic Gyroscope. *IEEE Photonics J*, 2018, vol. 10(2), p. 1-10.
[7] A.R. Spielvogel, L.L. Whitcomb. A Stable Adaptive Attitude Estimator on SO(3) for True-North Seeking Gyrocompass Systems: Theory and Preliminary Simulation Evaluation. Singapore, *ICRA 2017 IEEE*, 2017.
[8] JY Pan, CX Zhang, YX Niu, Z Fan. Accurate Calibration for Drift of Fiber Optic Gyroscope in Multi-position North-Seeking Phase. *Optik*, vol. 125, Iss. 24, p. 7244-7246. 2014.
[9] L Wang, Y Wang, YQ Deng, A Noureldin, et al. Drilling Trajectory Survey Technology Based on 3D RISS with a Single Fiber Optic Gyroscope. *Optik*, vol 203, Iss 8. 2020.
[10] Y.I. Binder, A.E. Eliseenkov, A.S. Lysenko, et al. MEMS-Based Downhole Tool Incorporated in Mobile Inclinometric Station for Trajectory Survey of Grouped Wellbores. *Gyroscopy Navig*, 2013, vol. 4, p, 85-91.
[11] L Zhu, JX Yang, M Smith. The Filtering Method of MEMS Gyro Signal Based on Sparse Decomposition. *J.Comput. Method Sci. Eng.*, 2020, vol. 20, Iss. 4, p, 1-12.
[12] D Bulz, S Weidlich, S Konietzka, T Motl, A Shaporin, et al. Compact Standalone North-finding Device Based on MEMS Gyroscope and Maytagging. *2021 Smart Systems Integration*, Grenoble, France, 2021.
[13] A.P Manjari, S Sanchana, P Vijayakumar, B.N Jyothi.V. Model Aided Simulation and Demonstration of True North Seeking Algorithm Using MEMS Based Gyroscope. *2021 IEEE Madras Section Conference*, Chennai, India, 2021

Round Traffic Sign Detection Algorithm

Yinrong Huang[1,*], Bing Wang[1,2] and Xiemin Yuan[1]

[1]School of Electronics and Information Engineering, Guang'an Vocational and Technical College, Sichuan, china
[2]School of Communication and Information Engineering, Chongqing University of Posts and Telecommunications, Sichuan,china

*Corresponding author email: 178600251@qq.com

Abstract. Traffic signs are an important part of autonomous driving and intelligent transportation. It provides instructions for pedestrians and vehicles and is critical to road traffic safety. However, existing detection algorithms cannot achieve real-time high-precision detection. Therefore, this paper proposes an algorithm that combines traditional methods with deep learning to detect circular traffic signs. Based on the HSV color space, the red and blue channel images are separated, and the candidate regions of the original image are extracted using the Hough transform. The shallow convolutional neural network (CNN) classifier rejects not traffic signs and classifies traffic signs. Experiments show that the algorithm is real and effective. On the CPU platform, the average accuracy rate is 96.2%, and the detection speed reaches 0.3 s / frame. Under the condition of ensuring the average accuracy rate, the detection speed is greatly reduced. The algorithm achieves the fastest speed, which makes real-time high-precision detection possible. The algorithm is more suitable for vehicle embedded systems.
Keywords: Traffic signs detection; HSV; Hough transform; CNN algorithm;OTSU algorithm.

1. Introduction

Circular traffic sign identification occupies an important position in traffic sign identification, according to statistics in the world mainstream traffic signs circular traffic signs occupy nearly half of [1], but there are a variety of factors interference problems, characteristics are not obvious, rainy days, sign aging, etc., in order to ensure road safety, to consider the accuracy of recognition and to ensure identification speed, so the actual application to achieve fast and high precision recognition.

At present, there are many methods for traffic sign identification. Due to the unique shape and color of traffic signs, some scholars have proposed color space [2-3], a method [4] of identifying traffic signs according to shape characteristics, or [5] of combining color and shape characteristics, or extracting candidate regions under color space to support vector machine (SVM) classifier identification [6-7].In recent years, due to the rise of the convolutional neural network [8], some scholars have proposed the method of detecting traffic signs based on the convolutional neural network [9], which extracts the candidate regions through the pre-selected box and is sent to the classification network for identification.But these methods all have some drawbacks,According to the existing publicly available traffic sign dataset, we found that the world's mainstream traffic signs, whether German traffic sign data set (GTSDB),Belgian traffic sign data set (Belgium TS), or Chinese traffic sign dataset (CTSDB),

are single and fixed, warning signs are circular shape of red outer circle, and blue background circular shape.
Therefore, according to the color and shape characteristics of circular traffic signs, this paper proposes an algorithm that separates red and blue color channels under HSV color space, uses Hough transformation to extract the original image candidates under each channel, and sends it to the constructed shallow convolutional neural network (CNN) classifier to detect circular traffic signs. The problem of real-time is solved by extracting a small number of candidate region images with the constructed shallow CNN classifier, and the accuracy and versatility of the algorithm are guaranteed. Due to the dataset, the algorithm divides circular traffic signs into two categories according to the characteristics of traffic signs:coercion and prohibition. After experimental test, the proposed method is true and effective. On the CPU platform, the average accuracy reaches 96.2%, and the detection speed is 0.3s / frame. In the same period of traffic sign recognition algorithm, the detection accuracy of the algorithm reaches the fastest. The combination of traditional methods and deep learning methods can make full use of the role of deep learning methods. The subsequent content is divided into three parts: extracting candidate regions, CNN classifiers, and experiments.

2. Candidate Regions were Extracted

The extracted original image candidates are sent to the CNN classifier for judgment and identification, and the non-traffic signs are removed and classified. The first thing to be done is the extraction of the candidate regions, and for this part, the text was extracted using Hough changes under the separated red and blue channel images.

2.1. HSV Color Space

HSV color space is a human visual perception system, using color (H), saturation (S) and brightness (V).Hue is usually used to distinguish a color from the macro level, the brighter a color, the higher the saturation value, the brightness describes the light and shade of a color, the higher the brightness value, the brighter the color. The threshold of the candidate region to be determined under the HSV color space to separate the blue channel from the red channel, and then do binary and median filter operations to obtain the image to be detected, as shown in Figure 1.

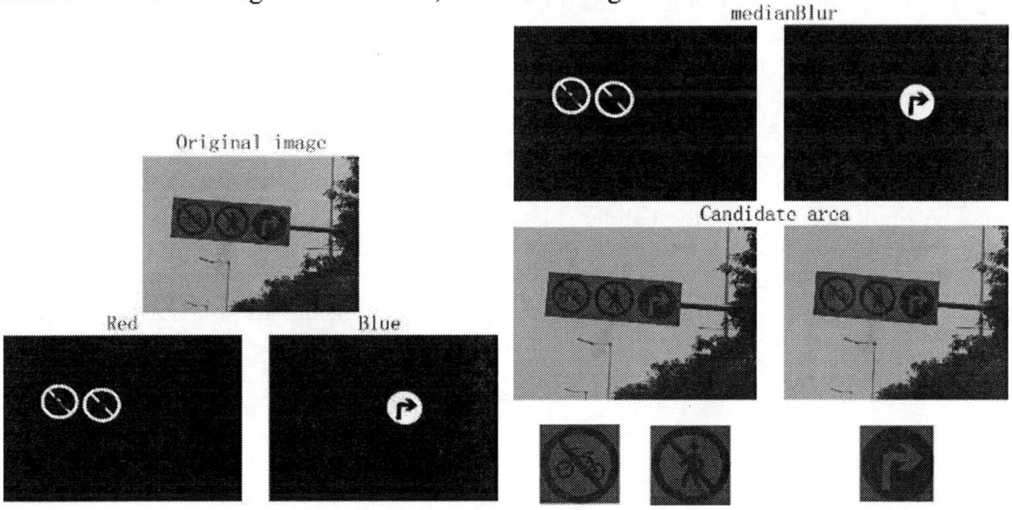

Figure 1. Extracting candidate area images.

2.2. Hough Transform

The Hough transformation is one of the basic methods of identifying geometry from an image in image processing. Assuming the line equation expression as formula (1), turn the formula to a parameter-based expression as formula (2), where ρ is the length of the line , θ is the angle of the line and the axis x.Assuming that N points are taken in the plane, and if three of the points are on one of the lines, then the ρ and θ of the lines where the three points are equal, it is ultimately determined by matching the line formed of the points in the plane.

$$y = kx + b \tag{1}$$

$$\rho = x\cos\theta + y\sin\theta \tag{2}$$

For circular detection, assuming the radius r of the circle has been determined, for the center (a,c), the point (x,y) on the line is known, according to the circular calculation formula (3), will form the center (x,y) of the circle, for the radius (a,c) of the circular curve, multiple points (x_i,y_i) on the line will constitute countless circle, the circle will eventually intersect at a point, that is the center (a,c), and determine the circle, as shown in Figure 2, different radius down into a conical floating. After the circle is established, the blank area with 2 pixels takes the minimum extangent rectangle, and finally determines the candidate region image, by operating as above, finally obtains the flow chart of Figure 1.

$$(x-a)^2 + (y-c)^2 = r^2 \tag{3}$$

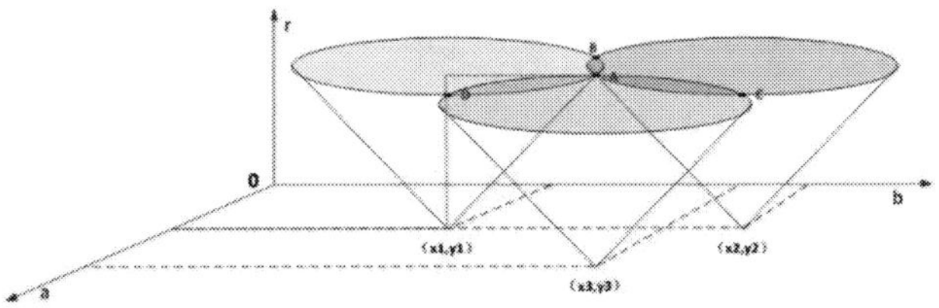

Figure 2. Establishing the Hough transform.

3. CNN Classifier

We used five convolutional layers with two fully connected layers to construct the CNN classifier. Its network structure diagram is shown in Figure 3. The input image size is all 32x32 pixels. Because the traffic signs contain rich edge information, the bright color contrast parts are mostly concentrated on the edges, which is very important for the traffic sign classification. Therefore, for each convolutional layer, for the convolutional supplement 0 around the image of the upper channel, the convolutional kernel size is 3x3, and the edge information is extracted by setting a different step length, the step length is 2 and the other is 1 for the downsampling operation. Since the background color of the traffic logo may be similar to its color, the extracted candidate area may not be the traffic sign image or the background image. For this end, we adjusted the extracted candidate area images to 32x32 resolution size to the CNN classifier, eliminate the non-traffic sign image, and will be the image classification of the traffic signs, as illustrated in Figure 4.With the CNN classifier, the non-traffic sign candidates can be effectively removed.

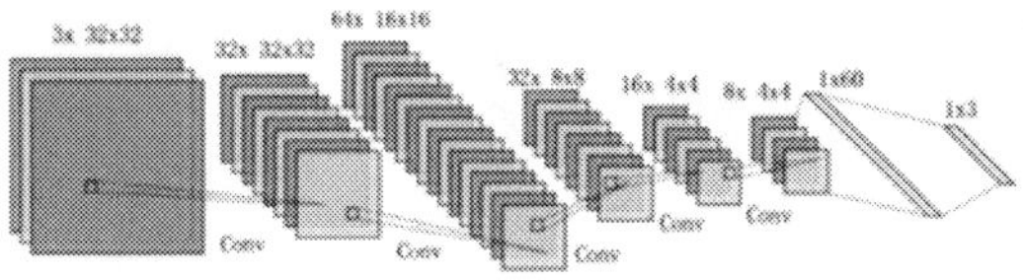

Figure 3. CNN classifier.

Figure 4. CNN classifier removes non-traffic sign.

4. Experiment

4.1. Experimental Platform

The hardware platform of this experiment is shown in Table 1, and the corresponding software platform is built above: python3.7, pytorch1.3.1, cuda9.2, cudnn7.5, cv2, all code is written in python, for CNN classifier training, this article is conducted under the GPU platform and will be The trained weight values were converted into the format of the CPU platform for testing.

Table 1. Hardware configuration.

name	model	number
computer	Laptop HP Shadow Elf 5	1
CPU	i5-9300H	1
GPU	GTX 1050	1
hardpan	512G PCIE solid state	1

4.2. CNN Classifier

The CNN classifier divided the categories into three categories: compulsory, prohibited and negative samples. In the ratio of 1:1:8,2,000 images were forced and prohibited, and the negative samples were 16,000 images, which were taken from the Chinese traffic sign public data set provided by Zhang and other [19] people.To test the performance of the classifier, we tested 1500 images, each classification of 500 images, 32x32 pixels in size to the CNN classifier and selected accuracy (P), recall (R), CPU classification speed (S) as validation criteria, as shown in formula (4) and formula (5), representing real positive samples, false positive samples, false negative samples, as shown in Table 2, test average classification accuracy reached 98.5%, average recall reached 97.3%, average classification rate of 12.5ms/ frames.

$$P = \frac{TP}{TP + FP} \tag{4}$$

$$R = \frac{TP}{TP + FN} \tag{5}$$

Table 2. CNN classifier comparison.

pavement markings	accuracy rate /%	Recall/%	velocity /ms
constraint	98.3	97.3	13.2
prohibit	99.1	96.3	12.2
prohibit	98.2	98.4	12.3
prohibit	98.5	97.3	12.5

4.3. Algorithmic Comparison

To test the overall algorithm performance, This paper sends 500 images containing circular traffic signs into different networks for comparison, Average accuracy (mAP) was also selected, The CPU detection speed (S) is used as the validation standard, As shown in Figure 5, We found that, when compared to YOLOv3 with the SSD algorithm, Our proposed algorithm, While the guaranteed average accuracy is not reduced, Significatly reduce the detection speed, On the CPU platform, The average accuracy rate reached 96.2%, While the detection speed was only 0.3s / frame, This is 68 times shorter than the YOLOv3, This is 67 times shorter than the SSD algorithm, In the simultaneous detection algorithm, The detection speed is optimal.

We set the minimum center of circle distance of the Hough change extraction candidate region to 30 pixels and selected different classification thresholds of the CNN classifier for testing, as shown in Figure 6, we found that the threshold size was 0.45 with an average accuracy of 96.2%.

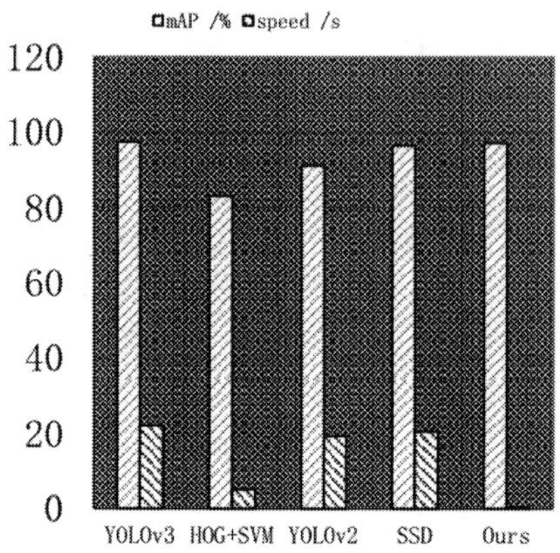

Figure 5. Algorithm comparison.

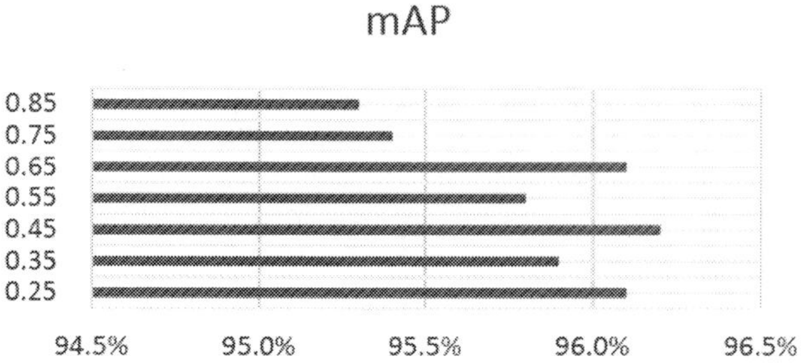

Figure 6. Detection threshold comparison.

To test the reliability of the candidate region extraction, we set the classification threshold of the CNN classifier to 0.45 to test the effect of the minimum center distance on the mean accuracy in the Hough transformation, as shown in Figure 7, we found that the effect was optimal when the minimum center distance is 30 pixels.

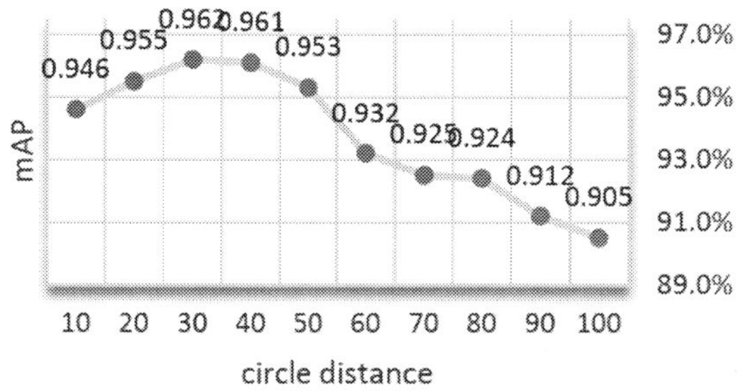

Figure 7. Comparison of minimum center distance.

Figure 8. Detection effect diagram.

Figure 9. Detection effect diagram.

Figure 8 and 9 are the recognition effects of the algorithm. It can be seen from the figure that for the large targets, it can be effectively identified, and the algorithm can also be effectively identified for the small targets. The areas framed by the algorithm can more accurately identify the traffic signs. Due to the CNN classifier, the algorithm can effectively remove the non-traffic sign image.

5. Conclusion

This paper proposes a novel algorithm for traffic sign detection combined with deep learning, which separates red and blue channels in HSV color space and uses Hough transform to select the constructed CNN classifier and eliminate the non-traffic signs and traffic signs. The experiment shows that the highest average accuracy is 96.2% and the detection accuracy is 0.3s / frame, ensuring high accuracy, making the algorithm more applicable to realistic scenarios and close to the embedded system.

Acknowledgements

Fund Project: National Natural Science Foundation (No.61671095, No.61702065, No.61701067, No.61771085); Signal and Information Processing Chongqing Municipal Key Laboratory Construction Fund Funding Project (No.CSTC2009CA2003); Chongqing Graduate Scientific Research and Innovation Fund Grant Project (No.CYS17219); Scientific Research Fund of Chongqing Education Commission (No.KJ1600427,No.KJ1600429)

References

[1] Mogelmose, A.: Trivedi,M.M.Moeslund,T.B. Vision-based traffic sign detection and analysis for intelligent driver assistance systems[C]. Perspectives and survey. IEEE Trans. Intell. Transp. Syst.2012,12,1484-1497.

[2] J Greenhalgh,M.Mirmehdi.Real-time detection and recognition of road traffic signs[J].Intelligent Transportation Systems.2012,13(4):1498-1506.

[3] I. M. Creusen,L. Hazelhoff, P.H.N.de With. Color transformation for improved traffic sign detection[C]. 2012 19th IEEE International Conference on Image Processing, 2012-9-30.

[4] S.Houben, J.Stallkamp, J.Salmen, etal.Detection of Traffic Signs in Real-World Images. The German Traffic Sign Detection Benchmark[C].International Joint Conference on Neural Networks,2013.

[5] H Pazhoumand-dar,M.Yaghoobi.A new approach in road sign recognition based on fast fractal coding[J].Neural Computing and Applications.2013,22(3-4):615-625.

[6] A Gudigar,B N Jagadale,P K Mahesh,etal.Kernel Based Automatic Traffic Sign Detection and Recognition Using SVM[C].Eco-friendly Computing and Communication Systems,Berlin,Heidelberg:153-161,2012.

[7] R Kumaraswarmy,L V Prabhu,K Suchithra,etal.SVM based classification of traffic signs for realtime embedded platform[C].Advances in Computing and Communications.Berlin,Heidelberg:339-348,2011.

[8] Rumin Zhang,Yifeng Yang,Wenyi Wang,etal.An Algorithm for Obstacle Detection based on YOLO and Light Filed Camera[C].Limerick,Ireland:2018 12th International Conference on Sensing Technology(ICST),2018-12-04.

[9] Xiong Changzhen, Wang Cong, Ma weixin, Shan Yanmei. A traffic sign detection algorithm based on deep convolutional neural network[C].Beijing, China: 2016 IEEE International Conference on Signal and Image Processing(ICSIP),2016-8-14.

A Region Preference-based Optimization Algorithm for Satellite Mission Planning Problem

Minghui Xiong* **and Wei Xiong**

Science and Technology on Complex Electronic System Simulation Laboratory, Space Engineering University, China

*Corresponding author email: xtkxxmh@163.com

Abstract. For the earth observation satellite mission planning problem, objectives such as observed target quantity, observation profit, energy consumption, image quality should be considered simultaneously, which is a many-objective optimization problem. Classical optimization-based mission planning algorithms obtain a set of non-dominated solutions in the entire search space, while only a single satisfy final plan is desired by decision maker. In this paper, a five-objective optimization model for satellite mission planning problem is constructed, then a region preference-based evolutionary algorithm, HMOEA-T, is applied to obtain the desired solutions. The decision makers describe the preference on each objective in target region form, then the algorithm guides a more detailed search within the preference region rather than the entire Pareto front. Comparative studies with preference-based methods (T-NSGA-III) and classical methods (NSGA-III) are conducted. We have exemplified the proposed method manage to obtain the solutions satisfying the mission planning preference and achieve better performance in convergence and diversity.
Keywords: Optimization algorithm; Preference articulation; Satellite mission planning; Many-objective optimization model.

1. Introduction

The original optimization objective of earth observation satellite mission planning problem is to maximize total profit of observed target, or to maximize the number of planned observation mission. However, in context of different application background, multiple objectives such as energy consumption, image quality should be optimized at the same time [1]. Multi-objective optimization problems with more than three objectives are defined many-objective optimization problems (MaOPs) [2]. Because of the NP-hard nature of the problem [3], heuristic [4] and meta-heuristic approaches are used to speed up the process of finding a satisfy solution, such as constructive heuristic [5], ant colony optimization [6], Tabu search [7], local search [8] and so on. Due to the simplicity and efficiency, evolutionary algorithms such as genetic algorithm [9] have been extensively applied in earth observation satellite mission planning.

In evolutionary optimization process, the algorithms attempt to describe the true Pareto front with a set of non-dominated solutions. With the increasement of objectives, discovering the entire would be difficult due to the dominance loss [10]. In addition, the decision makers only require one non-dominated solution as the final observation mission plan, which means the decision maker only requires part of the Pareto front. As a decision maker, the acquisition of uninterested solutions would increase the burden of choosing satisfy mission plan.

In many-objective optimization, introducing preference information into optimization process have been proved as an effective way to solve the problem of dominance loss [2]. In [11], we proposed an evolutionary algorithm based on region preference of the decision maker, HMOEA-T. The algorithm guides the population to search towards the predefined region, then all final solutions manage to reach the preferred region. The problem of dominance loss and acquisition of unnecessary solutions are thus solved.

In this paper, the previous work is extended to satellite mission planning. A five-objective optimization model is constructed. The preference information is described by target region. HMOEA-T is utilized to obtain preferred solutions. Comparative studies are conducted with a similar region preference-based algorithm T-NSGA-III [12]. A classical many-objective optimization algorithm NSGA-III [13] is selected as the baseline. Three mission planning cases reflecting different mission planning preferences are examined to verify the effectiveness of the proposed method.

2. Problem Formulation

Resources, missions, constraints and objectives are factors should be considered in earth observation satellite mission planning. It can be formulated by

$$\{S,T,C,O\} \tag{1}$$

S, *T*, *C*, *O* denotes the satellite resource set, observation mission set, constraint set and objective set respectively. Generally, earth observation satellite mission planning means under the condition of meeting visible time window, satellite storage capacity, satellite energy and other constraints, schedule various resources to meet the observation missions as much as possible, and maximize the optimization objectives. A five-objective optimization model is constructed as follows. For better visual presentation in experimental section, objective 1, 3 and 5 are normalized to [0, 1]:

- **Observed Target Quantity**: the ratio of observed targets to total candidate observation targets.

$$OTQ = \frac{\|OT\|}{\|CT\|} \tag{2}$$

where $\|OT\|$ and $\|RT\|$ are the number of observed targets and candidate observation targets.

- **Energy Consumption**: the energy consumed by observing the scheduled targets is represented by the average working time of each satellite.

$$EC = \frac{\sum_{i=1}^{\|S\|} \sum_{j\in OT_S} |ot_{i,j}|}{\|S\|} \tag{3}$$

where $|ot_{i,j}|$ is observation time of satellite *i* on target *j*, OT_S is the observed target set of satellite *s*.

- **Observed Target Profit**: the average profit of the observed targets.

$$OTP = \frac{\sum_{i=1}^{\|OT\|} pr_i}{\sum_{i=1}^{\|OT\|} pr_{\max}} \tag{4}$$

where pr_i is the profit of observed target *i*, and $pr_{\max}$ is the maximum profit of all targets, which is introduced to normalize the function value to [0, 1].

- **Mission Balance**: balance the mission load of each satellite.

$$MB = \sqrt{\frac{\sum_{i=1}^{\|S\|}(EC_i - EC_{ave})^2}{\|S\|-1}} \tag{5}$$

where EC_{ave} is the average energy consumption of the satellite.

- **Image Quality**: it is related to the distance and angle from satellite to the target. Since smaller maneuver angle means the shorter distance, the image quality function is defined as follows:

$$IQ = \frac{\sum_{i=1}^{\|OT\|}\theta_i}{\sum_{i=1}^{\|OT\|}\theta_{\max}} \tag{6}$$

where θ_i is the maneuver angle to target *i*, and $\theta_{\max}$ is the maximum maneuver angle.

Summary of objective functions is shown in table 1.

Table 1. Summary of objective functions.

$f(x)$	Description	Optimization direction
f_1	Observed Target Quantity (OTQ)	maximize
f_2	Energy Consumption (EC)	minimize
f_3	Observed Target Profit (OTP)	maximize
f_4	Mission Balance (MB)	minimize
f_5	Image Quality (IQ)	maximize

The constrains and assumptions are summarized as follows:

- Each target should not be repeatedly observed;
- The satellite can only process one task at any time;
- The satellite can start the mission after the attitude maneuver in consecutive observation.

3. Proposed Method

In earth observation satellite mission planning problem, the desired observation plan varies under different application conditions. In [11], we proposed an evolutionary algorithm based on region preference, HMOEA-T, which shows reliable performance on obtaining preferred solutions. In this paper, we extend this algorithm to solve the mission planning problem and propose a region preference-based earth observation satellite mission planning method.

3.1. Overview

The framework of proposed method for satellite mission planning is shown in figure 1, which includes three main parts: region-preference based search, overall performance evaluation and decision makers. Firstly, the preferred region is specified by the decision maker, then region-preference based search is utilized to obtain the preferred solutions, and the fitness of each individual is calculated by the overall performance evaluation part. Detailed procedure are as follows:

Step1: Initialization. A population P_0 of size N is randomly generated in decision space;

Step2: Offspring generation. Generate N offspring solutions using crossover and mutation operators;

Step3: Decoding. Acquire the mission plan coded by the individuals by decoding strategy;

Step4: Overall performance evaluation. Evaluate the overall performance of each individual after the decoding;

Step5: Region preference-based search. Guide the population search towards the predefined target region, and perform a more detailed search after reach into the region, loop Step 2~5 until the termination criteria is met;

Step6: Selection and determine. The decision maker selects a satisfy solution as final observation mission plan or determine an undated preference region if unsatisfied.

The overall performance evaluation of each individual has been described in section 2. More detailed information of region-preference based search method can be found in [11]. Other key points of the proposed method would be discussed in the following sections.

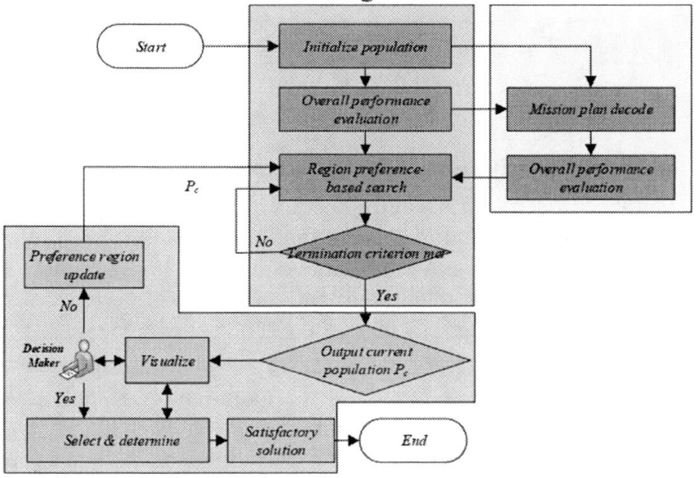

Figure 1. Framework of the proposed method for satellite mission planning.

3.2. Population Initialization

The mission execution strategy is represented by random integer numbers ranges $0\sim n_i$, where n_i is the number of executable time window for target i. The exact value j means the target would be observed in j_{th} executable time window, and 0 represents the target would not be observed. For instance, an individual representing the execution strategy of 6 candidate observation targets is shown in figure 2.

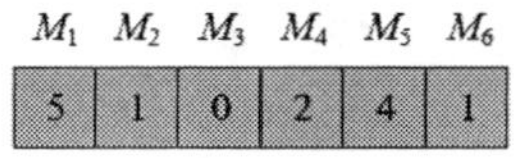

Figure 2. Individual representation.

3.3. Variation Operator

The mutation operator permutate the gene with a random integer number between [0 n_i], and the value of different target i varies. Hence, partial-mapped crossover [14] is utilized in this paper. Figure 3 fives an example of partial-mapped crossover over 6 candidate tasks.

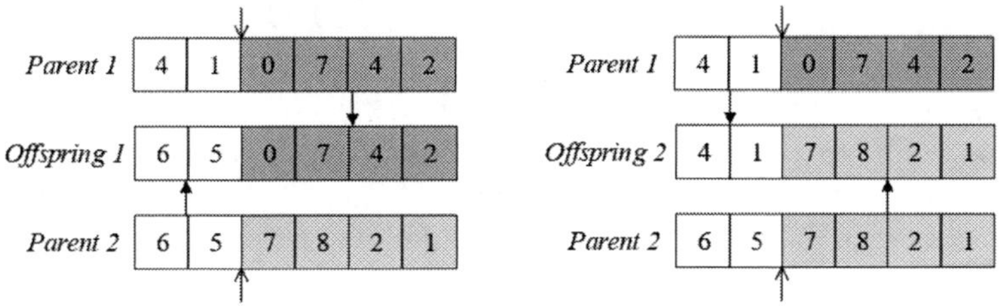

Figure 3. Examples of partial-mapped crossover.

4. Experimental Studies

4.1. Experimental Settings

In order to verify the effectiveness of proposed method, we set three problem cases with different mission planning preferences. For better visual presentation, all the objective values are normalized by [0 150 0.3 2 0.3] and [1 260 0.5 200 0.8]. Three different predefined preference regions are shown in table 2.

Table 2. Problem cases.

Case No.	Preference region
1	{[0 0 0 0 0.6], [1 1 1 1 1]}
2	{[0.95 0.75 0 0 0], [1 1 1 1 1]}
3	{[0 0.16 0 0 0.2], [1 0.35 0.43 1 0.58]}

Case 1 is relatively simple problem, which only specify preference requirements for image quality, aiming to test the capability of maintaining a well distribution within the preference region; case 2 describes tendence on increasing the task completion rate under high resource consumption; case 3 requires low energy consumption rate and maintain load balance to obtain medium image quality.

The simulation time is from 14 May 2021 12:00:00.000 UTCG to 15 May 2021 12:00:00.000 UTCG; the maximum half angle of the sensor view field is $10°$. Five low orbit earth observation satellite and 75 targets are inserted into the scenario, and the targets are randomly located on the mainland of China. We consider a similar region preference-based method T-NSGA-III [12] and a classical many-objective optimization algorithm NSGA-III [13] as the comparison. Other experimental settings are shown in table 3.

Table 3. Other experimental settings.

	Population	Generation	Crossover rate	Mutation rate
value	210	150000	0.6	0.01

4.2. Results and Discussion

The goal of solving satellite observation mission planning problem with preference-based algorithm is to obtain a mission planning plan that satisfies the preference requirements, which means obtaining the individuals located within the target region. In addition, a well distributed solution set within the target region would provide more satisfy alternatives for the decision maker to choose from. Therefore, in this section, the experimental results of different algorithms would be analyzed from the aspects of the effectiveness, convergence and distribution.

The distributions of final solutions the three algorithms obtained solving case 1,2,3 are visualized based on the visual clustering method [15], as shown in figure 4. The red and blue lines represent the preferred and unpreferred solutions, respectively.

For case 1, comparing Figures 4(a-1) and 4(a-2), the population obtained by HMOEA-T has better diversity on each objective, indicating more varieties of preferred solutions are obtained for the decision maker to choose from; the classical algorithm shows more random performance on finding preferred solutions (Figure 4(a, b, c-3)), while both HMOEA-T and T-NSGA-III shows steady performance obtaining more preferred solutions. In addition, the proportion of non-preferred solutions T-NSGA-III obtained reaches 61.4% and 98.5% on case 2 and 3, while HMOEA-T still shows stable performance. It means that HMOEA-T is better at guiding the population to converge to the preferred region.

The experimental results shows that HMOEA-T only obtains the final solutions satisfy the preferences of decision makers, avoiding the burden of selection caused by non-preference solutions for the final decision. Therefore, when solving the earth observation mission planning problem with HMOEA-T, the final mission plan would all satisfy the preference of decision makers. Table 4 shows three example solution selected from three final solution set, the normalized objective value is in the brackets. The normalized objective values are all located within the predefined preference region, the effectiveness of the proposes is thus numerically examined.

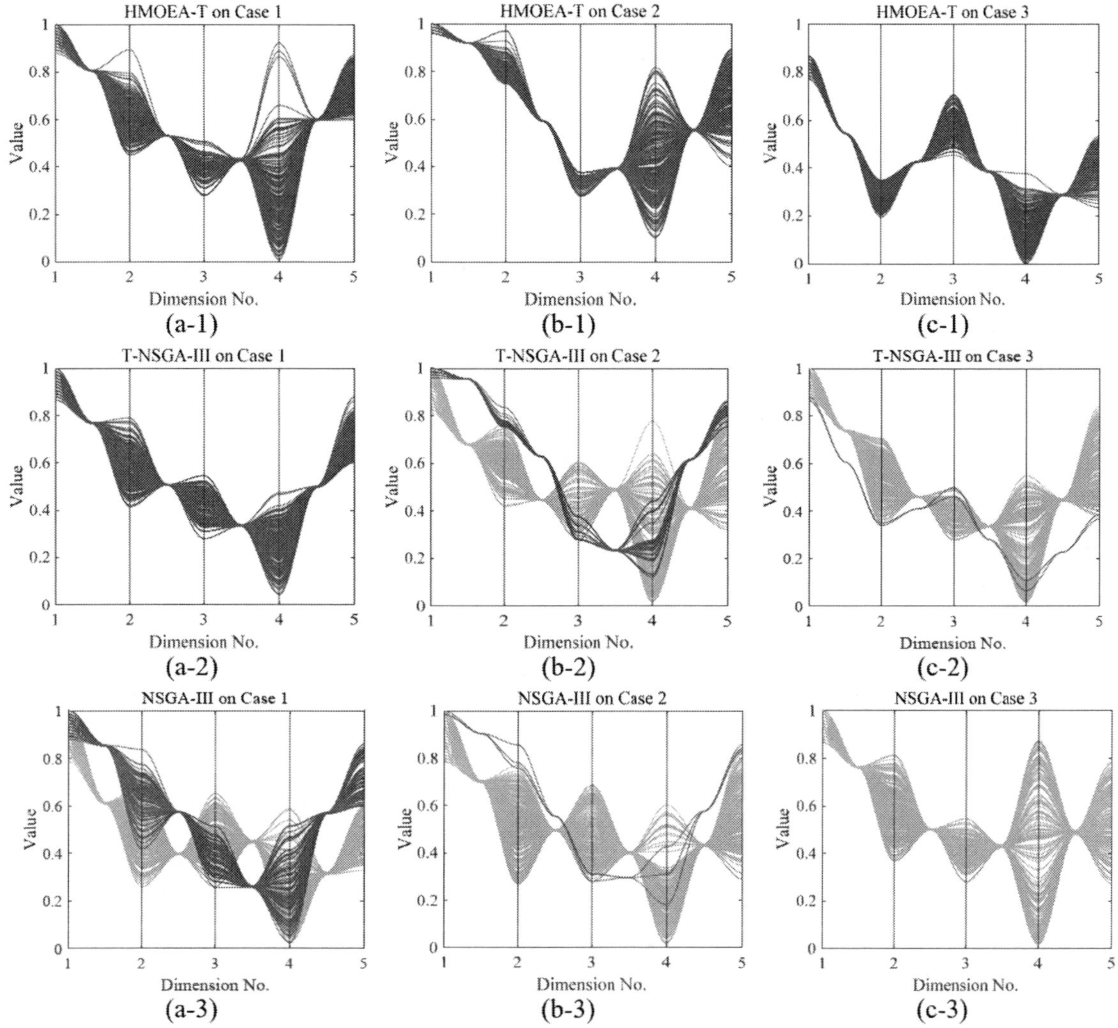

Figure 4. The distributions of final solutions the three algorithms obtained solving case 1,2,3.

Table 4. Examples of final solution.

	Optimization Objective				
	OTQ	**EC**	**OTP**	**MB**	**IQ**
Case 1	1.000 (1.000)	233.44 (0.759)	0.3560(0.280)	74.078 (0.364)	0.724 (0.848)
Case 2	0.973 (0.973)	213.05 (0.823)	0.3653 (0.342)	29.050 (0.510)	0.634 (0.634)
Case 3	0.827 (0.827)	179.13 (0.265)	0.4097 (0.548)	35.286 (0.168)	0.533 (0.467)

5. Conclusion

This paper studies satellite mission planning problem, constructs a five-objective optimization model, and propose a region preference-based evolutionary algorithm to optimize the mission plan. Different from the traditional heuristic methods, the population would not search the entire problem Pareto front, but search towards the preference region and obtain the solutions satisfy the optimization preference of the decision makers.

Experimental results have proved the effectiveness of our method. Compared with the non-preference information guided approach, the proposed is more efficient in finding satisfy solutions.
In future studies, we would perform more parameters analysis and numerical comparison on the performance of proposed method.

Acknowledgment
The authors acknowledge the support from Science and Technology on Complex Electronic System Simulation Laboratory (Funding No.DXZT-JC-ZZ-2019-002).

References

[1] Tangpattanakul P, Jozefowiez N, Lopez P. Multi-objective optimization for selecting and scheduling observations by agile earth observing satellites[C]//International Conference on Parallel Problem Solving from Nature. Springer, Berlin, Heidelberg, 2012: 112-121.
[2] Li K, Wang R, Zhang T, et al. Evolutionary many-objective optimization: A comparative study of the state-of-the-art[J]. IEEE Access, 2018, 6: 26194-26214.
[3] Wang X, Wu G, Xing L, et al. Agile earth observation satellite scheduling over 20 years: Formulations, methods, and future directions[J]. IEEE Systems Journal, 2020.
[4] Zhao L, Wang S, Hao Y, et al. Energy-dependent mission planning for agile earth observation satellite[J]. Journal of Aerospace Engineering, 2019, 32(1): 04018118.
[5] Wang X, Han C, Zhang R, et al. Scheduling multiple agile earth observation satellites for oversubscribed targets using complex networks theory[J]. IEEE Access, 2019, 7: 110605-110615.
[6] Cui K, Xiang J, Zhang Y. Mission planning optimization of video satellite for ground multi-object staring imaging[J]. Advances in Space Research, 2018, 61(6): 1476-1489.
[7] Lin Z. Mission planning for electromagnetism environment monitors satellite based on heuristic tabu search algorithm[C]//The 27th Chinese Control and Decision Conference (2015 CCDC). IEEE, 2015: 5698-5702.
[8] He L, Liu X, Laporte G, et al. An improved adaptive large neighborhood search algorithm for multiple agile satellites scheduling[J]. Computers & Operations Research, 2018, 100: 12-25.
[9] Li L, Yao F, Jing N, et al. Preference incorporation to solve multi-objective mission planning of agile earth observation satellites[C]//2017 IEEE Congress on Evolutionary Computation (CEC). IEEE, 2017: 1366-1373.
[10] Tian Y, Wang H, Zhang X, et al. Effectiveness and efficiency of non-dominated sorting for evolutionary multi-and many-objective optimization[J]. Complex & Intelligent Systems, 2017, 3(4): 247-263.
[11] Xiong M, Xiong W, Liu C. A hybrid many-objective evolutionary algorithm with region preference for decision makers[J]. IEEE Access, 2019, 7: 117699-117715.
[12] Li L, Chen H, Li J, et al. Preference-based evolutionary many-objective optimization for agile satellite mission planning[J]. IEEE Access, 2018, 6: 40963-40978.
[13] Deb K, Jain H. An evolutionary many-objective optimization algorithm using reference-point-based nondominated sorting approach, part I: solving problems with box constraints[J]. IEEE transactions on evolutionary computation, 2013, 18(4): 577-601.
[14] Varun Kumar S G, Panneerselvam R. A study of crossover operators for genetic algorithms to solve VRP and its variants and new sinusoidal motion crossover operator[J]. International Journal of Computational Intelligence Research, 2017, 13(7): 1717-1733.
[15] Xiong M, Xiong W, Jian P. Visualization of the Non-dominated Solutions in Many-objective Optimization[C]//2019 IEEE Fourth International Conference on Data Science in Cyberspace (DSC). IEEE, 2019: 188-195.

Variable Screening Optimization Algorithm for Mahalanobis-Taguchi System

Xiaohong Peng[1], Rui Zheng[2] and Jiufu Liu[2,*]

[1]Department of Information Engineering ,Jiangsu Maritime Institute, Nanjing 211170, China
[2]College of Automation, Nanjing University of Aeronautics and Astronautics, Nanjing 210016, China

*Corresponding author email: Liujiufu2@126.com

Abstract. This paper proposes a Mahalanobis-Taguchi system variable screening optimization method based on binary quantum behavior particle swarm.The main procedures and methods are as follows, Firstly, the Mahalanobis distance value is calculated by the Gram-Schmidt orthogonalization method.We build the multi-objective mixed planning model. The binary quantum behavior particle swarm optimization algorithm is used to solve the optimal combination. A new prediction system based on Mahalanobis-Taguchi metric is established, and the task of accurate discrimination is accomplished.
Keywords: Mahalanobis-Taguchi system; Variable screening; Binaryquantum behavior particle swarm; Optimization.

1. Introduction

Malalanobis-taguchi system (MTS) is a new method for pattern recognition[1-3]. Since quantum particle swarm optimization(QPSO) algorithm has advantages in variable optimization, this paper uses MTS model and QPSO to reduce the feature variables.

In this paper, a multi-objective mixed dimension reduction model is constructed using misclassification rate. The mahalanobis-Taguchi system variables are optimized by quantum behavior particle swarm optimization with stronger global search ability and better convergence,and the binary coding of quantum behavior particle swarm is improved. The optimization of feature variables is completed by iterative calculation.

2. Mahalanobis-Taguchi System

The Mahalanobis-Taguchi system uses the Mahalanobis distance (MD), On the basis of SNR(signal-to-noise ratio) and orthogonal table, the diagnosis and prediction of the system are completed.

Step 1: Build the reference space

N samples of the normal group were selected, the number of characteristic variables was set as K, and the vector group was normalized according to Formula (1). The Mahalanobis distance of each normal sample is calculated in the formula (2).

$$z_{ij} = \frac{1}{s_j}(x_{ij} - \overline{x}_j) \quad i = 1,2,\ldots,n, \quad j = 1,2,\ldots,k \tag{1}$$

$$MD_i = d_i^2 = \frac{1}{k} Z_i^T S^{-1} Z_i \qquad i = 1,2,\ldots,n \tag{2}$$

Here, $Z_i = [z_{i1}, z_{i2}, \ldots, z_{ik}], i = 1,2,\ldots,n$ is the normalized vector of the ith sample. $\bar{x}_j$ represents the data mean of all samples of the $j$thdimension variable. s_j representsthe data standard deviation of all samples of the jth dimension variable. S^{-1} represents the inverse matrix of the correlation matrix of the sample population.
Step 2: Optimization of the reference space
Equations (2) and (3) are used to calculate the MD and SNR values of the anomaly samples, respectively.

$$SN_i = -10\log\left[\frac{1}{n}\sum_{i=1}^{n}(1/MD_i^2)\right] \quad (3)$$

For the variable X_j ,It's calculated by the SNR average of all the screened variables, denoted as SN_j^+ .Then it's calculated by the SNR average value of the variable that is not screened,denoted as SN_j^- .
Step 3: thresholds are determined and classifications are predicted

3. Build a Dimension Reduction Model

3.1. Mahalanobis Distance
The calculation process is as follows[4-5]:

$$\begin{aligned} U_1 &= Z_1 \\ U_2 &= Z_2 - \frac{Z_2^T U_1}{U_1^T U_1} U_1 \\ &\cdots \\ U_k &= Z_k - \frac{Z_k^T U_1}{U_1^T U_1} U_1 - \cdots \frac{Z_k^T U_{k-1}}{U_{k-1}^T U_{k-1}} U_{k-1} \end{aligned} \quad (4)$$

Here $Z_j = (z_{1j}, z_{2j}, \ldots, u_{nj})^T$, $j = 1,2,\ldots,k$ is the $j$thdimensional normalized vector of the n samples. $U_k = (u_{1k}, u_{2k}, \ldots, u_{nk})^T$ is the mutual vertical vector of the $k$thdimension.
The Mahalanobis distance of the ith sample is given by:

$$MD_i = \frac{1}{k}\left(\frac{u_{i1}^2}{s_1^2} + \frac{u_{i2}^2}{s_2^2} + \cdots + \frac{u_{ik}^2}{s_k^2}\right) \quad (5)$$

Where u_{ik} is the element of the orthogonal vector U_k, s_k is the standard deviation of each element in the vector U_k, and k is the total number of variables.

3.2. Multi-objective Optimization Model
Goal 1: The variable screening standard adopts the concept of misclassification rate, and the variable combination to be optimized can be defined as:

$$x_i = \begin{cases} 1, & \textit{If the ith variable is selected} \\ 0, & \textit{If the ith variable is not selected} \end{cases}$$

The misclassification rate is as follows:

$$\begin{cases} f_1(X) = \frac{n_1^e}{n_1}, \\ f_2(X) = \frac{n_2^e}{n_2} \end{cases} \tag{6}$$

Here, n_1^e represents the number of normal observations, here misclassified as anomalies. n_2^e represents the total number of normal group samples. n_1 represents the total number of normal group samples. n_2 represents the total number of abnormal group samples.
The function input parameters $MD^1 = \{MD_1^1, MD_2^1, \ldots, MD_{n_1}^1\}$ and $MD^2 = \{MD_1^2, MD_2^2, \ldots, MD_{n_2}^2\}$ are the Mahalanobis distance. T is the threshold of the system prediction before optimization. The optimal threshold point of the system was determined by ROC curve method.
Goal 2: The way to minimize the number of system characteristic variables is to use dimensionality reduction, as follows:

$$f_3 = \frac{p_{selected}}{p} \tag{7}$$

The multi-objective optimization model is listed below:

$$min\ \ F(X) = \omega_1 \frac{n_1^e}{n_1} + \omega_2 \frac{n_2^e}{n_2} + \omega_3 \frac{p_{selected}}{p} \tag{8}$$

s.t.

$$\sum_{i=1}^{p} x_i \le p \tag{9}$$

$$\sum_{i=1}^{p} x_i = p_{selected} \tag{10}$$

$$f_1(X) \le f_1^{\max}, f_2(X) \le f_2^{\max} \tag{11}$$

In this paper, the above expression (8-11) represents reducing the dimension of the MTS variable.

4. Optimized Solution

4.1. On Quantum Behavior Particle Swarm Optimization(QPSO)

We can see from the following formula:

$$p_{ij}(\mathrm{t}) = \varphi_j(\mathrm{t}) P_{ijbest}(\mathrm{t}) + (1 - \varphi_j(\mathrm{t})) G_{jbest}(\mathrm{t}) \tag{12}$$

$$\varphi_j(\mathrm{t}) = \frac{c_1 r_1}{c_1 r_2 + c_2 r_3} \tag{13}$$

Where $p_{ij}(\mathrm{t})$ is the attractor of the jth dimension of the $i$thparticle,and $P_{ijbest}(\mathrm{t})$ represents the local optimal position of the jth dimension of the ith particle. $\varphi_j(\mathrm{t})$ consists of random numbers r_1, r_2, r_3 and normal numbers c_1, c_2 uniformly distributed over the interval $\{0,1\}$. Using the Monte Carlo method, the particle position update is obtained by the following formula:

$$x_{ij}(t+1) = \begin{cases} p_{ij}(\mathrm{t}) + \beta \left| M_{jbest}(\mathrm{t}) - x_{ij}(\mathrm{t}) \right| \ln(1.0/u_{ij}), & \text{if } r_6 \ge 0.5 \\ p_{ij}(\mathrm{t}) - \beta \left| M_{jbest}(\mathrm{t}) - x_{ij}(\mathrm{t}) \right| \ln(1.0/u_{ij}), & \text{otherwise} \end{cases} \tag{14}$$

Where $x_{ij}(t+1)$ is the position of the *j*th dimension of the *i*th particle, and u and r_6 are random numbers uniformly distributed over the interval $\{0,1\}$. β is the contraction-expansion coefficient, which controls the convergence speed of the algorithm and can be calculated by the following equation:

$$\beta = \beta_0 + \frac{(t_{\max} - t) \times (\beta_1 - \beta_0)}{t_{\max}} \tag{15}$$

Where t represents the current number of iterations and $t_{\max}$ represents the maximum number of iterations. The linear reduction of β from 1.0 to 0.5 can achieve better optimization results for the QPSO algorithm. In general, β_1 is 1 and β_0 is 0.5. The average optimal position is introduced,denoted as $M_{best}(\mathrm{t})$, which is defined as the average of the individual optimal positions and can be obtained from:

$$\begin{aligned} M_{best}(\mathrm{t}) &= (M_{1best}(\mathrm{t}), M_{2best}(\mathrm{t}), \ldots, M_{nbest}(\mathrm{t})) \\ &= (\frac{1}{N}\sum_{i=1}^{N} P_{i1best}(\mathrm{t}), \frac{1}{N}\sum_{i=1}^{N} P_{i2best}(\mathrm{t}), \ldots, \frac{1}{N}\sum_{i=1}^{N} P_{inbest}(\mathrm{t})) \end{aligned} \tag{16}$$

4.2. Particle Coding

The particle position value needs to be converted to binary form for calculation. The formula is based on the following:

$$x_{ij}(t+1) = \begin{cases} 0, & \mathrm{sig}(x_{ij}(t)) < \mathrm{rand}() \\ 1, & \mathrm{sig}(x_{ij}(t)) \geq \mathrm{rand}() \end{cases} \tag{17}$$

Here rand() is a random number, and the sigmoid function is as follows:

$$sig(x) = \frac{1}{1+e^{-x}}$$

4.3. Improved BinaryQuantum Behavior Particle Swarm Algorithm

This section proposes an improved variable reduction algorithm for Improved Binary Quantum behaved Particle Swarm Optimization (IBQPSO).

Algorithm: Quantum behavior binary particle swarm optimization

Input: sequence of characteristic variables $X = (x_1, x_2, \ldots, x_k)$, population size N, dimension of particles Dim, Upper and lower bounds of particles $x_{\max}$ and $x_{\min}$

Output: the global optimal fitness value F_{gbest}, the global optimal particle sequence X_{Gbest}

Randomly initializes the current position of each particle in the particle swarm x_i, determines the maximum number of iterations $MaxIter$

The particle position $x_{ij}(t)$ is binary coded according to equation (17), and the fitness function value of each particle is calculated, determines the individual optimal position $P_{best}(t)$ and the global optimal position $G_{best}(t)$.

Repeat

Calculate the average best position of the particles according to equation (16)

for $(i=1; i \leq N; i=i+1)$ **do**

  for $(j=1; j \leq Dim; j=j+1)$ **do**

Calculate the local attractor $p_{ij}(t)$ according to equation (12)

  Update the position of the particle $x_{ij}(t)$ according to equation (14)

  Limit particles to search intervals $x_{\max}$ and $x_{\min}$

  End for

Binary encoding the particle position $x_{ij}(t)$ according to equation (17)

Calculate the fitness function value of the current particle and update the individual optimal position of the particle $P_{best}(t)$
Compare $P_{best}(t)$ with the value of global optimal position $G_{best}(t)$ and update $G_{best}(t)$
End for
Until the maximum number of iterations $MaxIter$
Output F_{gbest} and X_{Gbest}

5. Case Study

5.1. Experiment Data

A fetal health diagnostic dataset measured for Cardiotocography (CTG) is collected by the University of Porto School of Medicine in Portugal. CTG is a technique for detecting fetal heart rate and uterine contractions in pregnant women using an electronic fetal monitor. It can be used to assess the health of the fetus. This data set has been labeled by the obstetrician based on morphological patterns and fetal status with 18 characteristic variables. It is necessary to reduce the dimension of the system. The data set is a post-diagnostic sample and the diagnosis results are divided into three types, including health, suspected conditions, and confirmed cases. In this paper, two types of healthy and confirmed cases are selected for subsequent data processing. The health category is taken as the normal sample, and the confirmed case category is used as the abnormal sample.512 groups of samples have been selected as the experimental data, including 336 normal samples and 176 abnormal samples.

5.2. Data Processing and Analysis

The initial characteristic variable sequence is used to calculate the MD of the remaining 232 sets of samples in the normal sample and the MD of the abnormal sample, as seen in Fig. 1 below.
As shown in Figure 1, the MD value of the normal group was smaller, with an average of 1.18. The MD value of abnormal group fluctuated greatly, with an average of 5.37. On the whole, MD has certain discriminant ability, but there are many overlaps, which makes further classification difficult.
In TABLE I , although BQPSO has the fastest spatial search speed under different weight ratios, the average number of iterations is the least, but its fitness value is not the best. The fitness values of the weight combinations (1) to (3) are all greater than BPSO and IBQPSO.

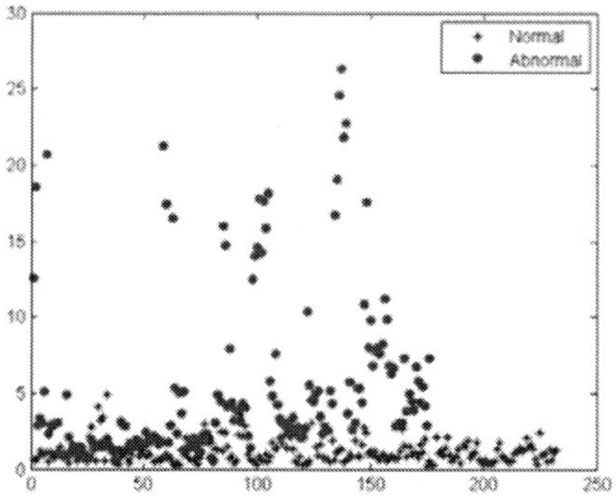

Figure 1. MD value distribution before optimization.

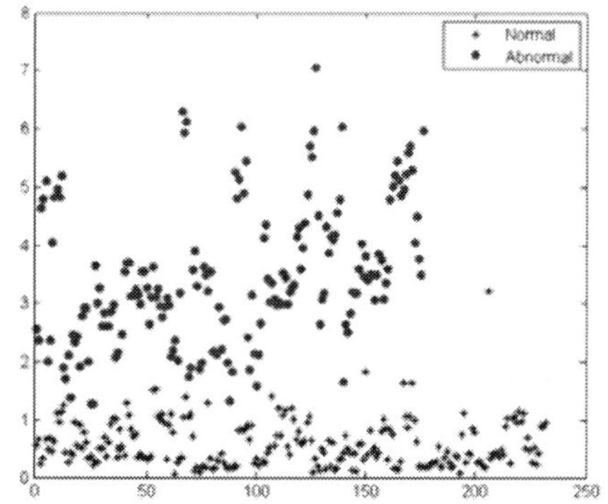

Figure. 2. MD value distribution after optimization.

In Figure 2, the optimized variables make the difference of MD values between normal samples and abnormal samples more obvious. Compared with before optimization, the overlapped parts of the system are reduced and the system identification efficiency is improved.

Table 1. Fitness value and optimal variables combination.

Weight combinations	Optimization algorithm	Fitness valucs	Iterations	optimal variables combination

ω_1=0.5,	BPSO	0.057	58	$X_3,X_5,X_6,X_{13},X_{15}$
ω_2=0.48,	BQPSO	0.063	7	X_3,X_5,X_6,X_7,X_{15}
ω_3=0.02	IBQPSO	0.052	20	$X_3,X_5,X_6,X_{13},X_{15}$
ω_1=0.3,	BPSO	0.043	73	X_3,X_6,X_{17}
ω_2=0.68,	BQPSO	0.048	8	$X_3,X_4,X_6,X_{10},X_{14},X_{15}$
ω_3=0.02	IBQPSO	0.042	20	$X_3,X_4,X_6,X_{11},X_{15},X_{16}$
ω_1=0.8,	BPSO	0.042	50	$X_3,X_5,X_6,X_{13},X_{15}$
ω_2=0.18,	BQPSO	0.045	6	$X_1,X_4,X_5,X_6,X_{13}, X_{15}$
ω_3=0.02	IBQPSO	0.039	16	$X_3,X_5,X_6,X_{13},X_{15}$

Table 2 lists the accuracy of the system predictions by selecting the optimized feature variables to reconstruct the measurement model. In terms of overall average accuracy, IBQPSO reached 94%, which was higher than 92% and 92.7% of BPSO and BQPSO. It can reflect that IBQPSO has fewer redundant variables in the system after effective reduction of characteristic variables. The accuracy of the discrimination of the result has been greatly improved.

Table 2. Accuracy comparison after optimization.

	Normal	Abnormal	Overall
BPSO	92.9%	91.1%	92.0%
BQPSO	93.8%	91.6%	92.7%
IBQPSO	94.7%	93.3%	94.0%

6. Conclusion

A new variable optimization method for MTS is proposed. In this paper, the misclassification weighted sum based on Mahalanobis distance calculation is proposed as a dimensionality reduction model, and the IBQPSO algorithm is used to solve the optimal variable combination. Through the example analysis, the iterative speed and classification success rate of variable dimension reduction are significantly improved by the method of this paper, and the efficiency of the method is verified.

References

[1] YazidMAM, Rijal JK, Awaluddin MS, Sari Emelia.Pattern Recognition on Remanufacturing Automotive Component as Support Decision Making Using Mahalanobis-taguchi System[J].Procedia CIRP,2015, 26:258-263

[2] Reséndiz E, Moncayo-Martínez, LA, SolísG. Binary ant colony optimization applied to variable screening in the Mahalanobis–Taguchi System[J]. Expert Systems with Applications, 2013, 40(2):634-637.

[3] Hadighi SA et al.Mahalanobis-Taguchi System-based criteria selection for strategy formulation a case in a training institution[J].Journal of industrial engineering international, 2013, 9(26):9-17.

[4] Sikder Sagar et al.A synergistic Mahalanobis–Taguchi system and support vector regression based predictive multivariate manufacturing process quality control approach[J].Journal of manufacturing system, 2013, 57: 323-337.

[5] Sikder Sagar et al. An integrated approach for multivariate statistical process control using mahalanobis-taguchi system and andrews function[J].International Quality Reliability Management, 2017, 34 (8): 186-1208.

A Passive Navigation Algorithm Based on Derivative Dynamic Time Regularization Geomagnetic Matching

Yiteng Wang[1,2*], Jinhong Bao[3] and Zehua Li[3]

[1]Science and Technology on Communication Information Security Control Laboratory, China
[2]No.36 Research Institute of China Electronics Technology Group Corporation, China
[3]College of Telecommunication Engineering, Hangzhou Dianzi University Hangzhou 310018, China

*Corresponding author email: yitengwang@bjtu.edu.cn

Abstract. As satellite positioning has shortcomings such as susceptibility to multipath interference, we look forward to using a new algorithm to solve these problems. The method of derivative dynamic time regularization is used as the matching method of geomagnetic navigation. Due to the factors of the inertial device itself, a certain cumulative error will occur in the positioning solution of the inertial navigation (INS). Therefore, an outdoor error model needs to be established, and the positioning error is corrected by filtering technology. To solve this problem, the Unscented Kalman Filter is used to fuse the positioning results of the two to obtain a more accurate positioning, with an average accuracy within 1.5 meters.
Keywords: Integrated navigation algorithm; Derivative dynamic time warping algorithm; Unscented Kalman filtering.

1. Introduction

In recent years, new navigation methods, especially the passive positioning navigation applications have been vigorously developed. As one of the important research objects in the domain of passive positioning navigation, the geomagnetic navigation has attracted the attention of many researchers in the domestic and abroad. Since the interior building steel structure and electrical wiring circuit will produce relatively strong constant geomagnetic field interference, geomagnetic navigation is more widely used indoors. Minor outdoor environmental changes can easily influence the geomagnetic field so strongly that the two intensity sequences, which are separately sampled from a pre-stored off-line geomagnetic database and an online collector, probably have difference both in local areas and sequence size. Therefore, it's not expected to use a point-to-point matching algorithm to match the influenced geomagnetic field intensity. In thesis, the derivative dynamic time warping (DDTW) method is used for trend matching. Compared with the point-to-point geomagnetic matching algorithm, the trend matching algorithm Performs better in accuracy. Nowadays, geomagnetic navigation is playing a great role in missile guidance, underwater vehicle areas. In thesis, the combination of geomagnetic navigation and inertial navigation (INS)[1][2] is used to navigate and locate vehicles in outdoor scenes. We use the Unscented Kalman Filter(UKF)[3][4] to correct the cumulative error which is commonly seen when using inertial device. At the same time, the motion trajectory of inertial navigation is used to assist the matching of geomagnetic navigation to improve the matching accuracy of geomagnetic navigation. Therefore, for different application environments, corresponding geomagnetic matching navigation and positioning methods are proposed according to geomagnetic characteristics. Through experiments, we prove that the method in thesis can make the vehicle obtain effective positioning accuracy in outdoor scenes, and the average accuracy can reach within 1.5 meters.

2. Geomagnetic Matching Algorithm for Outdoor Navigation

Dynamic time warping algorithm[5][6][7] is a method to figure out the similarity of two sets of time sequences of different lengths. In the field of geomagnetic navigation, if the environment constantly influences the geomagnetic field, the method can perfectly align the two geomagnetic field sequences of different lengths and finally get good matching results. But when additional factors from local environment affect the geomagnetic field, one of the method's defects is obviously shown. When a part of a sequence is obviously different from that corresponding part of the other, matching results may have local differences, especially in matching two troughs of different depth. As can we see in figure 1.

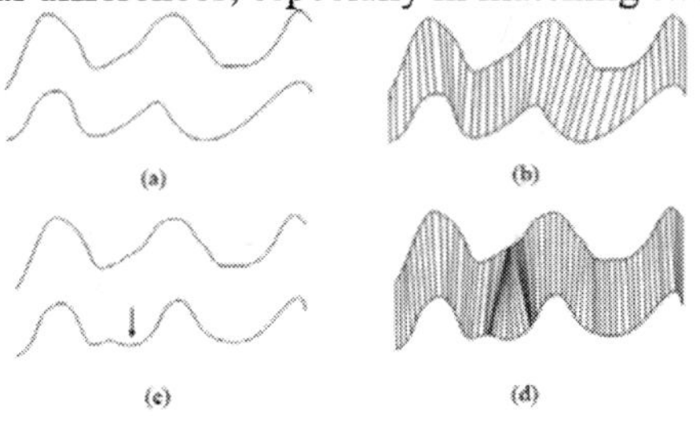

Figure 1. The impact of changes in local features of time sequence.

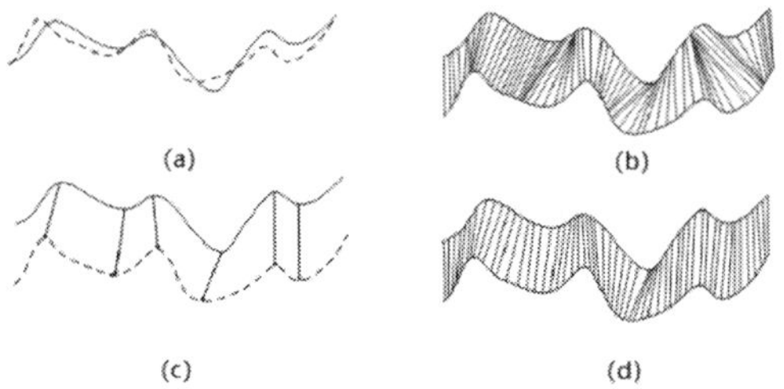

Figure 2. The comparison of DTW and DDTW.

It can be seen from the figure 1 that using DTW can map well from the sequence above to the sequence below when they are almost the same. However, when we change local characteristic of the sequence below (turning a trough of the sequence below into two), the DTW method would mapping them incorrectly. Due to the more complicated interference factors outdoors, it is not suitable to use the DTW method directly. To solve this problem, in the actual process of geomagnetic matching, "form", as a characteristic, should be taken into consideration. Therefore, the first derivative of the sequence is considered to obtain the relevant information of the shape, and the derivative dynamic time warping method is adopted as the geomagnetic matching method in the geomagnetic navigation matching of outdoor scenes in thesis. The DTW method is implemented by constructing a matrix of $N \times M$, the parameter (i^{th}, j^{th}) of matrix include the Euclidean distance $d(i, j)$ between the point p_i and q_j on two sets of time sequences. For DDTW, it is not the Euclidean parameter value between the two points in $d(i, j)$, but the square of the difference value between the point p_i and q_j corresponding derivatives of $D_x[p]$, $D_x[q]$. The following formula is usually used for calculation:

Local distance calculation based on derivative estimation is as follows:

$$d(i, j) = (D(p_i) - D(q_j))^2 \quad (1)$$

$$D_x[p] = 0.5(p_i - p_{i-1}) + 0.25(p_{i+1} - p_{i-1}) \quad , \quad 1<i<m \quad (2)$$

From the above equation, the derivative estimate $D_x[p]$ is the average of the straight slope passing through the point p_i and left neighbor and the straight slope passing through the point left and right neighbors.

From theoretical analysis, this derivative estimate is more robust to outliers than any estimate between two data nodes considered separately. Compared with DTW method, DDTW does not estimate the first and last data nodes of time sequences but uses the second node and the penultimate node. In the process of measurement, the short measuring accuracy of geomagnetic sampling module can lead to certain error to geomagnetic data. Therefore, before the geomagnetic matching process, we use the generalized continuation method[8][9] to smooth the sequences to be matched. The time complexity of DDTW is $O(mn)$, which is the same as that of DTW.

The similarity between sequence P and sequence Q obtained by DDTW method is:

$$L(p,q)=d(p_i,q_j)+\min\{L(i-1,j-1),L(i-1,j),L(i,j-1)\} \tag{3}$$

The specific comparison of DTW and DDTW in time sequences is shown in figure 2. In figure 2, figure a is two sets of time sequences to be matched, figure b is an intuitive display of the warpage alignment of two sets of time sequences, figure c is the alignment between time sequences generated by DTW method, and figure d is the alignment between time sequences generated by DDTW method. Therefore, DDTW method has better effect on the overall alignment of time sequences.

3. Outdoor Passive Integrated Navigation Based on Geomagnetic Matching

3.1. Subsection DDTW Matching Method

In the real environment, it's not an efficient way to real time positioning that using DDTW method alone to collect geomagnetic data for long period. Sliding window is used to integrate geomagnetic sequence, whose length is determined by experience. Wheel count calculates mileage information that is useful in narrowing the range of geomagnetic field strength sequence to match which is restored in offline geomagnetic database. The algorithm can reduce much calculating complexity while matching.

The process of sliding window matching on both online geomagnetic sequence and geomagnetic database field strength sequence is shown in the figure 3.

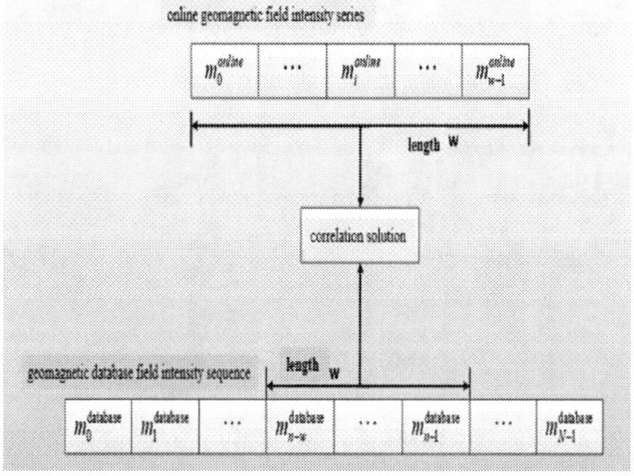

Figure 3. Geomagnetic field intensity sequence's slide match processing is measured by on-line.

As can we see from figure 3, w stands for sliding window length, N stands for the length of the geomagnetic field strength sequence corresponding off-line geomagnetic database, L stands for different sequences matching similarity, λ stands for matching similarity threshold. λ is an empirical value calculated by matching similarity of two sequences. Output the matching positioning results whenever $L < \lambda$ is observed. If not, the window continues sliding.

3.2. UKF Filtering

To fix the defects of EKF[10][11], Uhlmann and Julier thought *it is less difficult to approximate the probability distribution than the nonlinear function*, and proposed UKF. UKF[12][13], a type of frame, is mainly based on UT transform, distinguished from EKF by using Kalman linear filter, using UT transform to solve problems that covariance and mean may be transferred nonlinearly. First of all, UKF inherits the advantages of KF. The main advantage is that there is no need to store redundant historical information, we need this feature that can be used for real-time estimation. UKF does not approximate the nonlinear function, but approximates the probability density distribution of the nonlinear function, so that the equations of the nonlinear functions don't need to be known. Even if the system to be processed is nonlinear or approximately non-linear, the accuracy of all nonlinear systems is comparable

to the model using the second-order Taylor series approximation, and the system is still assumed to obey the Gaussian distribution. UT transform can produce a group of weighted sampling points(sigma), and non-linearly transform them. All operators are constrained by equations. The mean and variance of state parameters' nonlinear transform can be approached by those of transformed sampling points. UKF specific process is as follows.
Assuming state equation and observation equation:

$$X_k = F(X_{k-1}) + W_{k-1} \tag{4}$$

$$Z_k = H(X_k) + V_k \tag{5}$$

The mean and variance of the state vector X are denoted as $\bar{X}$, P_X. According to UT transform, we transform the 2n+1 sampling points surrounding $\bar{X}$ using following rules:

$$\chi_{j,k-1} = \begin{cases} \bar{X}_{k-1} & \mathrm{j=0} \\ \bar{X}_{k-1} + (\sqrt{(n+\lambda)P_{X,k-1}})_j & \mathrm{j=1,2,\ldots,}n \\ \bar{X}_{k-1} - (\sqrt{(n+\lambda)P_{X,k-1}})_j & \mathrm{j=n+1,n+2,\ldots,2n} \end{cases} \tag{6}$$

We define $\lambda = \alpha^2(n+k) - n$. k stands for semidefinite variance matrix parameters. Define the weight corresponding χ_{j} as:

$$w_j = \begin{cases} w_0^m = \lambda(n+\lambda)^{-1} & j=0 \\ w_0^c = \lambda(n+\lambda)^{-1} + (1+\alpha^2+\beta) & j=1,2,\ldots,n \\ w_j^m = w_j^c = ((n+\lambda)*2)^{-1} & j=n+1,n+2,\ldots,2n \end{cases} \tag{7}$$

In equation 7, α is used to adjust $\bar{X}$ and sampling distance, which is generally a decimal greater than 0. β contains prior information of X, adjust β to promote variance accuracy. It's prime to set β 2 for Gaussian distribution. We transform the sampling points non-linearly in each sampling period through nonlinear state function $\mathrm{F}(.)$ as:

$$\chi^*_{j,k/k-1} = \mathrm{F}(\chi_{j,k/k-1}) \tag{8}$$

Then we weight and estimate the results from non-linear transform:

$$\hat{X}_{k/k-1} = \sum_{j=0}^{2n} w_j^m \chi^*_{j,k/k-1} \tag{9}$$

$$P_{k/k-1} = \sum_{j=0}^{2n} w_j^m (\chi^*_{j,k/k-1} - \hat{X}_{k/k-1})(\chi^*_{j,k/k-1} - \hat{X}_{k/k-1})^T \tag{10}$$

We predict next value of vector using sampling points transform, therefore, the measured value of the estimator is:

$$z_{j,k/k-1} = H(\chi^*_{i,k/k-1}) \tag{11}$$

$$\hat{Z}_{k/k-1} = \sum_{j=0}^{2n} w_j^m z_{j,k/k-1} \tag{12}$$

$$\hat{X}_{k/k-1} = \sum_{j=0}^{2n} w_j^m \chi^*_{j,k/k-1} \tag{13}$$

Therefore, the UKF calculation step process can be expressed as:
Step 1 Collecting sampling points using UT transform:

$$\chi_{j,k-1} = \begin{cases} \bar{X}_{k-1} & \text{j}=0 \\ \bar{X}_{k-1}+(\sqrt{(n+\lambda)P_{X,k-1}})_j & \text{j}=1,2,\ldots,n \\ \bar{X}_{k-1}-(\sqrt{(n+\lambda)P_{X,k-1}})_j & \text{j=n+1,n+2,...,2n} \end{cases} \tag{14}$$

$$w_j = \begin{cases} w_0^m = \lambda(n+\lambda)^{-1} & j=0 \\ w_0^c = \lambda(n+\lambda)^{-1} + (1+\alpha^2+\beta) & j=1,2,\ldots,n \\ w_j^m = w_j^c = (2*(n+\lambda))^{-1} & j=n+1,n+2,\ldots,2n \end{cases} \tag{15}$$

Step 2 Predicting state:

$$\hat{X}_{k/k-1} = \sum_{j=0}^{2n} w_j^m \chi^*_{j,k/k-1} \tag{16}$$

$$P_{k/k-1} = \sum_{j=0}^{2n} w_j^m (\chi^*_{j,k/k-1} - \hat{X}_{k/k-1})(\chi^*_{j,k/k-1} - \hat{X}_{k/k-1})^T \tag{17}$$

$$z_{j,k/k-1} = H(\chi^*_{i,k/k-1}) \tag{18}$$

$$\hat{Z}_{k/k-1} = \sum_{j=0}^{2n} w_j^m z_{j,k/k-1} \tag{19}$$

Step 3 Measurement update:

$$P_{zz} = \sum_{j=0}^{2n} w_j^m (\hat{Z}_{k/k-1} - z_{j,k/k-1})(Z_{k/k-1} - z_{j,k/k-1})^T + \text{R}_k \tag{20}$$

$$P_{xz} = \sum_{j=0}^{2n} w_j^m (\chi^*_{j,k/k-1} - \hat{X}_{k/k-1})(\chi^*_{j,k/k-1} - z_{j,k/k-1})^T \tag{21}$$

$$K_k = P_{xz} P_{zz}^{-1} \tag{22}$$

$$\hat{X}_k = \hat{X}_{k/k-1} + K_k (Z_k - \hat{Z}_{k/k-1}) \tag{23}$$

$$P_k = P_{k/k-1} - K_k P_{zz} K_k^T \tag{24}$$

Step 4 repeat the above steps.

In real applying process, we define $\hat{X}_o = EX_o$ to accelerate calculation.

Journal of Physics: Conference Series **2179** (2022) 012037 doi:10.1088/1742-6596/2179/1/012037

The filter should be specifically designed. We first setting the parameters like INS longitude and latitude error as state quantity. The observation variable stands for the difference between INS solution and geomagnetic matching navigation solution. The error from INS can be corrected through integrated navigation system. Using filter to estimate this system's error, whose state equation can be written as:

$$\dot{X}(t) = F(t)X(t) + G(t)W(t) \tag{25}$$

$F_{13\times13}$ stands for State matrix, $W(t)$ stands for system noise, $G(t)$ stands for coefficient matrix, $X_{13\times1}$ stands for state variable.

$$X(t) = [\phi_E, \phi_N, \phi_U, \delta\lambda, \delta\varphi, \delta V_x, \delta V_y, \ell_{cx}, \ell_{cy}, \ell_{cz}, \alpha_x, \alpha_y, \alpha_z] \tag{26}$$

ϕ_E, ϕ_N, ϕ_U stand for direction attitude angle error including east-toward, north-toward and sky-toward, $\delta\lambda, \delta\varphi$ stand for longitude and latitude error, $\delta V_x, \delta V_y$ stands for east-toward and north-toward speed error, $\ell_{cx}, \ell_{cy}, \ell_{cz}$ stand for the drift of stable value of east-toward and north-toward gyroscope, $\alpha_x, \alpha_y, \alpha_z$ stand for the error from three accelerometer (east-toward, north-toward, up-toward).
The measurement equation of the system is expressed as:

$$Z(t) = [Z_1, Z_2]^T = HX(t) + V(t) \tag{27}$$

Z_1、Z_2 stand for longitude and latitude difference from geomagnetic matching navigation positioning and INS positioning, $H_{2\times13}$ stands for measurement matrix, $V(t)$ stands for measurement noise.
The schematic diagram of outdoor passive integrated navigation based on geomagnetic matching algorithm is shown in figure 4.

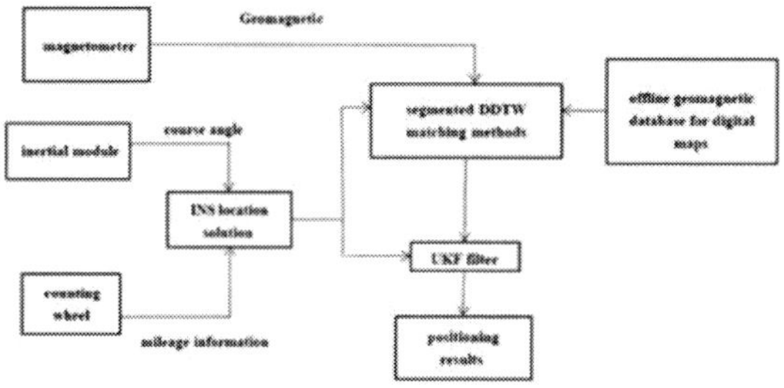

Figure 4. The principle schematic diagram of outdoor geomagnetic matching combined navigation and positioning.

There are two main steps in outdoor geomagnetic matching navigation positioning. Firstly, establishing the offline geomagnetic database for digital maps, establishing the offline geomagnetic database with GPS482 and magnetometer. The accuracy of GPS482 is centimeter level outdoors (except areas containing large buildings and forest). In special situations, wheel count and inertial device module are used to locate standard position and transform local coordinates into global ones (longitude and latitude). Secondly, online matching. Using segmented DDTW method to match geomagnetic sequences sampled online and that from the offline database for digital map. Defining INS speed error, attitude error, Gyro drift as state quantity, and the difference between positioning information calculated by INS and that of geomagnetic matching navigation positioning as observation. Using UKF to estimate INS error to correct INS. Finally we get the positioning result.

4. Experimental Simulation and Result Analysis

In this section, we design the filtered geomagnetic matching integrated navigation. The unfiltered geomagnetic matching integrated navigation corrects the positioning error of the INS, and calculates the trajectory of the carrier by using the method of geomagnetic matching navigation in most circumstances. In the filtered geomagnetic matching integrated navigation, we firstly build an offline database for digital maps, during which we select a cruising car equipped with a GPS482 module as well as a JY901 module in a magneto resistive sensor to collect three-axis geomagnetic field strength data. The output frequency of the GPS482 module, the JY901 module, round count of the cruising car is 5, 20, 5Hz. The demonstration environment locates at the Suzhou Geographical Science and Cultural Industry Building of Zhongke. As can be seen from figure 5 and 6.

Figure 5. A cruise vehicle with data acquisition module which is used to test.

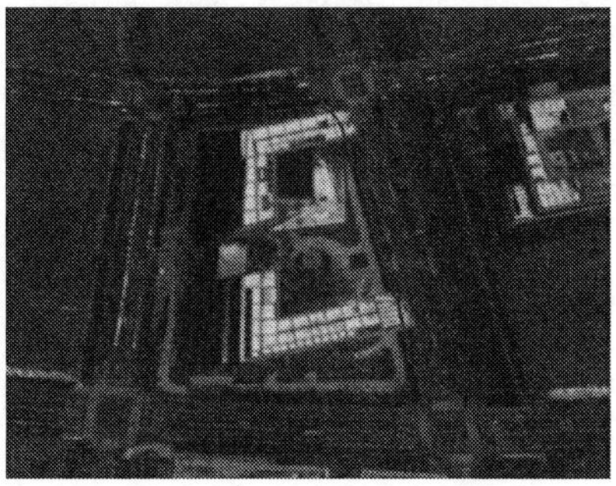

Figure 6. The route of cruise vehicle test.

Experimental parameters are shown in this table:

Table 1. Experimental parameters.

parameters	parameter values
original position	31.32525°,120.4278°
original speed	1m/s
initial horizontal attitude	0°
constant drift of gyroscope (three axes equal)	$0.03^o / h$
gyroscope random drift (three axes equal)	$0.005^o / h$
accelerometer stable offset (three axes equal)	0.001g
accelerometer random offset (three axes equal)	0.0005g
measure noise standard deviation of magnetic anomaly	1.4076mG
on- line sampling reluctance sensor sampling frequency	20Hz
accuracy of geomagnetic reference map in the test area	$0.000001^o \times 0.000001^o$
eastward velocity error of the cruise vehicle starting position	0.1m/s
north velocity error of the cruise vehicle starting position	0.1m/s
horizontal attitude error of the cruise vehicle starting position	1°
heading and attitude error of the cruise vehicle starting position	1°
initial latitude error of cruise vehicle	0.0001^o
initial longitude error of cruise vehicle	0.00005^o

According to experimental measurement, λ is 5.8701. w is the matching length of the sliding window. Let w Numerically equal to 20 times the Geomagnetic field strength. Let the cycle of location updating be 1 second while geomagnetic matching navigation and positioning. In the process of positioning solution, we save the last result of positioning that corresponds the digital maps, in which there is a geomagnetic field strength sequence of the offline geomagnetic field database. The sequence would be transferred at the next process for sliding match. location updating would be stopped when the wheel counter speed is 0 for over 2 seconds.

After building up the offline geomagnetic field database for digital maps, we start with matching the sampled geomagnetic sequence and geomagnetic field strength sequence of the offline geomagnetic field database for digital maps to resolve geomagnetic matching navigation and positioning trajectory by using the cruising car equipped with JY901 module containing Nine-axis gyroscope and Magneto resistive sensor, by combining the assist of history information of wheel count. We secondly UKF filter this trajectory with the help of INS position. Finally, we compare the algorithm with the Geomagnetic matching navigation method based on correlation coefficient matching algorithm (CC) and EKF filter processing[14].

The calculation formula of the correlation coefficient method can be seen as:

$$CC(X,Y)=\frac{\sum_{i=1}^{N}(x_i-\bar{x})(y_i-\bar{y})}{\sqrt{\sum_{i=1}^{N}(x_i-\bar{x})^2}\sqrt{\sum_{i=1}^{N}(y_i-\bar{y})^2}} \tag{28}$$

The correlation coefficient method is high-precision but complex. Therefore, segment processing is necessary while concerning time complexity. geomagnetic matching navigation and positioning trajectories calculated by different methods can be seen from figure 7.

As can be seen from figure 7, the red curve formed by the symbol "*" stands for the actual trajectory of the cruising car, the blue curve formed by the symbol "×" stands for the trajectory of the unfiltered geomagnetic matching integrated navigation based on DDTW, the green curve formed by the symbol "+" stands for the trajectory of the filtered integrated navigation based on CC and EKF, the purple curve formed by the symbol "⋄" stands for the trajectory of the filtered geomagnetic matching integrated navigation based on DDTW and UKF, which is proposed by thesis. figure 8 uses the same curve reference relationship.

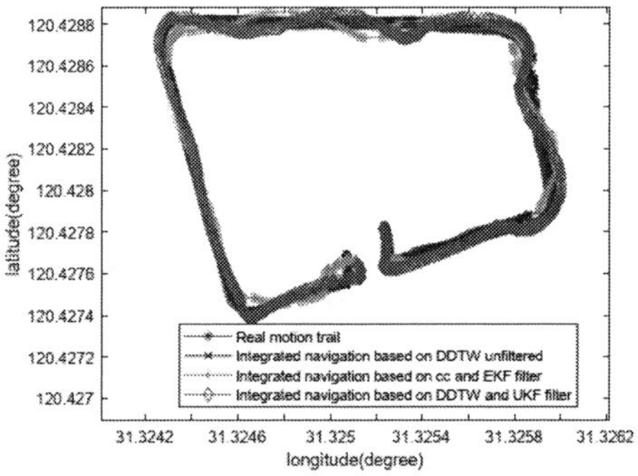

Figure 7. Different geomagnetic matching combined navigation and positioning trajectories.

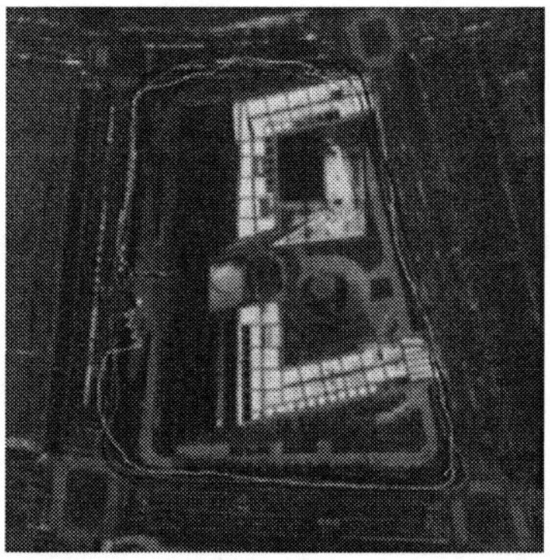

Figure 8. Navigation and positioning track with different methods on Google map.

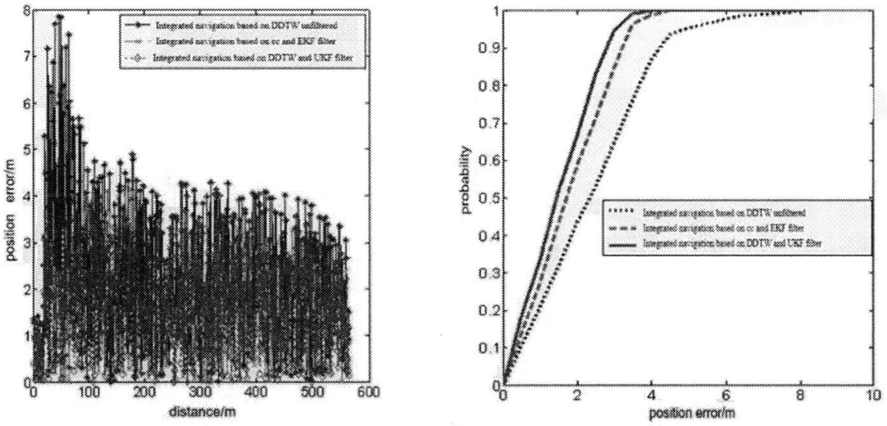

(a) (b)

Figure 9. Comparison of positioning errors of different geomagnetic matching combined navigation methods.

Table 2. Comparison of positioning errors of different geomagnetic matching combined navigation methods.

positioning error	1m	2m	3m	4m	average accuracy
integrated navigation based on DDTW unfiltered	23.76%	44.33%	67.73 %	90.07 %	2.31m
integrated navigation based on CC and EKF filter	27.13%	59.04%	84.75 %	98.76 %	1.83m
integrated navigation based on DDTW and UKF filter	34.93%	66.13%	95.92 %	100%	1.51m

As can be seen from figure 9, the maximum positioning error of the unfiltered geomagnetic matching integrated navigation based on DDTW is 7.857m, that of the filtered integrated navigation based on CC and EKF is 4.1897m, while that of the filtered geomagnetic matching integrated navigation based on DDTW and UKF is 3.6586m. As can be seen from table 4.2, the positioning error of the unfiltered geomagnetic matching integrated navigation based on DDTW ranges from 0 to 4m with 90.07% and average is 2.31m. the positioning error of the filtered integrated navigation based on CC and EKF ranges from 0 to 3m with 84.75% and average is 1.83m. the positioning error of the filtered geomagnetic matching integrated navigation based on DDTW and UKF ranges from 0 to 3m with 95.92% and average is 1.51m. In 2018, Namkyoung Lee et al[15]. used AMID and obtained a positioning accuracy of 0.8m in the corridor and 2.3m in the atrium.

From above experimental comparison, we see the ways, properly compensating and suppressing the errors, optimizing geomagnetic matching method to suppress the effect of local geomagnetic field interference, can be effective while many outdoor disturbing factors including complex terrain affect the accuracy of INS and geomagnetic matching. It can be seen that the filtered geomagnetic matching integrated navigation based on DDTW and UKF which is proposed by thesis has the feature of good accuracy.

5. Conclusion

Complex outdoor environment brings interference to INS and geomagnetic field. To solve this problem, we design a specific filter through using fusion positioning method to analyze and select filters. The state variables of the filter mainly contain initial positioning error of latitude and longitude, while the observed variables contain the difference between two results respectively calculated by geomagnetic matching method and INS method. The error from INS method can be fixed through using UKF filter to estimate error from the integrated navigation. In thesis, a filtered geomagnetic matching integrated navigation based on DDTW and UKF is proposed, which has the feature of good accuracy within 1.5m.The positioning and navigation scenarios studied in this paper are mainly for the study of geomagnetic matching navigation and positioning in local area scenarios. There are still deficiencies in the wide range of applications such as drones and underwater navigation, and the accuracy of the discrete database in the digital map needs to be further improved. . The research on the error correction of the geomagnetic navigation module itself is not deep enough, and it is hoped that a better method can be used to correct the error of the geomagnetic navigation module itself to improve the accuracy of geomagnetic information sampling. The algorithm we propose belongs to a gear that drives the times. Any drone or autonomous vehicle that uses this algorithm to operate will benefit. In the establishment of a big data network and the realization of smart city, the beneficial effect of any step is the meaning of our work.

References

[1] HUANG Liping. Research on inertial/geomagnetic integrated navigation matching algorithm [D]. Harbin Institute of Technology, 2017.
[2] LOU Xizhong, FANG Jun, ZHI li. An indoor magnetic navigation method based on Viterbi algorithm,2016,27(04):429-434.
[3] S. Wang et al., "A Data Fusion Method of Indoor Location Based on Adaptive UKF," 2018 7th International Conference on Digital Home (ICDH), Guilin, China, 2018, pp. 257-263.
[4] M. Wu and Jian Yao, "Adaptive UKF-SLAM based on magnetic gradient inversion method for underwater navigation," 2015 International Conference on Unmanned Aircraft Systems (ICUAS), Denver, CO, 2015, pp. 839-843.
[5] ZHOU Jiapeng, WANG Yunjia, LI Xin, CAO Xiaoxiang, CAO Hongji. Research on geomagnetic indoor positioning technology [J].Bulletin of Surveying and Mapping,2019(01):18-22.
[6] Yao Zhao, Lei Lin, Wei Lu and Yu Meng, "Landsat time series clustering under modified Dynamic Time Warping," 2016 4th International Workshop on Earth Observation and Remote Sensing Applications (EORSA), Guangzhou, 2016, pp. 62-66.
[7] QIAO Meiying, LIU Yuxiang, TAO Hui. A similarity metric algorithm for multivariate time series based on information entropy and DTW [J]. Journal of Shanxi Agricultural University (Natural Science Edition),2019,58(02):1-8.
[8] HUA Xirui, Ll Zhongqin, Ll Zhenchang, KOU Ruixiong. Application of generalized extension interpolation method in BDS precise clock correction [J]. GNSS World of China,2019,44(04):96-101.
[9] ZHANG Jie, MA Guanyi, Ll Jinghua, WANG Xiaolan. Generalized extended approximation GNSS code discriminator algorithm with maximum constraint [J]. Journal of Systems Engineering and Electronics,2017,(04):714-720.
[10] J. Lee, J. Lim and J. Lee, "Compensated Heading Angles for Outdoor Mobile Robots in Magnetically Disturbed Environment," in IEEE Transactions on Industrial Electronics, vol. 65, no. 2, pp. 1408-1419, Feb. 2018.
[11] X. Shuiqing, L. Xiangyuan, L. Zhengyi, Z. Hui and Y. Dandan, "Design of low cost pedestrian location system based on inertial navigation," 2017 International Conference on Progress in Informatics and Computing (PIC), Nanjing, 2017, pp. 471-475.
[12] Liang Ying. Research on 'Technologies of INS/Geomagnetic Matching Integrated Navigation System [D]. Harbin: Harbin Engineering University,2010.
[13] S. Wang et al., "A Data Fusion Method of Indoor Location Based on Adaptive UKF," 2018 7th International Conference on Digital Home (ICDH), Guilin, China, 2018, pp. 257-263.
[14] JI Xinchun, WEI Dongyan, LI Wen, GONG Peiwen, LAI Qifeng, YUAN Feng. A vehicle positioning system and method assisted by mileage information and geomagnetic matching [P]. CN107907134A,2018-04-13.
[15] Lee Namkyoung, Ahn Sumin, Han Dongsoo. AMID: Accurate Magnetic Indoor Localization Using Deep Learning[J]. Sensors, 18(5):1598-.

HISAT2 Parallelization Method Based on Spark Cluster

Jiannan Guo, Jing Gao* and Zhenyu Liu

[1]College of Computer and Information Engineering, Inner Mongolia Agricultural University, Hohhot, 010018, Inner Mongolia Autonomous Region, China
[2]Inner Mongolia Autonomous Region Key Laboratory of Big Data Research and Application of Agriculture and Animal Husbandry, Hohhot, 010018, Inner Mongolia Autonomous Region, China

*Corresponding author email: gaojing@imau.edu.cn

Abstract. Sequence alignment is one of the most important components in the Bioinformatics research field. It is of great significance to discover the functional structure and genetic information of nucleic acids and protein. With the rapid development and gradual maturity of high-throughput sequencing technology, the scale of gene data which have been discovered by that is going to increasingly large. Due to the gene sequence alignment calculation has high complexity and the sequencing gene data has large scale, the process of comparison computing will cause a plenty waste of computing time. HISAT2 is the one of most popular sequence comparison software. HISAT2 has better sensitivity and accuracy than other software, at the same time, the speed of process also has highly improved. According to those reasons, this passage implements the HISAT2 parallelization method based on Apache Spark cluster. Through the comparison experiment between single and cluster machine, the parallelization computing speed of HISAT2 parallelization method based on Spark cluster has increased obviously to 3.69 times, with the high rate of accuracy meanwhile.
Keywords: Spark, Big-data, Parallelization, HISAT2, Gene sequence alignment.

1. Introduction
With the development of high-throughput sequencing technology[1], the scale of gene sequencing data is getting larger. From the first generation of sequencing technology in the 1970s, the read length of the base sequence can reach 1000bp, to the mature high-throughput sequencing technology, the read length can be increased to more than 10,000, which is an order of magnitude leap, which makes the scale of bioinformatics data has greatly increased.
In the analysis of gene sequencing data, gene sequence comparison is one of the important links. Currently, commonly used gene sequence comparison software are: HISAT2[2], GSNAP10[3], TopHat2[4], STAR[5], Blast[6], Blat[7], BWA[8], Clustal[9], Muscle[10], Bowtie[11]. As one of the currently popular sequence comparison software, HISAT2 uses the improved BWT algorithm to obtain higher sensitivity and operation speed. Compared with programs of GSNAP10 and TopHat2, HISAT2 has a much faster processing speed than GSNAP10 and TopHat2. Compared with STAR11 and GSNAP10, HISAT2 has non-super-high sensitivity and can basically find all matching sequences. Although HISAT2 is slower than STAR11 in processing speed, the accuracy and sensitivity of HISAT2 are much higher than STAR[12]. The BLAST algorithm analysis speed is slow, the comparison result is not intuitive enough, the comparison can not display gene introns, exons and gene location. The Blat algorithm is not accurate when used in the comparison of nucleic acid sequences between distantly related species, and it produces too many meaningless sequence comparison fragments when repeatedly searching for

short matching fragments. BWA contains three different algorithms for the comparison of long and short clips, but the quality of the comparison will be reduced when multiple points are matched in the comparison. Clustal comparison adopts a progressive comparison method, which leads to the inability to guarantee the optimal comparison. Muscle uses an iterative method for comparison, and continuously uses dynamic programming algorithm rearrangement to achieve comparison, so it has extremely high memory requirements and low accuracy. Bowtie is more suitable for short sequence alignment tasks. In summary, the current gene sequence comparison tools mainly suffer from the problems of long time and low accuracy. Compared with other comparison software, HISAT2 has faster processing speed and higher accuracy, and the index of HISAT2 occupies less memory space than other software, and can support the operation of gene sequencing data of different lengths.

Since the gene sequence alignment process is very time-consuming, reducing time consumption has become one of the important issues that need to be resolved. There are currently three commonly used optimization methods. The first one is a single-machine optimization solution based on GPU, the second one is a MapReduce optimization solution based on Hadoop, and the third one is an optimization solution based on Spark clusters.

The first is a single-machine optimization solution based on GPU. With the concept of GPU[13] and the development of graphics processors, more and more industry projects combine GPU to improve the computing efficiency of tasks. In 2017, Si Yujia implemented a parallelized algorithm for profile-profile sequence alignment based on GPU[14]. By using a data set containing 53,646 sequences to test the algorithm, he analyzed the influence of the number of sequences participating in the sequence alignment, the sequence length and the assigned block on the acceleration effect. The results show that the running time of the algorithm on the GPU is on average 10 times that on the CPU, up to 23 times. In 2019, Chan Yuandong based on the many-core architecture Xeon Phi designed and implemented an efficient parallel DNA sequence comparison algorithm MyPhi[15], the Institute of Industry and Commerce East proposed a new filtering algorithm PUNAS, and gave a scalable platform based on multi-core and many-core platforms. Implementation of parallelization. In 2019, Zhu Xiangyuan solved the problem of large-scale sequence alignment based on multi-core computers and CPU+GPU heterogeneous systems, and proposed a general sequence alignment algorithm (DCPA) based on the divide-and-conquer method[16]. The experimental results show that compared with the classic sequence alignment algorithm MUSCLE, DCPA has obtained nearly 81 times the performance acceleration, and maintains better alignment accuracy. In 2019, Qiu Shenglin and Liu Chengzhong optimized the parameters of the support vector machine through the parallel method of GPU and Drosophila optimization algorithm[17], and performed comparison and classification experiments on gene sequences. In summary, although the GPU-based stand-alone optimization effect is very significant, the demand for hardware is also higher. Due to the high cost of GPU servers, general research institutions and laboratories cannot afford it, so acceleration by GPU is not a popular method.

The second scheme is based on H0adoop MapReduce optimization scheme. The popularity of Apache Hadoop[18] cluster and its distributed computing framework MapReduce[19] has given parallelization a new direction. Halvade[20] is a parallel processing framework for sequencing data based on MapReduce, which implements the sequence alignment process in the map stage. Through experiments on a cluster of 15 nodes with 360 CPU cores, and an Amazon Elastic MapReduce cluster of 16 worker nodes with 512 CPU cores, the computing time can be shortened from 2.5 to 5 days in a single-machine multi-threaded mode to 3 hour. In 2015, Meng Ming built a 16-node Hadoop environment on the vSphere virtualization platform, and realized the Blast algorithm based on MapReduce distributed parallelization through three methods[21]. The final results showed that MapReduce can effectively improve computing efficiency. In 2019, Wang Xiaoxia and Sun Decai proposed a local similarity connection algorithm based on Q-sample by using the MapReduce framework. The local similarity connection speed of this algorithm on large data sets is faster than that of LS-Jion, thereby optimizing the comparison of gene sequences. Locally similar connection operation[22]. In 2020, Wang Xiaoxia and Sun Decai proposed a self-join algorithm based on the MapReduce framework[23], which solved the local similar self-join localization problem. Experimental results show that the speed of self-join in this way is faster than the LS-Join algorithm.

The MapReduce framework has the characteristics of strong applicability and wide application fields. It can usually be used to process large-scale data. However, the MapReduce framework still has problems such as high I/O overhead, high calculation delay, and limited and inflexible programming methods.

The third solution is an optimization solution based on the Spark cluster. The AMP (Algorithms Machines and People) laboratory of the University of California at Berkeley proposed the Spark computing framework[24]. As a memory-based distributed computing framework, it can reduce I/O overhead very well. By using RDD to cache data across cluster nodes, which is versatile and easy to operate[25]. In 2016, Abuın J.M. et al. proposed SparkBWA[26]. Improve the performance of Burrows-Wheeler Aligner (BWA) by using big data technology Apache Spark technology. BWA is a software that reads DNA sequences into a reference genome. BWA is composed of optimized algorithms and is mainly used to process short-read comparison tasks. On the premise of not changing the source code, the designed SparkBWA can improve computing efficiency and has better scalability and ease of use while ensuring accuracy. By selecting the experimental data sets of 3.4GB, 11.8GB, and 48.3GB, carried out the comparison experiments of BWA, pBWA, SEAL, BigBWA, and Halvade, it was found that SparkBWA runs 1.9 times and 1.4 times faster than SEAL and pBWA. Compared with BigBWA and Halvade, the average speed is up to 1.4 times. In 2017, de Castro MR, Tostes CD, Davila AMR and others used the computing resources provided by cloud computing and the Apache Spark computing framework to speed up the process of gene similarity analysis of radionuclide-resistant bacteria, and proposed a Spark-based SparkBLAST algorithm[27]. Mainly through comparative experiments with CloudBLAST, two different comparative experiments were carried out by using the number of nodes and the number of cores as variables, and the data sets of 805M and 11G were selected for experiments. When two experiments are with 1 core and 2 cores, the acceleration effect of SparkBLAST is not as good as that of CloudBLAST. But when the number of cores continues to increase, the acceleration effect of SparkBLAST is significantly better than CloudBLAST. Finally, an acceleration of 41.78 was achieved by deploying a high distribution of 64 nodes and 13 partitions. In 2017, Zhou W, Li RL, Yuan S and others proposed MetaSpark[28] due to it is difficult for traditional biological information analysis tools to process massive data. By selecting 1 million reads as the experimental data set and 0.75GB of reference genome as the index, the experiment found that compared to SOAP2, BWA, and BLAST, MetaSpark can get more reads, and the number of reads read increased by 4%. Different tests show the scalability and high-performance advantages of MetaSpark. In 2017, Dong Gaifang, Fu Xueliang, and Li Honghui parallelized the design and implementation of the improved star comparison algorithm in the Spark cluster[29], using Spark Yarn-Client to run experiment on human mitochondrial data, and improve the performance of the algorithm. They also analyze and improve the performance of the algorithm. In 2019, Liu Zhenyu realized the parallelization of the Spark_BLAST comparison method in a 5-node cluster[30]. Through comparative experiments, it was proved that the speedup ratio of the Spark_BLAST algorithm can reach about 4 without changing the accuracy of the comparison results. In 2019, Li Leixiao, Liu Yanfeng, and Gao Jing improved and optimized the gene sequence alignment algorithm Smith-Waterman, and carried out a parallel design based on the Spark platform[31]. The experimental results show that implementing the optimized algorithm through a parallel method can greatly improve the running speed and scalability of the algorithm under the premise of ensuring the accuracy.

Compared with the GPU-based single-machine optimization solution, the Spark cluster-based optimization solution consumes less cost. Compared with the MapReduce optimization solution based on Hadoop, the optimization solution based on the Spark cluster has higher processing efficiency. The Spark cluster implements an efficient DAG execution process, which can efficiently process data through memory.

The comparison of gene sequences is an important step in the field of biological information. It is very important to shorten the processing time to obtain effective information, and the optimization solution based on the Spark cluster solves the problem of time-consuming comparison. Therefore, this paper chooses to parallelize the HISAT2 gene sequence alignment based on the Spark distributed framework.

2. HISAT2 Parallelization Method

This method implements HISAT2 parallelization method based on Spark cluster. The parallelization process is divided into two stages, data preparation stage and the parallel computing stage.

In the data preparation stage, the gene sequencing data in fastq format is copied to each worker node of the cluster and stored in the same path. The reference genome file (HISAT2 index) is constructed by the combining of HISAT2 algorithm and the fna file. Then the remote copy mechanism of the cluster is used to distribute the reference genome file to each worker node, as shown in Figure 1. Ensuring the path stored on each worker node is the same.

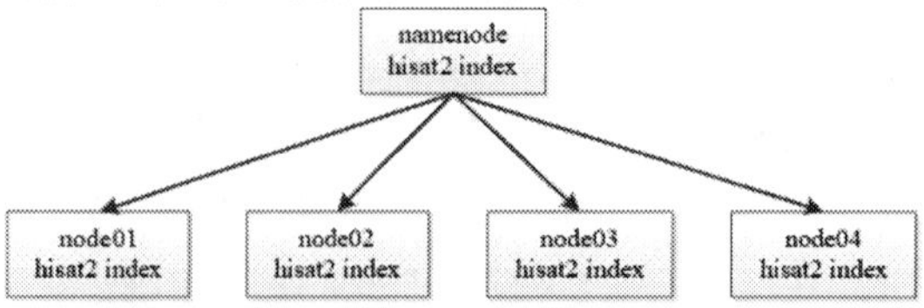

Figure 1. Distribution of HISAT2 index.

In the parallel computing stage, there are mainly 4 parts: tasks reading and division, tasks and resource allocation, parallel calculation, result combined and output.

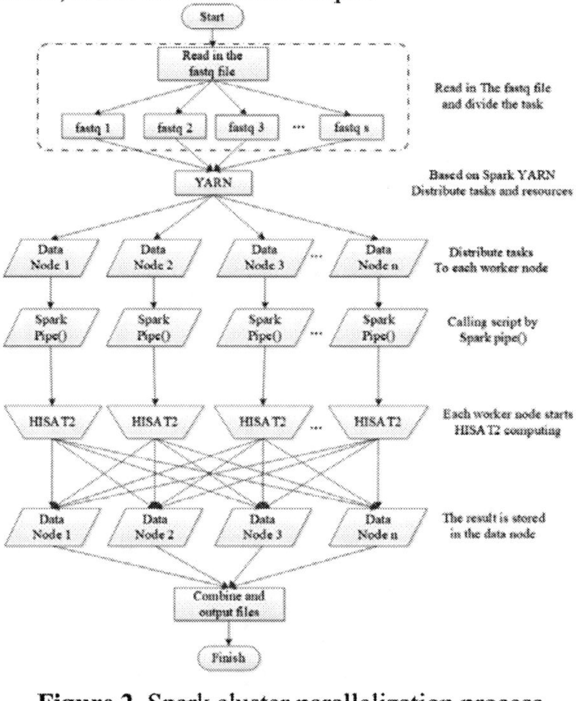

Figure 2. Spark cluster parallelization process.

As shown in Figure 2, the Spark cluster first generates an RDD through the makeRDD() operator, and the RDD contains the fastq format gene sequencing data content. In fastq gene sequencing data, each gene is divided by a specific identifier "@". Each gene consists of 4 lines of character strings. The contents of the 4 lines of character strings are the sequence ID, sequence information, sequence name and sequence quality information. Next, divide the read fastq file into S tasks. In general, in order to maximize the utilization of the cluster and prevent resource waste due to uneven task allocation, the number of division S is the multiple of worker nodes in the cluster. When the tasks are divided, the total number of genes can be equally divided by the number of tasks, the calculation formula for the number of genes in each task is shown in formula 1.

$$N = M / S \tag{1}$$

In the formula, M represents the number of genes in the experiment file, S is the number of divided tasks, N is the number of genes in each divided task. When the total number of genes cannot be divided equally by the number of divided tasks, the formula for calculating the number of genes in each divided task is shown in formula 2.

$$N = \begin{cases} floor(M / S) & N \leqslant R \\ floor(M / S) + 1 & N > R \end{cases} \quad (2)$$

In the formula, R is the remainder of M/S, and floor() is rounded down. It can be seen that the number of genes in the No.1 task to No. R task are "floor(M/S)+1". The number of genes in the No.(R+1) task to the No. S task are "floor(M/S)".
In the process of task and resource allocation, this experiment uses Hadoop Yarn resource manager. The Spark Yarn model has the advantage of flexible resource management, which can adjust resource allocation according to the current application pressure. The Spark cluster submits tasks through the Spark-submit command, and uses parameters to determine the start-up mode of the Spark cluster. In the Spark-submit command, set the parameter "--master" to Yarn-client. After starting the Yarn-client mode, the resource manager distributes tasks and resources to each worker node to prepare for the parallel computing steps.
In the process of parallel computing, after each worker node receives the execution command, it calls the HISAT2 algorithm script through the pipe() operator to start the computing task. The content of the HISAT2 algorithm script is the execution command of HISAT2, and the command content is as follows: "hisat2 -p *p -x /*/genome -U *U.fastq -S *S.sam". Among the command parameters, the "-p" parameter sets the number of threads started by the task. The "-x" parameter specifies the specific path of the reference genome genome file. The "-U" parameter specifies the path of single-ended fastq sequencing data. The "-S" parameter specifies the save path of the generated sam file. When the parallel calculation is completed, multiple sam files are generated in each worker node, and the multiple result files are transmitted to the specified path of the specified node through the "scp" command to obtain the final result file directory.

3. Text Environment

3.1. Configuration of Single Machine and Spark Cluster
In this experiment, the VMware virtualization platform of the big data processing center laboratory is used as the basic condition of the experiment. The cluster is composed of 5 virtual machine nodes, one node is the master node of Spark, and the other 4 nodes are used as the Spark worker node. The specific information of the cluster environment is shown in Table 1. The single machine configuration is the same as the Spark cluster.

Table 1. Spark cluster configuration information.

Name	System	RAM	CPU	HISAT2	Hadoop	Spark
Namenode	CentOS7	128	4	2.0.4	2.7.2	2.1.0
Node01	CentOS7	128	4	2.0.4	2.7.2	2.1.0
Node02	CentOS7	128	4	2.0.4	2.7.2	2.1.0
Node03	CentOS7	128	4	2.0.4	2.7.2	2.1.0
Node04	CentOS7	128	4	2.0.4	2.7.2	2.1.0

3.2. Experimental Data Source
To make sure the reliability of the experiment, the source of the experimental data is the open-source human gene data released by NCBI (National Center for Biotechnology Information). Five single-ended sequencing data of human genes of different sizes were selected as the experimental objects, and the experimental data was decompressed into fastq format.
The length of each sequence in the experimental data directly affects the calculation efficiency, the sequence length of the experimental data is controlled between 100 and 101. The specific data information is shown in Table 2. Selecting human reference genome file, which name is "GCF_002742125.1_Oar_rambouillet_v1.0_genomic.fna". The data size is about 2.7GB.

Table 2. Gene sequencing data information.

No.	Name of data	Size of file(MB)	Number of bases
1	SRR15204114	124.8	101
2	SRR2172077	411.2	101
3	ERR4057573	1228.8	101
4	ERR4057635	4096	101
5	ERR1078327	7270.4	100

4. Stand-alone Experiment

First, this experiment analyzes the relationship between threads and running speed. Selecting the SRR2172077 (411.2MB) file as the input data. When processing the sequencing data with dufferebt threads, the information of CPU usage, running time and speed are shown in Table 3.

Table 3. Single machine multi-thread running.

Number of Threads	1	2	4	6	8
CPU Usage (%)	100	200	398.8	398.9	396.4
Time (s)	1106	538	259	257	263
Speed (MB/s)	0.4	0.82	1.72	1.72	1.68

When the number of thread is 1, the CPU usage is 100%, and the speed is 0.4MB/s. With the increase of threads, when the number of thread is 4, the CPU usage reaches 398.8%, and the speed reaches 1.72MB/s. At this time, the utilization of CPU resources reaches the maximum. When the number of threads exceeds 4, the CPU usage does not increase significantly, and the running time and speed are basically stable. As shown in Figure 3, within the range of CPU resources, when a single task is processed, the thread is in direct proportion to the running speed. With the number of threads increase, the CPU usage and the running speed are increasing. When the upper limit of the single machine resource is reached, continue to increase the number of threads, the CPU usage does not change significantly, and the running speed tends to stabilize. Therefore, within the scope of CPU resources, when processing a single task, increasing the thread of the processing task can improve the operating efficiency. When the thread is set to 4, the best operating efficiency can be achieved, and the operating efficiency of continuing to increase the thread remains basically unchanged.

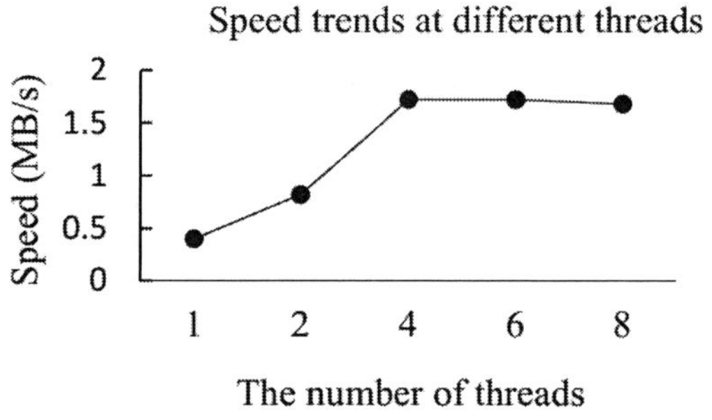

Figure 3. Speed trend at different threads.

During the experiment, the number of tasks running at the same time is also an important factor that affects the results of the experiment. Select the SRR2172077 (411.2MB) file to copy, and generate 5 task sets which is combined with different number of tasks, the number of tasks sets are 1, 2, 4, 6, 8. Set the threads of the calculation task to 1. CPU usage, running time and speed are shown in Table 4.

Table 4. Single machine multi-task running.

Number of tasks	1	2	4	6	8
CPU Usage (%)	100	200	398.8	397.4	396.4
Time (s)	1106	1091	1103	1552	2265
Speed (MB/s)	0.4	0.81	1.60	1.59	1.56

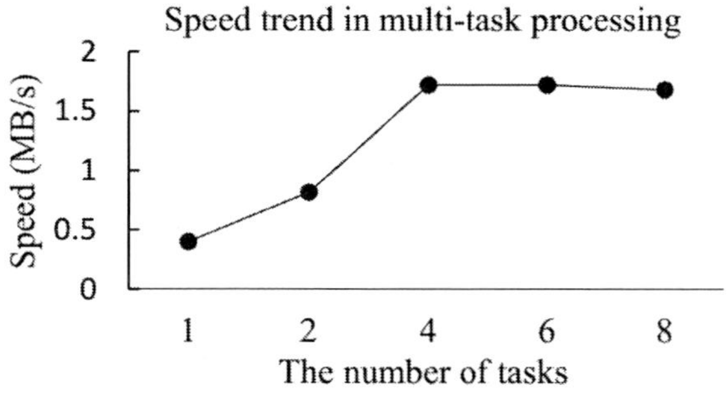

Figure 4. Speed trend in multi-task processing.

When processing 1 task, the CPU usage is 100%, and the speed is 0.4MB/S. With the increase in the number of processing tasks, when processing 4 tasks at the same time, the CPU usage reaches 398.8, the processing speed at this time is 1.6MB/s, and the utilization of CPU resources reaches the maximum. Continuing to increase the number of tasks, the CPU usage did not increase significantly, and the running speed was basically stable. As shown in Figure 4, within the range of CPU resources, when the threads are set to be the same, the number of parallel tasks is proportional to the running speed. As the number of parallel tasks increases, the CPU usage and the running speed are both increasing. After reaching the upper limit of single machine resources, continue to increase the number of parallel tasks, the CPU usage does not change significantly, and the running rate tends to stabilize. Therefore, within the range of CPU resources and when the thread used by the task is 1, increasing the number of processing tasks can improve the operating efficiency. When 4 tasks are processed in parallel, the best operating efficiency can be achieved, and the operating efficiency will not continue to be affected.

During the experiment, the number of running tasks and the number of using threads are both important factors that affect the efficiency of the operation. The SRR2172077 (411.2MB) file is selected, and a total of 9 experiments are carried out by setting different numbers of parallel tasks and threads. The running speed is as shown in the table. 5.

Table 5. Impact of the number of tasks and threads on rate.

Number of Tasks \ Speed (MB/s) \ Number of threads	1	2	4
1	0.4	0.82	1.72
2	0.81	1.57	1.60
4	1.60	1.58	1.61

When processing 1 task with 4threads, the efficiency of the task is best, and the speed can reach 1.72MB/s. Compared processing 2 tasks with 2 threads and 2 tasks with 4 threads, the computational efficiency is close, respectively, 1.57MB/s and 1.6MB/s. When 4 tasks are processed in parallel, the operating speed remains basically unchanged at 1.6MB/s. Therefore, in the range of CPU resources, increasing the number of tasks and the number of threads can effectively improve the efficiency of the comparison process, and the acceleration effect is basically the same.

A stand-alone experiment was performed on 5 gene sequencing data of different sizes, and the running time is shown in Table 6.

Table 6. Stand-alone processing time.

No.	1	2	3	4	5
Time (s)	157	259	432	1273	2353

5. Spark Cluster Experiment

In the Spark cluster experiment, the number of divided tasks is an important factor that affects the efficiency of the cluster. Copy the SRR3508721 file (411.2MB) to get different numbers of task sets, the number of task sets are 1, 2, 4, 6, 8. When the Spark cluster processes the comparison task, the thread is set to 4, and the running time and speed are shown in Table 7.

Table 7. Time and rate of the cluster processing multi-tasks.

Number of tasks	1	2	4	6	8
Time (s)	157	161	170	339	354
Speed (MB/s)	2.62	5.11	9.68	7.78	9.29

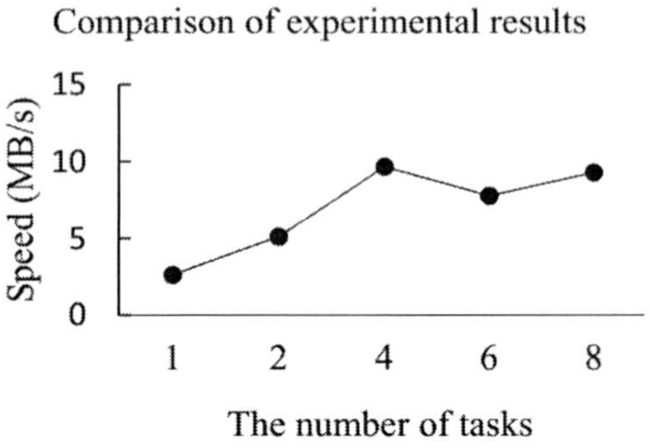

Figure 5. Spark Cluster multi-task rate trend.

As shown in Figure 5, as the number of tasks increases, the processing speed of the Spark cluster continues to increase. When 4 tasks are processed in parallel, the speed reaches 9.68MB/s. When processing 6 tasks, the processing speed drops to 7.78MB/s, and when processing 8 tasks, the processing speed increases to 9.29MB/s. When the number of parallel tasks is not evenly divisible by the number of cluster worker nodes, there will be a problem of uneven task distribution, resulting in a decrease in the running speed of the cluster. In this experiment, the Spark cluster is composed of 4 worker nodes. When the number of tasks is an integral multiple of the number of worker nodes, the utilization of cluster resources reaches the maximum. Therefore, in the Spark cluster experiment, the task is divided into 4 for parallelization, which can improve the operating efficiency to the best.

The Spark cluster experiment was performed on 5 different sizes of sequencing gene data, and the running time is shown in Table 8.

Table 8. Spark cluster processing time.

No.	1	2	3	4	5
Time (s)	50	110	195	426	636

The Spark cluster took 50 seconds to process the SRR15204114 file, and 636 seconds to process the ERR1078327 file. Compared with the stand-alone experiment, the running time is significantly reduced. Therefore, processing large-scale data through a Spark cluster can effectively reduce time consumption.

6. Comparative Experiment

Combining the running time of the stand-alone machine and the Spark cluster in different files, the running speed is further calculated, and the information obtained is shown in Table 9.

Table 9. The time and rate of contrast experiment.

No.	1	2	3	4	5
Time of Stand-alone (s)	76	216	432	1273	2353
Time of Spark cluster (s)	50	110	195	426	636
Speed of Stand-alone (MB/s)	1.64	1.90	2.84	3.21	3.10
Speed of Spark cluster (MB/s)	2.50	3.74	6.30	9.62	1143

When processing SRR15204114 files, the Spark cluster running time is 26 seconds faster than stand-alone machine. With the increase of gene sequencing data files, when processing ERR1078327 files, the Spark cluster running time is 1717 seconds faster than a stand-alone machine, and the single machine's computing speed has increased from 1.64MB/ to 3.21MB/s, and the computing speed of the cluster can be increased from 2.5MB/s to 11.43MB/s. Therefore, the computing efficiency of the Spark cluster is higher than single machine when processing large-scale data sets.

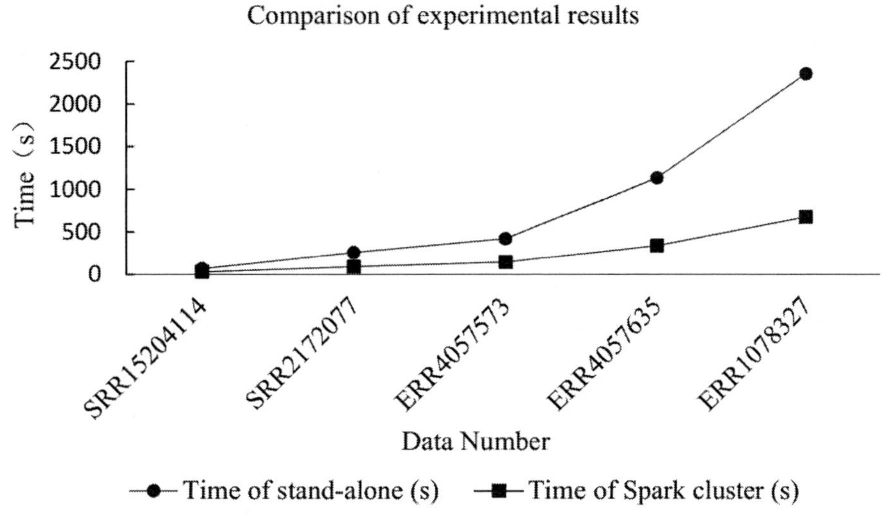

Figure 6. Comparison of experimental results.

As shown in Figure 6, when the processed data is small, the processing time of a single machine and a Spark cluster is almost the same. As the data increases, the processing time of the Spark cluster is significantly lower than single machine. Therefore, when processing small data, the acceleration effect of Spark cluster is not obvious, and when processing larger data, the Spark cluster processing method has more advantages.

Calculating the acceleration ratio of the Spark cluster with running different sequencing data, the information is as shown in Table 10.

Table 10. Spark cluster acceleration ratio.

No.	1	2	3	4	5
Acceleration ratio	1.52	1.97	2.21	3.0	3.69

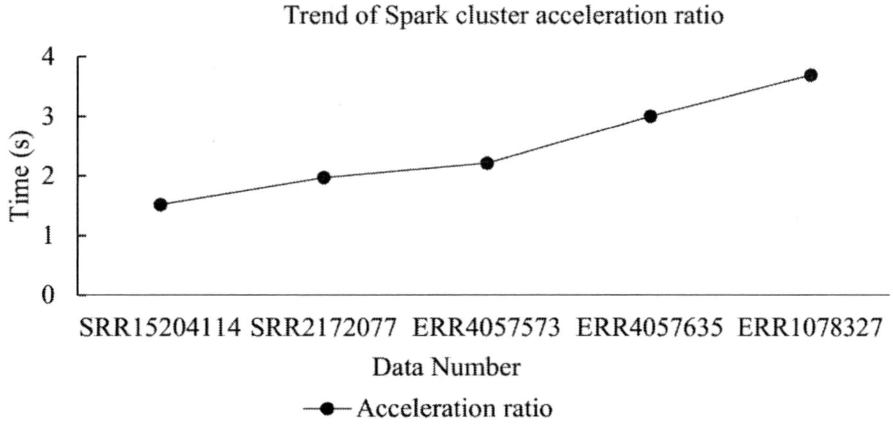

Figure 7. Trend of Spark cluster acceleration ratio.

As shown in Figure 7, with the increase of the experimental data, the acceleration ratio has increased. When calculating the ERR1078327 file, the acceleration ratio can reach 3.69, which verifies that the computing efficiency of the Spark cluster is better than that single machine.

7. Conclusion

In view of the current time-consuming problem of gene sequence comparison, this paper implements a parallel processing method of HISAT2 based on Spark cluster, and draws conclusions based on the results of comparative experiments. On the premise of not affecting the accuracy, compared with the stand-alone processing method, the processing method based on the 5 nodes Spark cluster can increase the acceleration ratio to 3.69.

Because the cluster start-up and the distribution of resource tasks have time consuming, the speed advantage is not obvious when processing small-scale data, which is similar to the processing time of a single machine. When processing large-scale data, the computational efficiency is significantly improved. At the same time, we analyzed the computing efficiency impact of the number of divided-tasks. By using 5-node cluster which is composed of 1 master node and 4 worker nodes, when the number of submitted tasks is an integral multiple of 4, the calculation efficiency is the best. In the field of bioinformatics, gene sequencing data is mostly large-scale data. Therefore, this article implements the HISAT2 parallelization method based on the Spark cluster, which can effectively solve the time-consuming problem.

Acknowledgments

When this paper is completed, I first express my heartfelt gratitude and sincere respect to the instructor Gao Jing. Teacher Gao Jing's persevering scientific research spirit, profound knowledge, fact-seeking work attitude, magnanimous and modest behavior, and selfless dedication have always inspired me to continuously learn knowledge and pursue the realm of life. In addition, I would also like to thank Dr. Liu Zhenyu for his patient guidance and put forward many useful suggestions for modification during the work process.

At the same time, I would also like to thank the following laboratories and projects for supporting this experiment, they are Inner Mongolia Autonomous Region Key Laboratory of Big Data Research and Application of Agriculture and Animal Husbandry, Inner Mongolia Natural Science Foundation-Research on efficient parallelization algorithms and software for mutation detection and transcriptome differential expression analysis of whole-genome resequencing data(2019MS03014), Development and application of private cloud operating system based on OpenStack(2019ZD15), Inner Mongolia Autonomous Region Science and Technology Major Project-Cloud Computing Application Technology Research and Development(2019ZD016), Inner Mongolia Autonomous Region Science and Technology Planning Project-Research and Application of Key Technology of Discipline Inspection and Supervision Big Data(2019GG372), Inner Mongolia Autonomous Region Science and Technology Department-Mongolian Plateau livestock genetic resource database and information platform construction and germplasm resource development and utilization(2020ZD0007).

There are too many people who have helped me. I sincerely thank all those who care about and help me. Thank them for giving me the power to overcome myself, surpass myself, and keep pursuing. Thank you.

References

[1] Wang X C, Yang Z R. High-throughput Sequencing Technology and Its Application[J]. China Biotechnology, 2012, 32(1):109-114.
[2] D Kim, Langmead B, Salzberg S L. HISAT: A fast spliced aligner with low memory requirements[J]. Nature Methods, 2015, 12(4):357-360.
[3] FA Buske, French H J, Smith M A, et al. NGSANE: A Lightweight Production Informatics Framework for High Throughput Data Analysis[J]. Bioinformatics, 2014, 30(10):1471-1472.
[4] Kim D, Pertea G, Trapnell C, et al. TopHat2: accurate alignment of transcriptomes in the presence of insertions, deletions and gene fusions[J]. Genome Biology, 2013, 14(4): r36.1-13.
[5] Valencia A. STAR: ultrafast universal RNA-seq aligner[J]. Bioinformatics, 2014, 22(1):1.
[6] Altschul S F. Basic local alignment search tool (BLAST)[J]. Journal of Molecular Biology, 2012, 215(3):403-410.
[7] Kent W J. Blat-the BLAST-like alignment tool[J]. Genome Research, 2001, 12(4):656-664.
[8] Heng,Li,Richard,et al. Fast and accurate short read alignment with Burrows-Wheeler

transform.[J]. Bioinformatics, 2009, 25(14):1754-1760.
[9] Thompson J D, Higgins D G, Gibson T J. CLUSTAL W: improving the sensitivity of progressive multiple sequence alignment through sequence weighting, position-specific gap penalties and weight matrix choice[J]. Nucleic Acids Research, 1994, 22(22):1673-1680.
[10] Edgar R C. MUSCLE: multiple sequence alignment with high accuracy and high throughput[J]. Nucleic Acids Research, 2004,32(5):1792-1797.
[11] Langmead B. Ultrafast and memory-efficient alignment of short DNA sequences to the human genome[J]. Genome biology, 2009, 10(3):1-10.
[12] Sahraeian S, Mohiyuddin M, Sebra R, et al. Gaining comprehensive biological insight into the transcriptome by performing a broad-spectrum RNA-seq analysis[J]. Nature Communications, 2017, 8(1):1-15.
[13] Nickolls J, Dally W J. The GPU Computing Era[J]. IEEE Micro, 2010, 30(2):56-69.
[14] Si Y J. Design and implementation of a fast sequence alignment software based on GPU[D]. Huazhong University of Science and Technology, 2017.
[15] Chan Y D. Parallel DNA sequence alignment algorithm based on multi-core and multi-core platforms [D]. Shandong University, 2019.
[16] Zhu X Y. Large scale biological sequence alignment algorithm and its parallelization[D]. Hunan University, 2019.
[17] Qiu S L, Liu C Z. Application of GPU based FOA parallel computing in gene classification[J]. Information & Communications, 2019(09):270-271, 273
[18] White T. Hadoop: The Definitive Guide[J]. Orlly Media Inc Gravenstn Highway North, 2012, 215(11):1-4.
[19] Dean J. MapReduce: Simplified Data Processing on Large Clusters[C]. Symposium on Operating System Design & Implementation. 2004,51(1):107-113.
[20] Decap D, Reumers J, Herzeel C, et al. Halvade: scalable sequence analysis with MapReduce[J]. Bioinformatics, 2015, 31(15):2482-2488.
[21] Meng M. Research and implementation of distributed parallelization of Genetic sequence alignment algorithm Blast based on Hadoop[D]. Inner Mongolia Agricultural University, 2015.
[22] Wang X X, Sun D C. Q-sample-based Local Similarity Join Parallel Algorithm[J]. Computer Science, 2019,46(12): 38-44
[23] Wang X X, Sun D C. A MapReduce-based Local Similarity Self-join Algorithm[J]. Computer Technology and Development, 2020, 30(02):88-93
[24] Zaharia M, Chowdhury M, Franklin M J, et al. Spark: Cluster computing with working sets[C]//HotCloud'10: Proceedings of the 2nd USENIX conference on Hot topics in cloud computing,2010,1-7.
[25] Awan A J, Brorsson M, Vlassov V, et al. Architectural Impact on Performance of In-memory Data Analytics: Apache Spark Case Study[J]. 2016. arXiv:1604.08484 [cs.DC].
[26] JM Abuín, Pichel J C, Pena T F, et al. SparkBWA: Speeding Up the Alignment of High-Throughput DNA Sequencing Data[J]. PLoS ONE, 2016,11(5): e0155461.
[27] Castro M, CDS Tostes. SparkBLAST: scalable BLAST processing using in-memory operations[J]. Bmc Bioinformatics, 2017,18(1):318.
[28] Zhou W, Li R, Yuan S, et al. MetaSpark: a spark-based distributed processing tool to recruit metagenomic reads to reference genomes[J]. Bioinformatics, 2017,33(7):1090-1092.
[29] Dong G F, Fu X L, Li H H. Research of Improvement and Parallelization for Sequence Assembly and Multiple Sequence Alignment[J]. Computer Science, 2017, 44(10): 55-58, 84.
[30] Liu Z Y. The Parallelization Research of Genomics Data Comparison Algorithm and the Construction Platform Based on Spark[D]. Inner Mongolia Agricultural University, 2019
[31] Li L X, Liu Y F, Gao J. Research on Optimization Improvement and Spark Parallelization of Smith-Waterman Algorithm[J]. Journal of Inner Mongolia Agricultural University (Natural Science Edition), 2019, 40(05):76-85.

Research on the Selection of Path Planning Algorithm: A Case Study in Leeds

Pang Yan*

University of Leeds, Southwest Jiaotong University, Sichuan Chengdu 610031, China

*Corresponding author email: sc18y2p@leeds.ac.uk

Abstract. Path planning algorithms are the core of path planning technology, and different path planning algorithms have different performances in terms of search efficiency and planning efficiency in different application areas. The traditional path planning algorithm selection mainly relies on personal knowledge and experience and takes a qualitative analysis method. With the variety of path planning algorithms and the increase of environmental complexity, the limitations of this empirical method are more and more obvious, and it is difficult to meet the actual needs of path planning and achieve the ideal path planning effect. The selection of path planning algorithms is an evaluation problem involving multiple factors and requires comprehensive algorithm selection and analysis. This paper addresses the problems of the selection method of the path planning algorithm, applies uncertainty theory and multi-attribute decision theory, constructs an evaluation system for path planning algorithm selection, proposes a multi-criteria EV-VIKOR decision model and decision method for path planning algorithm selection, and combines the results show the effectiveness of the decision-making method proposed in this paper.

1. Introduction

With the development of information technology and artificial intelligence technology, many new industries and application areas have arisen along with them, such as unmanned vehicles, mobile robots, unmanned aerial vehicles (boats), GPS navigation, communication routing problems, takeaway delivery, fire escape, etc. And these industries and applications all involve a common problem that needs to be solved, which is the path planning problem, so path planning algorithms have attracted the attention and interest of many research scholars. In recent years, a series of path optimization algorithms, such as the A* algorithm and Floyd's algorithm, have been proposed for different application areas. However, these algorithms are not perfect and have their strengths and weaknesses in terms of search time, path planning efficiency, algorithm time complexity, space complexity, and directivity. In the past, people only chose algorithms based on their personal experience and knowledge, and the results of such choices would be difficult to achieve the desired results. To select path planning algorithms with better performance in all aspects, it is necessary to compare these algorithms and to compare the advantages and disadvantages of various algorithms, it is necessary to conduct a comprehensive analysis of multiple index factors, which is a multi-criteria decision analysis problem.

For a long time, the research related to path planning algorithms focuses on the improvement and optimization of algorithms and is dedicated to proposing better algorithms. Most of the literature only focuses on and pursues the improvement and optimization of algorithms in terms of individual metrics such as search performance, while there is little discussion on the comparison of the differences in overall effectiveness between algorithms and how to select algorithms for path search and planning.

Some scholars have only done some review analysis on the analysis of the characteristics and comparative selection of various path planning algorithms. For example, Li et al.[1] reviewed traditional algorithms, intelligent optimization algorithms, reinforcement learning-based algorithms, and hybrid path planning algorithms for unmanned vehicles, and compared and analyzed the performance of each algorithm in terms of real-time, robustness, and time complexity of the algorithms. Huo et al.[2] divided the mobile robot path planning into two categories, global planning and local planning, summarized the advantages and disadvantages of the relevant algorithms. Gu et al.[3] conducted a comparative analysis of the genetic algorithm, the RRT algorithm, and the A* algorithm, pointing out the shortcomings and deficiencies of these three algorithms, and improved them according to the specific problem to be solved. Zhao et al.[4] systematically introduced the algorithmic principles and algorithmic processes of Dijkstra's algorithm, Floyd's algorithm, Bellman-Ford's algorithm, and SPFA's algorithm, and conducted a comprehensive comparative analysis of the time complexity and applicability of these path planning algorithms. Liang et al.[5]divided path planning algorithms into evolutionary and non-evolutionary algorithms and subdivided the non-evolutionary algorithms into two categories based on mathematical features: classical mathematics and geometric graph theory. Chen et al.[6]divided the UAV path planning algorithms into classical algorithms and intelligent reactive algorithms. They also analyzed the basic principles and problems of various algorithms in depth. Yang et al.[7]classified path planning algorithms into exact methods, heuristic algorithms and intelligent optimization algorithms. Zhang et al.[8] classified the unmanned vehicle path planning algorithms into two categories, global planning and local planning.

However, the aforementioned research on the selection of path planning algorithms is still lacking systematically, and only qualitative comparisons are made on the performance of various path algorithms, as well as on the improvement of the algorithms and other aspects. This paper addresses the shortcomings of path planning algorithm selection and proposes the EV-VIKOR multi-attribute decision method for path planning algorithm selection based on uncertainty multi-attribute decision theory.

2. Evaluation Criteria

2.1. Time Complexity

The essence of algorithm time complexity is the execution time of the algorithm, i.e. the sum of the frequencies of all statements in the algorithm. The statement frequency is the number of times a statement is executed and is closely related to the size of the problem that the algorithm is solving. Assuming that for a given algorithm, the current problem size is n, the statement frequency can be expressed as a function $T(n)$ of the problem size, and the algorithm time complexity can then be expressed as $T(n)$.

Common time complexity scales are constant order $O(1)$, logarithmic order $O(logn)$, linear order O(n), linear logarithmic order $O(nlogn)$, square order $O(n^2)$, cubic order $O(n^3)$, kth order $O(n^k)$, and exponential order $O(2^n)$.

2.2. Space Complexity

Space complexity is a measure of the amount of temporary storage space occupied by an algorithm during its run, denoted as $S(n) = O(f(n))$. The amount of temporary storage space occupied during runtime varies from path algorithm to path algorithm. Some algorithms require only a small amount of temporary space and do not change with the size of the problem, so they are said to be "in-situ" and are storage-efficient. Some algorithms require several temporary work units that are related to the size of the problem, n, and increases with n. When n is large, more storage units will be used.

2.3. Search Efficiency

Global path planning algorithms plan paths based on a priori information, while local path planning algorithms plan by acquiring information about the environment through a perceptron. Regardless of the algorithm, it must have the ability to respond quickly to environmental information, have a low computational effort and a short search time, and have good search efficiency. Different algorithms perform differently in terms of search efficiency. For example, the A* algorithm has the advantage of responding quickly to the environment and is a straightforward search algorithm.

2.4. Adaptability
There are many different path planning algorithms, and the range of adaptation varies from algorithm to algorithm due to the design of their algorithm principles and their computational performance. If the algorithm can adapt to various situations such as dense or dense relaxed graphs, graphs with positive or negative edge weights, and shortest paths with single or multiple sources, the wider the adaptation range, the better the algorithm. For example, Floyd's algorithm is suitable for multi-source shortest paths, which can calculate the shortest distance between any two nodes, is suitable for dense graphs, is more closely related to the vertices, and can solve the problem of negative-weighted edges, but the time complexity is relatively high and is not suitable for calculating a large amount of data.

3. Path Planning Algorithms

3.1. Depth-first Search(DFS)
Depth-first search is a graph traversal algorithm proposed by Hopcroft and Robert Tarjan. The algorithm prioritizes the expansion of the most recently discovered node until a certain depth limit is reached. If the target is not found or cannot be extended any further, it is then backtracked to another node to continue the extension.
Algorithm steps:

step1. Push the root node into the stack.

step2. Take the element from the top of the stack and find all the nodes in the adjacency table that are connected to him and have not been visited. Push these nodes into the stack. Mark the current node as the predecessor node of these nodes.

step3. Find the target node and exit.

step4. Traverse all nodes without finding the target node and exit the program.

Compared to Dijkstra's algorithm and Floyd's algorithm, depth-first search and breadth-first search are more oriented towards graph traversal operations, and the two algorithms are different forms of graph traversal and searching for shortest paths can be achieved by both algorithms, but their performance and efficiency are inferior to those of Dijkstra's algorithm and Floyd's algorithm and the A* algorithm. Depth-first search can find paths, but there is a great deal of uncertainty as to whether the path is the shortest. There is no perfect solution for the selection of the next node of the current node because of the choice of weights.

3.2. Breadth-first Search(BFS)
The breadth-first search was proposed by Edward Moore in 1959. The algorithm is an algorithm for traversing or searching a tree or graph data structure, starting at the root of the tree or some arbitrary node of the graph and exploring all neighboring nodes at the current depth before moving to the next node at the next depth level.
Algorithm steps:

step1. Add the root node to the queue.

step2. Each time you pop out the head element of the queue, look up all the nodes adjacent to it in the adjacency table and add the unvisited node to the queue. And set this node as the precursor of all new nodes added to the queue.

step3. Find the target node and exit.

step4. Iterate through all nodes without finding the target node, exit the program.

Dijkstra's algorithm is an improvement over breadth-first search in that for each node, the optimal solution is guaranteed to have been reached. Then, at the moment when the target node is finally found, the optimal solution must also be found. But the program will take longer to run.

3.3. Dijkstra Algorithm

Dijkstra algorithm[9] was proposed by the Dutch scientist Edsger W. Dijkstra in 1959. The key to this algorithm is that after each finding of a node that has the shortest path and has not been visited, the path is updated for all nodes to ensure that the optimal solution is obtained.

Algorithm steps:

step1. Create an array to mark whether a node is visited or not.

step2. Create a node to store the temporary distance of each node to the starting point. You can set the initial node to 0 and the other nodes to infinity. Set the current node as the start node.

step3. Set the current node to be accessible, e.g. mark it as "1" in the identity array. The array marked with "1" will not be accessed in the future.

step4. For the current node, iterate through all unvisited neighboring nodes and calculate a new distance from the current node. Compare the new distance with the value in the distance array and update or keep it as the smaller value. If the minimum value of the distance in the distance array is infinity, or if the endpoint has been marked as visited then stop. The procedure ends, otherwise, go to the next step.

Step 5. Continue to select the unvisited node marked with the minimum distance from the distance as the current node, continuing the loop from step 3.

In comparison with other algorithms, Dijkstra's algorithm is more accurate for shortest circuit search than the two graph traversal algorithms of DFS and BFS algorithms. In contrast to Floyd's algorithm, Dijkstra's time complexity is significantly better than Floyd's $O(n^3)$, but Floyd's algorithm is easier to understand and simpler to write in terms of program implementation.

3.4. Floyd Algorithm

Floyd algorithm[10] was proposed by Turing Award-winning professor Robert W. Floyd in 1959. Floyd's algorithm is a dynamic planning algorithm that works best for dense graphs, with positive or negative weights for paths, and is suitable for APSP (All Pairs Shortest Paths). The advantage of Floyd's algorithm is that it is simple to implement and, due to the compact structure of the triple loop, is more efficient than Dijkstra's algorithm for dense graphs. However, for searching single-source shortest paths, Floyd's algorithm will have a large amount of redundancy and will find a large number of shortest paths to unwanted nodes and take a lot of time.

Algorithm steps:

step1. Initialize a path matrix, adding the weights of the paths to the matrix if there is a path between two points, otherwise, set to infinity. Initialize a matrix of intermediate points, initializing each point's intermediate point to itself first.

step2. Loop for all nodes as intermediate nodes and update the value of the path matrix and the value of the intermediate point matrix if there is a node v that makes the path from node x to node y shorter.

step3. After three loops the algorithm ends and the desired value is obtained.

Compared to other algorithms, Floyd's algorithm has a space complexity of $O(n^2)$ and a time complexity of $O(n^3)$ compared to Dijkstra's algorithm because it relies on the path matrix and the intermediate point matrix to store the node information. However, the advantage of Floyd's algorithm is that it has only five lines of core code, which makes it extremely simple to implement. In terms of searching simple graphs, Floyd's algorithm performs essentially indistinguishably from Dijkstra's algorithm, and in terms of search results, it is similar to Dijkstra's algorithm, both outperforming DFS and BFS algorithms.

3.5. A Search Algorithm*

The A* algorithm[11,12] was co-published by Peter Hart and other scholars at Stanford Research Institute in 1968 and is a commonly used algorithm for path finding and graph traversal. The A* algorithm is based on Dijkstra's algorithm, which introduces heuristic functions and predicted costs, and is the most efficient direct search method for solving shortest paths in a static road network, as well as being a common heuristic for many other problems.

The core part of the algorithm lies in the design of the valuation function, $f(n) = g(n) + h(n)$, $f(n)$ is the estimated cost from the initial node via node n to the target node, $g(n)$ is the actual cost from the initial node to node n, and $h(n)$ is the estimated cost of the best path from node n to the target node.

Algorithm steps:

step1. Add the starting node to the priority queue.

step2. Traverse the current priority queue to find the node with the smallest F-value and make it the current node.

step3. Mark it as visited and process it for the adjacent nodes.

step4. If the neighboring node has not been visited, add it to the queue and set the current node as the parent of the newly added node, recording the F, H, and G values of the new node. If the neighboring node has already been visited, check if there is a better path. Check if the current node has a smaller G-value. If there is no smaller G value, no action is taken. If the G value is smaller, set the parent node of that node to the current node and re-set the G and H values of that node.

step5. When it happens that the target node is marked, and the path has been found. Or the priority queue is empty, terminate the algorithm. Otherwise, jump back to step 2.

step6. When the path is found, move from the endpoint to the start node along with the parent node. The A* algorithm is an advanced version of Dijkstra's algorithm compared to other algorithms. In terms of time complexity, the A* algorithm outperforms Dijkstra's algorithm and Floyd's algorithm, and can be asymptotically advanced to the $O(n)$ case.

4. Methodology; the Integrated EV–VIKOR

Figure 1 explains procedure of the integrated EV–VIKOR as follows:

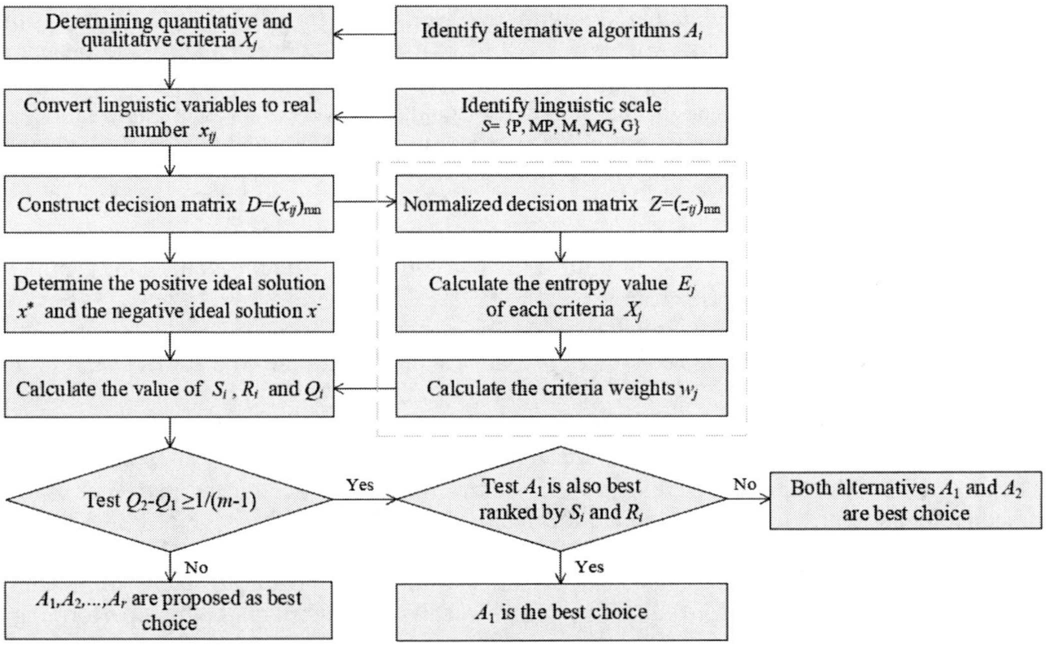

Figure 1. Procedure of the integrated EV–VIKOR.

Consider a multi-attribute decision problem, let the set of options be $A_i(i = 1,2,\cdots,m)$, the set of attributes be $X_j(j = 1,2,\cdots,n)$, and the values of the attributes given for each objective for multiple data types be x_{ij}, assuming that the attribute weight information $w_j(j = 1,2,\cdots,n)$ is known and $w_j \geq 0, j = 1,2,\cdots,n$, $\sum_{j=1}^{n} w_j = 1$, that the weights of each criteria are derived by the Entropy Value method (EV)[13,14]. The steps of the EV-VIKOR method are as follows:

step 1.Constructing multi-attribute decision matrices $D = (x_{ij})_{m\times n}$;

The decision-maker gives a value for each option under each attribute, both quantitative and qualitative, where the qualitative attribute can be given a linguistic value by the decision-maker based on his or her knowledge and experience.

step 2. Normalization of objective attribute values:

Before normalization, the linguistic variable data types will be transformed according to the linguistic evaluation set S= {P, MP, M, MG, G}. Here, for the quantification of qualitative attributes represented by linguistic variable data, the rating values are given on a scale of 1 to 5, with 1 being the lowest and 5 being the highest.

Let the normalized decision matrix be $Z = (z_{ij})_{m\times n}$. The common attributes are benefit-based and cost-based indicators, such that $I_j(j = 1,2)$ denotes the set of subscripts for benefit-based and cost-based, $= \{1,2,\cdots,m\}$, For real data, the normalized data are, after normalization,

Benefit-based indicators:

$$z_{ij} = x_{ij}/max(x_{ij}) \quad ,i \in M; j \in I_1 \tag{1}$$

Cost-based indicators:

$$z_{ij} = min(x_{ij})/x_{ij} \quad ,i \in M; j \in I_2 \tag{2}$$

step 3. Calculate the criteria weights w_j, using the Entropy Value (EV) method;

The entropy value of the real numbers under the set of criteria $X_j(j = 1,2,\cdots,n)$ is that,

$$E(X_j) = -\frac{1}{\ln m}\sum_{i=1}^{m} z_{ij}\ln z_{ij} \quad ,j = 1,2,\cdots,n \tag{3}$$

The weights w_j for each criteria $X_j(j = 1,2,\cdots,n)$ are,

$$w_j = -\frac{1-E(X_j)}{n-\sum_{j=1}^{n} E(X_j)} \quad ,j = 1,2,\cdots,n \tag{4}$$

step 4. Determine the positive ideal solution x_j^* and the negative ideal solution x_j^-, based on the decision information $D = (x_{ij})_{m\times n}$;

There are different types of attributes, which are generally divided into benefit and cost types.

$$x_j^* = \left\{\left(\max_i x_{ij} \middle| j \in J_1\right), \left(\min_i x_{ij} \middle| j \in J_2\right)\right\} = (x_1^*, x_2^*, \cdots, x_n^*) \tag{5}$$

$$x_j^- = \left\{\left(\min_i x_{ij} \middle| j \in J_1\right), \left(\max_i x_{ij} \middle| j \in J_2\right)\right\} = (x_1^-, x_2^-, \cdots, x_n^-) \tag{6}$$

In the equation, x_{ij} is the attribute value of option A_i concerning attribute X_j, J_1 and J_2 are the sets of subscripts for the benefit-based and cost-based objectives respectively.

step 5. Calculation of group benefits of options S_i, individual regret values R_i and trade-off values Q_i.

$$S_i = \sum_{j=1}^{n} \frac{w_j d(x_j^*, r_{ij})}{d(x_j^*, x_j^-)} \quad ,i = 1,2,\cdots,m \tag{7}$$

$$R_i = \max_j \left(\frac{w_j d(x_j^*, r_{ij})}{d(x_j^*, x_j^-)}\right) \quad ,i = 1,2,\cdots,m \tag{8}$$

$$Q_i = v\frac{S_i - S^*}{S^- - S^*} + (1-v)\frac{R_i - R^*}{R^- - R^*} \quad ,i = 1,2,\cdots,m \tag{9}$$

In the formula, $S^* = \min_i S_i$, $S^- = \max_i S_i$, $R^* = \min_i R_i$, $R^- = \max_i R_i$, the smaller the value of S_i , Q_i and R_i, the better. v is the decision mechanism coefficient, and $v = [0,1]$, indicates the decision-maker's preference for group opinion and individual regret values. A value of $v > 0.5$ indicates that the decision-maker prefers to make decisions based on maximizing the weight of group benefits, a value of $v < 0.5$ indicates that the decision-maker prefers to minimize individual regret values, and $v = 0.5$ indicates that there is no clear preference for the decision-maker to take a balanced compromise approach. In the VIKOR method, $v = 0.5$, the equilibrium trade-off to maximize group benefits and minimize individual regrets, is generally taken.

step 6.The values of S_i , R_i and Q_i are sorted in descending order, with smaller values representing better solutions;

step 7.Determining the compromise solution.

When the following two conditions hold, the solutions can be ranked according to the magnitude of the value of $Q_i(i = 1,2,\cdots,m)$. A smaller value means that the solution is better. Let the alternative corresponding to the smallest value of A_i be Q_i. If the smallest value Q_i satisfies both of the following two conditions C_1 and C_2, then solution is A_i the best compromise solution.

Condition C_1: acceptability advantage $Q_2 - Q_1 \geq 1/(m-1)$, where Q_1 and Q_2 are the smallest and second smallest combined values ranked by $Q_i(i = 1,2,\cdots,m)$, respectively, and m is the number of options.

Condition C_2: The best option A_i also has the smallest value of S_i or R_i in order Q_i.

If the above two conditions do not hold simultaneously, compromise solutions are obtained, divided into two types: if the condition C_1 is not satisfied, then $A_1, A_2, \cdots, A_r$ are a compromise solutions, where the maximum value r of is determined by $Q_r - Q_1 \geq 1/(m-1)$. These solutions are close to the ideal solution; if the condition C_2 is not satisfied, then the alternatives A_1 and A_2 are compromise solutions.

5. Case Studies

Figure 2 below shows a road map of part of Leeds city, with the upper end of the path as the starting point A and the lower end as the endpoint Z.

Five path planning algorithms are proposed, using $A_i(i = 1,2,\cdots,5)$, which are the DFS, BFS, Dijkstra, Floyd and A* search algorithms. There are four evaluation metrics considered, using $X_j(j = 1,2,3,4)$, which are time complexity X_1, space complexity X_2, search efficiency X_3 and adaptability X_4. Metrics X_1 and X_2 are given evaluation values based on time consumption and memory size, metrics X_3 and X_4 are qualitative attributes and are given linguistic evaluation values by the decision-maker, metrics X_1 and X_2 are cost-based metrics, metrics X_3 and X_4 are benefit-based metrics, and the path planning algorithm decision information matrix is shown in Table 1 below.

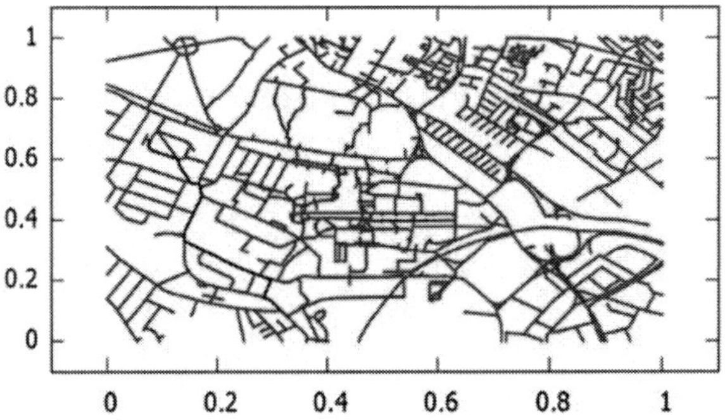

Figure 2. Map of part of Leeds city roads.

Table 1. Formatting sections, subsections and subsubsections.

	X_1	X_2	X_3	X_4
A_1	$O(n^2)$	$O(n^2)$	low	poor
A_2	$O(n^2)$	$O(n^2)$	medium low	medium poor
A_3	$O(n^2)$	$O(n^3)$	medium high	medium
A_4	$O(n^3)$	$O(n^2)$	medium	medium good
A_5	$O(n)$	$O(n^2)$	high	good

The initial decision matrix is mixed information, where all data is converted to a uniform real data type due to the inconvenience of calculation.
The indicator attribute values are quantified according to the algorithm consumption time according to a scale of 1 to 9 levels, where 1 corresponds to the time complexity $O(1)$ and 9 corresponds to $O(n^n)$; the indicator attribute value quantization processing is related to the temporary memory space occupation of the coding implementation, quantified by the actual value and processed regarding a scale of 1 to 9 levels, while, at the same time, allowing the quantified value of the attribute to maintain some variability; the indicator X_3 and X_4 are quantified according to the language evaluation sets are quantified according to a scale of 1 to 5, with poor (low) corresponding to 1 and high (good) corresponding to 5.
Normalized attribute matrix $(z_{ij})_{5\times4}$, as shown in Table 2 below.

Table 2. Normalized attribute matrix.

	X_1	X_2	X_3	X_4
A_1	0.2083	0.1939	0.0667	0.0667
A_2	0.2083	0.1939	0.1333	0.1333
A_3	0.2083	0.2091	0.2667	0.2000
A_4	0.2500	0.2015	0.2000	0.2667
A_5	0.1250	0.2015	0.3333	0.3333

Based on the normalized decision matrix, the weights of each attribute calculated according to the entropy method are $w_j(j = 1,2,3,4)$:

$$w = (0.0860, 0.0015,0.4563,0.4563)$$

The positive ideal solution x_j^* and the negative ideal solution x_j^- are:

$$x_j^* = (3,5.1,5,5)\ ,\ x_j^- = (6,5.3,1,1)$$

The group utility values S_i, individual regret values R_i and trade-off values Q_i for each algorithm , and are sorted in descending order as shown in Table 3 below, where is taken $v = 0.5$.

Table 3. Group utility values S_i, individual regret values R_i and trade-off values Q_i.

	S_i		R_i		Q_i	
	Calculated values	Sorting	Calculated values	Sorting	Calculated values	Sorting
A_1	0.4213	1	0.2082	1	1.0000	1
A_2	0.3172	2	0.1561	2	0.7515	2
A_3	0.1611	4	0.1041	3	0.4412	4
A_4	0.1635	3	0.1041	3	0.4441	3
A_5	0	5	0	4	0	5

From the calculated values Q_i, the path planning algorithm is ranked as $A_5 \succ A_3 \succ A_4 \succ A_2 \succ A_1$, where the algorithm corresponding to the smallest value of Q_i is A_5, and then check whether the algorithm A_5 satisfies the other two conditions C_1 and C_2. For the condition C_1, $Q(A_3) - Q(A_5) = 0.4412 > 0.25$, the condition C_1 is satisfied; for the condition C_2, the value S_i of for algorithm A_4 is the smallest and the condition C_2 is satisfied. Therefore, the algorithm A_5 is the best algorithm for path planning for urban vehicles in Leeds.

6. Conclusions and Further Discussion

This paper addresses the arbitrariness and unreasonableness of the current practice of choosing path planning algorithms based on experience alone, and attempts to propose a decision analysis method for choosing path planning algorithms, which provides a scientific decision-making tool for people to choose the appropriate algorithm for path planning.

Although the case study in this paper is focused on the selection of path planning algorithms for urban vehicles in Leeds, evaluation metrics can be added or subtracted as needed to apply to the selection of path planning algorithms in different application areas and complex environments.
There are also some shortcomings in the research presented in this paper. The selection of path planning algorithms is a multi-attribute decision problem, which requires a comprehensive analysis of various evaluation factors and a complete and comprehensive evaluation index system. influencing factors are not comprehensive enough, which is the shortcoming of this paper and a topic worthy of further research in the future.

References

[1] Li, Y.D.; Ma, T.L.; Chen, C.B.; Wei, H.L.; Yang, Q.G. A review of path planning algorithms for unmanned vehicles. *Foreign Electronic Measurement Technology* 2019,38, pp. 72-79.
[2] Huo, F.C.; Chi, J.; Huang, Z.J.; Ren, L.; Sun, Q.J.; Chen, J.L. A survey of path planning algorithms for mobile robots. *Journal of Jilin University (Information Science Edition)* 2018,36, pp. 639-647.
[3] Gu Lei. Vehicle routing algorithm and its application review. *Logistics Engineering and Management* 2019,41, 100-101+33.
[4] Zhao, W.J.; Gong, Z.Y.; Wang, W.; Fan, S.F. Comparison and analysis of several classical shortest path algorithms. *Journal of Chifeng University (Natural Science Edition)* 2018,34, pp. 47-49.
[5] Liang, X.H.; Mu, Y.H.; Wu, B.H.; Jiang, Y. A review of correlation algorithms for path planning. *Value Engineering* 2020, 39, pp. 295-299.
[6] Chen, Q.J.; Jin, Y.Q.; Han, L. Review of UAV path planning algorithm. *Flying Missiles* 2020,pp. 54-58.
[7] Yang, X.; Wang, R.; Zhang, T. Review of intelligent optimization algorithms for path planning of UAV cluster. *Control Theory & Applications* 2020,37, pp. 2291-2302.
[8] Zhang,K.Review of path planning algorithms for unmanned vehicles. *Equipment Manufacturing Technology* 2021,6, pp.111-113.
[9] Dijkstra, E.W. A note on two problems in connexion with graphs. *Numerische mathematic* 1959,1, pp. 269-271.
[10] Floyd, R.W. Shortest path. *Communications of the Association for Computing Machinery* 1962,15, pp. 245.
[11] Hart, P., Nilsson, N., Raphael, B. A formal basis for the heuristic determination of minimum cost paths. *IEEE Transactions on Systems Science and Cybernetics* 1968,4, pp. 100-1
[12] Dieter, F. A probabilistic approach to collaborative robot localization. *Autonomous Robots* 2008,8,pp. 325-344.
[13] Opricovic, S. Multi-criteria optimization of civil engineering systems. *Faculty of Civil Engineering*, Belgrade 1998,2, pp. 5-21.
[14] Zhu, Y.; Tian, D.; Yan, F. Effectiveness of entropy weight method in decision-making. *Mathematical Problems in Engineering* 2020, 3564835.

Outlier Detection of Leaf Images Based on One-Class Support of Vector Machine

Ming Yin and Liwei Wang*

Faculty of Electronic and Information, The Southwest Minzu University, Chengdu, 6100041, China

*Corresponding author email:wangliwei@swun.edu.cn

Abstract. In this paper, a novel approach was proposed to use the One Class Support Vector Machine (OCSVM). The growth status of fruit tree's leaves reflects the health level of the plant. Using computer system to detect the abnormal condition of plant leaves will help producers take measures to maintain the health of plants. We focus on the outlier leaf images which are obviously different from with the health leaf. A modified One Class Support Vector Machine is presented as a very efficient method to detect the outlier data from massive data of the leaves image dataset. The results show it worked with the test dataset and with very good performance.
Keywords: Leaf image; Image scoring; Data acquisition system; Leaf disease; Outlier detection; Anomaly score.

1. Introduction

With the development of the Internet of things and intelligent computer system. Modern information technology has been applied in the fields of agriculture and plant industry. The people acquire the data from the practical production field. The wireless sensor network collects and transfer data to the data to central control unit or to the cloud platform. Meanwhile the monitor system captures and records the image or video data to the specific database and to could data store center. It was rather difficult and low-efficiency to dealt with the mass data to aid the researcher. The consumption of data analysis system is too large to make it very inconvenient to leverage the image data. Machine learning and artificial intelligence grow up very fast in past twenty year which have very strong impaction to every corner of the modern society [1][2].

The plant leaf generally was classified as health leaf and poor growth leaf since the leaf affected by disease and by pest [3][4]. Though there are many different kinds of disease and pest which cause different symptoms of leaves. And the different plants also present different symptoms according the same damage source. In [5][6], a kinds of CNN method have been proposed to classify the specific plant leaf. Some algorithms focus on the traditional outlier detection [7][8][9]. However, from the perspective of the dataset of leaf images. The diversity of the leaf character shown as texture, shape, color and its period of growth. Therefore, the dataset was limited as the specific dataset which come from the very limited data source.

In this paper, we investigate the problem about the evaluation the leaf growth status. We leverage the modified one class support vector machine method to detect the outlier leaf image and use the score of anomaly degree to evaluate the leaf growth status, and we certainly can evaluate the plant growth by leaves status.

2. Background and Methods

2.1. The Image Data and Leaves Manual Scoring

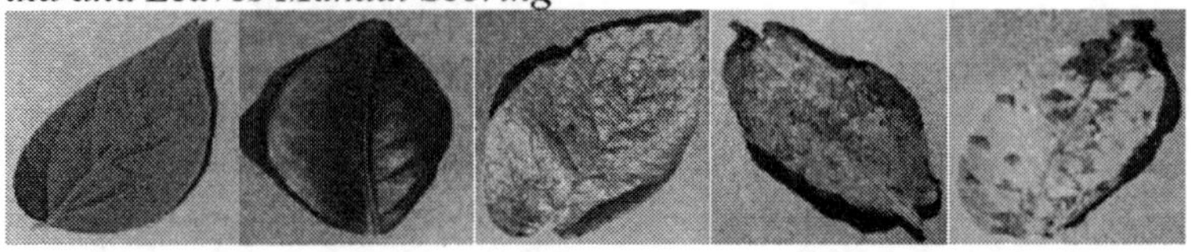

Figure 1. The example of the scored image descendent from 5 to 1.

We using the dataset from web resource, and we label and manual scoring the leaves image with the help of the agricultural expert. The score listed as 1 to 5. The picture 1 shows the example of the leaves image according the score. The leftmost score is 5 points, which is reduced to 1 point from left to right. We establish a labelled dataset which consist by labelled as "Positive" and "Negative". The Positive labelled dataset composed of images with a score of 5 and 4. On the contrary, The Negative labelled dataset composed of images with a score of 2 and 1. To maintain the diversity, the images scored 3 are divided into both two groups.

2.2. Outlier Detection

Outlier detection is an unsupervised classification scheme which also called as anomaly detection. Outlier detection is a method used to distinguish data in a dataset that is clearly different from other data. The dataset are divided into two categories: normal data and abnormal data. Normally, the volume of outlier data is small. In generally, the outliers have 3 types which include

- Point Outlier
- Contextual Outlier
- Collective Outlier.

The outlier is depended on dataset properties. In most cases, we rarely know outliers during training, outlier detection addresses the problem of modeling normal data distribution and defining measured values in this space, so as to classify samples as abnormal or normal. In high-dimensional data (such as images), the distance in the original space quickly loses the ability to describe because of dimension curse, and needs to be mapped to a more appropriate space.

The outlier detection about the leaf images is more complicated compare with the normal outlier detection problem since the dataset of leaves image have very wide distribution of the leaves sample. The composition of data sets is very diverse, and there are many kinds of samples. The accurate classification of diseased leaves and the evaluation of leaf health status are an important goal of anomaly detection.

2.3. The OCSVM and Application on Outlier Detection

OCSVM (one class support vector machine) is a single class support vector machine. Normally, it estimates the support of high-dimensional distribution. The model maps the data samples to the high-dimensional feature space through the kernel function, so that it has better aggregation. OCSVM also solves the optimal hyperplane in the feature space, and realizes the maximum separation between the target data and the coordinate origin. The equation of the OCSVM is shown below.

$$\hat{y} = \begin{cases} 0 & \text{if} \quad \mathbf{w}^T \cdot \mathbf{x} + b < 0 \\ 1 & \text{if} \quad \mathbf{w}^T \cdot \mathbf{x} + b \geq 0 \end{cases} \tag{1}$$

$$\begin{gathered} \hat{y} = \boldsymbol{w}^T \cdot \boldsymbol{x} + b \\ f(x) = sign(y) \end{gathered} \tag{2}$$

$$\begin{aligned} &\underset{\mathbf{m},b}{minimize} && \frac{1}{2}\mathbf{w}^T \cdot \mathbf{w} \\ &\text{subject to} && f_{(i)}\left(\mathbf{w}^T \cdot \mathbf{x}^{(i)} + b\right) \geq 1 \qquad \text{for } i = 1,2,\ldots,m \end{aligned} \tag{3}$$

The task is to define a boundary around the target class so that it can accept as many target objects as possible and minimize the chance of accepting exception objects. Therefore, in pattern recognition or machine learning, people try to infer classifiers from a set of training examples. Normally, the goal of classification is to distinguish a set of target objects from all other possible objects. Outlier detection is a specific classification which concern the minority of the dataset. It is mainly used to detect new objects that resemble a known set of objects [11]]. After experimental verification, the effect of direct application of SVM classification is not good. We exploit the OCSVM as a scoring framework to handle the dataset.

2.4. The Proposed Method

We have use a modified OCSVM method to detect the outlier of the leaf images [12]. Firstly, we acquired and check the input dataset. Then, with help of the agriculture expert, we design and label our dataset with scores. The following steps of algorithm flow chart is shown in the Figure 2.

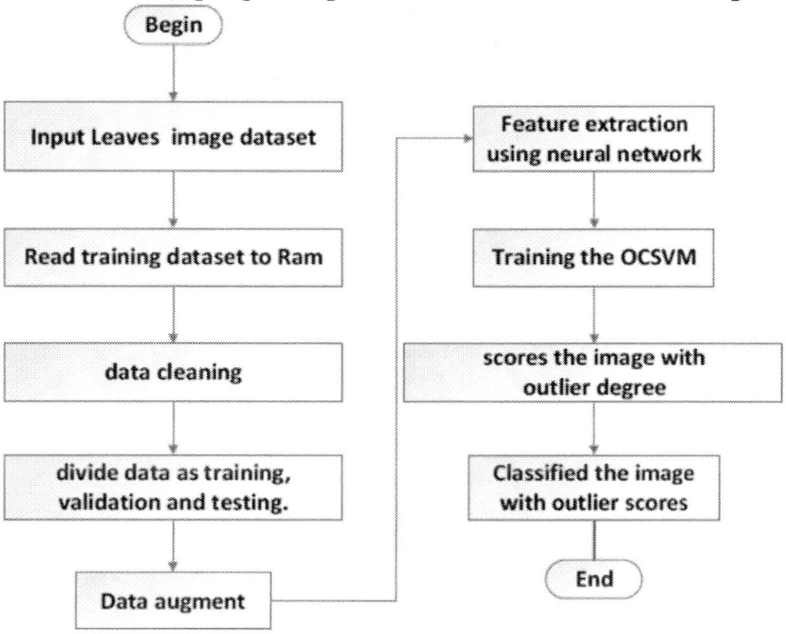

Figure 2. The algorithm flow chart.

3. Experiment and Results

We using Matlab2021a to perform the code and carry on our experiment, and the system include Intel i5 cpu,8G RAM. Figure 3 shows the results of our algorithm. The 100 leaves images listed by the score of anomaly degree from least to maximum. We can infer and evaluate the leaf's health status by the results. The Figure 3 lists the anomaly score which have the corresponding relationship with Figure 1. We also compare the accurate of the algorithm with different feature extraction. The Figure 4 show the sorted outlier degree. The results show that the outlier distribution of leaf image data set follows a similar half edge normal distribution. We use the outlier degree is more accurate than the outlier classification methods.

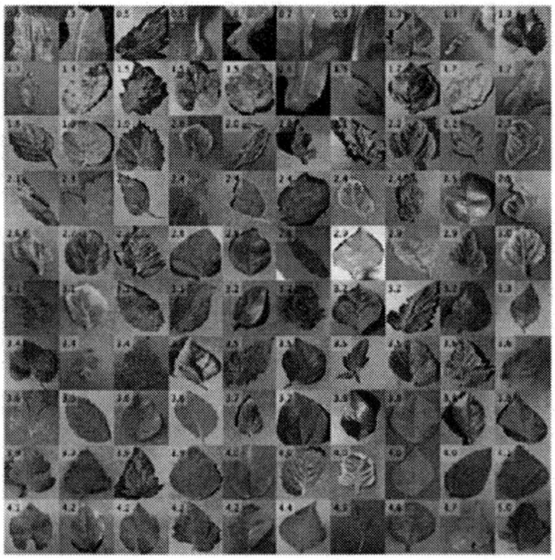

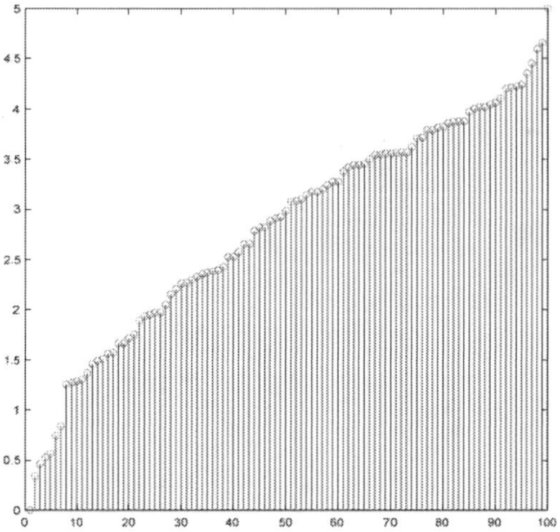

Figure 3. The outlier degree scores of the dataset. **Figure 4.** The sorted outlier degree.

4. Conclusion

By evaluate the outlier degree of the leaf images, we find a way to evaluate the plant growth status and the health status. Compare with the traditional classify methods or the anomaly detection methods. We leverage the Modified OCSVM not only find out the outlier data which represent the leaves with poor growth and health but also score the leaves image which represent the health degree of every leaf. The algorithm has good generalization ability and robustness.

Under the limit of the dataset source, some samples were labelled with fuzzy. In the future, we will refine the dataset and using the specific dataset to improve the algorithm accuracy.

Acknowledgement

This work supported by the Fundamental Research Funds for the Central Universities ,Southwest Minzu University, No:2018NQN09.

References

[1] N. Ahmed, H. Asif, and G. Saleem, "Leaf Image-based Plant Disease Identification using Color and Texture Features," 2021.

[2] Thakur, N. K. Kulkarni, T. Rajwal, and V. Deshpande, Smart Leaf Disease Detection Using Image Processing. Machine Learning and Information Processing, 2021.

[3] S. Raina and A. Gupta, "A Study on Various Techniques for Plant Leaf Disease Detection Using Leaf Image," in 2021 International Conference on Artificial Intelligence and Smart Systems (ICAIS), 2021.

[4] H. S. Ham and H. C. Cho, "Tomato Leaf Disease Classification with Image Augmentation Methods," Transactions of the Korean Institute of Electrical Engineers, vol. 70, no. 1, pp. 184-189, 2021.

[5] M. I. Dinata, S. Nugroho, and R. F. Rachmadi, "Classification of Strawberry Plant Diseases with Leaf Image Using CNN," in 2021 International Conference on Artificial Intelligence and Computer Science Technology (ICAICST), 2021.

[6] X. Du, C. M. Zhai, and Q. P. Wang, "Recognition of plant leaf image based on fractal dimension features," Neurocomputing, vol. 116, pp. 150-156, 2013.

[7] M. V. Applalanaidu and G. Kumaravelan, "A Review of Machine Learning Approaches in Plant Leaf Disease Detection and Classification," in 2021 Third International Conference on Intelligent Communication Technologies and Virtual Mobile Networks (ICICV), 2021.

[8] J. Yang, T. Deng, and S. Ran, "An Adaptive Weighted One-Class SVM for Robust Outlier Detection," Springer Berlin Heidelberg, 2016.

[9] S. Sinaga, "Analysis and Comparison of Sobel Edge Detection and Prewit Methods on Srilangka Leaf Image Edge Detection," CSRID (Computer Science Research and Its Development Journal), vol. 13, no. 1, pp. 13-22, 2021.
[10] H. Qu, J. Zhou, J. Qin, and X. Tian, "Anomaly Detection for Industrial Control Networks Based on Improved One-Class Support Vector Machine," International Journal of Pattern Recognition and Artificial Intelligence, 2020.
[11] Y. Aoki, F. Ehara, and Y. Muto, "Comparison of Acoustical Features for Abnormal Sound Detection by OCSVM," in International Conference on Industrial Application Engineering 2020, 2020.
[12] [J. N. Maloof, K. Nozue, M. R. Mumbach, and C. M. Palmer, "LeafJ: an ImageJ plugin for semi-automated leaf shape measurement," J Vis Exp, vol. 71, no. 71, pp. -, 2013.

Design of Wind Turbine Parameter Monitoring and Early Warning Software System Based on Deep Learning Modeling

Jiechang Wang*, Zhannan Chen, Jianfei Guo, Yan Dong, Jinglong Yu and Runan Zhou

Huaneng Jilin Renewable Energy Branch, 399# Huanengdasha, Xinghuo Rd. Qianjing Avenue, Changchun City, Jilin Province, China

*Corresponding author email: 919729665@qq.com

Abstract. This paper adopts deep learning self-encoding algorithm, using JAVA, C++ and python language mixed programming computing technology to realize wind turbine parameter monitoring and early warning and software system design and development. This paper describes the database design, process management mode, fault early warning model, alarm value processing and human-computer interaction interface of the system, explains the software operation process and workflow, as well as the early warning mechanism and algorithm of the system, and realizes the early fault early warning function of wind turbine operation parameters.
Keywords: Condition monitoring; Parameter early warning; Deep learning; Wind turbine.

1. Preface

As a renewable energy industry, wind power has developed the fastest, and it plays a pivotal role in China's green and low-carbon transition and energy structure adjustment. Wind turbines are in a complex environment, facing severe weather conditions such as high temperature, high humidity, salt spray corrosion, sand and dust, typhoons, thunderstorms, etc. As the service life of wind turbines increases, the probability of fan abnormal warning continues to increase, which brings challenges to the operation safety of the turbines Therefore, the safety issues in the operation of wind turbines are the most concerned by enterprises and the country[1].

At present, large wind turbines are generally equipped with supervisory control and data acquisition (SCADA) system to monitor and record the performance parameters of wind turbines and their core components, environmental parameters such as temperature / wind speed, electrical parameters such as grid voltage / frequency, wind turbine fault shutdown and maintenance and other operating parameters in real time and remotely. However, the current SCADA monitoring alarm is only at the monitoring level [2]. When the parameter exceeds the set threshold, the alarm will be triggered. This method fails to take into account environmental changes and the aging of the unit during operation. When the unit warning triggers the alarm, the abnormal warning of the unit has deteriorated, and early warning of the abnormal warning cannot be realized[3].

With the continuous development of sensor technology, artificial intelligence technology, and control technology [4], big data analysis methods are used, and large amounts of operation data and status information of the wind turbine SCADA system are used to carry out wind turbine performance evaluation, abnormal operating conditions early warning and early abnormal warnings detection [5,6] has become a current research hotspot. Based on this, this paper adopts deep learning self-encoding algorithm, JAVA, C++ and python language mixed programming calculation technology to realize wind turbine parameter monitoring and early warning and software system design and development, and realize the early fault warning function of wind turbine operating parameters.

2. The Architecture of Wind Turbine Parameter Monitoring and Early Warning Software System based on Deep Learning

The wind turbine parameter monitoring and early warning software system based on deep learning transmits the real-time values of the parameters in the SCADA system to the upper computer through Gigabit Ethernet communication technology, and the upper computer uses data mining, self-encoding, deep learning [7, 8], and intelligent analysis method to predict the safety value of wind turbines, and timely warns the subsystems that exceed the threshold, and stores the data in the database to provide guidance for the management of operation and maintenance personnel.
The system architecture is shown in Figure 1, which can be divided into 3 layers. The first layer is the data acquisition layer which will use the SCADA system to collect data on the various subsystems of the wind turbine, and upload the obtained data to the server through Gigabit Ethernet communication, and store it in the time series table database. The second layer is the information integration and calculation layer. The main function of the calculation layer is to realize the processing and calculation of each signal, call algorithms and models to perform in-depth calculations on the original data obtained on the spot, and obtain the calculated value of the acquired signal, and the calculated value and the true value Perform threshold calculation to obtain residuals and store them in the database. The third layer is the display and application layer. The function of the application layer is to display the results and warnings. The calculation results and other data in the database are displayed in the front-end Vue framework through the Java Spring Boot framework, and the fan equipment is obtained in the front-end Operation information and model early warning information, intuitively obtain equipment operating status and perform fault early warning processing on it.

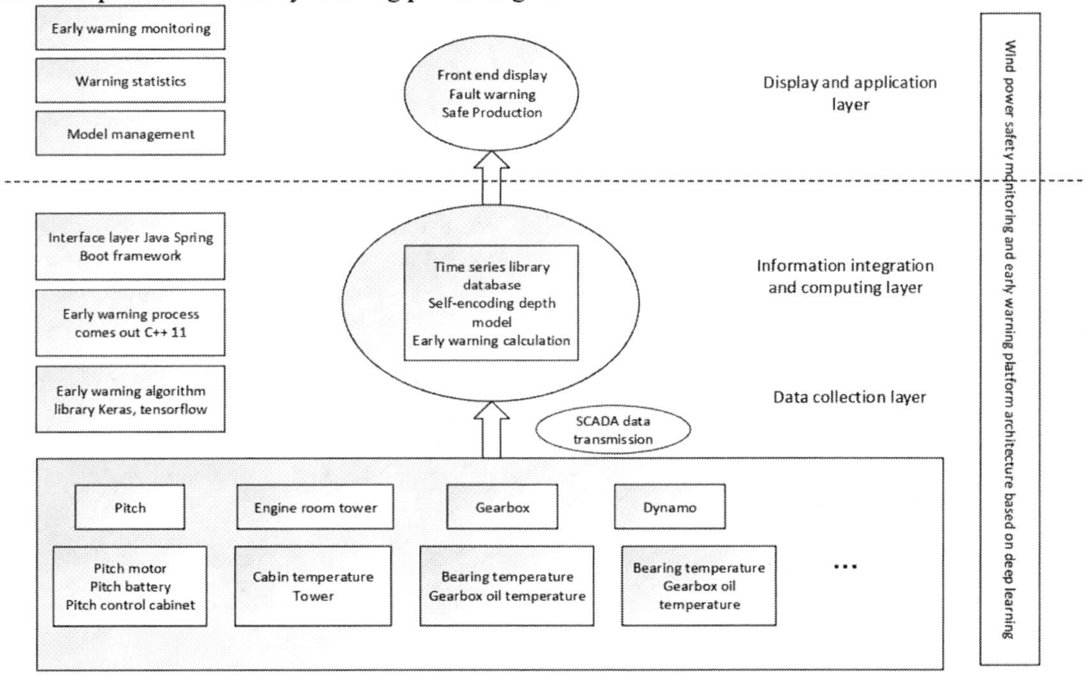

Figure 1. Wind turbine parameter monitoring and early warning software system architecture

3. Database Design

The database adopts the real-time historical database developed by Wuhan Yingfeng Energy Technology Co., Ltd. The database has excellent performance, efficient reading and writing, a single machine capacity supports 10 million measurement points, a storage compression ratio greater than 100:1, and features high security, high reliability, multi-platform support, and rich data access interfaces.

The real-time database uses shared memory and disk files in two ways. The memory/hard disk file management function has at least two processes: the generation of memory files or hard disk files and the function of providing file access.
The basic process of the database management function is as follows:
(1) Create shared memory according to the definition file progca.def, and set the file management information table fcb.
(2) Confirm the shared memory files and hard disk files logged in the file management information table.
(3) Load the initialization file under $HOME/semi/data/progca/ into the memory and set the shared memory. The basic structure of the real-time database is shown in Figure 2.

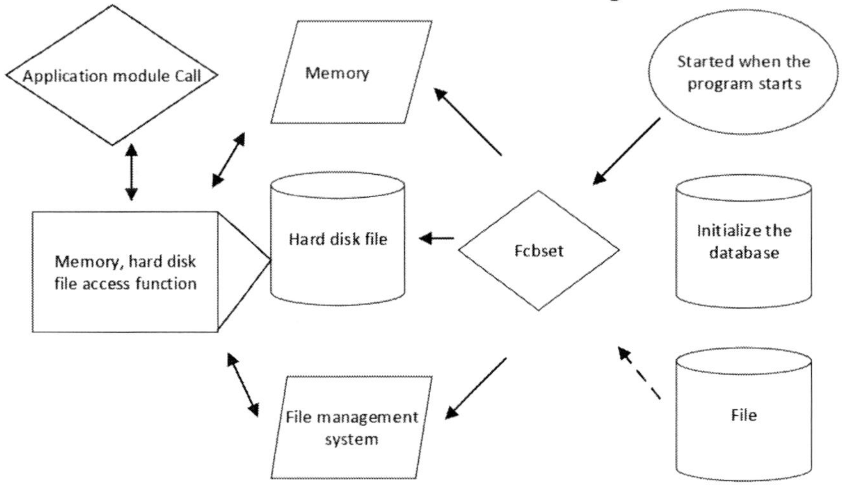

Figure 2. Real-time database composition

The real-time database of this system is implemented in a shared memory mode. Shared memory area is the fastest form of IPC. Once such a memory area is mapped to the address space of the processes that share it, the transfer of data between these processes no longer involves the kernel. However, some form of synchronization is usually required between processes that store information in or retrieve information from this memory area. The shmget function is usually used to create a shared memory area that does not yet exist, or to access an existing shared memory area. Usually shmat is called to attach it to the address space of the calling process.The data that the system needs to operate mainly include the following: communication data, early warning calculation data, picture representation data, monitoring data, etc. The shared memory information defined by it is shown below. This definition file is stored in the $home/semi/ aplgen folder. During initialization, the system will create shared memory according to the definition file. Fcb_ Path = user home directory/progca/(FCB data reading path name); File_ Max = 1024 (maximum number of shared memory files); File_ Size = 84 Mbytes (size of shared memory file); Shared memory files generally need to define the following information: file number, file lock flag bit, file type, data type, access type and file name.

4. The Design of the Process Module

The modules of the system process are mainly divided into four categories: basic service module, screen display module, SCADA data transmission module, and data processing module.The basic service module mainly provides the basic services of the system such as system process control, message control, module cycle control, etc.; the screen display module is mainly responsible for the action and message response processing of the window, and the display of each image interface; the SCADA data transmission module is mainly responsible for The device equipment performs SCADA data communication to obtain the raw data of real-time monitoring.The data processing module is mainly responsible for processing all kinds of collected data, judging whether the data is correct or not, and then calculating and making corresponding processing through the early warning model. As a running

program on Linux, good process management function can effectively manage the start and stop of processes and control the communication of processes. Its main functions are as follows:
(1) When the system is started, it can follow the process described in the initial start table fork;
(2) Start the process task according to the periodic process start table;
(3) In order to set / change the starting cycle, the cycle definition function is provided;
(4) It provides the functions of task process generation, termination and request.
The process startup process of the system is shown in Figure 3.

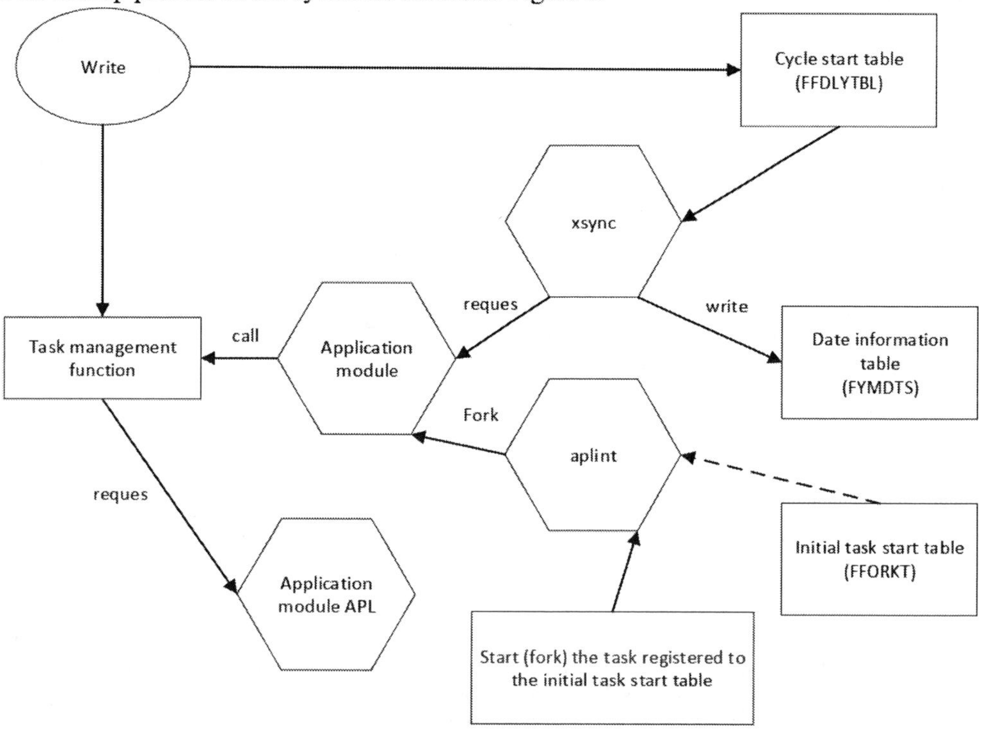

Figure 3. System process startup diagram

5. Abnormal Warning Early Warning Model

According to the working principle of wind turbine, the fault early warning system divides the wind turbine into subsystems such as wind speed and power, engine room temperature, pitch system, gearbox system and generator system. Based on the deep neural network of denoising auto-encoders (DAE), the parameter anomaly early warning model of each subsystem of wind turbine is established. On this basis, according to the actual operation data of SCADA system, the early warning platform adopts a complete set of self-developed data pre-processing methods, including null value, abnormal value, abnormal state data, normalization and other data processing methods.

The early warning method adopted by the fault early warning platform is the denoising auto-encoding (AE) algorithm, which is a neural network used to learn the nonlinear characteristics of input data and reconstruct it. It is a kind of The algorithm of non-linear relationship function learns the expression characteristics of the input data in an unsupervised way. The deep AE model of the wind power early warning platform based on deep learning sets three hidden layers, and performs grid search and Optimization on the number of units, learning rate and noise damage degree of each hidden layer. Finally, the number of units is 8 and 4 respectively, the damage degree is 0.4, the optimizer is Adam and the training loss function is MSE. The W, B and other parameters of the model are randomly initialized with uniform distribution at the beginning of training.

By inputting the SCADA data collected on site in real time into the trained deep AE model, the early warning results can be obtained. Comparing the results with the original input SCADA data, we can judge whether it deviates from the normal operation range, so as to judge whether it works normally.

6. Design of Alarm Processing

Considering the operation characteristics of wind turbine, the alarm processing design divides the alarm into parameter alarm and status alarm. Due to the large external interference during the operation of the wind turbine, the unit is prone to abnormal operation or abnormal warning, resulting in frequent alarm in parameter monitoring. In order to reflect the deeper alarm information, a certain logic is used to comprehensively consider the parameter alarm information, and then put forward the status alarm. There are the following processing methods for the realization of alarm function:

(1) Current value indicates processing: the equipment current value data is displayed on the display in real time, and the update cycle of the screen is the same as that of the monitoring processing.

(2) Current value sending processing: the monitoring and control computer sends the data of the existing value according to the periodic requirements.

(3) Parameter alarm monitoring and processing: calculate the predicted value through the depth network model according to the uploaded data of SCADA system. If the predicted value is different from the real value and exceeds the threshold, it will be regarded as a parameter alarm, which will be uploaded to the server database and automatically stored as an abnormal parameter change.

(4) status alarm monitoring and processing: when a parameter alarm record is generated, query the status alarm database table. When a parameter alarm that meets the status alarm logic, comprehensively raise multiple parameter alarms to status alarms, upload them to the server database, and automatically store them as abnormal status changes.

7. Design of Man-machine Interface

The main interface design refers to the design of the basic window displaying the graphical interface, on which all graphics will be displayed. Reasonable design is conducive to the display of future pictures, user operation and good visibility. As the system is a monitoring system, good graphical interface design is the key link in the system requirements.

1). Design of overall statistics interface for early warning

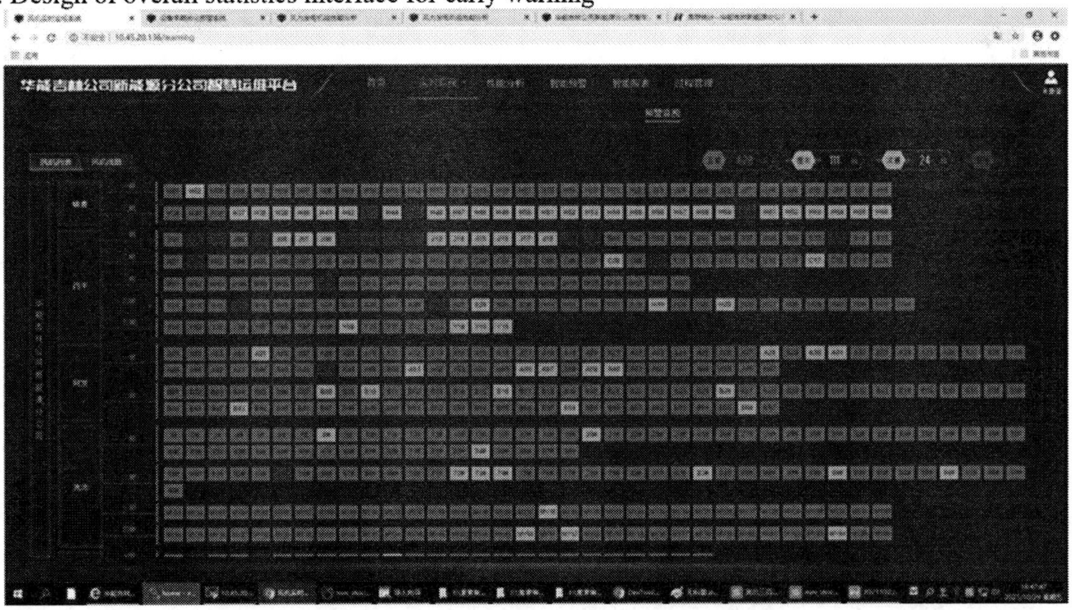

Figure 4. Overall alert statistics interface

The statistical interface can intuitively obtain fault statistical information. Green indicates that the equipment is normal, red indicates that there are early warnings, and the darker the color, the more alarms. The system can also carry out early warning statistics for fans, statistics by fan category, statistics by fan fault category, etc., so as to intuitively reflect the fault information of the overall unit operation status.

2). Design of early warning real-time monitoring interface

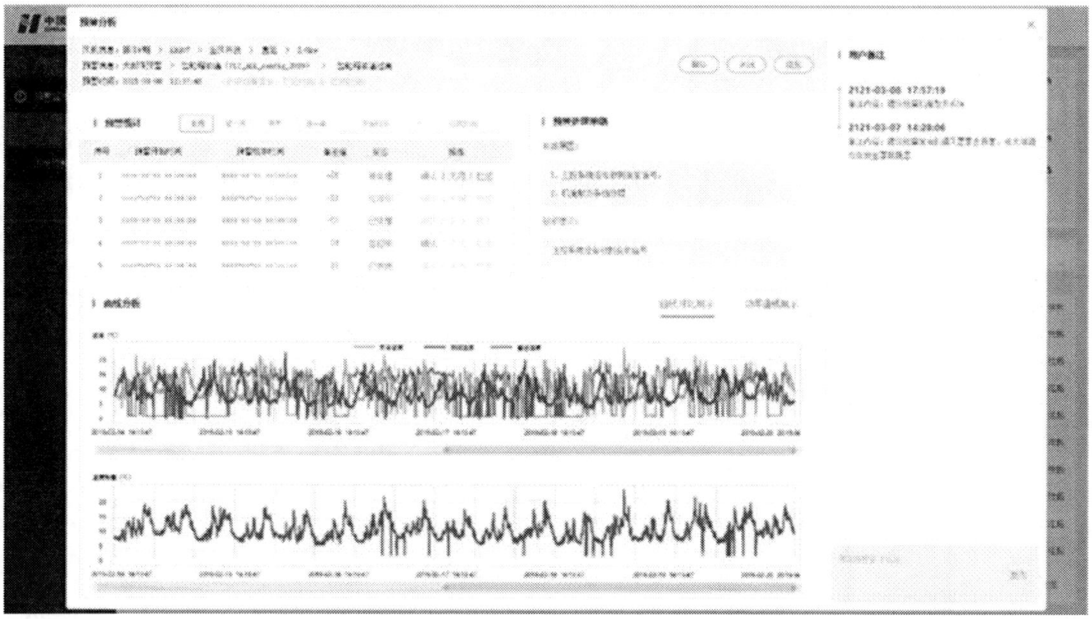

Figure 5. Alert real-time monitoring interface

The early warning real-time monitoring interface can intuitively obtain the real value of the real-time data returned by the on-site SCADA system. The predicted value and upper and lower threshold of the same depth model are also intuitively displayed in the drawing window of the early warning real-time monitoring interface. The drawing is intuitive, clear and beautiful.

3). Model management interface design

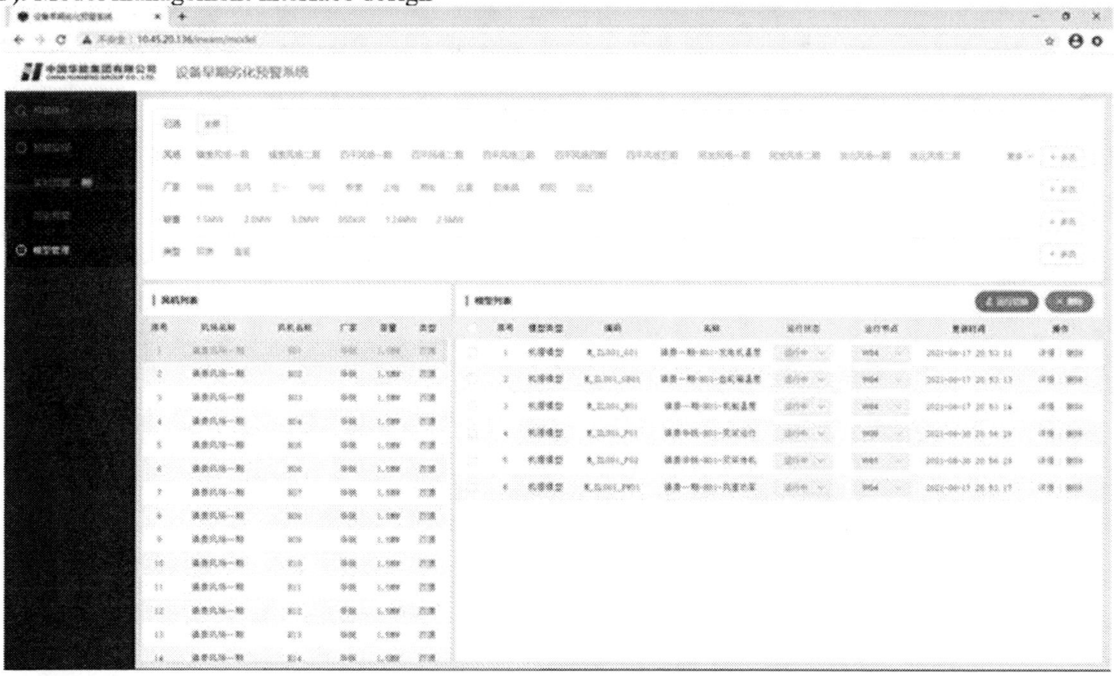

Figure 6. Model management interface

The model management interface can manage each sub model of each fan. In model management, you can operate the specific measurement points of each model, manage and display the predicted value, real value and measurement points of each model. The interface also provides an interface, which can directly modify the upper and lower threshold of each model, with simple and convenient management mode, flexible operation and friendly interface.

8. Summary

The platform architecture of wind power safety monitoring and early warning system based on deep learning is designed: data acquisition layer, calculation layer and application layer, and the functional requirements of the early warning system are analyzed. The database design, process management mode, fault early warning model, alarm value processing and human-computer interaction interface of the system are designed and completed. This paper expounds the software operation process of the deep learning safety monitoring and early warning system, designs the workflow of the wind power safety monitoring and early warning system based on deep learning, and designs the early warning mechanism of the system in detail.

References

[1] Huang Lingling, Fu Yang, Ren Haohan ect. Overview of research on wind turbine maintenance based on status information. Proceedings of the CSEE, 2020, v.40; No.656(21): 323-336.
[2] LI Dazhong, ZHANG Huaying. Wind turbine performance evaluation based on mutual information correlation of wind farm big data[J]. Science Technology and Engineering, 2018, 18(29): 211-215.
[3] Cheng Wenling. Research on Digital Coal Mine Safety Monitoring and Control Technology Based on Internet of Things[J]. Technological Innovation and Application,2020(26):160-161.
[4] Chen Ziqiang, Cheng Jian, Ji Wenqiang ect. Fault Diagnosis of Wind Turbine Pitch System Based on Deep Confidence Network[J]. Measurement and Control Technology, 2019, 38(5) : 18-22.
[5] LI Dazhong, ZHANG Huaying. Wind turbine performance evaluation based on mutual information correlation of wind farm big data[J]. Science Technology and Engineering, 2018, 18(29): 211-215.
[6] M. Patriksson, P. Bangalore. Analysis of SCADA data for early fault detection with application to the maintenance management of wind turbines[J]. Renewable Energy, 2018, 115:521-532.
[7] Fang Ronghua. Design and Realization of a Computer Monitoring System in a Hydropower Plant [D]. Huazhong University of Science and Technology, 2010.
[8] Hinton G E, Salakhutdinov R R. Reducing the dimensionality of data with neural networks[J]. science, 2006, 313(5786): 504-507.

A State Evaluation Method of Equipment Based on Safe Operation State Space Extraction with Zero-failure Data

Pan Li[1], Chaofan Wei[1], Runfeng Chang[2], Anying Liu[3] and Shenghan Zhou[1,*]

[1]School of Reliability and Systems Engineering, Beihang University, Beijing 100191, China
[2]School of Information Science and Technology, North China University of Technology, Beijing 100144, China
[3]School of Management Beijing Institute of Economics and Management, Beijing 100102, China

*Corresponding author email: zhoush@buaa.edu.cn

Abstract. For high reliability equipment and newly designed products, it is difficult to accurately evaluate the equipment status because almost no failure data will be generated during testing or using period. At present, the analysis of zero fault data mainly focuses on reliability estimation and unsupervised fault detection, and there is little research on the safe operation state of equipment to evaluate whether the equipment is in normal state. In this paper, we propose an equipment state evaluation method based on safe operation state space extraction, and verify the feasibility of this method through real data experiments. This method uses algorithms such as LOF (local outlier factor) and k-means algorithm to extract the parameter range of normal operation of the equipment from the historical zero-failure operation data, so as to judge whether the equipment is in a safe operation state. Compared with other methods, this method is simple and easy to use, and has wider applicability and stronger interpretability.
Keywords: Safe operation state space; Zero-failure data; Local outlier factor; k-means algorithm.

1. Introduction

With the continuous progress of production technology, the reliability of merchandise has been improved rapidly, and the failure rate decreases very quickly. In many industrial scenarios, the reliability of the equipment is so high that no failure will occur in the testing process of the equipment. These highly reliable products often need to be tested or used for a long time before it fails and no failure data will be generated at the initial stage of use, so it is difficult to monitor and predict the failure and guide the upgrading of products. Without failure information, it is difficult to judge and evaluate the work state of the equipment, because we can't get the abnormal state information of the equipment. Therefore, how to obtain the normal parameter range of product operation from zero-failure data is a problem worthy of study.

For high-reliability industrial merchandise, it is difficult to obtain the life of the product through testing and estimate the reliability of the products, and obtaining the sample fault data requires a long time of quality test, which not only costs a lot, but also prolongs the product design cycle. When the equipment life satisfies the exponential distribution, to solve the problem of zero fault equipment life test, Yin et al. [1] proposed an E-Bayesian method to predict the products' reliability, by integrating the zero-failure information with the hierarchical Bayesian estimation. Xu et al.[2] proposed a two-sided Modified Bayesian method to estimate the interval of reliability, and verified the effectiveness of the model with real data. When the life of equipment satisfies Weibull distribution, Li et al.[3] established a confidence limit method to predict reliability from zero-failure data, this method can estimate both the unilateral confidence interval and optimal confidence interval of equipment reliability. Since maximum likelihood estimation cannot be used for parameter estimation when there is no fault data, Jiang et al[4] proposed a modified MLE method to estimate the distribution parameters of product life, and combined the prior information obtained from similar products to predict the reliability; Jiang et al [5] presents a method to estimates the fault probability range of equipment directly from the fault data by using the properties of distribution function.

For zero-failure data, because there are only normal data, many supervised learning methods cannot be used. In addition to the equipment reliability prediction mentioned above, zero-failure data can also be used for unsupervised learning for fault detection. Costa et al [6] monitors the equipment fault in real time by calculating the sample density of the data space; Principi et al[7] used autoencoders to automatically diagnose motor fault; Michau[8] proposed a new Unsupervised Feature Alignment Network (UFAN), using incremental learning technology to modify the model continuously to adapt to the changes of equipment operating conditions and dynamically predict equipment faults.

In the above description, zero-failure data is mainly used for equipment reliability estimation and unsupervised fault detection. Although these methods describe some states of equipment, these methods have many problems: 1) The process of calculation is too complex and difficult to use; 2) These methods do not give the parameter range of safe operational state of an equipment, so it is difficult to accurately define the real state of the equipment; 3) These methods often require some prior knowledge, which is difficult to obtain in practical use. To solve these problems, we propose a new method to obtain the safe working state range of products directly from zero-failure data without any other external information.

As far as we know, there is no relevant research on extracting the normal working range of equipment from zero-failure data. Therefore, we propose a safe operation state space extraction (SOSSE)method, which directly obtains the safe operation state range of equipment by using the zero-failure data obtained from equipment testing or operation, and decide whether the equipment is in safe operation state without any fault information. The key contributions of this paper are as follows:

1) As far as we know, this is the first time to obtain the parameter space of equipment safe operation state by using zero-fault data;

2) We propose a simple and widely applicable method to extract the safe operation space of equipment, which can obtain the safe operation state space of equipment from the zero-failure operation data, and decide whether the equipment is in the safe operation state.

The rest of this paper is organized as follows. The second segment describes the safe operation state space extraction (SOSSE)method. In the third part, we use the equipment operation data in the real scene to test and verify the feasibility of the model. The fourth section summarizes the overall text.

2. Methodology

In this section, we propose a new safe operation state space extraction (SOSSE)method from the historical zero-failure data, so as to judge whether the new data is abnormal.

2.1. Model

As shown in Figure 1, the input of the model is historical zero-failure data. These data usually contain many features. These features may be continuous variables or discrete variables, and their values can be very different. Firstly, the model preprocesses the data by removing the monotonic or fixed features with time; Secondly, the data is split and analyzed respectively, and then the data points with abnormal distribution are deleted; Thirdly, the model cluster the data into different categories of each feature dimension with different number of clusters, and each category contains a group of value range; Finally, the clustering results are synthesized to obtain the final output.

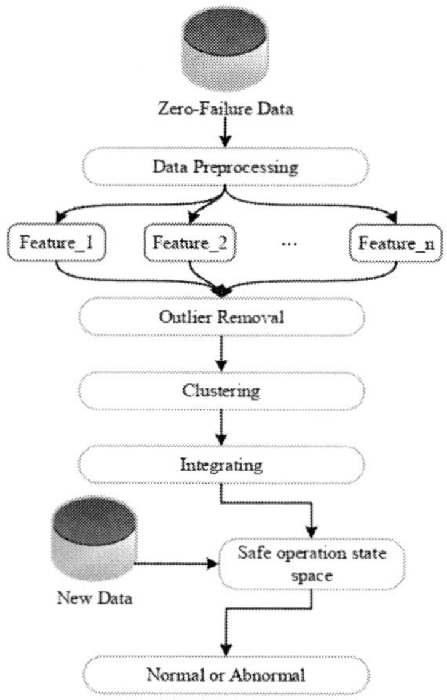

Figure 1. Model Structure.

2.2. Outlier Removal

In the actual scene, due to the disturbance of the environment, the daily detection parameters of the equipment will fluctuate in a certain range. Generally, the fluctuation of these parameters will follow a certain distribution, such as normal distribution, t-distribution, and some parameters may have long tail distribution. Most of the data are in a small value range, but a small amount of data will be distributed in a large range. If the actual range of the parameter is used as a limit, it is difficult to pick out the abnormal data. In order to solve this problem, we choose to use LOF (Local outlier factor) method[9] to eliminate the long tail data and low probability data in each feature.

The abnormal degree of a data point can be measured by calculating the LOF value for each feature of the data. The greater the LOF value, the greater the abnormal degree. By calculating the average value and standard deviation of the abnormal degree of all abnormal points, the data points beyond a certain range of standard deviation are regarded as abnormal points.

$$Outlier(p) = \{p \mid LOF(p) > mean(LOF) + z\text{-}std \times STD(LOF)\} \quad (1)$$

z-std in equation (1) can control the proportion of original data points in the value space. The larger *z-std* is, the less sensitive it is to abnormal data. The smaller z is, the more sensitive it is to abnormal data.

2.3. Clustering

After deleting the abnormal data points of each feature, we will cluster the data of each feature using k-means algorithm[10], and obtain the value range of each feature according by the clustering results, and form the final feature space. Suppose that after k-means clustering of feature *f*, *k* clustering centers are obtained. For category C_{mi} of feature *m*, the corresponding value range is:

$$range(C_{mi}) = [\min(p), \max(p)] \quad p \in cluster(C_i) \quad (2)$$

In the model, the accuracy of the value space can be controlled by *k-cluster*. We specify the range of k-cluster, use the data for model training, and select the best value of k-cluster according to the accuracy requirements of the actual situation. The higher the detection accuracy, the greater the value of k-cluster。 We have also tried contour coefficient and other methods, but the model performance is poor and the model accuracy cannot be controlled manually.

For category variables, when the value number num is less than *k-cluster*, *num-1* is used as the cluster number. For other features, *k-cluster* is used as the cluster number. The larger the value *k-cluster* is, the smaller the safe operation state space is, and the less sensitive it is to abnormal data detection.

2.4. Integrating

After the value range under each feature are obtained, the model combines consecutive intervals of each feature into one interval. Then the proportion of historical data in this space is calculated to decide whether it meets the requirements. If it does not meet the requirements, the input parameters *z-std* and *k-cluster* need to be adjusted.

3. Experiment

In this part, we use the equipment operation data in the real scene to test, and verify the feasibility of the model.

3.1. Data Description

The data we selected comes from the actual operation data of a navigation equipment. In the historical use of this type of product, only two failures have occurred. We select 50000 historical normal data records for model training, and select 100 fault data records (from one failure) to detect the effectiveness of the model. The original data format is shown in Table 1, including 38 feature variables, and the feature names are treated anonymously.

Table 1. Original data structure.

A	B	C	…	AJ	AK	AL
182	26887	1	…	44	-1	2270
192	26888	1	…	43	-1	2269
202	26889	1	…	44	-1	2269

In data preprocessing, we delete some monotonous and fixed features over time, and the number of remaining features is 33. The distribution of data is shown in Figure 2.

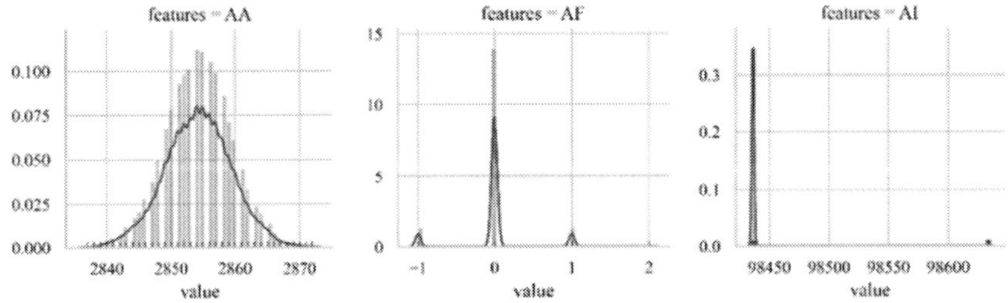

Figure 2. Distribution of different features of data.

3.2. Model Training

During the training, we selected 30 groups of different parameters with different value of *k-cluster* and *z-std*. Firstly, the model calculates the abnormal degree value of each feature in the data, and eliminates the data points whose abnormal degree is greater than z-std standard deviation. Then the k-means method is used to cluster the data of the feature with k-cluster as the number of clusters, and the set of parameter ranges of each cluster is taken as the parameter range of the feature. According to the detection accuracy requirements of specific scenes, the model parameters and training results that meet the requirements are selected.

Table 2. Safe Operation Space State.

Features	Value Range
A	[0, 255]
B	[0, 0], [1, 1], [2, 2]
…	…
M	[17, 16044], [21480, 65521],

After model training, the value ranges of all parameters are combined to form the safe operation state space of normal data. The value space is shown in Table 2. The value ranges of all features together constitute the safe operation parameter space.

3.3. Model Evaluation

We select different value of *k-cluster* and *z-std* to test the effect of the model. The evaluation results are shown in the Table 3. The statistical indicators are the proportion of historical normal data and the proportion of abnormal data in safe operation state space. For example, when *k-cluster* is 5 and *z-std* is 6, 99.516% of normal data falls in the safe operation state space and 0% of abnormal data falls in the safe operation space.

Table 3. Experimental Result.

k-cluster \ z-std	1	2	3	4	5	6
5	99.362/0	99.422/0	99.448/0	99.472/0	99.492/0	**99.516/0**
10	99.200/0	99.406/0	99.444/0	99.466/0	99.486/0	99.514/0
15	99.208/0	99.334/0	99.432/0	99.46/0	99.486/0	99.514/0
20	98.960/0	99.316/0	99.392/0	99.436/0	99.462/0	99.514/0
25	99.058/0	99.282/0	99.374/0	99.404/0	99.426/0	99.484/0

As shown in Figure 3 and Figure 4, when *z-std* remains unchanged, with the gradual increase of the number of n-clusters, the proportion value (percentage of normal data in value space) shows a downward trend as a whole; With the *k-cluster* unchanged, the normal proportional value increases gradually with the increase of *z-std*. We can control the accuracy of safe operation state space by controlling the size of z-std and k-cluster. The larger the *z-std*, the larger the state space and the wider the detection range. The larger the *k-cluster*, the smaller the state space and the more sensitive to abnormal data.

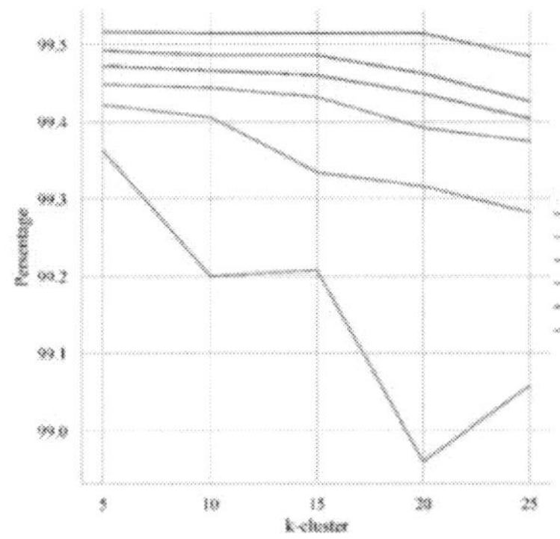

Figure 3. Percentage of normal data in value space with different z-std.

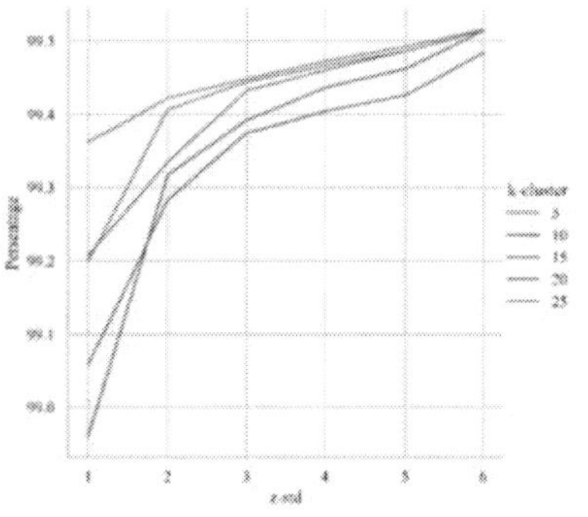

Figure 4. Percentage of normal data in value space with different k-cluster.

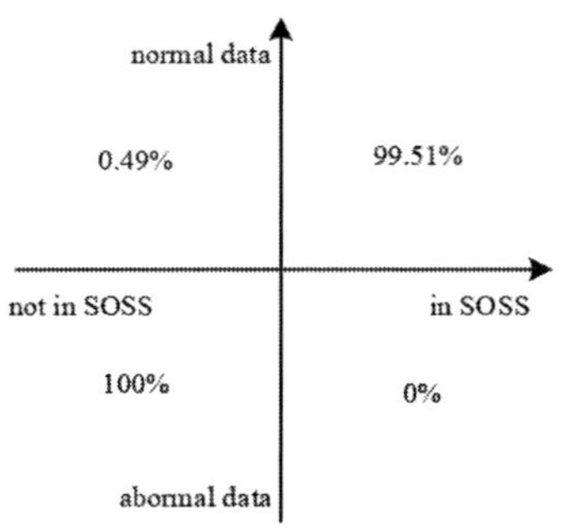

Figure 5. Experimental Result.

The visual display of the best experimental results is shown in Figure 5. When the data point falls on the right side of the horizontal axis, we can have a great probability that the equipment is in a safe state. When the data point falls on the left side of the horizontal axis, we think that the equipment may be abnormal and need to be diagnosed.
When the data point falls in the safe operation state space, we can think that the equipment is in the safe operation state at this time. Otherwise, we think that the equipment may have a fault and need to strengthen monitoring and further diagnosis.

4. Conclusion

In this paper, we propose a new algorithm to extract the safe operation space of equipment from zero-failure data, and verify the effectiveness of the algorithm by using real data. The algorithm can be widely used in various real scenes, such as high reliability equipment or products newly designed and put into use soon. Compared with other methods, this method can judge the safe working state of the equipment more simply and quickly, and give the probability that the equipment is in the safe working state.

Acknowledgments

This research was funded by the National Natural Science Foundation of China (Grant No. 71971013) and the Fundamental Research Funds for the Central Universities (YWF-20-BJ-J-943). The study was also sponsored by the Scientific Research Project of Beijing Institute of Economics and Management (21BSA04) and the Graduate Student Education & Development Foundation of Beihang University.

References

[1] Y. C. Yin, H. Z. Huang, W. Peng, Y. F. Li, and J. Mi, "An E-Bayesian method for reliability analysis of exponentially distributed products with zero-failure data," *Eksploatacja i Niezawodność,* vol. Vol. 18, 2016-01-01 2016.

[2] T. Xu and Y. Chen, "Two-sided M-Bayesian credible limits of reliability parameters in the case of zero-failure data for exponential distribution," *Applied Mathematical Modelling,* vol. 38, pp. 2586-2600, 2014-01-01 2014.

[3] H. Li and Z. Zheng, "Reliability Estimation for Zero-Failure Data Based on Confidence Limit Analysis Method," *Mathematical Problems in Engineering,* vol. 2020, p. e7839432, 2020-01-01 2020.

[4] P. Jiang, J. Lim, M. J. Zuo, and B. Guo, "Reliability estimation in a Weibull lifetime distribution with zero-failure field data," *Quality and Reliability Engineering International,* vol. 26, pp. 691-701, 2010-01-01 2010.

[5] P. Jiang, Y. Xing, X. Jia, and B. Guo, "Weibull Failure Probability Estimation Based on Zero-Failure Data," *Mathematical Problems in Engineering,* vol. 2015, p. e681232, 2015-01-01 2015.

[6] B. S. J. Costa, P. P. Angelov and L. A. Guedes, "Fully unsupervised fault detection and identification based on recursive density estimation and self-evolving cloud-based classifier," *Neurocomputing,* vol. 150, pp. 289-303, 2015-01-01 2015.

[7] E. Principi, D. Rossetti, S. Squartini, and F. Piazza, "Unsupervised electric motor fault detection by using deep autoencoders," *IEEE/CAA Journal of Automatica Sinica,* vol. 6, pp. 441-451, 2019-01-01 2019.

[8] G. Michau and O. Fink, "Unsupervised Fault Detection in Varying Operating Conditions," in *2019 IEEE International Conference on Prognostics and Health Management (ICPHM)*, 2019, pp. 1-10.

[9] M. M. Breunig, H. Kriegel, R. T. Ng, and J. Sander, "LOF: identifying density-based local outliers,", 2000, pp. 93-104.

[10] K. Alsabti, S. Ranka and V. Singh, "An efficient k-means clustering algorithm," *Electrical Engineering and Computer Science,* 1997-01-01 1997.

AUTHOR INDEX